Logic Colloquium 2006

The Annual European Meeting of the Association for Symbolic Logic, also known as the Logic Colloquium, is among the most prestigious annual meetings in the field. The current volume, *Logic Colloquium 2006*, with contributions from plenary speakers and selected special session speakers, contains both expository and research papers by some of the best logicians in the world. The most topical areas of current research are covered: valued fields, Hrushovski constructions (from model theory), algorithmic randomness, relative computability (from computability theory), strong forcing axioms and cardinal arithmetic, and large cardinals and determinacy (from set theory), as well as foundational topics such as algebraic set theory, reverse mathematics, and unprovability.

This volume will be invaluable for experts as well as those interested in an overview of central contemporary themes in mathematical logic.

LECTURE NOTES IN LOGIC

A Publication of
The Association for Symbolic Logic

This series serves researchers, teachers, and students in the field of symbolic logic, broadly interpreted. The aim of the series is to bring publications to the logic community with the least possible delay and to provide rapid dissemination of the latest research. Scientific quality is the overriding criterion by which submissions are evaluated.

Editorial Board

Anand Pillay, Managing Editor
Department of Pure Mathematics, School of Mathematics, University of Leeds

Lance Fortnow
Department of Computer Science, University of Chicago

Shaughan Lavine
Department of Philosophy, The University of Arizona

Jeremy Avigad
Department of Philosophy, Carnegie Mellon University

Vladimir Kanovei
Institute for Information Transmission Problems, Moscow

Steffen Lempp
Department of Mathematics, University of Wisconsin

See end of book for a list of the books in the series. More information can be found at http://www.aslonline.org/books-lnl.html

Logic Colloquium 2006

Edited by

S. BARRY COOPER
University of Leeds

HERMAN GEUVERS
Radboud University Nijmegen

ANAND PILLAY
University of Leeds

JOUKO VÄÄNÄNEN
University of Amsterdam

ASSOCIATION FOR SYMBOLIC LOGIC

CAMBRIDGE
UNIVERSITY PRESS

CAMBRIDGE UNIVERSITY PRESS
Cambridge, New York, Melbourne, Madrid, Cape Town, Singapore,
São Paulo, Delhi, Dubai, Tokyo

Cambridge University Press
32 Avenue of the Americas, New York, NY 10013-2473, USA

www.cambridge.org
Information on this title: www.cambridge.org/9780521110815

Association for Symbolic Logic
Richard Shore, Publisher
Department of Mathematics, Cornell University, Ithaca, NY 14853
http://www.aslonline.org

First published 2009

Printed in the United States of America

A catalog record for this publication is available from the British Library.

ISBN 978-0-521-11081-5 Hardback

CONTENTS

INTRODUCTION

These are the proceedings of the Logic Colloquium 2006, which was held July 27–August 2 at the Radboud University of Nijmegen in the Netherlands.

The Logic Colloquium is the annual European conference on logic, organized under the auspices of the Association for Symbolic Logic (ASL). The program of LC2006 consisted of a mixture of tutorials, invited plenary talks, special sessions, and contributed talks. Finally, there was a plenary discussion on Gödel's legacy, on the occasion of the 100th birthday of the great logician Kurt Gödel, moderated by William Tait. The program gave a good overview of the recent research developments in logic.

The tutorial speakers were Downey, Moerdijk, and Veličković. The invited plenary speakers were Abramsky, Arslanov, Friedman, Goldstern, Hrushovski, Koenigsmann, Lewis, Montalbán, Palmgren, Pohlers, Schimmerling, Steel, Tait, and Wagner. The five special sessions were devoted to computability theory, computer science logic, model theory, proof theory and type theory, and set theory.

For these proceedings we have invited the tutorial and plenary invited speakers—as well as one invited speaker from each of the special sessions—to submit a paper. All papers have been reviewed by independent referees. This has given rise to these proceedings, which give a good overview of the content and breadth of the Logic Colloquium 2006 and of the state of the art in logic at present.

The Editors
S. Barry Cooper
Herman Geuvers
Anand Pillay
Jouko Väänänen

DEFINABILITY AND ELEMENTARY EQUIVALENCE IN THE ERSHOV DIFFERENCE HIERARCHY

M. M. ARSLANOV

Abstract. In this paper we investigate questions of definability and elementary equivalence in the Ershov difference hierarchy. We give a survey of recent results in this area and discuss a number of related open questions. Finally, properties of reducibilities which are intermediate between Turing and truth table reducibilities and which are connected with infinite levels of the Ershov hierarchy are studied.

§1. Introduction. In this paper we consider the current status of a number of open questions concerning the structural organization of classes of Turing degrees below $\mathbf{0}'$, the degree of the Halting Problem. We denote the set of all such degrees by $\mathcal{D}(\leq \mathbf{0}')$.

The Ershov hierarchy arranges these degrees into different levels which are determined by a quantitative characteristic of the complexity of algorithmic recognition of the sets composing these degrees.

The finite level $n, n \geq 1$, of the Ershov hierarchy constitutes n-c.e. sets which can be presented in a canonical form as

$$A = \bigcup_{i=0}^{[\frac{n-1}{2}]} \{(R_{2i+1} - R_{2i}) \cup (R_{2i} - R_{2i+1})\}$$

for some c.e. sets $R_0 \subseteq R_1 \subseteq R_2 \subseteq \cdots \subseteq R_{n-1}$. (Here if n is an odd number then $R_n = \emptyset$.)

A (Turing) degree $\boldsymbol{a}$ is called an n-c.e. degree if it contains an n- c.e. set, and it is called a *properly n-c.e. degree* if it contains an n-c.e. set but no $(n-1)$-c.e. sets. We denote by $\mathcal{D}_n$ the set of all n- c.e. degrees. $\mathcal{R}$ denotes the set of c.e. degrees.

Degrees containing sets from different levels of the Ershov hierarchy, in particular the c.e. degrees, are the most important representatives of $\mathcal{D}(\leq \mathbf{0}')$. Investigations of these degree structures pursued in last two-three decades

The author is supported by RFBR Grant 05-01-00830.

Logic Colloquium '06
Edited by Barry Cooper, Herman Geuvers, Anand Pillay, and Jouko Väänänen
Lecture Notes in Logic, 32

show that the c.e. degrees and the degrees from finite levels of the Ershov hierarchy have similar properties in many respects.

The following theorem, which states that the classes of c.e. and n-c.e. degrees for $1 < n < \omega$ are indistinguishable from the point of view of their ability to compute fixed-point free functions, is a remarkable confirmation of this observation.

THEOREM 1. (*Arslanov* [1] *for* $n = 1$; *Jockusch, Lerman, Soare, Solovay* [17] *for* $n > 1$). *Suppose that* A *is a set which is* n*-c.e. for some* $n \geq 1$. *Then* A *has degree* $\mathbf{0}'$ *if and only if there is a function* f *computable in* A *with no fixed point, i.e.* $(\forall e)(W_{f(e)} \neq W_e)$.

Nevertheless, the elementary theories of the c.e. and the n-c.e. degrees for every $n > 1$ are different even at the Σ^0_2-level. This was shown by Downey [13] (the diamond lattice is embeddable in 2-c.e. degrees preserving $\mathbf{0}$ and $\mathbf{0}'$) and later by Cooper, Harrington, Lachlan, Lempp and Soare [10] (there is a 2-c.e. degree $\mathbf{d} < \mathbf{0}'$ which is maximal in the partial ordering of all n-c.e. degrees for all $n > 1$). Previous to this, a difference between the elementary theories of these degree structures at the Σ^0_3-level was exhibited in Arslanov [2, 3] (every n-c.e. degree for any $n > 1$ can be cupped to $\mathbf{0}'$ by a 2-c.e. degree $< \mathbf{0}'$). (Since any Σ^0_1-sentence satisfies $\mathcal{R}$ and $\mathcal{D}_n$ for any $n > 1$ if and only if it is consistent with the theory of partial orderings, $\mathcal{R}$ and $\mathcal{D}_n$ are indistinguishable at the Σ^0_1-level.)

These results initiated an intensive study of the properties of the n-c.e. degree structures for several $n > 1$. For the main part these investigations have concentrated on the following questions:

- Is the relation "$\mathbf{x}$ is c.e." definable in $\mathcal{D}_n$ for each (some) $n \geq 2$? Are there nontrivial definable in $\mathcal{D}_n, n \geq 2$, sets of c.e. degrees?
- Is the relation "$\mathbf{x}$ is m-c.e." definable in $\mathcal{D}_n$ for each (some) pair $n > m \geq 2$?
- Are $\{\mathcal{D}_2, \leq\}$ and $\{\mathcal{D}_3, \leq\}$ elementarily equivalent? (The famous *Downey's conjecture* states that they are elementarily equivalent.)
- Is $\{\mathcal{D}_2, \leq\}$ an elementary substructure of $\{\mathcal{D}_3, \leq\}$?
- Are $\{\mathcal{D}_m, \leq\}$ and $\{\mathcal{D}_n, \leq\}$ elementarily equivalent for each (some) $n \neq m, m, n \geq 2$?
- Is $\{\mathcal{D}_m, \leq\}$ an elementary substructure of $\{\mathcal{D}_n, \leq\}$ for $1 \leq m < n$?
- Is $\{\mathcal{D}_m, \leq\}$ a Σ_k-substructure of $\{\mathcal{D}_n, \leq\}$ for some $1 \leq m < n$ and some $k \geq 1$?
- Is the elementary theory of $\{\mathcal{D}_n, \leq\}$ undecidable for each (some) $n \geq 2$?

Known results on the definability of the relation "$\mathbf{x}$ is c.e." in $\{\mathcal{D}_n, \leq\}$, $n > 1$:

The following definition is from Cooper and Li [11].

DEFINITION 1. A Turing approximation to the class of c.e. degrees $\mathcal{R}$ in the n-c.e. degrees is a Turing definable class $\mathcal{S}_n$ of n-c.e. degrees such that

- either $\mathcal{R} \subseteq \mathcal{S}_n$ (in this case we say that $\mathcal{S}_n$ is an approximation to $\mathcal{R}$ from above), or
- $\mathcal{S}_n \subseteq \mathcal{R}$ ($\mathcal{S}_n$ is an approximation to $\mathcal{R}$ from below).

Obviously, $\mathcal{R}$ is definable in the n-c.e. degrees if and only if there is a Turing definable class $\mathcal{S}_n$ of n-c.e. degrees which is a Turing approximation to the class $\mathcal{R}$ in the n-c.e. degrees simultaneously from above and from below.

Non-trivial approximations from above:

First consider the following set of n-c.e. degrees: $\mathcal{S}_n = \{\mathbf{x} \in \mathcal{D}_n | (\forall \mathbf{z} > \mathbf{x})(\exists \mathbf{y})(\mathbf{x} < \mathbf{y} < \mathbf{z})\}$. The following two theorems show that $\mathcal{S}_n$ contains all c.e. degrees but does not coincide with the set of all n-c.e. degrees. Therefore, for any $n > 1$, $\mathcal{S}_n \neq \mathcal{D}_n$, and $\mathcal{S}_n$ is a *nontrivial approximation from above* of $\mathcal{R}$ in n-c.e. degrees.

THEOREM 2. (*Cooper, Yi* [12] *for* $n = 2$; *Arslanov, LaForte and Slaman* [6] *for* $n > 2$) *For any c.e. degree* $\mathbf{x}$ *and n-c.e. degree* $\mathbf{y}$, *if* $\mathbf{x} < \mathbf{y}$ *then* $\mathbf{x} < \mathbf{z} < \mathbf{y}$ *for some n-c.e. degree* $\mathbf{z}$.

THEOREM 3. [10] *There is a* 2-*c.e. degree which is maximal in each* $(\mathcal{D}_n, \leq)$, $n > 1$, *therefore* $\mathcal{S}_n \neq \mathcal{D}_n$ *for each* $n > 1$.

Using the Robinson Splitting Theorem technique (see Soare [20]) it is not difficult to construct a properly n-c.e. degree belonging to $\mathcal{S}_n$. Therefore $\mathcal{S}_n$ does not coincide with the class of all c.e. degrees and is not a Turing definition for the class $\mathcal{R}$.

Recently the following refinement of this approximation for the class $\mathcal{D}_2$ was obtained by Cooper and Li [11].

THEOREM 4. *For every c.e. degree* $\boldsymbol{a} < \boldsymbol{0}'$ *every* 2-*c.e. degree* $\boldsymbol{b} > \boldsymbol{a}$ *is splittable in the* 2-*c.e. degrees above* $\boldsymbol{a}$.

And again the set of c.e. degrees with this property does not coincide with the set of all c.e. degrees. Namely, using Cooper's [9] strategy of splitting of d-c.e. degrees over $\boldsymbol{0}$, M. Jamaleev [16] constructed a properly d-c.e. degree $\boldsymbol{a}$ such that any d-c.e. degree $\boldsymbol{d} > \boldsymbol{a}$ is splittable in d-c.e. degrees over $\boldsymbol{a}$.

So far there were no non-trivial approximations from below to the class of c.e. degrees $\mathcal{R}$ in the n-c.e. degrees, $n \geq 2$. Later in this paper the first such example will be presented.

Known results on elementary differences among $\{\mathcal{D}_n, \leq\}$ *for different* $n \geq 1$:

It was already mentioned that the elementary theories of the c.e. and the n-c.e. degrees for every $n > 1$ are different at the Σ^0_2-level, and that for any Σ_1-sentence φ, $\mathcal{D}_n \models \varphi$ if and only if φ is consistent with the theory of partial orderings. Therefore,

- $\mathcal{D}_n \equiv_{\Sigma_1} \mathcal{D}_m \equiv_{\Sigma_1} \mathcal{D}(\leq 0')$ for all $m \neq n, 1 \leq m, n < \omega$,

but

- for any $n \geq 1$, $\mathcal{D}_n$ is not a Σ_1-substructure of $\mathcal{D}(\leq 0')$ (Slaman, unpublished), and
- $\mathcal{R}$ is not a Σ_1-substructure of $\mathcal{D}_2$ (Y. Yang and L. Yu [21])

§2. Questions of definability and elementary equivalence. Investigation into the problems listed above is motivated by a desire for better understanding of the level of structural similarity of classes of c.e. and n-c.e. degrees for different $n > 1$, as well as better understanding of the level of homogeneity for the notion of c.e. with respect to n-c.e. degrees in the sense of the level of similarity of orderings of c.e. degrees and of n-c.e. degrees which are c.e. in a c.e. degree $\boldsymbol{d}$ and $\geq \boldsymbol{d}$. We consider the following two questions as corresponding examples for these two approaches.

QUESTION 1. Let $\boldsymbol{0} < \boldsymbol{d} < \boldsymbol{e}$ be c.e. degrees. There is a c.e. degree $\boldsymbol{c}$ such that $\boldsymbol{c} < \boldsymbol{e}$ and $\boldsymbol{c}|\boldsymbol{d}$. Does this property of the c.e. degrees also hold in the n-c.e. degrees? That is, given 2-c.e. degrees $\boldsymbol{d}$ and $\boldsymbol{e}$ such that $\boldsymbol{0} < \boldsymbol{d} < \boldsymbol{e}$, is always there a 2-c.e. degree $\boldsymbol{c}$ such that $\boldsymbol{c} < \boldsymbol{e}$ and $\boldsymbol{c}|\boldsymbol{d}$?

QUESTION 2. A relativization of the above stated property of the c.e. degrees to a c.e. degree $\boldsymbol{x}$ allows to obtain (having c.e. in and above $\boldsymbol{x}$ degrees $\boldsymbol{d} < \boldsymbol{e}$) a c.e. in $\boldsymbol{x}$ degree $\boldsymbol{c} > \boldsymbol{x}$ such that $\boldsymbol{c} < \boldsymbol{e}$ and $\boldsymbol{c}|\boldsymbol{d}$. Does this property hold also in the realm of 2-c.e. degrees in the following sense: let $\boldsymbol{0} < \boldsymbol{x} < \boldsymbol{d} < \boldsymbol{e}$ be such degrees that $\boldsymbol{x}$ is c.e., $\boldsymbol{d}$ and $\boldsymbol{e}$ are 2-c.e. degrees such that both of them are c.e. in $\boldsymbol{x}$. Is there a c.e. in $\boldsymbol{x}$ 2-c.e. degree $\boldsymbol{c}$ such that $\boldsymbol{c} < \boldsymbol{e}$ and $\boldsymbol{c}|\boldsymbol{d}$?

The following two results show that we have a negative answer to the first question and an affirmative answer to the second question.

THEOREM 5. [5] *There are 2-c.e. degrees* $\mathbf{d}$ *and* $\mathbf{e}$ *such that* $\boldsymbol{0} < \mathbf{d} < \mathbf{e}$ *and for any 2-c.e. degree* $\mathbf{u} < \mathbf{e}$ *either* $\mathbf{u} \leq \mathbf{d}$ *or* $\mathbf{d} \leq \mathbf{u}$.

THEOREM 6. [5] *For every c.e. degree* $\mathbf{x}$ *and all 2-c.e. degrees* $\mathbf{d}$ *and* $\mathbf{e}$ *such that* $\mathbf{d}, \mathbf{e}$ *are both c.e. in* $\mathbf{x}$ *and* $\boldsymbol{0} < \mathbf{x} < \mathbf{d} < \mathbf{e}$, *there is a c.e. in* $\mathbf{x}$ *2-c.e. degree* $\boldsymbol{u}$ *such that* $\mathbf{x} < \boldsymbol{u} < \mathbf{e}$ *and* $\mathbf{d}|\boldsymbol{u}$.

The following theorem is a refinement of Theorem 5.

THEOREM 7. a) *In Theorem 5 the degree* $\mathbf{d}$ *is necessarily c.e. and*

b) *for each 2-c.e. degree* $\mathbf{e}$ *there is at most one c.e. degree* $\mathbf{d} < \mathbf{e}$ *with this property.*

PROOF. Every 2-c.e. degree $\boldsymbol{u} > \boldsymbol{0}$ is c.e. in a c.e. degree $\boldsymbol{u}' < \boldsymbol{u}$ (this is the so-called Lachlan's Proposition). Therefore $\boldsymbol{e}$ is c.e. in a c.e. degree $\boldsymbol{e}' < \boldsymbol{e}$. If $\boldsymbol{e}' > \boldsymbol{d}$, then by Sacks Splitting Theorem we split $\boldsymbol{e}'$ into two c.e. degrees $\boldsymbol{e}_0$ and $\boldsymbol{e}_1$ avoiding the upper cone of $\boldsymbol{d}$ (avoiding $\boldsymbol{d}$, for short). At least one of these degrees must be incomparable with $\boldsymbol{d}$, a contradiction.

If $\boldsymbol{e}' < \boldsymbol{d}$, then consider the c.e. degree $\boldsymbol{c} = \boldsymbol{e}' \cup \boldsymbol{d}'$, where $\boldsymbol{d}' < \boldsymbol{d}$ is a c.e. degree such that $\boldsymbol{d}$ is c.e. in $\boldsymbol{d}'$. Obviously, $\boldsymbol{c} \leq \boldsymbol{d}$. If $\boldsymbol{c} < \boldsymbol{d}$ then we obtain a contradiction with Theorem 6, since both of the 2-c.e. degrees $\boldsymbol{e}$ and $\boldsymbol{d}$ are c.e. in $\boldsymbol{c}$. Therefore, $\boldsymbol{d} = \boldsymbol{c}$. Similar arguments prove also the second part of the theorem. ⊣

Now consider the following set of c.e. degrees:

$$\mathcal{S}_2 = \{0\} \bigcup \left\{ \boldsymbol{x} > \boldsymbol{0} \mid (\exists \boldsymbol{y} > \boldsymbol{x})(\forall \boldsymbol{z})(\boldsymbol{z} \leq \boldsymbol{y} \rightarrow \boldsymbol{z} \leq \boldsymbol{x} \vee \boldsymbol{x} \leq \boldsymbol{z}) \right\}.$$

It follows from Theorems 5 and 7 that

Corollary 1. *$\mathcal{S}_2 \subseteq \mathcal{R}$ and $\mathcal{S}_2 \neq \{0\}$.*

Therefore, $\mathcal{S}_2$ is a nontrivial approximation from below to the class of c.e., degrees $\mathcal{R}$ in the class of 2-c.e. degrees. A small additional construction in Theorem 5 allows to achieve that $\mathcal{S}_2$ contains infinitely many c.e. degrees.

It is a natural question to ask whether $\mathcal{S}_2$ coincides with the class of all c.e. degrees (and, therefore establishes definability of the c.e. degrees in $\mathcal{D}_2$) or not? To give a negative answer to this question we consider the isolated 2-c.e. degrees, which were introduced by Cooper and Yi [12]:

Definition 2. A c.e. degree $\mathbf{d}$ is *isolated* by an isolating 2-c.e. degree $\mathbf{e}$ if $\boldsymbol{d} < \boldsymbol{e}$ and for any c.e. degree $\mathbf{c}$, if $\boldsymbol{c} \leq \boldsymbol{e}$ then $\boldsymbol{c} \leq \boldsymbol{d}$.

A c.e. degree is *non-isolated* if it is not isolated.

Obviously, each non-computable c.e. degree $\boldsymbol{x}$ from $\mathcal{S}_2$ is isolated by a 2-c.e. degree $\boldsymbol{y}$. In Arslanov, Lempp, Shore [7] we proved that the non-isolated degrees are downward dense in the c.e. degrees and that they occur in any jump class. Therefore, $\mathcal{S}_2 \neq \mathcal{R}$.

Open Question. Whether for every pair of c.e. degrees $\boldsymbol{a} < \boldsymbol{b}$ there is a degree $\boldsymbol{c} \in \mathcal{S}_2$ such that $\boldsymbol{a} < \boldsymbol{c} < \boldsymbol{b}$ (i.e. $\mathcal{S}_2$ is dense in $\mathcal{R}$)?

An affirmative answer to this question implies definability of $\mathcal{R}$ in $\mathcal{D}_2$ as follows: given a c.e. degree $\boldsymbol{a} > \boldsymbol{0}$ we first split $\boldsymbol{a}$ into two incomparable c.e. degrees $\boldsymbol{a}_0$ and $\boldsymbol{a}_1$, then using density of $\mathcal{S}_2$ in $\mathcal{R}$ find between $\boldsymbol{a}$ and $\boldsymbol{a}_i, i \leq 1$, a c.e. degree $\boldsymbol{c}_i, i \leq 1$, obtaining $\boldsymbol{a} = \boldsymbol{c}_0 \cup \boldsymbol{c}_1$. This shows that in this case a nonzero 2-c.e. degree is c.e. if and only if it is a least upper bound of two incomparable 2-c.e. degrees from $\mathcal{S}_2$.

Conjecture 1. Each c.e. degree $\boldsymbol{a} > \boldsymbol{0}$ is the least upper bound of two incomparable degrees from $\mathcal{S}_2$ and, therefore, the c.e. degrees are definable in $\mathcal{D}_2$.

Corollary 2 (From Theorem 5). *There are no 2-c.e. degrees $\boldsymbol{f} > \boldsymbol{e} > \boldsymbol{d} > \boldsymbol{0}$ such that for any u,*

(i) *if $\mathbf{u} \leq \boldsymbol{f}$ then either $\mathbf{e} \leq \mathbf{u}$ or $\mathbf{u} \leq \mathbf{e}$, and*
(ii) *if $\mathbf{u} \leq \mathbf{e}$ then either $\mathbf{d} \leq \mathbf{u}$ or $\mathbf{u} \leq \mathbf{d}$.*

PROOF. If there are such degrees $\boldsymbol{f} > \boldsymbol{e} > \boldsymbol{d} > \boldsymbol{0}$ then by Theorem 7a the degree $\boldsymbol{e}$ is c.e. and by the Sacks Splitting Theorem is splittable avoiding $\boldsymbol{d}$, which is a contradiction. ⊣

OPEN QUESTION. Are there 3-c.e. degrees $\boldsymbol{f} > \boldsymbol{e} > \boldsymbol{d} > \boldsymbol{0}$ with this property?

Obviously, an affirmative answer to this question refutes Downey's Conjecture on the elementarily equivalency of $\mathcal{D}_2$ and $\mathcal{D}_3$.

Though this question still remains open, we can weaken a little this property of degrees $(\boldsymbol{d}, \boldsymbol{e}, \boldsymbol{f})$ to carry out the mission imposed to these degrees to refute Downey's Conjecture. We consider triples of non-computable n-c.e. degrees $\{(\boldsymbol{d}, \boldsymbol{e}, \boldsymbol{f}) \mid \boldsymbol{0} < \boldsymbol{d} < \boldsymbol{e} < \boldsymbol{f}\}$ with the following (weaker) property: for any n-c.e. degree u,

(i) if $\boldsymbol{u} \leq \boldsymbol{f}$ then either $\boldsymbol{u} \leq \boldsymbol{e}$ or $\boldsymbol{e} \leq \boldsymbol{d} \cup \boldsymbol{u}$, and
(ii) if $\boldsymbol{u} \leq \boldsymbol{e}$ then either $\boldsymbol{d} \leq \boldsymbol{u}$ or $\boldsymbol{u} \leq \boldsymbol{d}$.

(In the first line the former condition $\boldsymbol{e} \leq \boldsymbol{u}$ is changed to the weaker condition $\boldsymbol{e} \leq \boldsymbol{d} \cup \boldsymbol{u}$.)

We still have the following corollary from Theorems 5 and 6:

COROLLARY 3. *There are no 2-c.e. degrees $\boldsymbol{f} > \boldsymbol{e} > \boldsymbol{d} > \boldsymbol{0}$ such that for any 2-c.e. degree u,*

(i) *if $\boldsymbol{u} \leq \boldsymbol{f}$ then either $\boldsymbol{u} \leq \boldsymbol{e}$ or $\boldsymbol{e} \leq \boldsymbol{d} \cup \boldsymbol{u}$, and*
(ii) *if $\boldsymbol{u} \leq \boldsymbol{e}$ then either $\boldsymbol{d} \leq \boldsymbol{u}$ or $\boldsymbol{u} \leq \boldsymbol{d}$.*

PROOF. Suppose that there are such degrees $\boldsymbol{f} > \boldsymbol{e} > \boldsymbol{d} > \boldsymbol{0}$. Let $\boldsymbol{f}' \leq \boldsymbol{f}$ and $\boldsymbol{e}' \leq \boldsymbol{e}$ be c.e. degrees such that $\boldsymbol{f}$ and $\boldsymbol{e}$ are c.e. in $\boldsymbol{f}'$ and $\boldsymbol{e}'$, accordingly. Consider the degree $\boldsymbol{x} = \boldsymbol{d} \cup \boldsymbol{e}' \cup \boldsymbol{f}'$. Obviously, $\boldsymbol{d} \leq \boldsymbol{x} \leq \boldsymbol{f}$.

Since $\boldsymbol{x}$ is c.e., we have $\boldsymbol{e} \not< \boldsymbol{x}$, otherwise $\boldsymbol{x}$ is splittable in c.e. degrees avoiding $\boldsymbol{e}$, which is a contradiction. Also $\boldsymbol{x} \neq \boldsymbol{e}$, since in this case we can split $\boldsymbol{x}$ avoiding $\boldsymbol{d}$, which is again a contradiction. At last, if $\boldsymbol{x} \not\leq \boldsymbol{e}$ then it follows from condition (i) that $\boldsymbol{e} \leq \boldsymbol{d} \cup \boldsymbol{x} = \boldsymbol{x}$, a contradiction. Therefore, $\boldsymbol{x} < \boldsymbol{e}$. Since $\boldsymbol{f}$ and $\boldsymbol{e}$ are both c.e. in $\boldsymbol{x}$, it follows now from Theorem 6 that there is a 2-c.e. degree $\boldsymbol{u}$ such that $\boldsymbol{x} < \boldsymbol{u} < \boldsymbol{f}$ and $\boldsymbol{u} | \boldsymbol{e}$, a contradiction. ⊣

THEOREM 8. [5] *There exists a c.e. degree $\boldsymbol{d} > \boldsymbol{0}$, a 2-c.e. degree $\boldsymbol{e} > \boldsymbol{d}$, and a 3-c.e. degree $\boldsymbol{f} > \boldsymbol{e}$ such that for any 3-c.e. degree u,*

(i) *if $\boldsymbol{u} \leq \boldsymbol{f}$ then either $\boldsymbol{u} \leq \boldsymbol{e}$ or $\boldsymbol{e} \leq \boldsymbol{d} \cup \boldsymbol{u}$, and*
(ii) *if $\boldsymbol{u} \leq \boldsymbol{e}$ then either $\boldsymbol{d} \leq \boldsymbol{u}$ or $\boldsymbol{u} \leq \boldsymbol{d}$.*

COROLLARY 4. *$\mathcal{D}_2 \not\equiv \mathcal{D}_3$ at the Σ_2- level.*

In Theorem 8 we have a c.e. degree $\boldsymbol{d} > \boldsymbol{0}$ and a 2-c.e. degree $\boldsymbol{e} > \boldsymbol{d}$ such that every 3-c.e. degree $\boldsymbol{u} \leq \boldsymbol{e}$ is comparable with $\boldsymbol{d}$. Can this condition be strengthened in the following sense: there exists a c.e. degree $\boldsymbol{d} > \boldsymbol{0}$ and a 2-c.e. degree $\boldsymbol{e} > \boldsymbol{d}$ such that every n-c.e. degree $\leq \boldsymbol{e}$ for every $n < \omega$ is comparable with $\boldsymbol{d}$?

Open Question. Does there exist a c.e. degree $\boldsymbol{d} > \boldsymbol{0}$ and a 2-c.e. degree $\boldsymbol{e} > \boldsymbol{d}$ such that for any $n < \omega$ and any n-c.e. degree $\boldsymbol{u} \leq \boldsymbol{e}$ either $\boldsymbol{u} \leq \boldsymbol{d}$ or $\boldsymbol{d} \leq \boldsymbol{u}$?

An affirmative answer to this question would reveal an interesting property of the finite levels of the Ershov difference hierarchy with far-reaching prospects. From other side, if the question has a negative answer, then let $\boldsymbol{d} > \boldsymbol{0}$ and $\boldsymbol{e} > \boldsymbol{d}$ be accordingly c.e. and 2-c.e. degrees and $n \geq 3$ be the greatest natural number such that every n-c.e. degree $\boldsymbol{u} \leq \boldsymbol{e}$ is comparable with $\boldsymbol{d}$ and there is a $(n+1)$-c.e. degree $\boldsymbol{v} \leq \boldsymbol{e}$ which is incomparable with $\boldsymbol{d}$. Now consider the following Σ_1-formula:

$$\varphi(x, y, z) \equiv \exists u(x < y < z \,\&\, u \leq z \,\&\, u \not\leq y \,\&\, y \not\leq u).$$

Let $\boldsymbol{d}$ and $\boldsymbol{e}$ be degrees and n be the integer whose existence is assumed by the negative answer to the previous question. Then we have $\mathcal{D}_{n+1} \models \varphi(\boldsymbol{0}, \boldsymbol{d}, \boldsymbol{e})$, and $\mathcal{D}_n \models \neg\varphi(\boldsymbol{0}, \boldsymbol{d}, \boldsymbol{e})$, which means that in this case $\mathcal{D}_n$ is not a Σ_1-substructure of $\mathcal{D}_{n+1}$, thus answering a well-known open question.

We see that an answer to this question in any direction leads to very interesting consequences.

Theorems 5 and 8 raise a whole series of questions whose study could lead to a better understanding of the inner structure of the ordering of the n-c.e. degrees. Below we consider some of these questions.

Definition 3. Let $n > 1$. An $(n+1)$-tuple of degrees $\boldsymbol{a}_0, \boldsymbol{a}_1, \boldsymbol{a}_2, \dots, \boldsymbol{a}_{n-1}, \boldsymbol{a}_n$ forms an *n-bubble* in $\mathcal{D}_m$ for some $m \geq 1$, if $\boldsymbol{0} = \boldsymbol{a}_0 < \boldsymbol{a}_1 < \boldsymbol{a}_2 < \cdots < \boldsymbol{a}_{n-1} < \boldsymbol{a}_n$, $\boldsymbol{a}_k$ is k-c.e. for each k, $1 \leq k \leq n$, and for any m-c.e. degree $\boldsymbol{u}$, if $\boldsymbol{u} \leq \boldsymbol{a}_k$ then either $\boldsymbol{u} \leq \boldsymbol{a}_{k-1}$ or $\boldsymbol{a}_{k-1} \leq \boldsymbol{u}$.

An $(n+1)$-tuple of degrees $\boldsymbol{a}_0, \boldsymbol{a}_1, \boldsymbol{a}_2, \dots, \boldsymbol{a}_{n-1}, \boldsymbol{a}_n$ forms a *weak n-bubble* in $\mathcal{D}_m$ for some $m \geq 1$, if $\boldsymbol{0} = \boldsymbol{a}_0 < \boldsymbol{a}_1 < \boldsymbol{a}_2 < \cdots < \boldsymbol{a}_{n-1} < \boldsymbol{a}_n$, $\boldsymbol{a}_k$ is k-c.e. for each k, $1 \leq k \leq n$, and for any m-c.e. degree $\boldsymbol{u}$, if $\boldsymbol{u} \leq \boldsymbol{a}_k$ then either $\boldsymbol{u} \leq \boldsymbol{a}_{k-1}$ or $\boldsymbol{a}_{k-1} \leq \boldsymbol{u} \cup \boldsymbol{a}_{k-2}$.

Obviously, every n-bubble is also an n-weak bubble for every $n > 1$, but we don't know if the reverse holds. Theorem 5 and Corollary 2 state that in the 2-c.e. degrees there are 2-bubbles, but in the 2-c.e. degrees there are no n-bubbles (and even n-weak bubbles) for every $n > 2$. Theorem 8 states that in the 3-c.e. degrees there are 3-weak bubbles. The existence of n-bubbles (and even n-week bubbles) in the n-c.e. degrees for $n > 3$, and the existence of n-bubbles in m-c.e. degrees for $2 < m < n$ are open questions.

Conjecture 2. For every n, $1 < n < \omega$, $\mathcal{D}_n$ contains an n-bubble, but does not contain m-bubbles for every $m > n$. (As we saw already this is true for $n = 2$.)

Obviously, if this conjecture holds for some $n > 1$ then this means that $\mathcal{D}_n$ is not elementarily equivalent to $\mathcal{D}_m, m > n$.

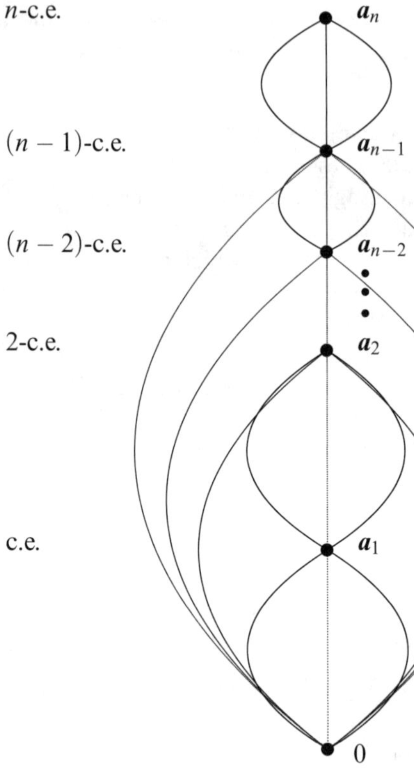

n-bubble

All known examples of sentences in the language of partial ordering, which are true in the n-c.e. degrees and false in the $(n+1)$-c.e. degrees for some $n \geq 1$, belong to the level $\forall\exists$ or to the higher levels of the arithmetic hierarchy. This and some other observations allow us to state the following plausible conjecture.

CONJECTURE 3. For all $n \geq 1$, for all $\exists\forall$-sentences φ, $\mathcal{D}_n \models \varphi \Rightarrow \mathcal{D}_{n+1} \models \varphi$. (The $\exists\forall$-theory of the n-c.e. degrees is a sub-theory of the $\exists\forall$-theory of the $(n+1)$-c.e. degrees).

The following question is posed in a number of publications (see, for instance, Y. Yang and L. Yu [21]):

QUESTION 3. Fix integers $n \geq 1$ and $m > n$. Is there a function $f : \mathcal{D}_n \to \mathcal{D}_m$ such that for each Σ_1-formula $\varphi(\bar{x})$ and each n-tuple $\bar{\boldsymbol{a}}$ from $\mathcal{D}_n$, $\mathcal{D}_n \models \varphi(\bar{\boldsymbol{a}})$ if and only if $\mathcal{D}_m \models \varphi(f(\bar{\boldsymbol{a}}))$?

If for some $n < m < \omega$ $\mathcal{D}_n$ is not a Σ_1-substructure of $\mathcal{D}_m$, then φ (if exists) cannot be the identity function.

How many parameters contain functions which are witnesses in the proof that $\mathcal{D}_1$ is not a Σ_1-substructure of $\mathcal{D}(\leq \mathbf{0}')$ and $\mathcal{D}_2$?

- Slaman's result ($\mathcal{R} \not\preceq_{\Sigma_1} \mathcal{D}(\mathbf{0}')$: 3 parameters;
- Yang and Yu ($\mathcal{R} \not\preceq_{\Sigma_1} \mathcal{D}_2$): 4 parameters.

QUESTION 4. Can be these numbers reduced?

§3. Generalized tabular reducibilities. In Arslanov [4] for each constructive ordinal $\alpha \leq \omega^\omega$ we defined the so-called generalized tabular reducibility $\leq_{gtt(\alpha)}$ such that $\leq_{gtt(\omega)}$ coincides with $\leq_{tt}$, for each $\alpha \leq \omega^\omega$ $gtt(\alpha)$-reducibility is intermediate between tt- and T-reducibilities, and for different α all $\leq_{gtt(\alpha)}$ are different. In Arslanov [4] we also outlined a proof that the $gtt(\alpha)$-reducibility carries out the following property of the tt-reducibility to other infinite levels of the Ershov hierarchy: a Δ^0_2-set is ω-c.e. if and only if it is tt-reducible to the creative set K. In this paragraph we give a corrected proof of this and some other related results, eliminating some inaccuracy in the argument.

The following definitions of infinite levels of the hierarchy are from Ershov [14, 15].

Let $P(x, y)$ be a computable predicate which on ω defines a partial ordering. (If $P(x, y)$ we write $x \leq_P y$.) A uniformly c.e. sequence $\{R_x\}$ of c.e. sets is a $\leq_P$-sequence, if for all x, y, $x \leq_P y$ implies $R_x \subseteq R_y$.

Hereinafter we will use the Kleene system of notation $(\mathcal{O}, <_0)$. For $a \in \mathcal{O}$ we denote by $|a|_0$ the ordinal α, which has $\mathcal{O}$-notation a. Therefore $|a|_0$ has the order type $\langle \{x | x <_0 a\}, <_0 \rangle$, and an "$a$-sequence of c.e. sets $\{R_x\}$" for $a \in \mathcal{O}$ is to be understood in the usual way. If α is a constructive ordinal and $a \in \mathcal{O}$ its notation, i.e. $|a|_0 = \alpha$, and $\lambda < \alpha$, then knowing a we can effectively find a notation b for λ, $|b|_0 = \lambda$.

An ordinal is even if it is either 0, or a limit ordinal, or a successor of an odd ordinal. Otherwise the ordinal is odd. Therefore if α is even, then α' (the successor of α) is odd and vise versa.

For the system of notation $\mathcal{O}$, the parity function $e(x)$ is defined as follows. Let $n \in \mathcal{O}$. Then $e(n) = 1$ if ordinal $|n|_0$ is odd, and $e(n) = 0$ if $|n|_0$ is even.

For any $a \in \mathcal{O}$ we first define operations S_a and P_a, which map a-sequences $\{R_x\}_{x <_0 a}$ to subsets of ω, as follows:

$S_a(R) = \{z | \exists x <_0 a(e(x) \neq e(a) \,\&\, z \in R_x \,\&\, \forall y <_0 x(z \notin R_y))\}$.

$P_a(R) = \{z | \exists x <_0 a(e(x) = e(a) \,\&\, z \in R_x \,\&$
$\& \forall y <_0 x(z \notin R_y))\} \cup \{\omega - \cup_{x <_0 a} R_x\}$.

It follows from these definitions that $P_a(R) = \overline{S_a(R)}$ for all $a \in \mathcal{O}$ and all a-sequences R.

The class Σ^{-1}_a (Π^{-1}_a) for $a \in \mathcal{O}$ is the class of all sets $S_a(R)$ (accordingly all sets $P_a(R)$), where $R = \{R_x\}_{x <_0 a}$ all a-sequences of c.e. sets, $a \in \mathcal{O}$. Define $\Delta^{-1}_a = \Sigma^{-1}_a \cap \Pi^{-1}_a$.

In particular, a set $A \subseteq \omega$ belongs to level Σ^{-1}_{ω} of the Ershov hierarchy (A is Σ^{-1}_{ω}-set), if there is a uniformly c.e. sequence of c.e. sets $\{R_x\}_{x\in\omega}$ such that $R_0 \subseteq R_1 \subseteq \cdots$ (ω-sequence of c.e. sets), and $A = \cup_{n=0}^{\infty}(R_{2n+1} - R_{2n})$.

Definition 4. A set A is ω-c.e. set if there is a computable function g of two variables s and x and a computable function f such that for all x $A(x) = lim_s g(s,x)$, $g(0,x) = 0$, and

$$|\{s | g(s+1,x) \neq g(s,x)\}| \leq f(x).$$

Theorem 9 (Ershov [14, 15]; Carstens [8], Selivanov [19]). *Let $A \subseteq \omega$. The following are equivalent*:

a) *A is ω-c.e.*;
b) *A is a Δ^{-1}_{ω}-set*;
c) *A is tt-reducible to a creative set K*;
d) *there is a uniformly c.e. sequence of c.e. sets $\{R_x\}_{x\in\omega}$, such that $\bigcup_{x\in\omega} R_x = \omega$, $R_0 \subseteq R_1 \subseteq \cdots$, and $A = \bigcup_{n=0}^{\infty}(R_{2n+1} - R_{2n})$.*

Theorem 10 (Ershov [14, 15]). *Let $a, b \in \mathcal{O}$ and $a <_0 b$. Then*

a) $\Sigma^{-1}_a \cup \Pi^{-1}_a \subset \Sigma^{-1}_b \cap \Pi^{-1}_b$ *and, therefore, for any $a \in \mathcal{O}$, $\Sigma^{-1}_a \subset \Delta^0_2$*;
b) $\cup_{a\in\mathcal{O}}\Sigma^{-1}_a = \cup_{a\in\mathcal{O},|a|_{\mathcal{O}}=\omega^2}\Sigma^{-1}_a = \Delta^0_2$.

For convenience we consider only ordinals $\leq \omega^{\omega}$, and for simplicity instead of notations we use ordinals meaning their representation in normal form

$$\alpha = \omega^m \cdot n_0 + \cdots + \omega \cdot n_{m-1} + n_m.$$

The material of this paragraph can, however, be extended also to all constructive ordinals (considering, for instance, Kleene system of ordinal notation).

We first define classes of generalized truth-table conditions ($gtt(\alpha)$-conditions) $\mathcal{B}_{\alpha}$, $\alpha \leq \omega^{\omega}$.

$\alpha = n > 1$: $\mathcal{B}_{\alpha}$ consists of all tt-conditions with norm $< n$;

$\alpha = \omega$: $\mathcal{B}_{\alpha}$ consists of all tt-conditions;

$\alpha = \omega^m \cdot n + \beta$, $\beta < \omega^m$ ($n > 1$; if $n = 1$ then $\beta > 0$): $\mathcal{B}_{\alpha}$ consists of all tt-conditions of the form

$$\sigma_1 \,\&\, \tau_1 \vee \cdots \vee \sigma_n \,\&\, \tau_n \vee \rho, \text{ or } \neg\big[\sigma_1 \,\&\, \tau_1 \vee \cdots \vee \sigma_n \,\&\, \tau_n \vee \rho\big],$$

where $\sigma_i \in \mathcal{B}_{\omega}$, $\tau_i \in \mathcal{B}_{\omega^m}$, $\rho \in \mathcal{B}_{\beta}$;

$\alpha = \omega^{m+1}$: $\mathcal{B}_{\alpha} = \bigcup_n \mathcal{B}_{\omega^m \cdot n}$;

$\alpha = \omega^{\omega}$: $\mathcal{B}_{\alpha} = \bigcup_n \mathcal{B}_{\omega^n}$.

It follows from these definitions that for each α, $\omega \leq \alpha \leq \omega^{\omega}$, gtt-conditions from $\mathcal{B}_{\alpha}$ are usual tt-conditions with a fixed inner structure of these conditions. Using this structure we define by induction on α an enumeration $\{\sigma^{\alpha}_n\}_{n\in\omega}$ of *$gtt(\alpha)$-formulas* related to gtt-conditions from $\mathcal{B}_{\alpha}$.

We denote by σ_n^ω the n-th tt-condition (which is a formula of propositional logic constructed from atomic propositions $\langle k \in X\rangle$ for several $k \in \omega$, and the norm of the tt-condition is the number of its atomic propositions).

For $\alpha = \omega^m \cdot n + \beta$, $m \geq 1, n \geq 1, \beta < \omega^m$ ($n > 1$; if $n = 1$ then $\beta > 0$) the $gtt(\alpha)$-formula $\sigma^\alpha_{\langle p,q,r\rangle}$ with index $\langle p, q, r\rangle$ is the formula

$$\left[\sigma^\omega_{\Phi_p(0)} \& \sigma^\gamma_{\Phi_q(0)} \vee \cdots \vee \sigma^\omega_{\Phi_p(n-1)} \& \sigma^\gamma_{\Phi_q(n-1)} \vee \sigma^\beta_r\right],$$

where $\gamma = \omega^m$, $\Phi_p(i)$ is the partial-computable function with index p, defined for all $i \leq n-1$, $\Phi_q(i)$ is the partial-computable function with index q.

Therefore, a $gtt(\alpha)$-formula σ_i^α with index $i = \langle p, q, r\rangle$ is a gtt-condition $\sigma \in \mathcal{B}_\alpha$, $\alpha = \omega^m \cdot n + \delta$, $m \geq 1, n \geq 1, \delta < \omega^m$, if and only if $\Phi_p(x)\downarrow$ for all $x \leq n-1$ and r is an index for some gtt-condition from $\mathcal{B}_\delta$.

For $\alpha = \omega^{m+1}$ and $\alpha = \omega^\omega$ the enumeration of $gtt(\alpha)$-formulas $\{\sigma^\alpha\}$ is defined using a fixed effective enumeration of all gtt-formulas from $\bigcup_{n,i}\sigma_i^{\omega^m\cdot n}$ (accordingly from $\bigcup_{n,i}\sigma_i^{\omega^n}$).

For convenience we add to the integers two additional objects *true* and *false*, for which σ^α_{true} is a tt-condition which is identically truth and σ^α_{false} is an inconsistent tt-condition.

From now on we identify the class $\mathcal{B}_\alpha$ with the class of all $gtt(\alpha)$-formulas.

Definition 5. We say that a gtt-formula σ from $\mathcal{B}_\alpha$ *converges* on a set $A \subseteq \omega$, if

1. $\alpha \leq \omega$, i.e. any tt-condition from $\mathcal{B}_\alpha$, $\alpha \leq \omega$, converges on any set $A \subset \omega$, or
2. σ is equal to $\left[\left(\bigvee_{i\leqslant m} \sigma^\omega_{\Phi_p(i)} \& \sigma^\gamma_{\Phi_q(i)}\right) \vee \sigma^\beta_j\right]$, and for any $i \leq m$ if A satisfies $\sigma^\omega_{\Phi_p(i)}$ (see the definition of "A satisfies $\sigma^\omega_{\Phi_p(i)}$" below), then $\Phi_q(i)\downarrow$ and $\sigma^\gamma_{\Phi_q(i)}$ converges on A.

Definition 6. A gtt-formula σ from $\mathcal{B}_\alpha$ is *satisfied* by a set $A \subseteq \omega$ (written as $A \models \sigma$), if σ converges on A and

- If $\sigma \in \mathcal{B}_\omega$, then A satisfies to the tt-condition σ,
- If σ is equal to $(\bigvee_{i\leqslant m} \sigma_i \& \tau_i) \vee \rho$, then $A \models \rho$ or there is an $i \leq m$ such that $A \models \sigma_i$ and $A \models \tau_i$.

$A \not\models \sigma$ means $A \models \neg\sigma$.

Definition 7. A set A is $gtt(\alpha)$-reducible to a set B (written as $A \leq_{gtt(\alpha)} B$), if there is a computable function f such that for any x

(i) gtt-formula $\sigma^\alpha_{f(x)}$ converges on B, and
(ii) $x \in A \leftrightarrow B \models \sigma^\alpha_{f(x)}$.

REMARK 1. Part (i) in this definition requires us to prove that the $\leq_{gtt(\alpha)}$-reducibility implies Turing reducibility for any $\alpha \leq \omega^{\omega}$ (see Theorem 11 below). From the other side, the presence of the condition (i) in the definition of the $gtt(\alpha)$-reducibility does not allow us to prove the main property which is incumbent on these reducibilities: a Δ_2^0-set A is α-c.e. if and only if $A \leq_{gtt(\alpha)} K$. (We say that a set A is α-c.e. for some infinite ordinal $\alpha \leq \omega^{\omega}$ if A belong to the level Δ_{α}^{-1} of the Ershov hierarchy.) In Theorem 12 below we prove a weaker version of this statement: if $A \leq_{gtt(\alpha)} K$ then A belongs to the level Σ_{α}^{-1} of the Ershov hierarchy, and if $A \in \Sigma_{\alpha}^{-1}$ then $A \leq_{gtt^{\star}(\alpha)} K$. Here the $gtt^{\star}(\alpha)$-reducibility is obtained from the the $gtt(\alpha)$-reducibility by removing part (i) in Definition 7.

If $\alpha \neq \omega^n$ for some $n \leq \omega$, then in general the reducibility $\leq_{gtt(\alpha)}$ is not transitive. For this reason we formulate the following theorem for ordinals $\omega^n, 1 \leq n \leq \omega$ only.

THEOREM 11. *For $\alpha = \omega, \omega^2, \ldots, \omega^{\omega}$ the reducibilities $\leq_{gtt(\alpha)}$ are reducibilities which are intermediate between the tt- and T-reducibilities, and for different α all $\leq_{gtt(\alpha)}$ are different.*

PROOF. From indexes i and j of σ_i^{α}, σ_j^{α} we can effectively compute an index k of the gtt-formula σ_k^{α}, which is obtained by substitution of σ_i^{α} into σ_j^{α}, which means that for $\alpha = \omega, \omega^2, \ldots, \omega^{\omega}$ the set $\mathcal{B}_{\alpha}$ is effectively closed on substitutions, and the relation $\leqslant_{gtt(\alpha)}$ in this case transitive. It is easy to see also that the relation $\leqslant_{gtt(\alpha)}$ is reflexive. To prove that $A \leqslant_{gtt(\alpha)} B \rightarrow A \leq_T B$ for all sets A, B, suppose that $x \in A$ if and only if $B \models \sigma_{f(x)}^{\alpha} = \left[\left(\bigvee_{i \leqslant m} \sigma_{\Phi_{g(x)}(i)}^{\omega} \& \sigma_{\Phi_{q(x)}(i)}^{\beta}\right) \vee \sigma_{r(x)}^{\gamma}\right]$ for some computable functions f, g, q and r. For each x using oracle of B we can list all $i \leq m$ such that $B \models \sigma_{\Phi_{g(x)}(i)}^{\omega}$. For each such i we have $\Phi_{q(x)}(i) \downarrow$ (condition (i) in the Definition 7), and we also check whether $B \models \sigma_{\Phi_{q(x)}(i)}^{\beta}$. Similarly for the $gtt(\gamma)$-formula $\sigma_{r(x)}^{\gamma}$. Now $x \in A$ if and only if for some $i \leq m$ we have $\models \sigma_{\Phi_{g(x)}(i)}^{\omega}$ and $B \models \sigma_{\Phi_{q(x)}(i)}^{\beta}$, or there is such i in $gtt(\gamma)$-formula $\sigma_{r(x)}^{\gamma}$.

It is well-known (see Selivanov [19]) that for all $a <_0 b$ the set of T-degrees of Δ_a^{-1}-sets is a proper subset of the set of Δ_b^{-1}-sets and, therefore, it follows from Theorem 12 below that at least the degrees of creative sets for these reducibilities are different. ⊣

DEFINITION 8. A set A is $gtt^{\star}(\alpha)$-reducible to a set B (written as $A \leq_{gtt^{\star}(\alpha)} B$), if there is a computable function f such that for any x

$\star$ $x \in A \leftrightarrow B \models \sigma_{f(x)}^{\alpha}$.

THEOREM 12. *Let $\omega \le \alpha \le \omega^\omega$. Then for any set $A \subseteq \omega$,*

(i) *If $A \in \Sigma^{-1}_\alpha$ then $A \le_{gtt^\star(\alpha)} K$;*
(ii) *If $A \le_{gtt(\alpha)} K$, then $A \in \Sigma^{-1}_\alpha$.*

PROOF. We will consider the case $\alpha = \omega \cdot 2$. After that it will be clear how to prove the theorem in the general case.

(i) Suppose that $R_0 \subseteq R_1 \subseteq \cdots \subseteq R_\omega \subseteq R_{\omega+1} \subseteq \cdots$ is a $\omega \cdot 2$-sequence such that $A = (\cup_{i=0}^\infty (R_{2i+1} - R_{2i})) \cup (\cup_{i=0}^\infty (R_{\omega+2i+1} - R_{\omega+2i}))$.

The proof of the following lemma is straightforward.

LEMMA 1. *Let A be a Σ^{-1}_ω set and $P_0 \subseteq P_1 \subseteq P_2 \subseteq \cdots$ be an ω-sequence such that $A = \bigcup_{i=0}^\infty (P_{2i+1} - P_{2i})$. For any $x \in \omega$, if $x \in P_n$ for some $n < \omega$, then there is a tt-condition σ^ω_m such that $x \in A$ if and only if $K \models \sigma^\omega_m$. The integer m can be found effectively from x and n.*

Given x first define $\sigma^\omega_{\Phi_{p(x)}(0)} = \sigma^\omega_{\Phi_{p(x)}(1)} =$ "true", leave $\Phi_{q(x)}(0)$ and $\Phi_{q(x)}(1)$ undefined, and wait for an $i < \omega$ such that either $x \in R_i$ or $x \in R_{\omega+i}$. Using the previous lemma in the first case find a tt-condition $\sigma^\omega_{x_0}$ such that $x \in A$ if and only if $K \models \sigma^\omega_{x_0}$, similarly in the second case find a tt-condition $\sigma^\omega_{x_1}$ (for the ω-sequence $R_\omega \subseteq R_{\omega+1} \subseteq \cdots$) such that $x \in A$ if and only if $K \models \sigma^\omega_{x_1}$. Now in the first case define $\sigma^\omega_{\Phi_{q(x)}(0)} = \sigma^\omega_{x_0}$, in the second case define $\sigma^\omega_{\Phi_{q(x)}(0)} = \sigma^\omega_{x_1}$.

Now it is clear that $\sigma^\omega_{\Phi_{q(x)}(0)} \& \sigma^\omega_{\Phi_{q(x)}(0)} \vee \sigma^\omega_{\Phi_{p(x)}(1)} \& \sigma^\omega_{\Phi_{q(x)}(1)}$ is the required $gtt^\star(\omega \cdot 2)$-formula which $gtt^\star(\omega \cdot 2)$-reduces A to K: if $x \notin (\cup_{i=0}^\infty R_i) \cup (\cup_{i=0}^\infty R_{\omega+i})$, then $x \notin A$ and $\Phi_{q(x)}(0)$ and $\Phi_{q(x)}(1)$ remain undefined. If $x \in (\cup_{i=0}^\infty R_i) \cup (\cup_{i=0}^\infty R_{\omega+i})$, then the claim follows directly from the lemma.

(ii) Now suppose that $A \le_{gtt(\omega \cdot 2)} K$, i.e. there is a computable function f such that $x \in A$ if and only if $K \models \sigma^{\omega \cdot 2}_{f(x)}$. By definition there are computable functions p and q such that for all x, $\sigma^{\omega \cdot 2}_{f(x)} = \sigma^\omega_{\Phi_{p(x)}(0)} \& \sigma^\omega_{\Phi_{q(x)}(0)} \vee \sigma^\omega_{\Phi_{p(x)}(1)} \& \sigma^\omega_{\Phi_{q(x)}(1)}$, and for $i \le 1$, if $K \models \sigma^\omega_{\Phi_{p(x)}(i)}$ then $\Phi_{q(x)}(i)\downarrow$.

Since $\sigma^\omega_{\Phi_{p(x)}(0)}$ is a usual truth-table condition, there are a Boolean function $\alpha_{0,x}$ and a finite set $F_0 = \{u_{0,1}, u_{0,2}, \ldots, u_{0,n_0(x)}\}$ such that $\sigma^\omega_{\Phi_{p(x)}(0)} = (F, \alpha_{0,x})$ (here $n_0(x)$ and $\alpha_{0,x}$ are computable functions on x). Similarly, if $\Phi_{q(x)}(0)\downarrow$, then let $\sigma^\omega_{\Phi_{q(x)}(0)} = (\{v_{0,1}, v_{0,2}, \ldots, v_{0,m_0(x)}\}, \beta_{0,x})$ for some functions $m_0(x)$ and $\beta_{0,x}$.

Similarly we define functions $n_1(x), m_1(x), \alpha_{1,x}$ and $\beta_{1,x}$ for tt-conditions $\sigma^\omega_{\Phi_{p(x)}(1)}$ and $\sigma^\omega_{\Phi_{q(x)}(1)}$ (if $\Phi_{q(x)}(1)\downarrow$).

We have $x \in A \leftrightarrow K \models \sigma^\omega_{\Phi_{p(x)}(0)} \& \Phi_{q(x)}(0)\downarrow \& K \models \sigma^\omega_{\Phi_{q(x)}(0)} \vee K \models \sigma^\omega_{\Phi_{p(x)}(1)} \& \Phi_{q(x)}(1)\downarrow \& K \models \sigma^\omega_{\Phi_{q(x)}(1)}$.

We construct an $\omega \cdot 2$-sequence $R_0 \subseteq R_1 \subseteq \cdots \subseteq R_\omega \subseteq R_{\omega+1} \subseteq \cdots$ such that $A = (\cup_{i=0}^{\infty}(R_{2i+1} - R_{2i})) \cup (\cup_{i=0}^{\infty}(R_{\omega+2i+1} - R_{\omega+2i}))$ as follows:

Given $x \in \omega$ we wait for a stage s such that either $K_s \models \sigma^{\omega}_{\Phi_{p(x)}(0)}$ & $\Phi_{q(x),s}(0)\downarrow$ & $K_s \models \sigma^{\omega}_{\Phi_{q(x)}(0)}$ or $K_s \models \sigma^{\omega}_{\Phi_{p(x)}(1)}$ & $\Phi_{q(x),s}(1)\downarrow$ & $K_s \models \sigma^{\omega}_{\Phi_{q(x)}(1)}$.

Suppose that at a stage s for some $i \leq 1$ we have $K_s \models \sigma^{\omega}_{\Phi_{p(x),s}(i)}$ & $\Phi_{q(x),s}(i)\downarrow$ & $K_s \models \sigma^{\omega}_{\Phi_{q(x),s}(i)}$.

Then we enumerate x into $R_{\omega+n_i(x)+m_i(x)}$, and wait for a stage $t > s$ when $K_t \not\models \sigma^{\omega}_{\Phi_{p(x),t}(i)} \vee K_t \not\models \sigma^{\omega}_{\Phi_{q(x),t}(i)}$. Then enumerate x into $R_{\omega+n_i(x)+m_i(x)-1}$ etc.

If later at a stage $l > s$ we obtain $\Phi_{q(x),l}(1-i)\downarrow$, then we enumerate x into $R_{r(x)+\varepsilon}$, where $r(x) = n_0(x) + n_1(x) + m_0(x) + m_1(x)$, and ε is defined as follows:

(i) $\varepsilon = 1$, if $K_l \models \sigma^{\omega}_{\Phi_{p(x),l}(0)}$ & $K_l \models \sigma^{\omega}_{\Phi_{q(x),l}(0)}$ or $K_l \models \sigma^{\omega}_{\Phi_{p(x),l}(1)}$ & $K_l \models \sigma^{\omega}_{\Phi_{q(x),l}(1)}$ and $r(x)$ is an even number, or

if $K_l \not\models \sigma^{\omega}_{\Phi_{p(x),l}(0)} \vee K_l \not\models \sigma^{\omega}_{\Phi_{q(x),l}(0)}$, and $K_l \not\models \sigma^{\omega}_{\Phi_{p(x),l}(1)} \vee K_l \not\models \sigma^{\omega}_{\Phi_{q(x),l}(1)}$ and $r(x)$ is an odd number.

(ii) $\varepsilon = 0$, if $K_l \models \sigma^{\omega}_{\Phi_{p(x),l}(0)}$ & $K_l \models \sigma^{\omega}_{\Phi_{q(x),l}(0)}$ or $K_l \models \sigma^{\omega}_{\Phi_{p(x),l}(1)}$ & $K_l \models \sigma^{\omega}_{\Phi_{q(x),l}(1)}$ and $r(x)$ is an odd number, or

if $K_l \not\models \sigma^{\omega}_{\Phi_{p(x),l}(0)} \vee K_l \not\models \sigma^{\omega}_{\Phi_{q(x),l}(0)}$, and $K_l \not\models \sigma^{\omega}_{\Phi_{p(x),l}(1)} \vee K_l \not\models \sigma^{\omega}_{\Phi_{q(x),l}(1)}$ and $r(x)$ is an even number.

Then as above we may enumerate x into $R_{r(x)+\varepsilon-1}$ (having for $\varepsilon = 1$ a negation of the condition (i), and for $\varepsilon = 0$ a negation of the condition (ii)) etc. As a result we obtain an $\omega \cdot 2$-sequence $R_0 \subseteq R_1 \subseteq \cdots \subseteq R_\omega \subseteq R_{\omega+1} \subseteq \cdots$ such that $A = \bigcup_{i=0}^{\infty}(R_{2i+1} - R_{2i}) \cup \bigcup_{i=0}^{\infty}(R_{\omega+2i+1} - R_{\omega+2i})$, which means that $A \in \Sigma^{-1}_{\omega\cdot 2}$. ⊣

Corollary 5. *Let $\omega \leq \alpha \leq \omega^\omega$. Then for any set A,*

if $A \leq_{gtt(\alpha)} K$ & $\bar{A} \leq_{gtt(\alpha)} K$ then $A \in \Delta^{-1}_\alpha$, and

if $A \in \Delta^{-1}_\alpha$, then $A \leq_{gtt^\star(\alpha)} K$ & $\bar{A} \leq_{gtt^\star(\alpha)} K$.

The following theorem shows that the Turing reducibility is not exhausted by any collection of $gtt(\alpha)$- and $gtt^\star(\alpha)$-reducibilities.

Theorem 13. *There is a set $A \leq_T \emptyset''$ such that for all α $A \not\leq_{gtt(\alpha)} \emptyset''$.*

Proof. The proof is based on the following lemma.

Lemma 2. *If $B \leqslant_{gtt(\alpha)} C$ for some α, then there exists a $\emptyset'$-computable function $\Phi_e^{\emptyset'}$ such that*

$$(\forall x)\left(x \in B \leftrightarrow C \models \sigma^{\omega}_{\Phi_e^{\emptyset'}(x)}\right).$$

PROOF. (of lemma). Let $B \leqslant_{gtt(\alpha)} C$ by a computable function f, i.e. for any x,

$$x \in B \leftrightarrow C \models \sigma^{\alpha}_{f(x)}.$$

From α and an index for f we can effectively find the following presentation

$$\sigma^{\alpha}_{f(x)} = \Big(\bigvee_{i \leqslant m} \sigma^{\omega}_{g(i,x)} \,\&\, \sigma^{\beta}_{\psi(i,x)} \Big) \vee \sigma^{\gamma}_{k(x)}$$

If $\beta > \omega$ then we find such a presentation also for $\sigma^{\beta}_{h(i,x)}$ via new β' and γ' and similarly for $\sigma^{\gamma}_{k(x)}$ etc. Finally, we obtain an extended presentation

$$\sigma^{\alpha}_{f(x)} = \bigvee_{i \leqslant m} \Big(\Big(\bigwedge_{j_i \leqslant n_i} \sigma^{\omega}_{p_{i,j_i}(x)} \Big) \,\&\, \sigma^{\omega}_{q_i(x)} \Big),$$

where all $p_{i,j_i}(x), 0 \leq i \leq m, 0 \leq j_i \leq n_i$, are defined, but some $q_i(x), 0 \leq i \leq m$, for some i can be undefined. Now let for $0 \leq i \leq m$,

$$\tau_s(x) = \begin{cases} \Big(\bigwedge_{j_i \leqslant n_i} \sigma^{\omega}_{p_{i,j_i}(x)} \Big) \,\&\, \sigma^{\omega}_{q_i(x)}, & \text{if } q_i(x) \downarrow; \\ \sigma^{\omega}_{false}, & \text{if } q_i(x) \uparrow. \end{cases}$$

We have $\sigma^{\alpha}_{f(x)} = \tau_0(x) \vee \tau_1(x) \cdots \vee \tau_k(x)$. This is obviously a tt-condition whose index in the enumeration of all tt-conditions can be computed using an oracle for $\emptyset'$. ⊣

Now let $B = \{x | (\exists y)[\varphi^{\emptyset'}_x(x) = y \,\&\, \emptyset'' \models \sigma^{\omega}_y]\}$ and let $A = \omega - B$.

The reducibility $A \leq_T \emptyset''$ is obvious. If $A \leqslant_{gtt(\alpha)} \emptyset''$ for some α, then there exists $\Phi^{\emptyset'}_e(x)$ from the lemma. Then

$$e \in A \leftrightarrow \emptyset'' \models \sigma^{\omega}_{\Phi^{\emptyset'}_e(x)} \leftrightarrow e \in B \leftrightarrow e \notin A \qquad ⊣$$

At last, the following theorem shows that the weak truth-table reducibility is a special case of the $gtt(\omega^2)$-reducibility.

DEFINITION 9. $A \leq_{wtt} B$, if $A = \Phi^B_e$ for some e and for all x $\varphi^B_e(x) \leq f(x)$ for some computable function f.

THEOREM 14. *If $A \leq_{wtt} B$, then $A \leq_{gtt(\omega^2)} B$.*

PROOF. Let $A = \Phi^B$ and let g be a computable function such that $\varphi^B(x) < g(x)$ for all x. There are $2^{g(x)}$ subsets $X_i \subseteq \{0, 1, \ldots, g(x) - 1\}$. For each of them we compose a tt-formula $\sigma^{\omega}_{p(i)}$, $i \leq 2^{g(x)}$, as follows:

$$X \models \sigma^{\omega}_{p(i)} \leftrightarrow X \lceil g(x) = X_i.$$

Now consider the formula

$$\sigma^{\omega^2}_{h(x)} \leqslant \sigma^{\omega}_{p(1)} \,\&\, \sigma^{\omega}_{q(1)} \vee \cdots \vee \sigma^{\omega}_{p(2^{g(x)})} \,\&\, \sigma^{\omega}_{q(2^{g(x)})},$$

where $\sigma^{\omega}_{p(i)}$ from above and $q(x)$ is defined as follows:

$$q(x) = \begin{cases} \text{true}, & \text{if } \Phi^{X_i}(x)\downarrow = 1; \\ \text{false}, & \text{if } \Phi^{X_i}(x)\downarrow = 0; \\ \uparrow. & \text{if } \Phi^{X_i}(x)\uparrow. \end{cases}$$

For any given x we can effectively compute an index $f(x)$ of the gtt-formula $\sigma^{\omega^2}_{h(x)}$. Now $\Phi^B(x) = 1 \leftrightarrow B \models \sigma^{\omega^2}_{f(x)}$, i.e. $A \leqslant_{gtt(\omega^2)} B$ by function $f(x)$. ⊣

The converse of this theorem is not true. Indeed, let $A \in \Delta^{-1}_{\omega^2} - \Delta^{-1}_{\omega}$. Then $A \leq_{gtt(\omega^2)} K$ but $A \not\leq_{wtt} K$, since $A \leq_{wtt} K$ if and only if $A \leq_{tt} K$ (see, for example, Rogers [18, exercise 9.45, page 159]) if and only if $A \in \Delta^{-1}_{\omega}$.

REFERENCES

[1] M. M. Arslanov, *Some generalizations of a fixed-point theorem*, ***Izvestiya Vysshikh Uchebnykh Zavedeniĭ. Matematika***, (1981), no. 5, pp. 9–16.

[2] ———, *Lattice properties of the degrees below* ***0′***, ***Doklady Akademii Nauk SSSR***, vol. 283 (1985), no. 2, pp. 270–273.

[3] ———, *The lattice of the degrees below* ***0′***, ***Izvestiya Vysshikh Uchebnykh Zavedeniĭ. Matematika***, (1988), no. 7, pp. 27–33.

[4] ———, *Generalized tabular reducibilities in infinite levels of the ershov difference hierarchy*, ***Logical Approaches of Computational Barriers*** (A. Beckmann, U. Berger, B. Lowe, and J. Tucker, editors), Swansea, UK, 2006, pp. 15–23.

[5] M. M. Arslanov, I. Sh. Kalimullin, and L. Lempp, *On Downey's conjecture*, to appear.

[6] M. M. Arslanov, Geoffrey L. LaForte, and Theodore A. Slaman, *Relative enumerability in the difference hierarchy*, ***The Journal of Symbolic Logic***, vol. 63 (1998), no. 2, pp. 411–420.

[7] M. M. Arslanov, S. Lempp, and R. A. Shore, *On isolating r.e. and isolated d-r.e. degrees*, ***Computability, Enumerability, Unsolvability***, London Mathematical Society Lecture Note Series, vol. 224, Cambridge University Press, Cambridge, 1996, pp. 61–80.

[8] H. G. Carstens, *Δ^0_2-Mengen*, ***Archiv für Mathematische Logik und Grundlagenforschung***, vol. 18 (1976/77), no. 1–2, pp. 55–65.

[9] S. B. Cooper, *A splitting theorem for the n-r.e. degrees*, ***Proceedings of the American Mathematical Society***, vol. 115 (1992), no. 2, pp. 461–471.

[10] S. B. Cooper, L. Harrington, A. H. Lachlan, S. Lempp, and R. I. Soare, *The d.r.e. degrees are not dense*, ***Annals of Pure and Applied Logic***, vol. 55 (1991), no. 2, pp. 125–151.

[11] S. B. Cooper and A. Li, *Turing definability in the Ershov hierarchy*, ***Journal of the London Mathematical Society. Second Series***, vol. 66 (2002), no. 3, pp. 513–528.

[12] S. B. Cooper and X. Yi, *Isolated d-r.e. degrees*, Preprint, 1993.

[13] R. Downey, *D.r.e. degrees and the Nondiamond theorem*, ***The Bulletin of the London Mathematical Society***, vol. 21 (1989), no. 1, pp. 43–50.

[14] Y. L. Ershov, *On a hierarchy of sets I*, ***Algebra i Logika***, vol. 7 (1968), no. 1, pp. 47–73.

[15] ———, *On a hierarchy of sets II*, ***Algebra i Logika***, vol. 7 (1968), no. 3, pp. 15–47.

[16] M. Jamaleev, *Splitting and cupping in the d-c. e. degrees*, submitted to ***Izvestiya Vysshikh Uchebnykh Zavedeniĭ. Matematika***.

[17] C. G. Jockusch, Jr., M. Lerman, R. I. Soare, and R. M. Solovay, *Recursively enumerable sets modulo iterated jumps and extensions of Arslanov's completeness criterion*, ***The Journal of Symbolic Logic***, vol. 54 (1989), no. 4, pp. 1288–1323.

[18] H. Rogers, Jr., ***Theory of Recursive Functions and Effective Computability***, McGraw-Hill, New York, 1967.

[19] V. L. Selivanov, *The Ershov hierarchy*, ***Akademiya Nauk SSSR. Sibirskoe Otdelenie. Sibirskiĭ Matematicheskiĭ Zhurnal***, vol. 26 (1985), no. 1, pp. 134–149.

[20] R. I. Soare, ***Recursively Enumerable Sets and Degrees***, Perspectives in Mathematical Logic, Springer-Verlag, Berlin, 1987.

[21] Y. Yang and L. Yu, *On differences among elementary theories of finite levels of Ershov hierarchies*, ***Theory and Applications of Models of Computation***, Lecture Notes in Computer Science, vol. 3959, Springer, Berlin, 2006, pp. 765–771.

KAZAN STATE UNIVERSITY
DEPARTMENT OF MATHEMATICS
KREMLEVSKAJA 18
420008, KAZAN, RUSSIA
E-mail: Marat.Arslanov@ksu.ru

A UNIFIED APPROACH TO ALGEBRAIC SET THEORY

BENNO VAN DEN BERG AND IEKE MOERDIJK

§1. Introduction. This short paper provides a summary of the tutorial on categorical logic given by the second named author at the Logic Colloquium in Nijmegen. Before we go into the subject matter, we would like to express our thanks to the organisers for an excellent conference, and for offering us the opportunity to present this material.

Categorical logic studies the relation between category theory and logical languages, and provides a very efficient framework in which to treat the syntax and the model theory on an equal footing. For a given theory T formulated in a suitable language, both the theory itself and its models can be viewed as categories with structure, and the fact that the models are models of the theory corresponds to the existence of canonical functors between these categories. This applies to ordinary models of first order theories, but also to more complicated topological models, forcing models, realisability and dialectica interpretations of intuitionistic arithmetic, domain-theoretic models of the λ-calculus, and so on. One of the best worked out examples is that where T extends the theory HHA of higher order Heyting arithmetic [24], which is closely related to the Lawvere-Tierney theory of elementary toposes. Indeed, every elementary topos (always taken with a natural numbers object here) provides a categorical model for HHA, and the theory HHA itself also corresponds to a particular topos, the "free" one, in which the true sentences are the provable ones.

The logic of many particular toposes shares features of independence results in set theory. For example, there are very natural constructions of toposes which model HHA plus classical logic in which the axiom of choice fails, or in which the continuum hypothesis is refuted. In addition, one easily finds topological sheaf toposes which model famous consistency results of intuitionistic logic, such as the consistency of HHA plus the continuity of all real-valued functions on the unit interval, and realisability toposes validating HHA plus "Church's thesis" (all functions from the natural numbers to itself are recursive). It took some effort (by Freyd, Fourman, McCarthy, Blass and Scedrov [13, 12, 6, 7] and many others), however, to modify the constructions

Logic Colloquium '06
Edited by Barry Cooper, Herman Geuvers, Anand Pillay, and Jouko Väänänen
Lecture Notes in Logic, 32

so as to provide models proving the consistency of such statements with HHA replaced by an appropriate set theory such as ZF or its intuitionistic counterpart IZF. This modification heavily depended on the fact that the toposes in question, namely various so-called Grothendieck toposes and Hyland's effective topos [18], were in some sense defined in terms of sets.

The original purpose of "algebraic set theory" [22] was to identify a categorical structure independently of sets, which would allow one to construct models of set theories like (I)ZF. These categorical structures were pairs $(\mathcal{E}, \mathcal{S})$ where $\mathcal{E}$ is a category much like a topos, and $\mathcal{S}$ is a class of arrows in $\mathcal{E}$ satisfying suitable axioms, and referred to as the class of "small maps". It was shown in loc. cit. that any such structure gave rise to a model of (I)ZF. An important feature of the axiomisation in terms of such pairs $(\mathcal{E}, \mathcal{S})$ is that it is preserved under the construction of categories of sheaves and of realisability categories, so that the model constructions referred to above become special cases of a general and "elementary" preservation result.

In recent years, there has been a lot of activity in the field of algebraic set theory, which is well documented on the web site www.phil.cmu.edu/projects/ast. Several variations and extensions of the the original Joyal-Moerdijk axiomatisation have been developed. In particular, Alex Simpson [30] developed an axiomatisation in which $\mathcal{E}$ is far from a topos (in his set-up, $\mathcal{E}$ is not exact, and is only assumed to be a regular category). This allowed him to include the example of classes in IZF, and to prove completeness for IZF of models constructed from his categorical pairs $(\mathcal{E}, \mathcal{S})$. This approach has been further developed by Awodey, Butz, Simpson and Streicher in their paper [3], in which they prove a categorical completeness theorem characterising the category of small objects in such a pair $(\mathcal{E}, \mathcal{S})$ (cf. Theorem 3.9 below), and identify a weak "basic" intuitionistic set theory BIST corresponding to the core of the categorical axioms in their setting.

In other papers, a variant has been developed which is adequate for constructing models of *predicative* set theories like Aczel's theory CZF [1, 2]. The most important feature of this variant is that in the structure $(\mathcal{E}, \mathcal{S})$, the existence of suitable power objects is replaced by that of inductive W-types. These W-types enabled Moerdijk and Palmgren in [28] to prove the existence of a model V for CZF out of such a structure $(\mathcal{E}, \mathcal{S})$ on the basis of some exactness assumptions on $\mathcal{E}$, and to derive the preservation of (a slight extension of) the axioms under the construction of sheaf categories. This result was later improved by Van den Berg [35]. It is precisely at this point, however, that we believe our current set-up to be superior to the ones in [28] and [35], and we will come back to this in some detail in Section 6 below. We should mention here that sheaf models for CZF have also been considered by Gambino [15] and to some extent go back to Grayson [17]. Categorical pairs $(\mathcal{E}, \mathcal{S})$ for weak predicative set theories have also been considered by Awodey-Warren [5] and

Simpson [31]. (Note, however, that these authors do not consider W-types and only deal with set theories weaker than Aczel's CZF.)

The purpose of this paper is to outline an axiomatisation of algebraic set theory which combines the good features of all the approaches mentioned above. More precisely, we will present axioms for pairs $(\mathcal{E}, \mathcal{S})$ which

- imply the existence in $\mathcal{E}$ of a universe V, which models a suitable set theory (such as IZF) (cf. Theorem 4.1 below);
- allow one to prove completeness theorems of the kind in [30] and [3] (cf. Theorem 3.7 and Theorem 3.9 below);
- work equally well in the predicative context (to construct models of CZF);
- are preserved under the construction of sheaf categories, so that the usual topological techniques automatically yield consistency results for IZF, CZF and similar theories;
- hold for realisability categories (cf. Examples 5.3 and 5.4 and Theorem Theorem 7.1).

Before we do so, however, we will recall the axioms of the systems IZF and CZF of set theory. In the next Section, we will then present our axioms for small maps, and compare them (in Subsection 3.4) to those in the literature. One of the main features of our axiomatisation is that we do not require the category $\mathcal{E}$ to be exact, but only to possess quotients of "small" equivalence relations. This restricted exactness axiom is consistent with the fact that every object is separated (in the sense of having a small diagonal), and is much easier to deal with in many contexts, in particular those of sheaves. Moreover, together with the Collection axiom this weakened form of exactness suffices for many crucial constructions, such as that of the model V of set theory from the universal small map $E \to U$, or of the associated sheaf of a given presheaf. In Section 4 we will describe the models of set theory obtained from pairs $(\mathcal{E}, \mathcal{S})$ satisfying our axioms, while Section 5 discusses some examples. Finally in Sections 6 and 7, we will discuss in some detail the preservation of the axioms under the construction of sheaf and realisability categories.

Like the tutorial given at the conference, this exposition is necessarily concise, and most of the proofs have been omitted. With the exception perhaps of Sections 6 and 7, these proofs are often suitable adaptations of existing proofs in the literature, notably [22, 30, 28, 3, 33]. A complete exposition with full proofs will appear as [37, 38, 39].

We would like to thank Thomas Streicher, Jaap van Oosten and the anonymous referees for their comments on an earlier draft of this paper, and Thomas Streicher in particular for suggesting the notion of a display map defined in Section 7.

§2. Constructive set theories. In this Section we recall the axioms for the two most prominent constructive variants of Zermelo-Fraenkel set theory, IZF and CZF. Like ordinary ZF, these two theories are formulated in first-order logic with one non-logical symbol ε. But unlike ordinary set theory, these theories are constructive, in that their underlying logic is intuitionistic.

In the formulation of the axioms, we use the following standard abbreviations: $\exists x \,\varepsilon\, a\,(\dots)$ for $\exists x\,(x \,\varepsilon\, a \wedge \dots)$, and $\forall x \,\varepsilon\, a\,(\dots)$ for $\forall x\,(x \,\varepsilon\, a \rightarrow \dots)$. Recall also that a formula is called *bounded*, when all the quantifiers it contains are of one of these two forms. Finally, a formula of the form $\forall x \,\varepsilon\, a\, \exists y \,\varepsilon\, b\, \phi \wedge \forall y \,\varepsilon\, b\, \exists x \,\varepsilon\, a\, \phi$ will be abbreviated as:

$$\mathrm{B}(x \,\varepsilon\, a, y \,\varepsilon\, b)\, \phi.$$

The axioms which both theories have in common are (the universal closures of):

Extensionality: $\forall x\,(x \,\varepsilon\, a \leftrightarrow x \,\varepsilon\, b) \rightarrow a = b$.
Empty set: $\exists x\, \forall y\, \neg y \,\varepsilon\, x$.
Pairing: $\exists x\, \forall y\,(y \,\varepsilon\, x \leftrightarrow y = a \vee y = b)$.
Union: $\exists x\, \forall y\,(y \,\varepsilon\, x \leftrightarrow \exists z \,\varepsilon\, a\; y \,\varepsilon\, z)$.
ε-induction: $\forall x\,(\forall y \,\varepsilon\, x\, \phi(y) \rightarrow \phi(x)) \rightarrow \forall x\, \phi(x)$
Bounded separation: $\exists x\, \forall y\,(y \,\varepsilon\, x \leftrightarrow y \,\varepsilon\, a \wedge \phi(y))$, for any bounded formula ϕ in which a does not occur.
Strong collection: $\forall x \,\varepsilon\, a\, \exists y\, \phi(x, y) \rightarrow \exists b\, \mathrm{B}(x \,\varepsilon\, a, y \,\varepsilon\, b)\, \phi$.
Infinity: $\exists a\,(\exists x\; x \,\varepsilon\, a) \wedge (\forall x \,\varepsilon\, a\, \exists y \,\varepsilon\, a\; x \,\varepsilon\, y)$.

One can obtain an axiomatisation for the constructive set theory IZF by adding to the axioms above the following two statements:

Full separation: $\exists x\, \forall y\,(y \,\varepsilon\, x \leftrightarrow y \,\varepsilon\, a \wedge \phi(y))$, for any formula ϕ in which a does not occur.
Power set axiom: $\exists x\, \forall y\,(y \,\varepsilon\, x \leftrightarrow y \subseteq a)$.

To obtain the predicative constructive set theory CZF, one should add instead the following axiom (which is a weakening of the Power Set Axiom):

Subset collection: $\exists c \forall z (\forall x \,\varepsilon\, a \exists y \,\varepsilon\, b \phi(x, y, z) \rightarrow \exists d \,\varepsilon\, c \mathrm{B}(x \,\varepsilon\, a, y \,\varepsilon\, d) \phi(x, y, z))$.

The Subset Collection Axiom has a more palatable formulation (equivalent to it relative to the other axioms), called Fullness (see [2]). Write $\boldsymbol{mv}(a, b)$ for the class of all multi-valued functions from a set a to a set b, i.e. relations R such that $\forall x \,\varepsilon\, a\, \exists y \,\varepsilon\, b\, (x, y) \,\varepsilon\, R$.

Fullness: $\exists u\,(u \subseteq \boldsymbol{mv}(a, b) \wedge \forall v \,\varepsilon\, \boldsymbol{mv}(a, b)\, \exists w \,\varepsilon\, u\,(w \subseteq v))$.

Using this formulation, it is also easier to see that Subset Collection implies Exponentiation, the statement that the functions from a set a to a set b form a set.

§3. Categories with small maps. Here we introduce the categorical structure which is necessary to model set theory. The structure is that of a category $\mathcal{E}$ equipped with a class of morphisms $\mathcal{S}$, satisfying certain axioms and being referred to as the *class of small maps*. The canonical example is the one where $\mathcal{E}$ is the category of classes in a model of some weak set theory, and morphisms between classes are small in case all the *fibres* are sets. More examples will follow in Section 5. In Section 4, we will show that these axioms actually provide us with the means of constructing models of set theory.

3.1. Axioms. In our work, the underlying category $\mathcal{E}$ is a Heyting category with sums. More precisely, $\mathcal{E}$ satisfies the following axioms (for an excellent account of the notions involved, see [20, Part A1]):

- $\mathcal{E}$ is cartesian, i.e. it has finite limits.
- $\mathcal{E}$ is regular, i.e. every morphism factors as a cover followed by a mono and covers are stable under pullback.
- $\mathcal{E}$ has finite disjoint and stable coproducts.
- $\mathcal{E}$ is Heyting, i.e. for any morphism $f : X \longrightarrow Y$ the functor $f^* : \mathrm{Sub}(Y) \longrightarrow \mathrm{Sub}(X)$ has a right adjoint $\forall_f$.

This expresses precisely that $\mathcal{E}$ is a categorical structure suitable for modelling a typed version of first-order intuitionistic logic with finite product and sum types.

We now list the axioms that we require to hold for a class of small maps, extending the axioms for a class of open maps (see [22]). We will comment on the relation between our axiomatisation and existing alternatives in Subsection 3.4 below.

The axioms for a class of open maps $\mathcal{S}$ are:

(A1) (Pullback stability) In any pullback square

$$\begin{array}{ccc} D & \longrightarrow & B \\ {\scriptstyle g}\downarrow & & \downarrow{\scriptstyle f} \\ C & \xrightarrow[p]{} & A, \end{array}$$

where $f \in \mathcal{S}$, also $g \in \mathcal{S}$.

(A2) (Descent) Whenever in a pullback square as above, $g \in \mathcal{S}$ and p is a cover, $f \in \mathcal{S}$.

(A3) (Sums) Whenever $X \longrightarrow Y$ and $X' \longrightarrow Y'$ belong to $\mathcal{S}$, so does $X + X' \longrightarrow Y + Y'$.

(A4) (Finiteness) The maps $0 \longrightarrow 1$, $1 \longrightarrow 1$ and $2 = 1 + 1 \longrightarrow 1$ belong to $\mathcal{S}$.

(A5) (Composition) $\mathcal{S}$ is closed under composition.

(A6) (Quotients) In any commutative triangle

$$\begin{array}{ccc} Z & \xrightarrow{p} & Y \\ & {\scriptstyle g}\searrow \quad \swarrow{\scriptstyle f} & \\ & X, & \end{array}$$

where p is a cover and g belongs to $\mathcal{S}$, so does f.

These axioms are of two kinds: the axioms (A1-3) express that the property we are interested in is one of the *fibres* of maps in $\mathcal{S}$. The others are more set-theoretic: (A4) says that the collections containing 0, 1 or 2 elements are sets. (A5) is a union axiom: the union of a small disjoint family of sets is again a set. Finally, (A6) is a form of replacement: the image of a set is again a set.

We will always assume that a class of small maps $\mathcal{S}$ satisfies the following two additional axioms, familiar from [22]:

(C) (Collection) Any two arrows $p : Y \longrightarrow X$ and $f : X \longrightarrow A$ where p is a cover and f belongs to $\mathcal{S}$ fits into a quasi-pullback diagram[1] of the form

$$\begin{array}{ccccc} Z & \longrightarrow & Y & \xrightarrow{p} & X \\ {\scriptstyle g}\downarrow & & & & \downarrow{\scriptstyle f} \\ B & & \xrightarrow[h]{} & & A, \end{array}$$

where h is a cover and g belongs to $\mathcal{S}$.

(R) (Representability, see Remark 3.4) There exists a small map $\pi : E \longrightarrow U$ (a "universal small map") such that for every small map $f : X \longrightarrow Y$ there is a diagram of the shape

$$\begin{array}{ccccc} X & \longleftarrow & A & \longrightarrow & E \\ {\scriptstyle f}\downarrow & & \downarrow & & \downarrow{\scriptstyle \pi} \\ Y & \xleftarrow[p]{} & B & \longrightarrow & U, \end{array}$$

where the left square is a quasi-pullback, the right square is a pullback and p is a cover.

The collection principle (C) expresses that in the internal logic it holds that for any cover $p : Y \longrightarrow X$ with small codomain there is a cover $Z \longrightarrow X$ with small domain that factors through p, while (R) says that there is a (necessarily class-sized) family of sets $(E_u)_{u \in U}$ such that any set is covered by one in this family.

[1] Recall that a commutative square in a regular category is called a quasi-pullback if the unique arrow from the initial vertex of the square to the inscribed pullback is a cover.

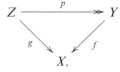

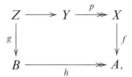

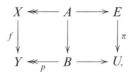

The next requirement is also part of the axioms in [22]. For a morphism $f : X \longrightarrow Y$, the pullback functor $f^* : \mathcal{E}/Y \longrightarrow \mathcal{E}/X$ always has a left adjoint Σ_f given by composition[2]. It has a right adjoint Π_f only when f is exponentiable.

(ΠE) (Existence of Π) The right adjoint Π_f exists, whenever f belongs to $\mathcal{S}$.

This intuitively means that for any set A and class X there is a class of functions from A to X.

When f is exponentiable, one can define an endofunctor P_f (the polynomial functor associated with f) as the composition:

$$P_f = \Sigma_Y \Pi_f X^*.$$

Its initial algebra (whenever it exists) is called the W-type associated to f. For extensive discussion and examples of these W-types we refer the reader to [27, 33, 16]. We impose the axiom (familiar from [27, 14]):

(WE) (Existence of W) The W-type associated to any map $f : X \longrightarrow Y$ in $\mathcal{S}$ exists.

In non-categorical terms this means that for a signature consisting of a (possibly class-sized) number of term constructors each of which has an arity forming a set, the free term algebra exists (but maybe not as a set).

The following two axioms are necessary to have bounded separation as an internally valid principle (see Remark 3.3). For this purpose we need a piece of terminology: call a subobject

$$m : A \rightarrowtail X$$

$\mathcal{S}$-bounded, whenever m belongs to $\mathcal{S}$; note that the $\mathcal{S}$-bounded subobjects form a submeetsemilattice of $\mathrm{Sub}(X)$. We impose the following axiom:

(HB) (Heyting axiom for bounded subobjects) For any small map $f : Y \longrightarrow X$ the functor $\forall_f : \mathrm{Sub}(Y) \longrightarrow \mathrm{Sub}(X)$ maps $\mathcal{S}$-bounded subobjects to $\mathcal{S}$-bounded subobjects.

In addition, we require that all equalities are bounded. Call an object X separated, when the diagonal $\Delta : X \longrightarrow X \times X$ is small. We furthermore impose (see [3]):

(US) (Universal separation) All objects are separated.

We finally demand a limited form of *exactness*, by requiring the existence of quotients for a restricted class of equivalence relations. To formulate this categorically, we recall the following definitions. Two parallel arrows

$$R \underset{r_1}{\overset{r_0}{\rightrightarrows}} X$$

---

[2] We will write X^* and Σ_X for f^* and Σ_f, where f is the unique map $X \longrightarrow 1$.

in category $\mathcal{E}$ form an *equivalence relation* when for any object A in $\mathcal{E}$ the induced function

$$\mathrm{Hom}(A, R) \longrightarrow \mathrm{Hom}(A, X) \times \mathrm{Hom}(A, X)$$

is an injection defining an equivalence relation on the set $\mathrm{Hom}(A, X)$. We call an equivalence relation bounded, when R is a bounded subobject of $X \times X$. A morphism $q : X \longrightarrow Q$ is called the *quotient* of the equivalence relation, if the diagram

$$R \underset{r_1}{\overset{r_0}{\rightrightarrows}} X \overset{q}{\longrightarrow} Q$$

is both a pullback and a coequaliser. In this case, the diagram is called *exact*. The diagram is called *stably exact*, when for any $p : P \longrightarrow Q$ the diagram

$$p^*R \underset{p^*r_1}{\overset{p^*r_0}{\rightrightarrows}} p^*X \overset{p^*q}{\longrightarrow} p^*Q$$

is also exact. If the quotient completes the equivalence relation to a stably exact diagram, we call the quotient stable.

In the presence of (US), any equivalence relation that has a (stable) quotient, must be bounded. So our last axiom imposes the maximum amount of exactness that can be demanded:

(BE) (Bounded exactness) All $\mathcal{S}$-bounded equivalence relations have stable quotients.

This completes our definition of a class of small maps. A pair $(\mathcal{E}, \mathcal{S})$ satisfying the above axioms now will be called *a category with small maps*.

When a class of small maps $\mathcal{S}$ has been fixed, we call a map f small if it belongs to $\mathcal{S}$, an object A small if $A \longrightarrow 1$ is small, a subobject $m : A \longrightarrow X$ small if A is small, and a relation $R \subseteq C \times D$ small if the composite

$$R \subseteq C \times D \longrightarrow D$$

is small.

We conclude this Subsection with some remarks on a form of exact completion relative to a class of small maps. As a motivation, notice that axiom (BE) is not satisfied in our canonical example, where $\mathcal{E}$ is the category of classes in a model of some weak set theory. To circumvent this problem, we will prove the following theorem in our companion paper [37]:

THEOREM 3.1. *The axiom* (BE) *is conservative over the other axioms, in the following precise sense. Any category $\mathcal{E}$ equipped with a class of maps $\mathcal{S}$ satisfying all axioms for a class of small maps except* (BE) *can be embedded in a category $\overline{\mathcal{E}}$ equipped with a class of small maps $\overline{\mathcal{S}}$ satisfying all the axioms, including* (BE). *Moreover, the embedding $\mathbf{y} : \mathcal{E} \longrightarrow \overline{\mathcal{E}}$ is fully faithful, bijective on subobjects and preserves the structure of a Heyting category with sums, hence* *preserves and reflects validity of statements in the internal logic. Finally, it also preserves and reflects smallness, in the sense that* $\mathbf{y}f$ *belongs to* $\overline{\mathcal{S}}$ *iff* f *belongs to* $\mathcal{S}$.

The category $\overline{\mathcal{E}}$ is obtained by formally adjoining quotients for bounded equivalence relations, as in [11, 9]. Furthermore, a map $g : B \longrightarrow A$ in $\overline{\mathcal{E}}$ belongs to $\overline{\mathcal{S}}$ iff it fits into a quasi-pullback square

$$\begin{array}{ccc} \mathbf{y}D & \twoheadrightarrow & B \\ {\scriptstyle \mathbf{y}f}\downarrow & & \downarrow{\scriptstyle g} \\ \mathbf{y}C & \twoheadrightarrow & A, \end{array}$$

with f belonging to $\mathcal{S}$ in $\mathcal{E}$.

3.2. Consequences. Among the consequences of these axioms we list the following.

REMARK 3.2. For any object X in $\mathcal{E}$, the slice category $\mathcal{E}/X$ is equipped with a class of small maps $\mathcal{S}/X$, by declaring that an arrow $p \in \mathcal{E}/X$ belongs to $\mathcal{S}/X$ whenever $\Sigma_X f$ belongs to $\mathcal{S}$. Any further requirement for a class of small maps should be stable under slicing in this sense, if it is to be a sensible addition. We will not explicitly check this every time we introduce a new axiom, and leave this to the reader.

REMARK 3.3. In a category $\mathcal{E}$ with small maps the following internal form of "bounded separation" holds. If $\phi(x)$ is a formula in the internal logic of $\mathcal{E}$ with free variable $x \in X$, all whose basic predicates are bounded, and contains existential and universal quantifications $\exists_f$ and $\forall_f$ only along small maps f, then

$$A = \{x \in X \mid \phi(x)\} \subseteq X$$

defines a bounded subobject of X. In particular, smallness of X implies smallness of A.

REMARK 3.4. It follows from the axioms that any class of small maps $\mathcal{S}$ is also representable in the stronger sense that there is a universal small map $\pi : E \longrightarrow U$ such that for every small map $f : X \longrightarrow Y$ there is a diagram of the shape

$$\begin{array}{ccccc} X & \twoheadleftarrow & A & \longrightarrow & E \\ {\scriptstyle f}\downarrow & & \downarrow & & \downarrow{\scriptstyle \pi} \\ Y & \underset{p}{\twoheadleftarrow} & B & \longrightarrow & U, \end{array}$$

where the left square is a pullback, the right square is a pullback and p is a cover. Actually, this is how representability was stated in [22]. We have chosen the weaker formulation (R), because it is easier to check in some examples.

REMARK 3.5. Using the axioms (ΠE), (R), (HB) and (BE), it can be shown along the lines of Theorem 3.1 in [22] that for any class of small maps the following axiom holds:

(PE) (Existence of power class functor) For any object C in $\mathcal{E}$ there exists a power object $\mathcal{P}_s C$ and a small relation $\in_C \subseteq C \times \mathcal{P}_s C$ such that, for any D and any small relation $R \subseteq C \times D$, there exists a unique map $\rho : D \longrightarrow \mathcal{P}_s C$ such that the square:

$$\begin{array}{ccc} R & \longrightarrow & \in_C \\ \downarrow & & \downarrow \\ C \times D & \xrightarrow[1 \times \rho]{} & C \times \mathcal{P}_s C \end{array}$$

is a pullback.

In addition, one can show that the object $\mathcal{P}_s C$ is unique (up to isomorphism) with this property, and that the assignment $C \mapsto \mathcal{P}_s C$ is functorial.

A special role is played by $\Omega_b = \mathcal{P}_s 1$, what one might call the object of bounded truth-values, or the bounded subobject classifier. There are a couple of observations one can make: bounded truth-values are closed under small infima and suprema, implication, and truth and falsity are bounded truth-values. A subobject $m : A \longrightarrow X$ is bounded, when the assertion "$x \in A$" has a bounded truth-value for any $x \in X$, as such bounded subobjects are classified by maps $X \longrightarrow \Omega_b$.

REMARK 3.6 (See [5]). When $\mathcal{E}$ is a category with a class of small maps $\mathcal{S}$, and we fix an object $X \in \mathcal{E}$, we can define a full subcategory $\mathcal{S}_X$ of $\mathcal{E}/X$, whose objects are small maps into X. The category $\mathcal{S}_X$ is a Heyting pretopos, and the inclusion into $\mathcal{E}/X$ preserves this structure; this was proved in [5]. This result can be regarded as a kind of categorical "soundness" theorem, in view of the following corresponding "completeness" theorem, which is analogous to Grothendieck's result that every pretopos arises as the coherent objects in a coherent topos (see [21, Section D.3.3]).

THEOREM 3.7. *Any Heyting pretopos $\mathcal{H}$ arises as the category of small objects $\mathcal{S}_1$ in a category $\mathcal{E}$ with a class of small maps $\mathcal{S}$.*

This theorem was proved in [5], where, following [3], the objects in $\mathcal{E}$ were called the *ideals* over $\mathcal{H}$.

3.3. Strengthenings. For the purpose of constructing models of important (constructive) set theories, we will consider the following additional properties which a class of small maps may enjoy.

(NE) (Existence of nno) The category $\mathcal{E}$ possesses a natural numbers object.
(NS) (Smallness of nno) In addition, it is small.

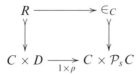

There is no need to impose (NE), as it follows from (WE). The axiom (NS) is necessary for modelling set theories with Infinity. The property (PE) in Remark 3.5 has a similar strengthening, corresponding to the Power set Axiom:

(PS) (Smallness of power classes) For each X the $\mathcal{P}_s$-functor on $\mathcal{E}/X$ preserves smallness of objects over X.

Both (NS) and (PS) were formulated in [22] for the purpose of modelling IZF.

REMARK 3.8 (Cf. [3]). Let X be an object in a category with small maps $(\mathcal{E}, \mathcal{S})$ satisfying (PS). The category $\mathcal{S}_X$ is a topos, and the inclusion into $\mathcal{E}/X$ preserves this structure. In fact, every topos arises in this way:

THEOREM 3.9. *Any topos $\mathcal{H}$ arises as the category of small objects $\mathcal{S}_1$ in a category equipped with a class of small maps satisfying* (PS).

Like Theorem 3.7, this is proved in [3] using the ideal construction.

We will also need to consider requirements corresponding to the axioms of Full Separation and Fullness. To Full Separation corresponds the following axiom, introduced in [22]:

(M) All monos are small.

A categorical axiom corresponding to Fullness was first stated in [36]. In order to formulate it, we need to introduce some notation. For two morphisms $A \longrightarrow X$ and $B \longrightarrow X$, we will denote by $M_X(A, B)$ the poset of multi-valued functions from A to B over X, i.e. jointly monic spans in $\mathcal{E}/X$,

$$A \twoheadleftarrow P \longrightarrow B$$

with $P \longrightarrow X$ small and the map to A a cover. By pullback, any $f : Y \longrightarrow X$ determines an order preserving function

$$f^* : M_X(A, B) \longrightarrow M_Y(f^*A, f^*B).$$

(F) For any two small maps $A \longrightarrow X$ and $B \longrightarrow X$, there exist a cover $p : X' \longrightarrow X$, a small map $f : C \longrightarrow X'$ and an element $P \in M_C(f^*p^*A, f^*p^*B)$, such that for any $g : D \longrightarrow X'$ and $Q \in M_D(g^*p^*A, g^*p^*B)$, there are morphisms $x : E \longrightarrow D$ and $y : E \longrightarrow C$, with $gx = fy$ and x a cover, such that $x^*Q \geq y^*P$.

Though complicated, it is "simply" the Kripke-Joyal translation of the statement that there is for any pair of small objects A and B, a small collection P of multi-valued relations between A and B, such that any multi-valued relation contains one in P.

3.4. Relation to other settings. The axioms for a category with small maps $(\mathcal{E}, \mathcal{S})$ as we have presented them are very close to the original axioms as presented by Joyal and Moerdijk on pages 6-8 of their book [22]. We only require the weak form of exactness of (BE) (instead of ordinary exactness), and added the axioms (WE), (HB) and (US).

Since the appearance of [22], various axiomatisations have been proposed, which can roughly be subdivided into three groups. To the first group belong axiom systems extending the original presentation in [22]. Already in [22], it is shown how to extend these axioms for the purpose of obtaining models for IZF, and this is followed up in [23]. In [14] Gambino introduces an extension of the original axiomatisation leading to models of predicative set theories.

A second group of papers starts with Simpson's [30] and comprises [3, 8, 31, 4, 5]. In these axiomatisations, the following axioms which are here taken as basic are regarded as optional features: the Collection Axiom (C), Bounded Exactness (BE), and also (WE) (although they all hold in the category of ideals). Instead, the existence of a $\mathcal{P}_s$-functor as in (PE) is postulated, as is a model of set theory, either in the form of a universe, or a universal object. In the approach taken here, these are properties derived from the existence of a universal small map $\pi : E \longrightarrow U$. Part of the purpose of this paper is to make clear that the results for axiom systems in [30, 3, 5] also hold for our axiomatisation. We list the achievements in order to make a comparison possible: in [30], Simpson obtained a set-theoretic completeness result for an impredicative set theory (compare Theorem 4.4). Then in [3], Awodey, Butz, Simpson and Streicher prove a categorical completeness result of which our Theorem 3.9 is variant. A predicative version of this result which does not involve W-types but is otherwise analogous to Theorem 3.7 above, was then proved by Awodey and Warren in [5].

The fact that our set-up contains the Collection Axiom (C) makes it less appropriate for modelling set theories based on the axiom of Replacement. However, in our theory this Collection Axiom plays a crucial rôle: for example, in the construction of the initial ZF-algebra from W-types (see Theorem 4.1 below), or in showing the existence of the associated sheaf functor.

A third group of papers starts with [28], and continues with [35, 34]. These axiomatisations have a flavour different from the others, because here the axioms for a class of small maps do not extend the axioms for a class of open maps, as the Quotient Axiom (A6) is dropped. The aim of Moerdijk and Palmgren in [28] was to find an axiomatisation related to Martin-Löf's predicative type theory which included the category-theoretic notion of a W-type, from which models of Aczel's CZF could be constructed. We will point out below that the same is true here (in fact, we can construct models of CZF proper, rather than of something less or more). Another concern of [28], which is also the topic of Van den Berg's paper [35], is the stability of the notion of category with small maps under sheaves. The earlier results in [28] and [35] concerning sheafification were less than fully satisfactory. For the notion of category with small maps explained here, the theory of sheaves can be developed very smoothly (see Section 6), using the combination of the axioms (BE) and (US). We consider this one of the main advantages of the present axiomatisation.

§4. Models of set theory. For the purpose of discussing models of set theory, we recall from [22] the notion of a ZF-*algebra* in a category with small maps $(\mathcal{E}, \mathcal{S})$. A ZF-algebra V is an object in $\mathcal{E}$ equipped with two independent algebraic structures: on the one hand, it is an (internal) poset with small (in the sense of $\mathcal{S}$) sups. On the other hand, it is equipped with an endomap $s : V \longrightarrow V$, called "successor". A morphism of ZF-algebras should preserve both these structures: the small suprema, and the successor.

A crucial result is the following:

THEOREM 4.1. *In any category with small maps* $(\mathcal{E}, \mathcal{S})$, *the initial* ZF-*algebra exists.*

This theorem can be proved along the lines of [28]. Indeed, one can consider the W-type associated to the universal small map $\pi : E \longrightarrow U$. One can then show that the equivalence relation given by bisimulation is bounded so that the quotient exists. This quotient is the initial ZF-algebra (more details will appear in [37]). This initial ZF-algebra has a natural interpretation as a model of set theory. We think of the order as inclusion, suprema as union, and sx as $\{x\}$. This suggests to define membership as:

$$x \,\varepsilon\, y := sx \leq y.$$

Since $\mathcal{E}$ is a Heyting category, one can ask oneself the question which set-theoretic statements the structure (V, ε) satisfies in the internal logic of $\mathcal{E}$. The answer is given by the following theorem, whose second part was proved in [22] (the first part can be proved in a similar manner):

THEOREM 4.2. *Let* $(\mathcal{E}, \mathcal{S})$ *be a category with small maps in which the natural numbers object is small* (*so* (NS) *holds*).

1. *If* $(\mathcal{E}, \mathcal{S})$ *satisfies the Fullness Axiom* (F), *then the initial* ZF-*algebra models* CZF.
2. *If* $(\mathcal{E}, \mathcal{S})$ *satisfies the Power Set Axiom* (PS) *and the Separation axiom* (M), *then the initial* ZF-*algebra models* IZF.

REMARK 4.3. To obtain models for classical set theories, one may work in Boolean categories. Initial ZF-algebras in such categories validate classical logic, and therefore model classical set theories.

As a counterpart to Theorem 4.2 we can formulate a completeness theorem:

THEOREM 4.4. *The semantics of Theorem* 4.2 *is complete for both* CZF *and* IZF *in the following strong sense.*

1. *There is a category with small maps* $(\mathcal{E}, \mathcal{S})$ *satisfying* (NS) *and* (F) *such that its initial* ZF-*algebra* V *has the property that, for any sentence* ϕ *in the language of set theory*:

$$V \models \phi \Leftrightarrow \mathrm{CZF} \vdash \phi.$$

2. *There is a category with small maps* $(\mathcal{E}, \mathcal{S})$ *satisfying* (NS), (M) *and* (PS), *such that its initial* ZF*-algebra* V *has the property that, for any sentence* ϕ *in the language of set theory*:

$$V \models \phi \Leftrightarrow \mathrm{CZF} \vdash \phi.$$

To prove this theorem one builds the syntactic category of classes and a ZF-algebra V such that validity in V is the same as derivability in the appropriate set theory. Problems concerning (BE) are, of course, solved by appealing to Theorem 3.1. The first person to prove a completeness result in this manner was Alex Simpson in [30] for an impredicative set theory. A predicative variation is contained in [5] and [14].

REMARK 4.5. Every (ordinary, classical) set-theoretic model (M, ε) is also subsumed in our account, because every such model there is an initial ZF-algebra V_M in a category with small maps $(\mathcal{E}_M, \mathcal{S}_M)$ having the property that for any set-theoretic sentence ϕ:

$$V_M \models \phi \Leftrightarrow M \models \phi.$$

$\mathcal{E}_M$ is of course the category of classes in the model M, with those functional relations belonging to $\mathcal{S}_M$ that the model believes to have sets as fibres, extended using Theorem 3.1 so as to satisfy (BE). One could prove completeness of our categorical semantics for classical set theories along these lines.

§5. Examples.

We recall from [22] the basic examples of categories satisfying our axioms.

EXAMPLE 5.1. The canonical example is the following. Let $\mathcal{E}$ be the category of classes in some model of set theory, and declare a morphism $f : X \longrightarrow Y$ to be small, when all its fibres are sets. If the set theory is strong enough, this will satisfy all our axioms, except for (BE), but an appeal to Theorem 3.1 will resolve this issue.

EXAMPLE 5.2. Let $\mathcal{E}$ be a category of sets (relative to some model of ordinary set theory, say), and let κ be an infinite regular cardinal. Declare $f : X \longrightarrow Y$ to be small, when all fibres of f have cardinality less than κ. This will validate all our basic axioms, as well as (M). When $\kappa > \omega$, (NS) will also hold, and when κ is inaccessible, (PS) and (F) will hold.

EXAMPLE 5.3. The following two examples are related to realisability, and define classes of small maps on the effective topos $\mathcal{E}ff$ (see [18]). Recall that there is an adjoint pair of functors $\Gamma \dashv \nabla$, where $\Gamma = \mathcal{E}ff(1, -) : \mathcal{E}ff \longrightarrow \mathcal{S}ets$ is the global sections functor. Fix a regular cardinal $\kappa > \omega$, and declare

$f : X \longrightarrow Y$ to be small, whenever there is a quasi-pullback square

$$\begin{array}{ccc} Q & \twoheadrightarrow & X \\ {\scriptstyle g}\downarrow & & \downarrow{\scriptstyle f} \\ P & \underset{p}{\twoheadrightarrow} & Y \end{array}$$

with p a cover, and g a morphism between projectives such that Γg is κ-small, in the sense of the previous example. This example was further studied by Kouwenhoven and Van Oosten in [23], and shown to lead to McCarty's realisability model of set theory for an inaccessible cardinal κ (see [26]).

EXAMPLE 5.4. Another class of small maps on $\mathcal{E}ff$ is given as follows. Call a map $f : X \longrightarrow Y$ small, whenever the statement that all its fibres are subcountable is true in the internal logic of $\mathcal{E}ff$ (a set is subcountable, when it is the quotient of a subset of the natural numbers). These maps were studied in [19] and dubbed "quasi-modest" in [22]. The first author showed they lead to a model of CZF in which all sets are subcountable, and therefore refutes the Power Set Axiom (see [34]). He also showed the model is the same as the one contained in [32] and [25].

EXAMPLE 5.5. Once again, fix an infinite regular cardinal κ, and let $\mathcal{C}$ be a subcanonical site which is κ-small, in the sense that every covering family has cardinality strictly less than κ. We say that a sheaf X is κ-small, whenever it is covered by a collection of representables whose cardinality is less than κ. Finally, a morphism $f : X \longrightarrow Y$ will be considered to be κ-small, whenever for any map $y : C \longrightarrow Y$ from a representable $C \in \mathcal{C}$ the pullback $f^{-1}(y)$ as in

$$\begin{array}{ccc} f^{-1}(y) & \longrightarrow & X \\ \downarrow & & \downarrow{\scriptstyle f} \\ C & \underset{y}{\longrightarrow} & Y \end{array}$$

is a κ-small sheaf. This can again be shown to satisfy all our basic axioms. Also, when $\kappa > \omega$, (NS) will hold, and so will (PS) and (F), when κ is inaccessible.

§6. **Predicative sheaf theory.** The final example of the previous Section, that of sheaves, can be internalised, in a suitable sense. Starting from a category with small maps $(\mathcal{E}, \mathcal{S})$, and an appropriate site $\mathcal{C}$ in $\mathcal{E}$, one can build the category $\mathrm{Sh}_{\mathcal{E}}(\mathcal{C})$ of internal sheaves over $\mathcal{C}$, which is again a Heyting category with stable, disjoint sums. Furthermore, there is a notion of small maps between sheaves, turning it into a category with small maps. In fact, stability of our notion of a category with small maps under sheaves is one of its main assets. Here we will limit ourselves to formulating precise statements, leaving the proofs for [39].

For the site $\mathcal{C}$ we assume first of all that the underlying category is small, in that the object of objects C_0 and of arrows C_1 are both small. By a *sieve* on $a \in C_0$ we mean a *small* collection of arrows into a closed under precomposition. We assume that the collection of covering sieves $\mathrm{Cov}(a)$ on an object $a \in C_0$ satisfies the following axioms:

(M) The maximal sieve $M_a = \{f \in C_1 \mid \mathrm{cod}(f) = a\}$ belongs to $\mathrm{Cov}(a)$.

(L) For any $U \in \mathrm{Cov}(a)$ and morphism $f : b \longrightarrow a$, the sieve $f^*U = \{g : c \longrightarrow b \mid fg \in U\}$ belongs to $\mathrm{Cov}(b)$.

(T) If T is a sieve on a, such that for a fixed $U \in \mathrm{Cov}(a)$ any pullback h^*T along a map $h : b \longrightarrow a \in U$ is an element of $\mathrm{Cov}(b)$, then $T \in \mathrm{Cov}(a)$.

The definition of an (internal) presheaf and sheaf is as usual.

Using the bounded exactness of $(\mathcal{E}, \mathcal{S})$ and assuming that the relation $S \in \mathrm{Cov}(a)$ is bounded, one can show the existence of the associated sheaf functor (the cartesian left adjoint for the inclusion of sheaves into presheaves). This functor can then be used to prove in the usual way that the sheaves form a Heyting category with stable and disjoint sums.

As the small maps between sheaves we take those that are "pointwise small". Observe that there is a forgetful functor $U : \mathrm{Sh}_{\mathcal{E}}(\mathcal{C}) \longrightarrow \mathcal{E}/C_0$, and call a morphism $f : B \longrightarrow A$ of sheaves *pointwise small*, when Uf is. To show that these morphisms form a class of small maps, we make two additional assumptions. First of all, we assume the Exponentiation Axiom in the "metatheory" $(\mathcal{E}, \mathcal{S})$:

(ΠS) For any small map $f : B \longrightarrow A$, the functor $\Pi_f : \mathcal{E}/B \to \mathcal{E}/A$ preserves small objects.

Furthermore, we also assume that our site has a basis, meaning the following: for any $a \in C_0$ there is a *small* collection of covering sieves $\mathrm{BCov}(a)$ such that

$$S \in \mathrm{Cov}(a) \Leftrightarrow \exists R \in \mathrm{BCov}(a) : R \subseteq S.$$

Note that the relation $S \in \mathrm{Cov}(a)$ is bounded, when the site has a basis.

THEOREM 6.1. *Let $(\mathcal{E}, \mathcal{S})$ be a category with small maps, and let $\mathcal{C}$ be an internal site with a basis. If the class $\mathcal{S}$ satisfies* (ΠS), *then* $\mathrm{Sh}_{\mathcal{E}}(\mathcal{C})$ *with the class of pointwise small maps is again a category with small maps satisfying* (ΠS). *Furthermore, all the axioms that we have introduced,* (NS), (PS), (F) *and* (M), *are stable in the sense that each of these holds in sheaves, whenever it holds in the original category.*

§7. **Predicative realisability.** In this Section we outline how the construction of [22] of a class of small maps in Hyland's effective topos (as in Example 5.3), can be mimicked in the context of a category with small maps $(\mathcal{E}, \mathcal{S})$ as introduced in Section 3. Our construction is inspired by the fact that the effective topos arises as the exact completion of the category of assemblies, as in [10].

Let us start with a category with small maps $(\mathcal{E}, \mathcal{S})$ satisfying (NS) (so the nno in $\mathcal{E}$ is small). The first observation is that we can internalise enough recursion theory in $\mathcal{E}$ for doing realisability. In fact, enough can already be formalised in Heyting Arithmetic HA, so certainly in a category with small maps. We then define the category of assemblies, as follows. An *assembly* consists of an object A in $\mathcal{E}$ together with a surjective relation $\alpha \subseteq \mathbb{N} \times A$. For pairs (n, a) belonging to this relation, we write $n \in \alpha(a)$, which we pronounce as "n realises (the existence of) a"; surjectivity of the relation then means that every $a \in A$ has at least one realiser. A morphism f of assemblies from (B, β) to (A, α) is given by a morphism $f : B \longrightarrow A$ in $\mathcal{E}$ for which the internal logic of $\mathcal{E}$ verifies that:

> there is a natural number r such that for all $b \in B$ and $n \in \beta(b)$, the Kleene application $r \cdot n$ is defined, and realises $f(b)$ (i.e. $r \cdot n \in \alpha(fb)$).

One can now prove that the category of assemblies $\mathcal{E}[\mathcal{A}sm]$ relative to $\mathcal{E}$ is a Heyting category with stable and disjoint sums (see [18], where the assemblies occur as the $\neg\neg$-separated objects in the effective topos).

In order to describe the relevant exact completion of this category of assemblies, we first outline a construction. Consider two assemblies (B, β) and (A, α) and a morphism $f : B \longrightarrow A$, not necessarily a morphism of assemblies. Then this defines a morphism of assemblies $(B, \beta[f]) \longrightarrow (A, \alpha)$ by declaring that $n \in \beta[f](b)$, whenever n codes a pair $\langle n_0, n_1 \rangle$ such that $n_0 \in \alpha(fb)$ and $n_1 \in \beta(b)$. In case f belongs to $\mathcal{S}$ and β is a bounded relation, a morphism of this form will be called a *standard display map* relative to $\mathcal{S}$ (this notion was pointed out to us by Thomas Streicher). A *display map* is a morphism that can be written as an isomorphism followed by a standard display map. These display maps do *not* satisfy the axioms for a class of small maps; in particular, they are not closed under Descent and Quotients. Another problem is that the category of assemblies is not exact, not even in the more limited sense of being bounded exact.

Both problems can be solved by appealing to Theorem 3.1. Or, to be more precise, they can be solved by constructing an exact completion for categories with a class of display maps, resulting in categories with small maps satisfying (BE) (how this is to be done will be shown in [38]). Recall that the small maps in the exact completion are precisely those g that fit into a quasi-pullback diagram

$$\begin{array}{ccc} yD & \twoheadrightarrow & B \\ {\scriptstyle yf}\downarrow & & \downarrow{\scriptstyle g} \\ yC & \twoheadrightarrow & A, \end{array}$$

where f is a small map in the original category. Therefore it is to be expected that the class of small maps in the exact completion of a category with display maps satisfies Descent and Quotients even when the class of display maps in the original category from which it is defined, does not satisfy these axioms. In fact, as it turns out, the display maps between assemblies have enough structure for the maps g in the exact completion of assemblies that fit into a square as above with f a display map, to form a class of small maps. In this way, both problems with the category of assemblies can be solved at the same time by moving to the exact completion. Therefore we define the realisability category $(\mathcal{E}[\mathcal{E}ff], \mathcal{S}[\mathcal{E}ff])$ to be this exact completion of the pair $(\mathcal{E}[\mathcal{A}sm], \mathcal{D})$, where $\mathcal{D}$ is the class of display maps in the category of assemblies.

THEOREM 7.1. *If* $(\mathcal{E}, \mathcal{S})$ *is a category with small maps satisfying* (NS), *then so is* $(\mathcal{E}[\mathcal{E}ff], \mathcal{S}[\mathcal{E}ff])$. *Furthermore, all the axioms that we have introduced,* (ΠS), (PS), (F) *and* (M), *are stable in the sense that each of these holds in the realisablity category, whenever it holds in the original category.*

The initial ZF-algebra in the realisability category should be considered as a suitable internal version of McCarty's realisability model [26] (see also [23]), which in our abstract approach is also defined for predicative theories like CZF (compare [29]).

REFERENCES

[1] P. ACZEL, *The type theoretic interpretation of constructive set theory*, ***Logic Colloquium '77 (Proc. Conf., Wrocław, 1977)***, Studies in Logic and the Foundations of Mathematics, vol. 96, North-Holland, Amsterdam, 1978, pp. 55–66.

[2] P. ACZEL and M. RATHJEN, *Notes on Constructive Set Theory*, Technical Report 40, Institut Mittag-Leffler, 2000/2001.

[3] S. AWODEY, C. BUTZ, A. K. SIMPSON, and T. STREICHER, *Relating topos theory and set theory via categories of classes*, Available from `http://www.phil.cmu.edu/projects/ast/`, June 2003.

[4] S. AWODEY and H. FORSSELL, *Algebraic models of intuitionistic theories of sets and classes*, ***Theory and Applications of Categories***, vol. 15 (2005/06), no. 5, pp. 147–163.

[5] S. AWODEY and M.A. WARREN, *Predicative algebraic set theory*, ***Theory and Applications of Categories***, vol. 15 (2005/06), no. 1, pp. 1–39.

[6] A. R. BLASS and A. SCEDROV, *Freyd's models for the independence of the axiom of choice*, ***Memoirs of the American Mathematical Society***, vol. 79 (1989), no. 404, p. 134.

[7] A.R. BLASS and A. SCEDROV, *Complete topoi representing models of set theory*, ***Annals of Pure and Applied Logic***, vol. 57 (1992), no. 1, pp. 1–26.

[8] C. BUTZ, *Bernays-Gödel type theory*, ***Journal of Pure and Applied Algebra***, vol. 178 (2003), no. 1, pp. 1–23.

[9] A. CARBONI, *Some free constructions in realizability and proof theory*, ***Journal of Pure and Applied Algebra***, vol. 103 (1995), pp. 117–148.

[10] A. CARBONI, P. J. FREYD, and A. SCEDROV, *A categorical approach to realizability and polymorphic types*, ***Mathematical Foundations of Programming Language Semantics (New Orleans, LA, 1987)***, Lecture Notes in Computer Science, vol. 298, Springer Verlag, Berlin, 1988, pp. 23–42.

[11] A. CARBONI and R. CELIA MAGNO, *The free exact category on a left exact one*, ***Australian Mathematical Society. Journal. Series A***, vol. 33 (1982), no. 3, pp. 295–301.

[12] M. P. Fourman, *Sheaf models for set theory*, ***Journal of Pure and Applied Algebra***, vol. 19 (1980), pp. 91–101.

[13] P. Freyd, *The axiom of choice*, ***Journal of Pure and Applied Algebra***, vol. 19 (1980), pp. 103–125.

[14] N. Gambino, *Presheaf models for constructive set theories*, ***From Sets and Types to Topology and Analysis***, Oxford Logic Guides, vol. 48, Oxford University Press, Oxford, 2005, pp. 62–77.

[15] ———, *Heyting-valued interpretations for constructive set theory*, ***Annals of Pure and Applied Logic***, vol. 137 (2006), no. 1-3, pp. 164–188.

[16] N. Gambino and J. M. E. Hyland, *Wellfounded trees and dependent polynomial functors*, ***Types for Proofs and Programs***, Lecture Notes in Computer Science, vol. 3085, Springer-Verlag, Berlin, 2004, pp. 210–225.

[17] R. J. Grayson, *Forcing in intuitionistic systems without power-set*, ***The Journal of Symbolic Logic***, vol. 48 (1983), no. 3, pp. 670–682.

[18] J. M. E. Hyland, *The effective topos*, ***The L.E.J. Brouwer Centenary Symposium (Noordwijkerhout, 1981)***, Studies in Logic and the Foundations of Mathematics, vol. 110, North-Holland, Amsterdam, 1982, pp. 165–216.

[19] J. M. E. Hyland, E. P. Robinson, and G. Rosolini, *The discrete objects in the effective topos*, ***Proceedings of the London Mathematical Society. Third Series***, vol. 60 (1990), no. 1, pp. 1–36.

[20] P. T. Johnstone, ***Sketches of an Elephant: A Topos Theory Compendium. Vol. 1***, Oxford Logic Guides, vol. 43, The Clarendon Press Oxford University Press, New York, 2002.

[21] ———, ***Sketches of an Elephant: A Topos Theory Compendium. Vol. 2***, Oxford Logic Guides, vol. 44, The Clarendon Press Oxford University Press, Oxford, 2002.

[22] A. Joyal and I. Moerdijk, ***Algebraic Set Theory***, London Mathematical Society Lecture Note Series, vol. 220, Cambridge University Press, Cambridge, 1995.

[23] C. Kouwenhoven-Gentil and J. van Oosten, *Algebraic set theory and the effective topos*, ***The Journal of Symbolic Logic***, vol. 70 (2005), no. 3, pp. 879–890.

[24] J. Lambek and P. J. Scott, ***Introduction to Higher Order Categorical Logic***, Cambridge Studies in Advanced Mathematics, vol. 7, Cambridge University Press, Cambridge, 1986.

[25] R. S. Lubarsky, *CZF and second order arithmetic*, ***Annals of Pure and Applied Logic***, vol. 141 (2006), no. 1-2, pp. 29–34.

[26] D. C. McCarty, ***Realizability and recursive mathematics***, Ph.D. thesis, Oxford University, Oxford, 1984.

[27] I. Moerdijk and E. Palmgren, *Wellfounded trees in categories*, ***Annals of Pure and Applied Logic***, vol. 104 (2000), no. 1-3, pp. 189–218.

[28] ———, *Type theories, toposes and constructive set theory: predicative aspects of AST*, ***Annals of Pure and Applied Logic***, vol. 114 (2002), no. 1-3, pp. 155–201.

[29] M. Rathjen, *Realizability for constructive Zermelo-Fraenkel set theory*, ***Logic Colloquium '03***, Lecture Notes in Logic, vol. 24, Association for Symbol Logic, La Jolla, CA, 2006, pp. 282–314.

[30] A. K. Simpson, *Elementary axioms for categories of classes (extended abstract)*, ***14th Symposium on Logic in Computer Science (Trento, 1999)***, IEEE Computer Society, Los Alamitos, CA, 1999, pp. 77–85.

[31] A.K. Simpson, *Constructive set theories and their category-theoretic models*, ***From Sets and Types to Topology and Analysis***, Oxford Logic Guides, vol. 48, Oxford University Press, Oxford, 2005, pp. 41–61.

[32] T. Streicher, *Realizability models for* CZF + ¬ *Pow*, Unpublished note, March 2005.

[33] B. van den Berg, *Inductive types and exact completion*, ***Annals of Pure and Applied Logic***, vol. 134 (2005), no. 2-3, pp. 95–121.

[34] ———, ***Predicative topos theory and models for constructive set theory***, Ph.D. thesis, University of Utrecht, The Netherlands, 2006.

[35] ———, *Sheaves for predicative toposes*, Accepted for publication in ***Archive for Mathematical Logic***. Available from `arXiv:math.LO/0507480`, July 2005.

[36] B. van den Berg and F. De Marchi, *Models of non-well-founded sets via an indexed final coalgebra theorem*, ***The Journal of Symbolic Logic***, vol. 72 (2007), no. 3, pp. 767–791.

[37] B. van den Berg and I. Moerdijk, *Aspects of predicative algebraic set theory I: Exact completion*, ***Annals of Pure and Applied Logic***, vol. 156 (2008), pp. 123–159.

[38] ———, *Aspects of predicative algebraic set theory II: Realizability*, Accepted for publication in ***Theoretical Computer Science***.

[39] ———, *Aspects of predicative algebraic set theory III: Sheaves*, Under construction, 2008.

TECHNISCHE UNIVERSITÄT DARMSTADT
FACHBEREICH MATHEMATIK
SCHLOSSGARTENSTR. 7
64289 DARMSTADT, GERMANY
E-mail: berg@mathematik.tu-darmstadt.de

MATHEMATICAL INSTITUTE
UTRECHT UNIVERSITY
PO BOX 80.010
3508 TA UTRECHT, THE NETHERLANDS
E-mail: moerdijk@math.uu.nl

BRIEF INTRODUCTION TO UNPROVABILITY

ANDREY BOVYKIN

Abstract. The article starts with a brief survey of Unprovability Theory as of summer 2006. Then, as an illustration of the subject's model-theoretic methods, we re-prove threshold versions of unprovability results for the Paris-Harrington Principle and the Kanamori-McAloon Principle using indiscernibles. In addition, we obtain a short accessible proof of unprovability of the Paris-Harrington Principle. The proof employs old ideas but uses only one colouring and directly extracts the set of indiscernibles from its homogeneous set. We also present modified, abridged statements whose unprovability proofs are especially simple. These proofs were tailored for teaching purposes.

The article is intended to be accessible to the widest possible audience of mathematicians, philosophers and computer scientists as a brief survey of the subject, a guide through the literature in the field, an introduction to its model-theoretic techniques and, finally, a model-theoretic proof of a modern theorem in the subject. However, some understanding of logic is assumed on the part of the readers.

The intended audience of this paper consists of logicians, logic-aware mathematicians and thinkers of other backgrounds who are interested in unprovable mathematical statements. We start with a brief survey, listing many important achievements and directions of the subject. Most of the results speak for themselves and we omit a discussion of how they are interrelated as well as the story of the subject's big questions, philosophies, goals, exciting conjectures and dreams which is presumed to be partly known to the readers. The survey is biased towards the Paris-Harrington Principle and its exact versions (understanding this topic is an excellent first step for anyone who decides to study unprovability). In the second part of the paper the reader will find full and very accessible unprovability proofs developed for teaching purposes: the optimal unprovability proof of the Paris-Harrington Principle, two abridged statements whose unprovability proofs are simplest possible and, finally, a model-theoretic proof of some threshold results (exact unprovability results).

This paper was written in autumn 2005 in Liverpool.

§1. Brief survey.

Peano Arithmetic. Peano Arithmetic (PA) is a first-order theory in the language of arithmetic $\mathcal{L} = \{+, \times, <, 0, 1\}$ that consists of the following axioms: associativity and commutativity of $+$ and $\times$, their neutral elements

Logic Colloquium '06
Edited by Barry Cooper, Herman Geuvers, Anand Pillay, and Jouko Väänänen
Lecture Notes in Logic, 32

are 0 and 1 respectively, distributivity, discrete linear order axioms for $<$ (total order, there is a first element 0, no last element, every element has an immediate successor, every nonzero element has an immediate predecessor), 1 is the successor of 0, $x < y \to x + z < y + z$ and the induction scheme: for every $\mathcal{L}$-formula $\varphi(x, \overline{y})$, we have an axiom $\forall \overline{y}\, [\, \varphi(0, \overline{y}) \wedge \forall x (\varphi(x, \overline{y}) \to \varphi(x + 1, \overline{y})) \to \forall x \varphi(x, \overline{y})\,]$. In this article we shall deal only with Peano Arithmetic and some of its subsystems but for further study of the subject, beyond this article, the readers will eventually need to explore the whole story surrounding the notions of logical strength and consistency strength of theories and understand the picture of the scale of consistency strength. Although not required to understand the current article, a list of theories that are important in the subject (stretching all way from $I\Delta_0$ and $I\Delta_0(\exp)$ to the strongest extensions of ZFC) can be found on pages 39-40 in [36].

All concrete mathematics of the past can be conducted in Peano Arithmetic. It is common to identify theorems of PA with 'finite mathematics', that is the world of mathematical theorems that can be formulated in $\mathcal{L}$ and whose proof does not require the use of any notion of 'infinite set' in an essential way. Theorems of finite mathematics include the table of derivatives, the table of integrals of elementary functions, for every arithmetically definable complex function f, "if f is analytic on $\mathbb{C}$ and bounded then f is a constant", "in $\mathbb{R}^{24}$, there is a way to place 196560 non-overlapping unit spheres that touch the unit sphere", "the sum $\sum_{\substack{p, p+2 \\ \text{both prime}}} \frac{1}{p}$ converges", etc. (It is an exercise to check that all these statements can be formulated in the language $\mathcal{L}$ and their usual proofs can be conducted in Peano Arithmetic.) We give these examples in order to illustrate the extent of what is meant by 'finite mathematics' and show that in this understanding, 'finite mathematics' embraces not only finite combinatorial manipulations but all imaginable mathematics whose objects can be somehow finitely approximated or finitely encoded, including everyday 'continuous' mathematics and many branches of mathematics that seem to use notions beyond that of a natural number but will usually have a way to avoid it by approximations and coding.

Paris-Harrington Principle. Since Gödel's Incompleteness Theorems, for almost half a century logicians did not have examples of PA-unprovable statements that would not refer to diagonalisation or other logicians' tricks. The first PA-unprovable statements of 'mathematical' character (not referring to arithmetisation of syntax and provability) appeared in 1976 in the work of J. Paris (building upon joint work with L. Kirby [57]) and led to the formulation in [72] of the Paris-Harrington Principle (denoted PH): "for any numbers m, n and c, there exists a number N such that for every colouring f of m-subsets of $\{0, 1, \ldots, N-1\}$ into c colours, there is an f-homogeneous $H \subset \{0, 1, \ldots, N-1\}$ of size n such that $|H| > \min H$". This statement PH is not provable in Peano Arithmetic.

Many statements equivalent to PH have been studied: the Hercules-Hydra battle and termination of Goodstein sequences by L. Kirby and J. Paris [73], the flipping principle of L. Kirby [56], the arboreal statement by G. Mills [68], P. Pudlák's Principle [74, 45], the kiralic and regal principles by P. Clote and K. McAloon [27].

An important PA-unprovable statement was introduced in [50] by A. Kanamori and K. McAloon. A function f in m arguments is called regressive if $f(x_0, x_1, \dots, x_{m-1}) \leq x_0$ for all $x_0 < x_1 < \dots < x_{m-1}$. For regressive functions of m arguments, we cannot guarantee existence of a homogeneous set of size $(m+1)$, e.g., for $f(x_0, x_1, \dots, x_{m-1}) = x_0 - 1$, every set of size $(m+1)$ is not homogeneous. However, we can talk about min-homogeneous sets: a set H is called min-homogeneous if for all $c_0 < c_1 < \dots < c_{m-1}$ and $c_0 < d_1 < \dots < d_{m-1}$ in H, $f(c_0, c_1, \dots, c_{m-1}) = f(c_0, d_1, \dots, d_{m-1})$. Now, KM is the following statement: "for any numbers m, a and n with $n \geq m$, there exists $b > a$ such that for every regressive function f defined on m-subsets of $[a, b]$, there is a min-homogeneous set $H \subset [a, b]$ of size at least n". The statement KM is unprovable in PA and is equivalent to PH.

Indicators. First independence results grew out of indicator theory (best references are [55], [57] and [71]). In a model $M \vDash I\Sigma_1$, an initial segment I is called semi-regular if for every $a \in I$ and every function $f \colon [0, a] \to M$ (i.e., an M-coded set of pairs), the image of f is bounded in I. Every semi-regular initial segment satisfies $I\Sigma_1$. An initial segment I is called strong if for every M-coded partition $P \colon [I]^3 \to 2$, there is an M-coded I-unbounded P-homogeneous subset. Every strong initial segment is a model of PA. Other important classes of initial segments (closed under certain externally-described combinatorial operations) include regular (satisfies $B\Sigma_2$), n-extendible (corresponds to $I\Sigma_{n+1}$) and n-Ramsey (corresponds to $I\Sigma_{n+1}$) initial segments. Now, if a Σ_1-definable in $I\Sigma_1$ Skolem function $Y(x, y)$ is such that for all $a < b$ in M, $Y(a, b) > \mathbb{N}$ if and only if there is a strong initial segment between a and b then the statement "for all a and c, there is $b > a$ such that $Y(a, b) > c$" is unprovable in Peano Arithmetic. The function $Y(x, y)$ is called an indicator for strong initial segments because it indicates whether the set $[a, b]$ is large enough to accommodate a strong initial segment. (Similarly for all other kinds of initial segments above.) All early Ramsey-style independence results can be described in this setting. In case of the Paris-Harrington Principle, the function $Y(x, y) =$ the maximal c such that for every colouring $P \colon [x, y]^c \to c$, there is a homogeneous subset H of size $2c$ such that $|H| > \min H$ is an indicator for strong initial segments.

Many early indicator proofs were conducted in terms of a game between two players where Player I tries to ensure that the final initial segment between a and b is, say, strong, and Player II tries to prevent it. The game of finite (nonstandard) length is determined, so it turns out that if a set is large enough then Player I has a winning strategy, otherwise Player II has a winning strategy. Original sources are [57] and [71]. For another example connecting games and independence results, see the general idea and the Peano Arithmetic section in a recent article by P. Pudlák [75]. There is much more left to say about games and independence results, e.g., about unprovability of existence of a winning strategy in certain games.

Indicator theory ideas are also useful in the model-theoretic approach to Reverse Mathematics, in the spirit of the model-theoretic proof by J. Paris and L. Kirby in [57] that RT^3_2 (the infinite Ramsey Theorem for triples and two colours) implies all of PA, thus providing an alternative to the recursion-theoretic approach to these matters. Modern-day examples of such model-theoretic proofs in the style of indicator theory can be found in the articles [15] and [16] by the author and A. Weiermann.

PH is an arithmetical version of large cardinals. The historical prototypes of the Paris-Harrigton Principle and the earlier PA-unprovable statements of [70] are large cardinal axioms. In the case of arithmetic, closedness properties postulated by large cardinal axioms correspond to closedness properties of initial segments of models of arithmetic under the (external) combinatorial properties described above: semi-regularity, regularity, strength, extendibility and Ramseyness. This analogy eventually led to the Paris-Harrington principle and was important in the early days of the subject but was later abandoned and almost forgotten. It may be very fruitful to have a fresh look at this analogy, especially having in mind the modern advances in the study of large cardinals. J. Ketonen's manuscripts [52], [53] provide an alternative way to view such analogy, using ordinals, a theme that is closely connected with the fundamental Ketonen-Solovay article [54].

Fragments of Peano Arithmetic. If we restrict the induction scheme to Σ_n-formulas, that is, arithmetical formulas of the form

$$\exists x_1 \forall x_2 \exists x_3 \ldots \varphi(x_1, x_2, \ldots, x_n, \overline{y}),$$

where φ is preceded by no more than n quantifiers and itself contains no unbounded quantifiers, then the theory obtained is denoted by $I\Sigma_n$ ("induction for Σ_n-formulas"). Clearly, for every n, $I\Sigma_n \subseteq I\Sigma_{n+1}$ and PA $= \bigcup_{n=1}^{\infty} I\Sigma_n$. Also, it is known that for every n, $I\Sigma_n \neq I\Sigma_{n+1}$ since $I\Sigma_{n+1}$ proves $\mathrm{Con}(I\Sigma_n)$.

A useful alternative axiomatisation of $I\Sigma_n$ uses an instance of the least-number principle $\forall \overline{y}\,[\exists x\varphi(x, \overline{y}) \rightarrow \exists z(\varphi(z, \overline{y}) \wedge \forall w < z \neg\varphi(w, \overline{y})]$ for every Σ_n-formula $\varphi(x, \overline{y})$.

It is widely believed that all $\mathcal{L}$-theorems of existing mathematics (apart from logicians' discoveries we are talking about in this article) can be proved not only in PA but even in $I\Sigma_2$. It would be surprising if someone managed to find an existing theorem in mathematics that can be formulated in $\mathcal{L}$ but does not have a proof formalisable in $I\Sigma_2$.

For every $k \in \mathbb{N}$, the statement $\mathrm{PH}^{(k+1)}$ defined as "for all n and c, there exists N such that for every colouring f of $(k+1)$-element subsets of $\{0, 1, \dots, N-1\}$ into c colours, there is an f-homogeneous $H \subset \{0, 1, \dots, N-1\}$ of size at least n such that $|H| > \min H$" is $I\Sigma_k$-unprovable [71] and is equivalent to $\mathrm{KM}^{(k+1)}$ (the Kanamori-McAloon Principle for $(k+1)$-element subsets) and to $\mathrm{RFN}_{\Sigma_1}(I\Sigma_k)$, the 1-consistency of $I\Sigma_k$: $\forall \varphi \in \Sigma_1\ (\mathrm{Pr}_{I\Sigma_k}(\varphi) \rightarrow \varphi)$. (It says "for all Σ_1-statements φ, if $I\Sigma_k$ proves φ then φ holds". In order to write it as a formula, it is necessary to use the satisfaction predicate for Σ_1-formulas. Unprovability of $\mathrm{RFN}_{\Sigma_1}(I\Sigma_k)$ in $I\Sigma_k$ easily follows from Gödel's Second Incompleteness Theorem: put φ to be $\exists x\ x \neq x$ to observe that $\mathrm{RFN}_{\Sigma_1}(I\Sigma_k)$ implies $\mathrm{Con}_{I\Sigma_k}$.) A good exposition of 1-consistency and reflection principles is in the old Smorynski's paper [86], a good exposition of the satisfaction predicate is in Kaye's textbook [51].

Threshold results for PH and KM. Let $\log^{(n)}(x) = \underbrace{\log_2(\log_2 \dots \log_2(x))}_{n \text{ times}}$ and the tower-function $2_n(x)$ be defined as $2_0(x) = x$, $2_{n+1}(x) = 2^{2_n(x)}$. Also, define $\log^*(m)$ as the minimal n such that $2_n(2) \geq m$. The Ramsey number $R(k, c, m)$ is defined as the minimal number such that for any colouring of k-subsets of $\{0, 1, 2, \dots, R(k, c, m) - 1\}$ in c colours, there is a monochromatic subset of size m. For any set X, we write $[X]^n$ for the set of its n-element subsets. The set of all n-element subsets of $\{a, a+1, \dots, b\}$ will be denoted by $[a, b]^n$, without any confusion. As usual, a natural number N is identified with the set of its predecessors $\{0, 1, \dots, N-1\}$.

For every function $F(x)$, define $\mathrm{PH}_F^{(k)}$ as the statement "for all n and c there exists N such that for every $f : [N]^k \rightarrow c$, there is a homogeneous $H \subseteq N$ of size at least n and such that $F(\min H) < |H|$". We say that f is F-regressive if for all $x_0 < x_1 < \cdots < x_{k-1}$, we have $f(x_0, x_1, \dots, x_{k-1}) \leq F(x_0)$. Now define $\mathrm{KM}_F^{(k)}$ as the statement "for all n there exists N such that for every F-regressive f defined on $[N]^k$, there is a min-homogeneous subset of N of size at least n". Also, define PH_F as $\forall k\ \mathrm{PH}_F^{(k)}$ and KM_F as $\forall k\ \mathrm{KM}_F^{(k)}$. It is easy to see that for every strictly increasing F, PH_F implies PH and KM_F implies KM thus making these statements PA-unprovable. A. Weiermann [90] proved that for every n, $\mathrm{PH}_{\log^{(n)}}$ is PA-unprovable but $\mathrm{PH}_{\log^*}$ is PA-provable. In the case of fixed dimension, the story is more complex. The following interesting result was first proved by Gyesik Lee [61] for all $n \neq k-1$ and later completed by L. Carlucci, G. Lee and A. Weiermann [23]: if $k \geq 2$ then

1. if $n \leq k-1$ then $\mathrm{KM}_{\log^{(n)}}^{(k+1)}$ is $I\Sigma_k$-unprovable;
2. if $n > k-1$ then $I\Sigma_1$ proves $\mathrm{KM}_{\log^{(n)}}^{(k+1)}$.

Similar theorems hold for the family $\mathrm{PH}^{(k)}_{\log^{(n)}}$:

1. if $n \le k$ then $I\Sigma_k$ does not prove $\mathrm{PH}^{(k+1)}_{\log^{(n)}}$ (A. Weiermann [90]);
2. if $n > k$ then $I\Sigma_1$ proves $\mathrm{PH}^{(k+1)}_{\log^{(n)}}$ (G. Lee [61]).

There are even slightly sharper threshold results for this case in [23]. Here is the picture in the case $k = 1$ (see [59]): if A^{-1} is the inverse of the Ackermann function and $\{F_m\}_{m\in\omega}$ is the Grzegorczyk hierarchy of primitive recursive functions then:

1. $I\Sigma_1 \nvdash \mathrm{PH}^{(2)}_{\frac{\log}{A^{-1}}}$;
2. for every $m \in \omega$, $I\Sigma_1 \vdash \mathrm{PH}^{(2)}_{\frac{\log}{F_m^{-1}}}$;
3. $I\Sigma_1 \nvdash \mathrm{KM}^{(2)}_f$, where $f(x) = x^{\frac{1}{A^{-1}(x)}}$;
4. for every $m \in \omega$, $I\Sigma_1 \vdash \mathrm{KM}^{(2)}_{f_m}$, where $f_m(x) = x^{\frac{1}{F_m^{-1}(x)}}$.

In particular, $\mathrm{KM}^{(2)}_{\log}$ is provable but $\mathrm{PH}^{(2)}_{\log}$ is unprovable.

The reason for $I\Sigma_1$-provability of $\mathrm{PH}^{(k+1)}_{\log^{(n)}}$ and $\mathrm{KM}^{(k+1)}_{\log^{(n)}}$ for large n comes from the Erdös-Rado theorem [30], which implies that an upper bound for the Ramsey number $R(m, k+1, m)$ is $2_n(m)$ for some large enough n depending only on k. Given m and c, let $\ell = \max\{m, c\}$. Consider any colouring $f : [0, 2_n(\ell)]^{k+1} \to c$. By the Ramsey Theorem, there is $H \subset [0, 2_n(\ell)]$ of size at least ℓ which is f-homogeneous. Also, $\log^{(n)}(\min H) < \log^{(n)}(2_n(\ell)) = \ell \le |H|$. Thus $I\Sigma_1 \vdash \mathrm{PH}^{(k+1)}_{\log^{(n)}}$.

A similar argument for $I\Sigma_1$-provability of $\mathrm{KM}^{(k+1)}_{\log^{(n)}}$ for large n goes as follows: consider n such that $R(m, k+1, m) < 2_n(m)$ for all m. Let f be a $\log^{(n)}$-regressive function defined on $[0, 2_n(m)]$. Then the image of f is contained in $[0, m]$. Hence there is a homogeneous (thus also min-homogeneous) subset $H \subseteq [0, 2_n(m)]$ of size at least m.

Kruskal's Theorem and well-quasi-orders. Very often an unprovable statement can be viewed as a 'miniaturisation' of an infinitary theorem. A spectacular example of miniaturisation is H. Friedman's Theorem [81] on unprovability of a finite version of Kruskal's Theorem. Define a (nonplane, rooted) tree as a partially ordered set with the least element and such that the set of all predecessors of every point is linearly ordered. Infinite Kruskal's Theorem says: if $T_1, T_2, \ldots$ is an infinite sequence of finite trees then there are $i < j$ such that $T_i \trianglelefteq T_j$, i.e., there is an inf-preserving embedding from T_i into T_j (that is, trees are well-quasi-ordered by $\trianglelefteq$). Friedman's Theorem says that neither the Infinite Kruskal Theorem nor its finite version ('miniaturisation') "for all k there is N such that whenever $\langle T_i \rangle_{i=1}^N$ is a sequence of finite trees such that for all $i \le N$ we have $|T_i| \le k + i$ then there are $i < j \le N$ such that $T_i \trianglelefteq T_j$" is provable in ATR_0, a theory stronger than Peano Arithmetic. Here, $|T|$ is the number of vertices in T. Also, Kruskal's theorem restricted to binary trees is unprovable in ACA_0 (and its finite version unprovable in PA) and Kruskal's theorem for binary trees and two labels unprovable in ATR_0. (There are also important results by L. Gordeev about the strength of Kruskal's theorem with gap-condition and other unprovability results in [42, 43, 44].) It was later shown by M. Loebl and J. Matoušek [62] that if the condition $|T_i| \leq k + i$ is replaced by $|T_i| \leq k + \frac{1}{2}\log_2 i$ then the statement becomes $I\Sigma_1$-provable but for the condition $|T_i| \leq k + 4\log_2 i$, the statement is PA-unprovable. What happens between $\frac{1}{2}$ and 4 was recently resolved by A. Weiermann [89]. Let α be Otter's constant ($\alpha = \frac{1}{\rho}$, where ρ is the radius of convergence of $\sum_{i=0}^{\infty} t_i z^i$, where t_i is the number of finite trees of size i), $\alpha \approx 2.955765\ldots$. Then for any primitive recursive real number r,

1. if $r \leq \frac{1}{\log_2 \alpha}$ then the statement with the condition $|T_i| \leq k + r \log i$ is $I\Sigma_1$-provable;
2. if $r > \frac{1}{\log_2 \alpha}$ then the statement is PA-unprovable.

Similar results have been proved by A. Weiermann for plane trees, binary plane trees, Higman's lemma [93], ordinal notations [91], [89]. Several general theorems have recently been proved by the author [12], giving an answer to the following question: if X is a well-quasi-ordered combinatorial class (i.e., there is a fixed notion of size of objects and for every n there are finitely-many objects of size n) such that for some function f, the statement "for all K there is N such that whenever $X_1, X_2, \ldots, X_N$ are in X and for all $i \leq N$, $|X_i| < K + f(i)$ then for some $i < j \leq N$, $X_i \trianglelefteq X_j$" is unprovable in some theory T, what is the exact threshold function that separates provable and T-unprovable instances of the corresponding well-quasi-orderedness statements for $\mathrm{Seq}(X)$ (the set of all finite sequences of elements of X), $\mathrm{Cyc}(X)$ (finite cycles of elements of X), $\mathrm{Mult}(X)$ (finite multisets of X), $\mathrm{Trees}(X)$ (plane rooted trees labeled by X) and some other compound combinatorial classes that inherit well-quasi-orderedness from X by a usual minimal bad sequence argument? The proofs use Weiermann-style compression techniques and ideas from analytic combinatorics. There should be ordinal-theoretic results closely connected to these theorems (roughly: how the multiset-, sequence-, cycle-, tree- and other constructions transform the maximal order-types of well-ordered linearisations of original well-quasi-orders) but these results, to the author's knowledge, haven't yet been written down by anyone.

Another important example is a theorem by H. Friedman, N. Robertson and P. Seymour on unprovability of the Graph Minor Theorem [40]. For multigraphs G and H, we say that H is a minor of G if H can be obtained from G by a succession of three elementary operations: edge removal, edge contraction and removal of an isolated vertex. The first-order version of their theorem states that the statement "for all k there is N such that whenever $G_1, \ldots, G_N$ is a sequence of finite multigraphs such that for all $i \leq N$ we have $|G_i| \leq k + i$ then for some $i < j \leq N$, G_i is a minor of G_j" is not provable in Π^1_1-CA_0, a very strong subsystem of second-order arithmetic. Multigraphs can well be replaced by simple graphs in this formulation.

For a function f, let the statement GM_f be "for all K there is N such that for any sequence of simple graphs $G_1, G_2, \ldots, G_N$ such that $|G_i| < K + f(i)$, there are $i < j$ such that G_i is a minor of G_j". It is now easy to conjecture that for every primitive recursive real number a, if $a \leq \sqrt{2}$ then $\mathrm{GM}_{a \cdot \sqrt{\log}}$ is $I\Sigma_1$-provable but if $a > \sqrt{2}$ then $\mathrm{GM}_{a\sqrt{\log}}$ is unprovable. So far it has been proved by the author in [12] that if $a \leq \sqrt{2}$ then $\mathrm{GM}_{a \cdot \sqrt{\log}}$ is $I\Sigma_1$-provable and $\mathrm{GM}_{7\log}$ is PA-unprovable. A certain lemma on graph enumeration is currently missing for the proof of the full conjecture to go through. There are similar conjectures about unprovability thresholds for Kruskal's theorem with gap condition and for graph minor theorem for subcubic graphs and for classes of graphs omitting certain minors. This enterprise has to use analytic combinatorics, graph minor theory and Pólya theory.

Well-quasi-order theory has been extensively studied in the framework of Reverse Mathematics [84], the study of logical strength and consistency strength of second-order arithmetical assertions. Reverse Mathematics is the closest relative of first-order unprovability theory and there is exchange of ideas flowing both ways. There are many results, methods and ideas in Reverse Mathematics that are very relevant to the subject but we omit them in this brief survey. It is difficult to draw a strict line between the two subjects and any future exposition of first-order unprovability may have to incorporate some discussion of relevant parts of Reverse Mathematics.

Single tree, single sequence and boolean relation theory. Here is a new way to obtain unprovable statements from existing statements that talk about long sequences of objects by assembling elements of sequences of objects into one single object to talk about, with order (or embeddability) relation between objects in the sequence becoming order (or embeddability) relation between chunks of this single object. For a tree T, let $T[i]$ be the tree of nodes of T of height $\leq i$. It has been proved by H. Friedman in [35] that the statement "for all n and k there is K such that whenever T is a rooted nonplane tree labeled by $\{1, 2, \ldots, n\}$ and every non-leaf has degree k then there are $i < j \leq K$ such that $T[i]$ is inf-preserving, label-preserving and leaf-preserving embeddable into $T[j]$" is equivalent to 1-consistency of Π^1_2-TI_0, a system much stronger than Peano Arithmetic, and even the statement with $k = 2$ already implies 1-consistency of ATR_0. There are also versions that involve a gap-condition. Another application of similar ideas deals with Higman's lemma. The statement "for all m there is K such that for every K-sequence $x_1, x_2, \ldots, x_K$ of elements of $\{1, 2, \ldots, m\}$, there are $i < j < \frac{K}{2}$

such that $x_i, x_{i+1}, \ldots, x_{2i}$ is a subsequence of $x_j, x_{j+1}, \ldots, x_{2j}$" is unprovable in $I\Sigma_2$ (see [34]).

Another series of H. Friedman's results under the general name of "boolean relation theory" can be found in [37]: if we list all (second-order) statements of a certain simple shape (all of them a priori equally simple and natural) and try to classify them according to their truth, some of them turn out to be unprovable in some theories stronger than ZF, e.g. the following statement from [37] is provably in ACA' equivalent to 1-consistency of the theory ZFC+ $\{$there is an n-Mahlo cardinal$\}_{n\in\omega}$: "for any two functions $f, g\colon \mathbb{N}^k \to \mathbb{N}$ such that there are two constants $c, d > 1$ with $c \cdot |\overline{x}| < f(\overline{x}), g(\overline{x}) < d \cdot |\overline{x}|$ for all but finitely many $\overline{x} \in \mathbb{N}^k$, there exist infinite sets $A, B, C \subseteq \mathbb{N}$ such that $A \cup .f(A) \subseteq C \cup .g(B)$ and $A \cup .f(B) \subseteq C \cup .g(C)$", where $|\overline{x}|$ is the maximal element of the k-tuple $\overline{x}$, $f(A)$ is the image of f on k-tuples of A and $A \cup .D$ means the union of A and D together with the statement that they are disjoint.

Sine, zeta-function, diophantine approximation and universality. The results in this section spring from H. Friedman's sine-principle [11]. For every $n \geq 1$ and every function F of one argument, let us introduce the statement SP^n_F: "for all m, there is N such that for any sequence $a_1, a_2, \ldots, a_N$ of rational numbers, there is $H \subseteq A$ of size m such that for any two n-element subsets $a_{i_1} < a_{i_2} < \cdots < a_{i_n}$ and $a_{i_1} < a_{k_2} < \cdots < a_{k_n}$ in H, we have $|\sin(a_{i_1} \cdot a_{i_2} \cdot \ldots \cdot a_{i_n}) - \sin(a_{i_1} \cdot a_{k_2} \cdot \ldots \cdot a_{k_n})| < F(i_1)$". For $n \geq 2$ and any function $F(x)$ eventually dominated by $(\frac{2}{3})^{\log^{(n-1)}(x)}$, the principle SP^{n+1}_F is not provable in $I\Sigma_n$. The proof in [11] uses the Rhin-Viola theorem on irrationality measure of π.

The sine-principle led to a series of exciting developments. What other functions can be taken instead of sine so that the statement would still be unprovable? The following theorem by the author and A. Weiermann [17] describes a large class of such functions. Let f, g and h be three functions such that for any $N \in \mathbb{N}$ and any small $\varepsilon > 0$,

1. h is a periodic function with period a, continuous on its period;
2. f is such that for any $b_1, b_2, \ldots, b_N$, linearly independent over $a\mathbb{Q}$ and any $c_1, c_2, \ldots, c_N \in [0, a)$, there is $x \in \mathbb{Q}$ such that for all $i \in \{1, 2, \ldots, N\}$, $|f(b_i \cdot x) \bmod a - c_i| < \varepsilon$;
3. g is any continuous function on a subset of $\mathbb{R}$ whose image contains $[d, +\infty)$ for some $d \in \mathbb{R}$.

Then the statement "for all m and n, there is N such that for any sequence $\langle a_i \rangle_{i=1}^N$ of rational numbers, there is $H \subseteq N$ of size m such that for any two n-sequences $i_1 < i_2 < \cdots < i_n$ and $i_1 < k_2 < \cdots < k_n$ of natural numbers smaller than N,

$$|h(f(g(a_{i_1} \cdot a_{i_2} \cdot \ldots \cdot a_{i_n}))) - h(f(g(a_{i_1} \cdot a_{k_2} \cdot \ldots \cdot a_{k_n})))| < 2^{-i_1}\text{"}$$

is unprovable in Peano Arithmetic and the versions for fixed n are unprovable in $I\Sigma_{n-1}$. This class of functions includes $\sin(p(x_1 \cdot x_2 \cdot \ldots \cdot x_n))$, $\{p(x_1 \cdot x_2 \cdot \ldots \cdot x_n)\}$, $\{\frac{1}{p(x_1 \cdot x_2 \cdot \ldots \cdot x_n)}\}$ for any non-constant polynomial p (using H. Weyl's theorem on simultaneous diophantine approximation). In particular, the proof of this general theorem shows that unprovability of the sine-principle can now be demonstrated using simultaneous diophantine approximation instead of an argument involving irrationality measure of π. We have several conjectures that suggest versions of these results in p-adic setting.

Another interesting example deals with the Riemann zeta-function [17]: for any $\sigma \in \mathbb{R}$, consider the statement "for all m and n, there is N such that for any sequence $\langle a_i \rangle_{i=1}^N$ of rational numbers, there is $H \subseteq N$ of size m such that for any two n-sequences $i_1 < i_2 < \cdots < i_n$ and $i_1 < k_2 < \cdots < k_n$ of natural numbers smaller than N, $|\zeta(\sigma + i \cdot a_{i_1} \cdot a_{i_2} \cdot \ldots \cdot a_{i_n}) - \zeta(\sigma + i \cdot a_{i_1} \cdot a_{k_2} \cdot \ldots \cdot a_{k_n})| < 2^{-i_1}$". For $\sigma > 1$, the unprovability proof is similar to that for $\sin(x)$, using almost periodicity. For $\sigma \in (\frac{1}{2}, 1]$, the unprovability proof uses a probabilistic argument.

The reason behind unprovability of these statements is that Ramsey-style assertions (in this case the Kanamori-McAloon Principle) can be concealed in this setting by replacing quantification over all possible colourings by the quantifier over all possible sequences of rational numbers. E.g. in the case of the sine-function, what we need is that every function is approximated by sine on some subset: for any $\varepsilon > 0, K, n$ and any function $g\colon [K]^n \to [-1, 1]$, there is a sequence of rational numbers $\langle a_1, a_2, \ldots, a_K \rangle$ such that for any $i_1 < \cdots < i_n \leq K$, $|g(i_1, \ldots, i_n) - \sin(a_{i_1} \cdot \ldots \cdot a_{i_n})| < \varepsilon$. This situation where all possible patterns (e.g. functions, colourings, etc.) are already present in one complex object is called universality and there is more to say about using universal objects to reformulate Ramsey-style statements.

The following unprovability results have been proved in [17], using discrete analogues of S. Voronin's universality theorem about the Riemann zeta-function. The statement "for all n there is N such that whenever $\langle a_i \rangle_{i=1}^N$ are natural numbers, there is a subset $H \subset N$ of size n such that for all $k < l < m$ in H,

$$\left|\zeta^{(a_k)}\left(\frac{3}{4} + i \cdot a_l\right) - \zeta^{(a_k)}\left(\frac{3}{4} + i \cdot a_m\right)\right| < \frac{1}{2^k}\text{''}$$

is unprovable in $I\Sigma_1$. The statement "for all n there is N such that whenever $\langle a_i \rangle_{i=1}^N$ are natural numbers, there is a subset $H \subset N$ of size n such that for all $k < l < m$ in H,

$$\left|\zeta\left(\frac{1}{a_k} + i \cdot a_l\right) - \zeta\left(\frac{1}{a_k} + i \cdot a_m\right)\right| < \frac{1}{2^k}\text{''}$$

is unprovable in $I\Sigma_1$. Similar results will hold for a wide range of zeta- and L-functions.

Universality, in the broad understanding of the word, will be a rich source of unprovable statements in the future. We want to encode and treat a random colouring of n-tuples, so assertions that involve all possible patterns (or some 'random' patterns) of finite configurations should attract our attention. Nowadays, it should be possible to show unprovability of some (versions of) already existing strong conjectures in, say, number theory. A promising example that begs for an independence proof is Schinzel's hypothesis H (that is rooted in the work of 19th century mathematicians, e.g. Bunyakovskiy): "for any finite collection of irreducible polynomials $P_1(x), P_2(x), \ldots, P_n(x)$ with integer coefficients and such that $\prod_{i \leq n} P_i$ has no fixed prime divisor, there exist infinitely-many integers m such that for all $i \leq n$, $P_i(m)$ are prime". This conjecture is extremely strong (implying the twin-prime conjecture and infinity of the set of primes of the form n^2+1) and its formulation already provides some necessary ingredients for the unprovability proof. ($I\Sigma_1$-unprovability of hypothesis H was already conjectured in the PhD thesis of A. Woods [94]).

Braids. Braids are very popular and interesting objects in today's mainstream mathematics. The n-strand braid group B_n is a group with the following presentation:

$$B_n = \langle \sigma_1, \ldots, \sigma_{n-1};\ \sigma_i\sigma_j = \sigma_j\sigma_i \text{ for } |i-j| \geq 2,$$
$$\sigma_i\sigma_j\sigma_i = \sigma_j\sigma_i\sigma_j \text{ for } |i-j| = 1 \rangle.$$

A braid is called positive if it has a representation without σ_i^{-1} for any i.

There are several independence results about braids, relying on P. Dehornoy's left-invariant ordering $\prec$ of positive braids as ω^{ω^ω} and S. Burckel's ordering of n-strand positive braids as $\omega^{\omega^{n-2}}$, see [28]. Some unprovability results immediately follow from $I\Sigma_2$-unprovability of transfinite induction up to ω^{ω^ω}. The statement "for every K there is N such that for any sequence $B_1, B_2, \ldots, B_N$ of positive braids such that $|B_i| < K + i$, there are $i < j \leq N$ such that $B_i \prec B_j$" is not provable in $I\Sigma_2$. (Here, $|B|$ is the smallest number of letters in a braid word representing B.) It is also possible to do a Friedman-style argument to turn a statement about a sequence of braids into a statement about comparing internal segments of one long braid.

At the dawn of modern logic, early logicians, for example R. Goodstein in 1957, wrote about ordinal descent through ω^{ω^ω} with justification why this is an acceptable mathematical principle. This ordinal was perceived by some people at that time as the biggest ordinal that allows for a convincing verbal "justification" of why the corresponding ordinal descent principle is true. This ordinal nowadays turns up in various natural mathematical contexts, see for example Higman's lemma and S. Simpson's article [83] on the Robson Basis Theorem.

One more family of independence results comes from braid-theoretic analogues of the hydra battle. Termination of the following game on positive braids is unprovable in $I\Sigma_2$. The following game was introduced by

Lorenzo Carlucci in [13]. In the rest of this section, we write i instead of σ_i. For numbers a, b, define a wave between a and b as the braid word $w_{a,b} = a(a+1)^2 \dots (b-1)^2 b^2 (b-1)^2 \dots (a+1)^2 a$ if $a < b$ and $w_{a,b} = a(a-1)^2 \dots (b+1)^2 b^2 (b+1)^2 \dots (a-1)^2 a$ if $a > b$. A braid word is even if all blocks of its consecutive equal letters are of even length and for neighbouring blocks consist of numbers i and j that differ by 1. (We are temporarily restricting ourselves to even braids in order not to worry about applications of braid relations.) Given a positive even braid word w, we define a reduct $w[k]$ for every $k > 0$ as follows. If w ends in 11 then $w[k]$ is obtained from w by deleting this 11. Otherwise let $b_i^{m_i}$ be the first block of equal letters (counting from right to left) of length $m_i > 2$. Let $\hat{b_i}$ be the closest occurrence of the letter b_i to the right of $b_i^{m_i}$ if it exists and the empty word if there is no such occurrence. Let us then write w as $\dots\ b_i^{m_i}\ a\, u\, \hat{b_i}\ \dots$. If $a < b_i$ then $w[k]$ is defined as

$$\dots\ b_i^{m_i-2}\ (b_i-1)^3\ w^k_{b_i-1,\min u}\ (b_i-1)\ u\ \hat{b_i}\ \dots,$$

otherwise $w[k]$ is $\dots\ b_i^{m_i-1}\ (b_i+1)^3\ w^k_{b_i+1,\max u}\ (b_i+1)\ u\ \hat{b_i}\ \dots$. The sequence $w[1][2]\dots[k]$ consists of positive even braid words and decreases with respect to the braid ordering $\prec$. Eventually it terminates but its termination is unprovable in $I\Sigma_2$. There are versions of this sequence for 3-strand braids [13]: a simple version, whose termination time is Ackermannian and another one, formulated as a game, whose termination is unprovable in $I\Sigma_2$.

There is some hope to translate statements about braids into geometrical, topological statements with some amount of 'physical' meaning, e.g. using the fact that braid groups are fundamental groups of certain configuration spaces. This is a very rich topic with many new results (all so far based on left-orderability of certain groups) being obtained these days by P. Dehornoy, A. Weiermann, L. Carlucci and the author (for a survey, see [14]).

Long term rewriting. Another interesting twist in the story is the rewrite systems whose termination is PA-unprovable. Good references are short articles by L. Beklemishev [6] and W. Buchholz [19]. An early rewrite system that imitated the hydra was proposed by Dershowitz. Recently, G. Moser [29] provided a full analysis of this system and related systems. Alternative systems have been formulated by W. Buchholz and by H. Touzet [88]. There are also relevant results by I. Lepper. Here is an example by L. Beklemishev [5] presented as a battle between a gardener and a worm. (It is not exactly a rewrite system but there are versions of it formulated as a rewrite system in [6].) A worm is a finite sequence of natural numbers $f : \{0, 1, \dots, m\} \to \mathbb{N}$. For every worm $w = f(0), f(1), \dots, f(m)$, define $w[n]$ as follows. If $f(m) = 0$ then chop it off, i.e., put $w[n] = f(0), f(1), \dots, f(m-1)$. Otherwise find the maximal place $k < m$ such that $f(k) < f(n)$, define two words $r = f(0), \dots, f(k)$ and $s = f(k+1), \dots, f(m-1), f(m) - 1$ and set $w[n]$ to be r followed by

$(n+1)$ copies of s. The statement $\forall w \exists n\ w[1][2]\ldots[n] = \varnothing$ ("every worm dies") is unprovable in Peano Arithmetic.

Other examples. Many other examples of unprovable statements and a discussion of the subject in its late 1970s – early 1980s state can be found in the four volumes [82], [80], [69] and [7], devoted entirely or partially to arithmetical unprovability results. In addition, I would like to mention unprovable statements obtained by translating consistency statements for various theories into non-solvability assertions about corresponding Diophantine equations (see [65]), unprovable statements about Diophantine games [65], the Schütte-Simpson treatment of finite sequences of natural numbers with gap-condition [85], K. McAloon's [66], [67] and Z. Ratajczyk's [76], [77] theories of iterations of the Paris-Harrington Principle and R. Sommer's related result on transfinite induction [87], independence results on pointwise induction by T. Arai [1], J. Avigad's statement with update procedures [2], P. Clote's PhD thesis and his anti-basis approach to independence results [25], Cichon's treatment of Goodstein sequences [24], Bigorajska-Kotlarski series of results on partitioning α-large sets (see e.g. [8], [9]), the Friedman-McAloon-Simpson early fundamental article [39] and the subsequent paper by S. Shelah [79], J.-P. Ressayre's statements unprovable in $I\Sigma_1$, PA and ZF that state existence of finite approximations of models of these theories [78], the treatment of infinitary strong statements by means of J. Paris-style density assertions in [15] and [16] by the author and A. Weiermann, the Kochen-Kripke proof with ultrapowers [58], the Buchholz Hydra [18], L. Carlucci's recent uniform treatment of hydras, worms and sequences with gap condition [22] and H. Friedman's many results on combinatorial statements unprovable in ZFC+ large cardinals [32], [33]. (Many new unprovable statements are announced regularly on H. Friedman's webpage, so we shall not even attempt to give an up-to-date account of these rapid developments here.)

The subject of first-order arithmetical unprovability has a big number of results, some of them masterpieces, produced by different authors, and is rapidly developing these days. Although it is one of the most central subjects in logic and is already seriously connected with many mathematical disciplines, there are currently no comprehensive monographs, no textbooks, not even surveys. This current sketchy and incomplete introduction is a mere glimpse at the subject and is intended to be a modest patch in this state of affairs, without any serious discussion of the subject's goals, its big fundamental questions, its history, its implications for philosophy of mathematics or its visions of the future and without any attempt to analyse and systematise its methods and ideas.

Methods of proving unprovability. Unprovability proofs so far usually fall into one of two broad categories. The first category consists of model-theoretic constructions that demonstrate how, assuming a statement, a model of our strong theory can be built directly, 'by hands'. Proofs of the first category work very well for Ramsey-style statements and for strong theories. The second category consists of combinatorial proofs springing from proof theory and the Ketonen-Solovay article [54] (which shows combinatorially that the function arising from PH eventually dominates every function of the Grzegorczyk-Schwichtenberg-Wainer hierarchy (since all PA-provably recursive functions occur in this hierarchy, the result follows)) or from the study of well-quasi-ordered sets [81]. Most of the proofs in the volume [82] are of this category, as well as the articles [63] and [64] by M. Loebl and J. Nešetřil. Proofs in this category work very well (often better than model-theoretic proofs) at the lower end of consistency strength and for well-quasi-orders, like trees, whose ordinal treatment is well-developed but the model-theoretic treatment is currently absent.

Apart from the original articles we mentioned above, other good sources reporting on proofs of the first category would be the book [46] by P. Hájek and P. Pudlák and the papers [3], [4] by J. Avigad and R. Sommer (proof-theoretic aspects). A recent book manuscript [60] by H. Kotlarski is an exposition of both approaches as well as of many different proofs of Gödel's Theorem. The second category is well explained in an early article by W. Buchholz and S. Wainer [20], in the Friedman-Sheard article [41], in the Fairtlough-Wainer paper [31] and in the recent article by A. Weiermann [92].

Apparently, there is also a third category of unprovability proofs, proofs that interpret directly independent statements as reduction strategies for proof systems. We have little to say about it and refer the reader to a recent article by L. Carlucci [21], an old paper by H. Jervell [49] and the articles [48] and [47] by M. Hamano and M. Okada.

Once we have one unprovable statement, it is usually possible to find a series of mathematical statements in very different contexts that imply our statement, sometimes in a very non-trivial way. It often happens that non-logical mathematical obstacles in proving such implications are difficult and deep. In this way, the original Ramsey-theoretic and well-quasi-order-theoretic reasons for unprovability may end up very well hidden, and new mathematical statements appear natural and still carry a big amount of unprovability and logical strength.

Some people somehow view research in unprovability in terms of *big numbers and fast-growing functions.* Indeed, big numbers and fast-growing functions often occur in our discourse. (For example, the Paris-Harrington Principle implies every Π_2 theorem of Peano Arithmetic, including totality of every PA-provably recursive function.) However, this simple perception of the subject eventually misleads when one reaches the higher strata of consistency strength where the philosophical status of unprovable statements is very different. It is better, at least in the context of this article (and especially in the context of statements unprovable in higher theories) to view unprovable statements as possessing a large "amount of logical strength" that allows us to build approximations of models of strong theories.

§2. Unprovability of PH. Nowadays unprovability of the Paris-Harrington Principle in Peano Arithmetic is presented via unprovability proof of KM followed by a proof that PH implies KM or by the combinatorial Ketonen-Solovay method using α-large sets or by a sequence of lemmas (gluing different colourings) as in the original article [72]. A model-theoretic proof was promised in the original article [72] but never appeared. This was a folklore understanding in the 1970s and 1980s that a model-theoretic proof can be written down. The closest it got is in the treatment of n-extendible initial segments by J. Paris on pages 324-328 in [71] and in the book by P. Hájek and P. Pudlák [46] but still far from being instantly accessible to a wide audience. We shall give a single simple proof (inspired by the Kanamori-McAloon proof [50]) that will be useful in logic courses. There will be one colouring which will yield a desired set of indiscernibles among its large homogeneous set. To our knowledge, the proof below is the first time a simple direct model-theoretic unprovability proof for PH appears in print.

Finally, let me mention that all statements below, unprovable in Peano Arithmetic, are easy consequences of the Infinite Ramsey Theorem, so will usually be labeled as "true".

Theorem 1 (Paris-Harrington). *The statement "for all a, k, m, c, there is $b > a$ such that for every $f : [a,b]^k \to c$, there is an f-homogeneous $H \subset N$ of size m such that $|H| > \min H$" is not provable in Peano Arithmetic.*

The readers who need a reminder of basic notions from logic (e.g., models of arithmetic, overspill and the usual way in which sets, functions and formulas are encoded and treated within a model of arithmetic) are referred to Kaye's textbook [51].

Proof. Let $M \models I\Sigma_1$ be nonstandard, $a, e \in M$ be nonstandard and $\varphi_1(z, x_1, x_2, \ldots, x_e), \ldots, \varphi_e(z, x_1, x_2, \ldots, x_e)$ be the first e Δ_0-formulas in at most the free variables shown. In particular, this list contains all Δ_0-formulas of standard size. Denote $R(2e+1, e+1, 5e+1)$ by $r(e)$. Let $b \in M$ be minimal such that for every $g : [a,b]^{4e+1} \to 3e+1$, there is a g-homogeneous $H \subset [a,b]$ of size $r(e)$ such that $|H| > \min H$. We shall build a model of PA between a and b.

First define a function $f : [a,b]^{2e+1} \to e+2$ as follows[1]: for $c < \overline{d_1} < \overline{d_2}$ in $[a,b]$, put

$$f(c, \overline{d_1}, \overline{d_2}) = \min i \leq e\ \exists p < c\ \Big(\varphi_i(p, \overline{d_1}) \not\leftrightarrow \varphi_i(p, \overline{d_2})\Big)$$

if such i exists and $e+1$ otherwise. (In order to write down this formula we

[1]we shall often shorten $c < a_1 < \cdots < a_e < b_1 < \cdots < b_e$ as $c < \overline{a} < \overline{b}$

should use a formula that represents the satisfaction relation for Δ_0-formulas.) Informally, $f(c,\overline{d_1},\overline{d_2})$ is the first formula (with a parameter smaller than c allowed) that distinguishes the tuples $\overline{d_1}$ and $\overline{d_2}$.

Define another function h, defined on $[a,b]^{2e+1}$, as follows:

$$h(c,\overline{d_1},\overline{d_2}) = \min p < c \ \left(\varphi_{f(c,\overline{d_1},\overline{d_2})}(p,\overline{d_1}) \not\leftrightarrow \varphi_{f(c,\overline{d_1},\overline{d_2})}(p,\overline{d_2})\right)$$

if $f(c,\overline{d_1},\overline{d_2}) \neq e+1$ and c otherwise. I.e., the value $h(c,\overline{d_1},\overline{d_2})$ is the first parameter p with which $\varphi_{f(c,\overline{d_1},\overline{d_2})}$ distinguishes $\overline{d_1}$ and $\overline{d_2}$.

Now let us introduce our colouring $g\colon [a,b]^{4e+1} \to 3e+1$ as follows:

$$g(c,\overline{d_1},\overline{d_2},\overline{d_3},\overline{d_4}) = \begin{cases} 0 \quad \text{if} & h(c,\overline{d_1},\overline{d_2}) = h(c,\overline{d_3},\overline{d_4}), \\ j \quad \text{if} & h(c,\overline{d_1},\overline{d_2}) \neq h(c,\overline{d_3},\overline{d_4}) \text{ and} \\ & 0 < j \leq 3e \text{ and } h(c,\overline{d_1},\overline{d_2}) \equiv j \pmod{3e}. \end{cases}$$

Choose a g-homogeneous set $H \subseteq [a,b]$ of size at least $r(e)$ and such that $|H| > \min H$. Let us show that the value of g on $[H]^{4e+1}$ is 0. Suppose that the value of g on $[H]^{4e+1}$ is $j \neq 0$. Notice that then there are not more than $[\frac{\min H}{3e}] < \frac{|H|}{2e}$ points below $\min H$ that can be values of $h(\min H,\overline{d_1},\overline{d_2})$ for $\overline{d_1} < \overline{d_2}$ in H. Hence, by the pigeonhole principle, there are $\overline{d_1} < \overline{d_2} < \overline{d_3} < \overline{d_4}$ such that $h(\min H,\overline{d_1},\overline{d_2}) = h(\min H,\overline{d_3},\overline{d_4})$, which is a contradiction. Hence the value of g on $[H]^{4e+1}$ is 0.

Since $|H| \geq r(e)$, there is a set $C \subseteq H$ of $5e+1$ points $c_0 < \cdots < c_{5e}$ in H such that for any two $(2e+1)$-tuples $c < \overline{d_1} < \overline{d_2}$ and $c' < \overline{d_1'} < \overline{d_2'}$ in C, we have $f(c,\overline{d_1},\overline{d_2}) = f(c',\overline{d_1'},\overline{d_2'})$. If the value of f on $[C]^{2e+1}$ is $i \neq e+1$ then define $p = h(c_0, c_{3e+1},\dots,c_{5e})$. Then for all $\overline{d_1} < \overline{d_2}$ among $\{c_1,\dots,c_{3e}\}$, we have $h(c_0,\overline{d_1},\overline{d_2}) = p$ because $g \equiv 0$ on $[H]^{4e+1}$. Hence

$$\varphi_i(p,c_1,\dots,c_e) \not\leftrightarrow \varphi_i(p,c_{e+1},\dots,c_{2e}) \not\leftrightarrow \varphi_i(p,c_{2e+1},\dots,c_{3e}) \\ \not\leftrightarrow \varphi_i(p,c_1,\dots,c_e),$$

which is impossible since we have only two truth values. Hence $i = e+1$ and we can conclude that the set $C' = \{c_0 < c_1 < \cdots < c_{2e}\}$ is our desired set of indiscernible elements (=indiscernibles): for any standard Δ_0-formula $\varphi(z,x_1,\dots,x_n)$ any $c \in C'$, any $d_1 < \cdots < d_n$ and $e_1 < \cdots < e_n$ above c and any $p < c$, we have $\varphi(p,d_1,\dots,d_n) \leftrightarrow \varphi(p,e_1,\dots,e_n)$. Now we shall repeat the usual argument that the initial segment $I = \sup_{k\in\mathbb{N}} c_k$ is a model of Peano Arithmetic. Once we prove that, observe that $I \models \mathrm{PA} + \neg\mathrm{PH}$ since $b > I$ was the minimal point for which a large homogeneous set exists for every colouring $g\colon [a,b]^{4e+1} \to 3e+1$.

To prove that I satisfies PA, we have to show that I is closed under addition and multiplication and satisfies the induction scheme. All other axioms are inherited from M. To show that I is closed under addition, it suffices to show that $c_1 + c_2 < c_3$. If it was the case that $c_1 + c_2 \geq c_3$ then for some $p < c_1$, $p + c_2 = c_3$. Then, by indiscernibility (using the formula $z + x_2 = x_3$ as φ), for any $c > c_2$ in H, we have $p + c_2 = c$, which is a contradiction. Hence $c_1 + c_2 < c_3$ and I is closed under addition. The same argument for multiplication: if $c_1 \cdot c_2 > c_3$ then there is $p < c_1$ such that $p \cdot c_2 < c_3 < p \cdot c_2 + c_2$. By indiscernibility this would now mean that $c_4 < p \cdot c_2 + c_2 < c_3 + c_2$, which is impossible since we already proved that $c_4 > c_2 + c_3$. Hence I is a structure in the language of arithmetic. Let us now show that the scheme of induction holds in I. Let $p \in I$ and $\varphi(p, x_1, x_2, \ldots, x_n, x)$ be such that $I \models \exists x\ \forall x_1 \exists x_2 \forall x_3 \ldots\ \varphi(p, x_1, x_2, \ldots, x_n, x)$, i.e., we take an arbitrary formula with a parameter $p \in I$ and one free variable x and assume that there is $x \in I$ which satisfies it. We want to find a minimal such x in I. For some $k \in \mathbb{N}$ and some $\ell \in I$, we have $\langle p, \ell \rangle < c_k$ and $I \models \forall x_1 \exists x_2 \forall x_3 \ldots \varphi(p, x_1, \ldots, x_n, \ell)$. By indiscernibility, for every $d \leq \ell$, $I \models \forall x_1 \exists x_2 \forall x_3 \ldots \varphi(p, x_1, \ldots, x_n, d)$ if and only if $M \models \forall x_1 < c_{k+1} \exists x_2 < c_{k+2} \forall x_3 < c_{k+3} \ldots \varphi(p, x_1, \ldots x_n, d)$. However, in M there is a minimal d such that $M \models \forall x_1 < c_{k+1} \exists x_2 < c_{k+2} \forall x_3 < c_{k+3} \ldots \varphi(p, x_1, \ldots x_n, d)$, hence this d is minimal such that $I \models \forall x_1 \exists x_2 \forall x_3 \ldots \varphi(p, x_1, \ldots, x_n, d)$. We have shown that $I \models \mathrm{PA} + \neg\mathrm{PH}$ and our proof is complete. $\dashv$

In this proof, we actually needed much less from our nonstandard ground model than that it satisfies $I\Sigma_1$, namely Δ_0-induction and existence of Ramsey numbers (i.e., closedness under the tower-function) but the assumption of existence of Ramsey numbers can be eliminated (by the overspill argument below), so Δ_0-induction plus totality of exponentiation is enough.

Also notice that the model of $\mathrm{PA} + \neg\mathrm{PH}$ was constructed from just one nonstandard instance of PH (the existence of $b \in M$ such that for any colouring $[a, b]^{4e+1} \to 3e + 1$ there exists a large homogeneous subset of size $r(e)$). Why would such an instance exist? We could of course start with a ground model $M \models I\Sigma_1 + \mathrm{PH}$ and not worry any more (this should be the pedagogically preferred option). However notice that for every standard $a, n \in M$ there is b_n such that for any colouring $[a, b_n]^{4n+1} \to 3n + 1$, there exists a homogeneous H of size $r(n)$ such that $|H| > \min H$ (because PH holds in the "standard model" part of M). Now do a Δ_0-overspill argument to find the nonstandard numbers a, e and b as needed in the beginning of the proof. (Notice also that this overspill argument immediately re-proves the famous fact that every nonstandard model of $I\Sigma_1$ has many initial segments satisfying PA. Probably this is a well-known observation.)

A good exercise for the reader at this stage would be to modify the proof of Theorem 1 to show that the number of colours in PH can be fixed as 2 and the statement stays unprovable. A stronger result (unprovability of the Paris-Harrington principle in fixed dimension and two colours) belongs to J. Paris and is Lemma 29 in [71]. Another exercise for the reader would be to modify the above proof and get unprovability with fixed $a = 0$.

§3. Adapted version of PH. The unprovability proof for PH is now easy (as easy as for KM) but we can formulate unprovable statements with yet simpler unprovability proofs if we sacrifice one exponent. I suggest that logic courses can use the proofs below instead of proofs for PH or KM.

First, we present an adapted version of PH, whose unprovability is established easily by one straightforward step, and the proof is rid of all unnecessary combinatorics.

Theorem 2. *The statement "for all m, n and c, there exists N such that for every $f\colon [N]^n \to c$, there is an f-homogeneous $H \subset N$, of size at least m and such that $|H| > n \cdot (2^{n \cdot \min H} + 1)$" is not a theorem of Peano Arithmetic.*

Proof. We shall prove unprovability already with $c = 2$. Let $M \models I\Sigma_1$ be nonstandard, $e \in M \setminus \mathbb{N}$, and $\varphi_1(x_1, \dots, x_e, y), \dots, \varphi_e(x_1, \dots, x_e, y)$ be the enumeration of the first e Δ_0-formulas in at most the free variables shown. Suppose $N \in M$ is the minimal point such that for every $f\colon [N]^{2e+1} \to 2$, there is an f-monochromatic $H \subseteq N$ such that $|H| > e \cdot (2^{e \cdot \min H} + 1)$. Define our colouring $f\colon [N]^{2e+1} \to \{0, 1\}$ as follows: if $\{a < b_1 < b_2 < \cdots < b_e < c_1 < \cdots < c_e\}$ is a $(2e+1)$-subset of N, put $f(a, b_1, \dots, b_e, c_1, \dots, c_e) = 0$ if for all $x < a$,

$$\{i \le e \mid \varphi_i(b_1, \dots, b_e, x)\} = \{i \le e \mid \varphi_i(c_1, \dots, c_e, x)\}$$

and 1 otherwise. Using the definition of N, extract an f-homogeneous set $H \subset N$ of size greater than $e \cdot (2^{e \cdot \min H} + 1)$.

For every e-tuple $b_1 < b_2 < \cdots < b_e$ in $H \setminus \{\min H\}$, define the following sequence of e-many subsets of $[0, \min H)$:

$$\langle \{x < \min H \mid \varphi_1(b_1, \dots, b_e, x)\}, \dots, \{x < \min H \mid \varphi_e(b_1, \dots, b_e, x)\} \rangle.$$

There can be no more than $2^{e \cdot \min H}$ such sequences, hence, by the pigeonhole principle, there are $b_1 < b_2 < \cdots < b_e < c_1 < \cdots < c_e$ in $H \setminus \{\min H\}$ such that

$$f(\min H, b_1, \dots, b_e, c_1, \dots, c_e) = 0.$$

Hence, by homogeneity, f is constant 0 on $[H]^{2e+1}$.

Let $d_1 < \cdots < d_e$ be the last e elements of H. Then for any $a < b_1 < \cdots < b_e$ and $a < c_1 < \cdots < c_e$ in $H \setminus \{d_1, \dots, d_e\}$, we have: for every $x < a$,

$$\begin{aligned}\{i \le e \mid \varphi_i(b_1, \dots, b_e, x)\} &= \{i \le e \mid \varphi_i(d_1, \dots, d_e, x)\} \\ &= \{i \le e \mid \varphi_i(c_1, \dots, c_e, x)\},\end{aligned}$$

which is the indiscernibility condition we need. Again define an initial segment I as the supremum of the first $\mathbb{N}$ points of H and notice that $I \models \mathrm{PA}$ and that I satisfies the negation of our statement. ⊣

§4. Adapted version of KM. For KM, there is also a short adapted version which we first presented in [10].

THEOREM 3. *The statement "for all* m, a *and* n *with* $n < m$, *there is* $b > a$ *such that for every* $f : [a, b]^n \to 2^{nb}$ *such that* $f(x_1, \ldots, x_n) < 2^{n \cdot x_1}$, *there is an* f*-min-homogeneous subset of* $[a, b]$ *of size* m*" is not provable in Peano Arithmetic.*

PROOF. Let $M \models I\Sigma_1$ be a nonstandard ground model, $e, a, b \in M \setminus \mathbb{N}$, $e < a < b$ and $b \in M$ be minimal such that for every function f defined on $[a, b]^{e+1}$ such that $f(x_0, x_1, \ldots, x_e) < 2^{e \cdot x_0}$, there is a min-homogeneous subset of $[a, b]$ of size $2e$.

Let $\varphi_1(x_0, x_1, \ldots, x_e), \ldots, \varphi_e(x_0, x_1, \ldots, x_e)$ be the first e Δ_0-formulas in not more than the $e + 1$ free variables shown. Define f as follows: for every $x_0 < x_1 < \cdots < x_e$ in $[a, b]$, put

$$f(x_0, x_1, \ldots, x_e) = \left\langle \begin{matrix} \{p < x_0 \mid \varphi_1(p, x_1, \ldots, x_e)\} \\ \vdots \\ \{p < x_0 \mid \varphi_e(p, x_1, \ldots, x_e)\} \end{matrix} \right\rangle.$$

i.e., $f(x_0, x_1, \ldots, x_e)$ is a code of a collection of e subsets of $[0, x_0 - 1)$, hence $f(x_0, x_1, \ldots, x_e) < 2^{e \cdot x_0}$. Extract an f-min-homogeneous subset $H = \{c_i\}_{i=1}^{2e} \subseteq [a, b]$ and notice that for every Δ_0-formula $\varphi(x_0, x_1, \ldots, x_n)$, any $i_0 < i_1 < \cdots < i_n$ and $i_0 < j_1 < \cdots < j_n$ and all $p < c_{i_0}$, we have $M \models \varphi(p, c_{i_1}, \ldots, c_{i_n}) \leftrightarrow \varphi(p, c_{j_1}, \ldots, c_{j_n})$, which is exactly our indiscernibility condition. Again define an initial segment I as the supremum of the first $\mathbb{N}$ points of H and notice that $I \models$ PA and that I satisfies the negation of our statement. ⊣

§5. Model-theoretic proof of a threshold result for PH. The following theorem was first proved by A. Weiermann in [90] using ordinals.

THEOREM 4. *For every* $n \in \omega$, *Peano Arithemtic does not prove* $\mathrm{PH}_{\log^{(n)}}$.

The combinatorial treatment of KM was done by G. Lee in [61], also using ordinals.

THEOREM 5. *For every* $n \in \omega$, *Peano Arithmetic does not prove* $\mathrm{KM}_{\log^{(n)}}$.

We shall give one model-theoretic proof of both of these theorems. It is also possible to show by a combinatorial argument how each of these statements implies PH, by a usual trade-off where increased dimension allows for weaker largeness restrictions on a homogeneous set. The reader can think of these possible combinatorial arguments as being already incorporated into the complete model-theoretic proof below.

For every number a, let $X(a)$ be the set coded by a. We fix a coding method such that every subset of $[0, a]$ has a code before $2^{a+1} + 1$. For example, let every $X \subseteq [0, a]$ be coded by $\sum_{i=0}^{a} \chi(i) \cdot 2^i$, where χ is the characteristic function of X. Let us first write down the proof for $n = 1$.

PROOF. As usual, take a nonstandard ground model $M \models I\Sigma_1$ and elements $a > e > \mathbb{N}$. Let $\varphi_1(z, x_1, x_2, \dots, x_e), \dots, \varphi_e(z, x_1, x_2, \dots, x_e)$ be the first e Δ_0-formulas in at most the free variables shown. In particular, this list will contain all Δ_0-formulas of standard size. Denote $R(2e+1, e+2, 10e+1)$ by $r(e)$. Let $b \in M$ be minimal such that for every $g\colon [a, b]^{8e+1} \to 5e+1$ there is $H \subset [a, b]$ of size at least $r(e)$ such that $|H| > \log(\min H)$. Define $f\colon [a, b]^{2e+1} \to e+2$, the minimal formula of disagreement, as before: for $c < \overline{d_1} < \overline{d_2}$ in $[a, b]$, put $f(c, \overline{d_1}, \overline{d_2}) = \min i \leq e\ \exists p < c\ (\varphi_i(p, \overline{d_1}) \not\leftrightarrow \varphi_i(p, \overline{d_2}))$ if such i exists and $e+1$ otherwise. Define a regressive function h, the minimal parameter of disagreement, as before: $h(c, \overline{d_1}, \overline{d_2}) = \min p < c\ (\varphi_{f(c,\overline{d_1},\overline{d_2})}(p, \overline{d_1}) \not\leftrightarrow \varphi_{f(c,\overline{d_1},\overline{d_2})}(p, \overline{d_2}))$ if $f(c, \overline{d_1}, \overline{d_2}) \neq e+1$ and c otherwise. Now, define a log-regressive function ℓ defined on $[a, b]^{4e+1}$. Let $c < \overline{d_1} < \overline{d_2} < \overline{d_3} < \overline{d_4}$ be a $(4e+1)$-subset in $[a, b]$. Let $p_1 = h(c, \overline{d_1}, \overline{d_2})$, $p_2 = h(c, \overline{d_3}, \overline{d_4})$. Put

$$\ell(c, \overline{d_1}, \overline{d_2}, \overline{d_3}, \overline{d_4}) = \min(\boldsymbol{X}(p_1) \Delta \boldsymbol{X}(p_2)) \text{ if } p_1 \neq p_2$$

and $\log(c)$ otherwise. Here Δ denotes the symmetric difference relation: $A \Delta B = (A \setminus B) \cup (B \setminus A)$. Clearly, $\ell(c, \overline{d_1}, \overline{d_2}, \overline{d_3}, \overline{d_4}) \leq \log(c)$.

Now, define the main colouring $g\colon [a, b]^{8e+1} \to 5e+1$ as follows:

$$g(c, \overline{d_1}, \dots, \overline{d_8}) = \begin{cases} 0 & \text{if} \quad \ell(c, \overline{d_1}, \dots, \overline{d_4}) = \ell(c, \overline{d_5}, \dots, \overline{d_8}), \\ j & \text{if} \quad \ell(c, \overline{d_1}, \dots, \overline{d_4}) \neq \ell(c, \overline{d_5}, \dots, \overline{d_8}) \text{ and} \\ & \quad 0 < j \leq 5e \text{ and } \ell(c, \overline{d_1}, \dots, \overline{d_4}) \equiv j \pmod{5e}. \end{cases}$$

Extract a g-homogeneous $H \subset [a, b]$ of size $r(e)$ such that $|H| > \log(\min H)$. Observe, as usual, that g is constant 0 on $[H]^{8e+1}$. Indeed, if $g|_{[H]^{8e+1}}$ is $j \neq 0$ then below $\log(\min H)$ there are at most $[\frac{\log(\min H)}{5e}]$ possible values of $\ell(\min H, \overline{d_1}, \overline{d_2}, \overline{d_3}, \overline{d_4})$, hence for $\frac{|H|}{4e}$ successive $4e$-subsets in $H \setminus \{\min H\}$, there are not enough separate spaces, hence for some $\overline{d_1} < \dots < \overline{d_8}$, $\ell(\min H, \overline{d_1}, \overline{d_2}, \overline{d_3}, \overline{d_4}) = \ell(\min H, \overline{d_5}, \overline{d_6}, \overline{d_7}, \overline{d_8})$. Hence $g|_{[H]^{8e+1}} = 0$.

Choose $H_1 \subset H$ of size $10e+1$, homogeneous for f. This can be done by our definition of $r(e)$. Let H_2 be $H_1 \setminus \{\text{the last } 4e \text{ elements of } H_1\}$.

Let us show that $f|_{[H_2]^{2e+1}} = e+1$. Suppose that $f|_{[H_2]^{2e+1}}$ has constant value $i \neq e+1$ and we shall get a contradiction. List H_2 as $c < \overline{d_1} < \dots < \overline{d_6}$. Denote the value of $\ell(c, \dots)$ on H_2 by p and let $h(c, \overline{d_1}, \overline{d_2}) = p_1$, $h(c, \overline{d_3}, \overline{d_4}) = p_2$, $h(c, \overline{d_5}, \overline{d_6}) = p_3$. Let us first notice that p_1, p_2 and p_3 are all different. If two of them coincided, say $p_1 = p_2$ then $\ell(c, \overline{d_1}, \overline{d_2}, \overline{d_3}, \overline{d_4}) = p = \log c$ and hence $p_1 = p_3$. So, let us assume $p_1 = p_2 = p_3 = q$. Since $i \neq e+1$, we know that $q < c$. Suppose without loss of generality that $\varphi_i(q, \overline{d_1})$, $\varphi_i(q, \overline{d_3})$ and $\varphi_i(q, \overline{d_5})$ are true (and $\varphi_i(q, \overline{d_2})$, $\varphi_i(q, \overline{d_4})$ and $\varphi_i(q, \overline{d_6})$ are false). Since $\ell(c, \overline{d_1}, \overline{d_2}, \overline{d_3}, \overline{d_5}) = \ell(c, \overline{d_1}, \overline{d_2}, \overline{d_3}, \overline{d_4})$, we conclude that $\overline{d_3}$ and $\overline{d_5}$ disagree on q, which contradicts our knowledge that $\varphi_i(q, \overline{d_3})$ and $\varphi_i(q, \overline{d_5})$ are both false. We arrived at a contradiction, hence $p_1 \neq p_2$, $p_1 \neq p_3$ and $p_2 \neq p_3$ and the sets $\boldsymbol{X}(p_1), \boldsymbol{X}(p_2)$ and $\boldsymbol{X}(p_3)$ are different.

Notice that by definition of ℓ, we have $p \in \boldsymbol{X}(p_1) \Leftrightarrow p \notin \boldsymbol{X}(p_2)$, $p \in \boldsymbol{X}(p_2) \Leftrightarrow p \notin \boldsymbol{X}(p_3)$ and $p \in \boldsymbol{X}(p_1) \Leftrightarrow p \notin \boldsymbol{X}(p_3)$, which are contradictory assertions. Hence $f|_{[H_2]^{2e+1}} = e + 1$ and H_2 is our desired set of indiscernibles. ⊣

On the way we also proved PA $\nvdash$ KM$_{\log}$: indeed the set H_1 is obtained as min-homogeneous for the log-regressive function ℓ.

Now the proof for arbitrary n. The idea is the same as in case $n = 1$. We shall make n steps downstairs and focus the contradiction below $\log^{(n)}(c)$.

Proof. As usual, take a nonstandard ground model $M \models I\Sigma_1$ and elements $a > e > \mathbb{N}$. Let $\varphi_1(z, x_1, x_2, \dots, x_e), \dots, \varphi_e(z, x_1, x_2, \dots, x_e)$ be the first e Δ_0-formulas in the free variables shown. In particular, this list will contain all Δ_0-formulas of standard size. Denote $R(2e + 1, e + 2, 5 \cdot 2^n \cdot e + 1)$ by $r(e)$. Let $b \in M$ be minimal such that for every $g\colon [a, b]^{e \cdot 2^{n+2}+1} \to 2^{n+2}e + 1$ there is $H \subset [a, b]$ of size at least $r(e)$ and such that $|H| > \log^{(n)}(\min H)$.

Define $f\colon [a, b]^{2e+1} \to e+2$, the minimal formula of disagreement as before: for $c < \overline{d_1} < \overline{d_2}$ in $[a, b]$, put $f(c, \overline{d_1}, \overline{d_2}) = \min i \leq e\ \exists p < c\ \varphi_i(p, \overline{d_1}) \not\leftrightarrow \varphi_i(p, \overline{d_2})$ if such i exists and $e + 1$ otherwise. Define a (regressive) function h, the minimal parameter of disagreement as before: $h(c, \overline{d_1}, \overline{d_2}) = \min p < c\ [\varphi_{f(c,\overline{d_1},\overline{d_2})}(p, \overline{d_1}) \not\leftrightarrow \varphi_{f(c,\overline{d_1},\overline{d_2})}(p, \overline{d_2})]$ if $f(c, \overline{d_1}, \overline{d_2}) \neq e + 1$ and c otherwise.

Now, define a $\log^{(n)}$-regressive function ℓ defined on $[a, b]^{2^{n+1} \cdot e+1}$. Let $c < \overline{d_1} < \cdots < \overline{d_{2^{n+1}}}$ be a $(2^{n+1} \cdot e + 1)$-subset in $[a, b]$. First, define for every $i = 1, 2, \dots, n + 1$ a sequence of points $\{p^i_k\}_{k=1}^{2^{i-1}}$ as follows. For all $k = 1, 2, \dots, 2^n$, put $p^{n+1}_k = h(c, \overline{d_{2k-1}}, \overline{d_{2k}})$. For every $1 \leq i < n + 1$ and every $k = 1, 2, \dots, 2^{i-1}$, put $p^{i-1}_k = \min(\boldsymbol{X}(p^i_{2k-1}) \Delta \boldsymbol{X}(p^i_{2k}))$. Now, put $\ell(c, \overline{d_1}, \dots, \overline{d_{2^{n+1}}}) = p^1_1$. Clearly, $\ell(c, \overline{d_1}, \dots, \overline{d_{2^{n+1}}}) \leq \log^{(n)}(c)$. Now, define the main colouring $g\colon [a, b]^{2^{n+2} \cdot e+1} \to 2^{n+2}e + 1$ as follows:

$$g(c, \overline{d_1}, \dots, \overline{d_{2^{n+1}}}, \overline{d_{2^{n+1}+1}} \dots, \overline{d_{2^{n+2}}}) =$$
$$= \begin{cases} 0 & \text{if} \quad \ell(c, \overline{d_1}, \dots, \overline{d_{2^{n+1}}}) = \ell(c, \overline{d_{2^{n+1}+1}}, \dots, \overline{d_{2^{n+2}}}), \\ j & \text{if} \quad \ell(c, \overline{d_1}, \dots, \overline{d_{2^{n+1}}}) \neq \ell(c, \overline{d_{2^{n+1}+1}}, \dots, \overline{d_{2^{n+2}}}) \text{ and} \\ & \quad 0 < j \leq 2^{n+2}e \text{ and } \ell(c, \overline{d_1}, \dots, \overline{d_{2^{n+1}}}) \equiv j \pmod{2^{n+2}e}. \end{cases}$$

Choose a g-homogeneous set $H \subseteq [a, b]$ of size $r(e)$ such that $|H| > \log^{(n)}(\min H)$. Let us show that the value of g on $(2^{n+2} \cdot e + 1)$-subsets of H

is 0. Suppose that the value of g on $(2^{n+2} \cdot e+1)$-subsets of H is $j \neq 0$. Notice that then there are not more than $[\frac{\log^{(n)}(\min H)}{2^{n+2}e}]$ points below $\log^{(n)}(\min H)$ that can possibly be values of $\ell(\min H, \overline{d_1}, \dots, \overline{d_{2^{n+1}}})$. Hence, by the pigeonhole principle (using the fact that $\frac{\log^{(n)}(\min H)}{2^{n+2}e} < \frac{|H|}{2^{n+1}e}$), there are $\overline{d_1} < \cdots < \overline{d_{2^{n+1}}} < \overline{d_{2^{n+1}+1}} < \cdots < \overline{d_{2^{n+2}}}$ such that $\ell(c, \overline{d_1}, \dots, \overline{d_{2^{n+1}}}) = \ell(c, \overline{d_{2^{n+1}+1}}, \dots, \overline{d_{2^{n+2}}})$, which is a contradiction. Hence the value of g on $(2^{n+2}e+1)$-subsets of H is 0.

Since $|H| \geq r(e)$, we can choose an f-homogeneous $H' \subseteq H$, of size $5 \cdot 2^n \cdot e + 1$.

Let us show that the value of f on $(2e+1)$-subsets of H' is $e+1$, thus producing our desired set of indiscernibles. Let $H'' = H' \setminus \{$the last $2^{n+1} \cdot e$ points of $H'\}$. List the set H'' as $c < \overline{d_1} < \cdots < \overline{d_{3 \cdot 2^{n+1}}}$.

Notice that the function $\ell(c, \dots)$ is constant on H''. Denote the value of ℓ on H'' by p. Again, show (by an argument similar to the argument above) that $p_1^n \neq p_2^n \neq p_3^n \neq p_1^n$ and notice that since $\ell(c, \overline{d_1}, \dots, \overline{d_{2^n}}, \overline{d_{2^n+1}}, \dots, \overline{d_{2^{n+1}}}) =$

$$= \ell(c, \overline{d_1}, \dots, \overline{d_{2^n}}, \overline{d_{2^{n+1}+1}}, \dots, \overline{d_{3 \cdot 2^n}}) = \ell(c, \overline{d_{2^n+1}}, \dots, \overline{d_{2^{n+1}}}, \overline{d_{2^{n+1}+1}}, \dots, \overline{d_{3 \cdot 2^n}})$$
$$= p,$$

we have $p \in \boldsymbol{X}(p_1^n) \Leftrightarrow p \notin \boldsymbol{X}(p_2^n) \Leftrightarrow p \in \boldsymbol{X}(p_3^n) \Leftrightarrow p \notin \boldsymbol{X}(p_1^n)$, which is a contradiction. Hence f is constant $e+1$ on H'' and H'' is our desired set of indiscernibles. ⊣

It is clear that on the way we proved that Peano Arithmetic does not prove $\mathrm{KM}_{\log^{(n)}}$: notice that an ℓ-min-homogeneous set of size $r(e)$ produces the same set of indiscernibles as in our proof.

We conjecture that the proof above can now be converted into a proof of the full A. Weiermann's threshold result from [90]: Peano Arithmetic does not prove PH_f, where $f(n) = \log^{H_{\varepsilon_0}^{-1}(n)}(n)$, where H_{ε_0} is the ε_0th function in the Hardy hierarchy (that eventually dominates all PA-provably recursive functions).

Another approach to proving threshold results for PH would be to use Lemma 3.3. of [50] (which gives KM-thresholds) and then prove combinatorial implications between PH_f and KM_g for different functions f and g. In the case of PH-thresholds, this would require the same amount of work as we did above.

REFERENCES

[1] T. Arai, *A slow growing analogue to Buchholz' proof*, ***Annals of Pure and Applied Logic***, vol. 54 (1991), no. 2, pp. 101–120.

[2] J. Avigad, *Update procedures and the 1-consistency of arithmetic*, ***Mathematical Logic Quarterly***, vol. 48 (2002), no. 1, pp. 3–13.

[3] J. Avigad and R. Sommer, *A model-theoretic approach to ordinal analysis*, ***The Bulletin of Symbolic Logic***, vol. 3 (1997), no. 1, pp. 17–52.

[4] ———, *The model-theoretic ordinal analysis of theories of predicative strength*, ***The Journal of Symbolic Logic***, vol. 64 (1999), no. 1, pp. 327–349.

[5] L. D. Beklemishev, *The worm principle*, ***Logic Colloquium '02***, Lecture Notes in Logic, vol. 27, ASL, La Jolla, CA, 2006, pp. 75–95.

[6] ———, *Representing Worms as a Term Rewriting System*, Oberwolfach Report 52/2006, 2007, pp. 7–9.

[7] C. Berline, K. McAloon, and J.-P. Ressayre (editors), ***Model Theory and Arithmetic***, Lecture Notes in Mathematics, vol. 890, Springer, Berlin, 1981.

[8] T. Bigorajska and H. Kotlarski, *A partition theorem for α-large sets*, ***Fundamenta Mathematicae***, vol. 160 (1999), no. 1, pp. 27–37.

[9] ———, *Partitioning α-large sets: some lower bounds*, ***Transactions of the American Mathematical Society***, vol. 358 (2006), no. 11, pp. 4981–5001.

[10] A. Bovykin, *Several proofs of PA-unprovability*, ***Logic and its Applications***, Contemporary Mathematics, vol. 380, AMS, Providence, RI, 2005, pp. 29–43.

[11] ———, *Unprovability of sharp versions of Friedman's sine-principle*, ***Proceedings of the American Mathematical Society***, vol. 135 (2007), no. 9, pp. 2967–2973.

[12] ———, *Exact unprovability results for compound well-quasi-ordered combinatorial classes*, ***Annals of Pure and Applied Logic***, vol. 157 (2009), pp. 77–84.

[13] A. Bovykin and L. Carlucci, *Long games on braids*, preprint, available online at `http://logic.pdmi.ras.ru/~andrey/research.html`, 2006.

[14] A. Bovykin, L. Carlucci, P. Dehornoy, and A. Weiermann, *Unprovability results involving braids*, preprint, 2007.

[15] A. Bovykin and A. Weiermann, *The strength of infinitary ramseyan statements can be accessed by their densities*, to appear in ***Annals of Pure and Applied Logic***.

[16] ———, *The strength of infinitary statements can be approximated by their densities*, preprint, 2007.

[17] ———, *Unprovable statements based on diophantine approximation and distribution of values of zeta-functions*, preprint, `http://logic.pdmi.ras.ru/~andrey/research.html`, 2007.

[18] W. Buchholz, *An independence result for Π^1_1-CA+BI*, ***Annals of Pure and Applied Logic***, vol. 33 (1987), no. 2, pp. 131–155.

[19] ———, *Another Rewrite System for the Standard Hydra Battle*, Oberwolfach Report 52/2006, 2007, pp. 13–15.

[20] W. Buchholz and S. Wainer, *Provably computable functions and the fast growing hierarchy*, ***Logic and Combinatorics (Arcata, Calif., 1985)***, Contemporary Mathematics, vol. 65, AMS, Providence, RI, 1987, pp. 179–198.

[21] L. Carlucci, *A new proof-theoretic proof of the independence of Kirby-Paris' hydra theorem*, ***Theoretical Computer Science***, vol. 300 (2003), no. 1-3, pp. 365–378.

[22] ———, *Worms, gaps, and hydras*, ***Mathematical Logic Quarterly***, vol. 51 (2005), no. 4, pp. 342–350.

[23] L. Carlucci, G. Lee, and A. Weiermann, *Classifying the phase transition threshold for regressive ramsey functions*, submitted, 2006.

[24] E. A. Cichon, *A short proof of two recently discovered independence results using recursion theoretic methods*, ***Proceedings of the American Mathematical Society***, vol. 87 (1983), no. 4, pp. 704–706.

[25] P. Clote, *Weak partition relation, finite games, and independence results in Peano arithmetic*, ***Model Theory of Algebra and Arithmetic (Proc. Conf., Karpacz, 1979)***, Lecture Notes in Mathematics, vol. 834, Springer, Berlin, 1980, pp. 92–107.

[26] ———, *Anti-basis theorems and their relation to independence results in Peano arithmetic*, ***Model Theory and Arithmetic***, Lecture Notes in Mathematics, vol. 890, Springer, 1981, pp. 115–133.

[27] P. Clote and K. McAloon, *Two further combinatorial theorems equivalent to the 1-consistency of Peano arithmetic*, ***The Journal of Symbolic Logic***, vol. 48 (1983), no. 4, pp. 1090–1104.

[28] P. Dehornoy, I. Dynnikov, D. Rolfsen, and B. Wiest, ***Why Are Braids Orderable?***, Panoramas et Synthèses [Panoramas and Syntheses], vol. 14, Société Mathématique de France, Paris, 2002.

[29] N. Dershowitz and G. Moser, *The hydra battle revisited*, ***Rewriting Computation and Proof***, Lecture Notes in Computer Science, vol. 4600, Springer, Berlin, 2007, pp. 1–27.

[30] P. Erdös and R. Rado, *Combinatorial theorems on classifications of subsets of a given set*, ***Proceedings of the London Mathematical Society. Third Series***, vol. 2 (1952), pp. 417–439.

[31] M. Fairtlough and S. Wainer, *Hierarchies of provably recursive functions*, ***Handbook of Proof Theory***, Studies in Logic and the Foundations of Mathematics, vol. 137, North-Holland, Amsterdam, 1998, pp. 149–207.

[32] H. Friedman, *On the necessary use of abstract set theory*, ***Advances in Mathematics***, vol. 41 (1981), no. 3, pp. 209–280.

[33] ———, *Finite functions and the necessary use of large cardinals*, ***Annals of Mathematics. Second Series***, vol. 148 (1998), no. 3, pp. 803–893.

[34] ———, *Long finite sequences*, ***Journal of Combinatorial Theory. Series A***, vol. 95 (2001), no. 1, pp. 102–144.

[35] ———, *Internal finite tree embeddings*, ***Reflections on the Foundations of Mathematics (Stanford, CA, 1998)***, Lecture Notes in Logic, vol. 15, ASL, Urbana, IL, 2002, pp. 60–91.

[36] ———, *Interpretations of set theory in discrete mathematics and informal thinking*, preprint, Tarski Lectures, available online, 2007.

[37] ———, *Interpreting set theory in discrete mathematics*, preprint, Tarski Lectures, available online, 2007.

[38] ———, ***Boolean Relation Theory and the Incompleteness Phenomena***, a 600-page book, to appear.

[39] H. Friedman, K. McAloon, and S. Simpson, *A finite combinatorial principle which is equivalent to the 1-consistency of predicative analysis*, ***Patras Logic Symposion (Patras, 1980)***, Studies in Logic and the Foundations of Mathematics, vol. 109, North-Holland, Amsterdam, 1982, pp. 197–230.

[40] H. Friedman, N. Robertson, and P. Seymour, *The metamathematics of the graph minor theorem*, ***Logic and Combinatorics (Arcata, Calif., 1985)***, Contemporary Mathematics, vol. 65, AMS, Providence, RI, 1987, pp. 229–261.

[41] H. Friedman and M. Sheard, *Elementary descent recursion and proof theory*, ***Annals of Pure and Applied Logic***, vol. 71 (1995), no. 1, pp. 1–45.

[42] L. N. Gordeev, *Generalizations of the one-dimensional version of the Kruskal-Friedman theorems*, ***The Journal of Symbolic Logic***, vol. 54 (1989), no. 1, pp. 100–121.

[43] ———, *Generalizations of the Kruskal-Friedman theorems*, ***The Journal of Symbolic Logic***, vol. 55 (1990), no. 1, pp. 157–181.

[44] ———, *A modified sentence unprovable in PA*, ***The Journal of Symbolic Logic***, vol. 59 (1994), no. 4, pp. 1154–1157.

[45] P. Hájek and J. Paris, *Combinatorial principles concerning approximations of functions*, ***Archive for Mathematical Logic***, vol. 26 (1986), no. 1-2, pp. 13–28.

[46] P. Hájek and P. Pudlák, ***Metamathematics of First-Order Arithmetic***, Perspectives in Mathematical Logic, Springer-Verlag, Berlin, 1993.

[47] M. Hamano and M. Okada, *A direct independence proof of Buchholz's Hydra game on finite labeled trees*, ***Archive for Mathematical Logic***, vol. 37 (1997), no. 2, pp. 67–89.

[48] ———, *A relationship among Gentzen's proof-reduction, Kirby-Paris' hydra game and Buchholz's hydra game*, ***Mathematical Logic Quarterly***, vol. 43 (1997), no. 1, pp. 103–120.

[49] H. Jervell, *Gentzen games*, ***Zeitschrift für Mathematische Logik und Grundlagen der Mathematik***, vol. 31 (1985), no. 5, pp. 431–439.

[50] A. Kanamori and K. McAloon, *On Gödel incompleteness and finite combinatorics*, ***Annals of Pure and Applied Logic***, vol. 33 (1987), no. 1, pp. 23–41.

[51] R. Kaye, ***Models of Peano Arithmetic***, Oxford Logic Guides, vol. 15, The Clarendon Press Oxford University Press, New York, 1991, Oxford Science Publications.

[52] J. Ketonen, *Set theory for a small universe, I. The Paris-Harrington axiom*, Unpublished manuscript, 1979.

[53] ———, *Some remarks on finite combinatorics*, Unpublished Manuscript, 1981.

[54] J. Ketonen and R. Solovay, *Rapidly growing Ramsey functions*, ***Annals of Mathematics. Second Series***, vol. 113 (1981), no. 2, pp. 267–314.

[55] L. Kirby, ***Initial Segments of Models of Arithmetic***, Ph.D. thesis, Manchester University, 1977.

[56] ———, *Flipping properties in arithmetic*, ***The Journal of Symbolic Logic***, vol. 47 (1982), no. 2, pp. 416–422.

[57] L. Kirby and J. Paris, *Initial segments of models of Peano's axioms*, ***Set Theory and Hierarchy Theory, V (Proc. Third Conf., Bierutowice, 1976)***, Lecture Notes in Mathematics, vol. 619, Springer, Berlin, 1977, pp. 211–226.

[58] S. Kochen and S. Kripke, *Nonstandard models of Peano arithmetic*, ***Logic and Algorithmic (Zurich, 1980)***, Monographie de l'Enseignment Mathematique, vol. 30, University of Genève, Geneva, 1982, pp. 275–295.

[59] M. Kojman, G. Lee, E. Omri, and A. Weiermann, *Sharp thresholds for the phase transition between primitive recursive and Ackermannian Ramsey numbers*, ***Journal of Combinatorial Theory. Series A***, vol. 115 (2008), no. 6, pp. 1036–1055.

[60] H. Kotlarski, ***A Model-Theoretic Approach to Proof Theory for Arithmetic***, unpublished book manuscript, 2003.

[61] G. Lee, ***Phase Transitions in Axiomatic Thought***, Ph.D. thesis, University of Münster, 2005.

[62] M. Loebl and J. Matoušek, *On undecidability of the weakened Kruskal theorem*, ***Logic and Combinatorics (Arcata, Calif., 1985)***, Contemporary Mathematics, vol. 65, AMS, Providence, RI, 1987, pp. 275–280.

[63] M. Loebl and J. Nešetřil, *Fast and slow growing (a combinatorial study of unprovability)*, ***Surveys in Combinatorics, 1991***, London Mathematical Society Lecture Notes Series, vol. 166, Cambridge University Press, Cambridge, 1991, pp. 119–160.

[64] ———, *An unprovable Ramsey-type theorem*, ***Proceedings of the American Mathematical Society***, vol. 116 (1992), no. 3, pp. 819–824.

[65] Yu. Matiyasevich, ***Hilbert's Tenth Problem***, Foundations of Computing Series, MIT Press, Cambridge, MA, 1993.

[66] K. McAloon, *Progressions transfinies de théories axiomatiques, formes combinatoires du théorème d'incomplétude et fonctions récursives à croissance rapide*, ***Models of Arithmetic***, Astérisque, vol. 73, Soc. Math. France, Paris, 1980, pp. 41–58 (in French).

[67] ———, *Paris-Harrington incompleteness and progressions of theories*, ***Recursion Theory (Ithaca, N.Y., 1982)***, Proceedings of Symposia in Pure Mathematics, vol. 42, AMS, Providence, RI, 1985, pp. 447–460.

[68] G. Mills, *A tree analysis of unprovable combinatorial statements*, ***Model Theory of Algebra and Arithmetic***, Lecture Notes in Mathematics, vol. 834, Springer, Berlin, 1980, pp. 248–311.

[69] L. Pacholski and J. Wierzejewski (editors), ***Model Theory of Algebra and Arithmetic***, Lecture Notes in Mathematics, vol. 834, Springer, Berlin, 1980.

[70] J. Paris, *Some independence results for Peano arithmetic*, ***The Journal of Symbolic Logic***, vol. 43 (1978), no. 4, pp. 725–731.

[71] ———, *A hierarchy of cuts in models of arithmetic*, ***Model Theory of Algebra and Arithmetic***, Lecture Notes in Mathematics, vol. 834, Springer, Berlin, 1980, pp. 312–337.

[72] J. Paris and L. Harrington, *A mathematical incompleteness in Peano arithmetic*, ***Handbook for Mathematical Logic*** (J. Barwise, editor), North-Holland, 1977.

[73] J. Paris and L. Kirby, *Accessible independence results for Peano arithmetic*, ***The Bulletin of the London Mathematical Society***, vol. 14 (1982), no. 4, pp. 285–293.

[74] P. Pudlák, *Another combinatorial principle independent of Peano's axioms*, unpublished, 1979.

[75] ———, *Consistency and games—in search of new combinatorial principles*, ***Logic Colloquium '03***, Lecture Notes in Logic, vol. 24, ASL, La Jolla, CA, 2006, pp. 244–281.

[76] Z. Ratajczyk, *Arithmetical transfinite induction and hierarchies of functions*, ***Fundamenta Mathematicae***, vol. 141 (1992), no. 1, pp. 1–20.

[77] ———, *Subsystems of true arithmetic and hierarchies of functions*, ***Annals of Pure and Applied Logic***, vol. 64 (1993), no. 2, pp. 95–152.

[78] J.-P. Ressayre, *Nonstandard universes with strong embeddings, and their finite approximations*, ***Logic and Combinatorics (Arcata, Calif., 1985)***, Contemporary Mathematics, vol. 65, AMS, Providence, RI, 1987, pp. 333–358.

[79] S. Shelah, *On logical sentences in* PA, ***Logic Colloquium '82***, Studies in Logic and the Foundations of Mathematics, vol. 112, North-Holland, Amsterdam, 1984, pp. 145–160.

[80] S. Simpson (editor), ***Harvey Friedman's Research on the Foundations of Mathematics***, Studies in Logic and the Foundations of Mathematics, vol. 117, North-Holland Publishing, Amsterdam, 1985.

[81] ———, *Nonprovability of certain combinatorial properties of finite trees*, ***Harvey Friedman's Research on the Foundations of Mathematics***, Studies in Logic and the Foundations of Mathematics, vol. 117, North-Holland, Amsterdam, 1985, pp. 87–117.

[82] S. Simpson (editor), ***Logic and Combinatorics***, Contemporary Mathematics, vol. 65, AMS, Providence, RI, 1987.

[83] ———, *Ordinal numbers and the Hilbert basis theorem*, ***The Journal of Symbolic Logic***, vol. 53 (1988), no. 3, pp. 961–974.

[84] ———, ***Subsystems of Second Order Arithmetic***, second ed., Perspectives in Mathematical Logic, Springer-Verlag, Berlin, 2007.

[85] S. Simpson and K. Schütte, *Ein in der reinen Zahlentheorie unbeweisbarer Satz über endliche Folgen von natürlichen Zahlen*, ***Archiv für Mathematische Logik und Grundlagenforschung***, vol. 25 (1985), no. 1-2, pp. 75–89, (in German) [A theorem on finite sequences of natural numbers that is unprovable in pure number theory].

[86] C. Smoryński, *The incompleteness theorems*, ***Handbook for Mathematical Logic***, 1977, pp. 821–865.

[87] R. Sommer, *Transfinite induction within Peano arithmetic*, ***Annals of Pure and Applied Logic***, vol. 76 (1995), no. 3, pp. 231–289.

[88] H. Touzet, *Encoding the hydra battle as a rewrite system*, ***Mathematical Foundations of Computer Science, 1998 (Brno)***, Lecture Notes in Computer Science, vol. 1450, Springer, Berlin, 1998, pp. 267–276.

[89] A. Weiermann, *An application of graphical enumeration to PA*, ***The Journal of Symbolic Logic***, vol. 68 (2003), no. 1, pp. 5–16.

[90] ———, *A classification of rapidly growing Ramsey functions*, ***Proceedings of the American Mathematical Society***, vol. 132 (2004), no. 2, pp. 553–561.

[91] ———, *Analytic combinatorics, proof-theoretic ordinals, and phase transitions for independence results*, ***Annals of Pure and Applied Logic***, vol. 136 (2005), no. 1-2, pp. 189–218.

[92] ———, *Classifying the provably total functions of PA*, ***The Bulletin of Symbolic Logic***, vol. 12 (2006), no. 2, pp. 177–190.

[93] ———, *Phase transition thresholds for some Friedman-style independence results*, ***Mathematical Logic Quarterly***, vol. 53 (2007), no. 1, pp. 4–18.

[94] A. R. Woods, ***Some Problems in Logic and Number Theory and Their Connections***, Ph.D. thesis, Manchester University, 1981.

DEPARTMENT OF MATHEMATICS
UNIVERSITY OF BRISTOL
BRISTOL, BS8 1TW, ENGLAND
E-mail: andrey@logic.pdmi.ras.ru

HIGHER-ORDER ABSTRACT SYNTAX IN TYPE THEORY

VENANZIO CAPRETTA AND AMY P. FELTY

Abstract. We develop a general tool to formalize and reason about languages expressed using higher-order abstract syntax in a proof-tool based on type theory (Coq). A language is specified by its signature, which consists of sets of sort and operation names and typing rules. These rules prescribe the sorts and bindings of each operation. An algebra of terms is associated to a signature, using de Bruijn notation. Then a higher-order notation is built on top of the de Bruijn level, so that the user can work with meta-variables instead of de Bruijn indices. We also provide recursion and induction principles formulated directly on the higher-order syntax. This generalizes work on the Hybrid approach to higher-order syntax in Isabelle and our earlier work on a constructive extension to Hybrid formalized in Coq. In particular, a large class of theorems that must be repeated for each object language in Hybrid is done once in the present work and can be applied directly to each object language.

§1. Introduction. We aim to use proof assistants (in our specific case Coq [9, 6]) to formally represent and reason about languages using higher-order syntax, i.e. object languages where binding operators are expressed using binding at the meta-level. This is an active and fertile field of research. Several methods contend to become the most elegant, efficient, and easy to use. The differences stem from the approach of the researchers and the characteristics of the proof tool used.

Our starting point was the work on the Hybrid tool in Isabelle/HOL by Ambler, Crole, and Momigliano [2]. We began by replicating their development step by step in Coq, but soon realized that the different underlying meta-theory (the Calculus of Inductive Constructions [10, 38], as opposed to higher-order logic) provided us with different tools and led us to diverge from a simple translation of their work. The final result [8] exploits the computational content of the Coq logic: we prove a recursion principle that can be used to program functions on the object language. These functions can be directly computed inside Coq. In the present work, we extend it and provide a general tool in which the user can easily define a language by giving its signature, and a set of tools (higher-order notation, recursion and induction principles) are automatically available.

Let us introduce the problem of representing languages with bindings in a logical framework. While a first-order language contains operation symbols that take just elements of the domain(s) as arguments, a second-order

Logic Colloquium '06
Edited by Barry Cooper, Herman Geuvers, Anand Pillay, and Jouko Väänänen
Lecture Notes in Logic, 32

language may have operations whose input consists of functions of arbitrary complexity. From a syntactic point of view these operations bind some of the free variables in their arguments. The simplest example of this phenomenon is the abstraction operation in λ-calculus. We can see the operation λ, syntactically, as a binder; its use is: $\lambda x.t$, where we indicate that the free occurrences of the variable x in t are now bound. Alternatively, we can see it as taking a function as input; its use is: $\lambda(f)$, where f maps λ-terms to λ-terms.

Both outlooks have their strong and weak points. First-order approaches are unproblematic from the logical point of view and most proof tools allow their formalization by direct inductive definitions. However, a series of problems arise. Terms are purely syntactical objects and therefore are considered equal only when they are syntactically the same; while terms that differ only by the name of bound variables should be considered identical (α-conversion). Substitution becomes problematic because binders may capture free variables, so we need to use some renaming mechanism. These approaches implement object languages by a first-order encoding with some bookkeeping mechanism to keep track of bound variables and avoid problems of variable capture and renaming. In general, many definitions and lemmas must be formalized for each implemented object language.

The higher-order approach has the advantage that the implementation of the meta-theory can be reused: In implementing the proof tool, the developers were already faced with the decision of how to represent functions and operators on functions. Higher-Order Abstract Syntax (HOAS) aims at reusing this implementation work by seeing arguments as functions at the meta-level, thus delegating the issues of α-conversion, substitution, and renaming to the logical framework. The main disadvantage is that higher-order data-types are simply not allowed in most systems. For example, a higher-order representation of the λ-calculus would be an inductive data-type with two constructors:

$$
\begin{aligned}
&\textbf{\textit{Inductive}}\ \Lambda := \\
&\quad \mathsf{abs} : (\underline{\Lambda} \to \Lambda) \to \Lambda \\
&\quad \mathsf{app} : \Lambda \to \Lambda \to \Lambda
\end{aligned}
$$

But the constructor for abstraction abs is not allowed in most systems, since it contains a negative occurrence (underlined) of the defined type. Changing the meta-theory to permit such definitions would just make the logic inconsistent. Some constraint must be imposed on negative occurrences to avoid the kind of circularity that results in inconsistency of the logical framework. Several ways of doing this are documented in the literature.

Another drawback of HOAS is the appearance of *exotic terms*. A term is called exotic if it results from the application of a binder to a function that is not uniform in its argument. For example $\lambda(X \mapsto F^{|X|}\,X)$, where F is any term and $|\_|$ is the length function on terms, is exotic: its body does not itself represent a term, but only reduces to a term for every concrete input. Exotic terms do not exist in the informal treatment of object languages and should be precluded.

The Hybrid strategy (which we appropriated) is a combination of the two approaches: It uses an internal first-order syntactic representation (de Bruijn syntax [12]) combined with a higher-order user interface.

The original elements of the present work with respect to [2] and [8] are:

- *Definition of a type of signatures for Universal Algebra with binding operations:* The general shape of formalizations of various case studies in [2] and [8] was informally explained, but the low level work had to be repeated for each new formal system. Our higher-order signatures generalize Plotkin's *binding signatures* [29, 15]. To our knowledge, there is no other formalization of this notion.
- *Automatic definition of the families of types of terms from every signature:* The user need not define explicitly the types of formal terms of the object language anymore. It is enough to specify the construction rules for terms and the higher-order syntax is automatically generated.
- *Generation of typed terms:* In [2] and [8] terms were generated in an untyped way, any term could be applied to any other term. The set of correct terms was defined by using a well-formedness predicate. This meant that proofs needed to explicitly manipulate proofs of well-formedness. In the present work, terms are well-formed from the start, thanks to the use of an inductive family that directly implements the typing rules given in the signature.
- *Generation of the recursion principle and proof of its correctness for every signature:* In our previous work we formalize recursion principles for specific object languages. Now we have a parameterized version of it; it can be used to define computable functions on the higher-order syntax of any signature.
- *General proof of the induction principle for all signatures:* This gives appropriate reasoning principles for every object language, derived automatically by the system from the specification of the signature. In our previous work, this principle had to be derived separately for every object language. Neither the recursion nor the induction principle were proved in the original Hybrid system, neither for the general case nor for specific case studies.
- *Higher-order recursive calls:* In the recursion principle, the recursive call is a function on all possible results on the bound variables, rather that just the result on the body of the abstraction. This means, for example, that in defining a recursive function f on the typed λ-calculus, in the abstraction case $f\ (\lambda x.b)$ the recursive call isn't just the value of f on b, but a function mapping all possible results for the bound variable x to the corresponding value for b.

In Section 2 we present our tool from the user point of view. We describe how to define an object language by giving its signature and how to obtain higher-order notation, recursion and induction principles on it. In Section 3 we describe the implementation of our method in Coq and explain the main ideas and problems involved. In Section 4, we discuss related work, and in Section 5, after a review of our work, we state our goals for future research.

Prerequisites: We assume that the reader is familiar with Dependent Type Theory and (Co)Inductive types. Good introductions are Barendregt [5] and the books by Luo [21] and Sørensen and Urzyczyn [34]. An introduction to inductive and coinductive types, as used here, is Part I of the first author's Ph.D. thesis [7].

More specifically, we formalized everything in the proof assistant Coq. A good introduction to this system is the book by Bertot and Castéran [6]. A complete formal description is in the Coq manual [9]. The Coq files of our development are available at: `http://www.cs.ru.nl/~venanzio/Coq/HOUA.html`.

In this paper we avoid giving specific Coq code. Instead we have tried to adopt notation and terminology accessible to people familiar with other versions of dependent type theory and (co)inductive definitions. In addition, we have tried to make our treatment comprehensible to users of other systems, like Lego [1], HOL [19], and Isabelle [24].

In this paper, Prop is the type of logical propositions, whereas Set is the type of data types. We use the standard logical connectives for formulas in Prop. We write $\{A\} + \{B\}$ for Coq's constructive disjunction (in Set). We write $:=$ for definitional equality.

We need several notations for lists and tuples. The standard type of lists of elements of set A is $\mathsf{list}\ A$ and a typical element of it is $[a_1, \ldots, a_n]$. A Cartesian product $A_1 \times \cdots \times A_n$ has as elements n-tuples of the form $\langle t_1, \ldots, t_n \rangle$. Beside these standard types, we need to use dependent lists, in which every element has a different type belonging to a class of types called *argument types:* the type of such a lists is denoted by $\{\!|A_1, \ldots, A_n|\!\}$, where the A_is belongs to the class of argument types, and its elements are n-tuples of the form $\{a_1, \ldots, a_n\}$. Braces in conjunction with :: denote the cons operator, e.g., if $\vec{a} = \{a_1, \ldots, a_n\}$, then $\{a_0 :: \vec{a}\} = \{a_0, a_1, \ldots, a_n\}$. Finally, we have a different notation $\langle\!\langle t_1, \ldots, t_n \rangle\!\rangle$ for list of *terms with bindings.* These types of lists are defined rigorously at the moment of their introduction in Section 3. The reader who is not interested in their specific implementation may think of them as being equivalent to Cartesian products.

The notation $(\mathsf{fun}\, x \mapsto b)$ denotes λ-abstraction: variable x is bound in body b. Abstraction over an n-tuple is often written $(\mathsf{fun}\, \langle x_1, \ldots, x_n \rangle \mapsto b)$. The keyword ***Inductive*** introduces an inductive type; the keyword ***Record*** introduces a record type and is syntactic sugar for a dependent tuple type.

We freely use infix notations, without explicit declarations. An underscore _ denotes an implicit type parameter that can be inferred from context. Other notations are described as they are introduced.

§2. **Higher-order universal algebra.** We describe our Coq development of Universal Algebra with bindings from a user point of view. In the next section we give the details of the implementation. We use the simply typed λ-calculus as a running example. We use Coq as a logical framework. An object language is specified by giving its signature, which is a tuple consisting of a set of names for sorts, a set of names for operations, and an assignment of a typing rule to every operation; in addition, we require equality on the sorts to be decidable. This notion is a multisorted extension of Plotkin's binding signatures [29, 15]. Our tool automatically generates the family of types of terms for the object language and provides programming and reasoning support for it.

Before describing the general development, let us illustrate how a user defines the simply typed λ-calculus in our tool. As stated, the user must give a signature, that is, a 4-tuple:

$$\mathsf{sig}_\lambda : \mathsf{Signature} := \mathsf{signature}\,\mathsf{type}_\lambda\,\mathsf{dec}_\lambda\,\mathsf{operation}_\lambda\,\mathsf{rule}_\lambda.$$

Below we explain the meaning of the components of the signature sig_λ.

First we must give the set type_λ of sorts, that are, in our case, codes for simple types: they are generated from the base type o_λ by applying the arrow constructor, if A and B are codes for types, then $(\mathsf{arrow}_\lambda\ A\ B)$ is also a code for a type. We call attention to the distinction between Coq types, like Signature and type_λ themselves, that represent sets in the metalanguage, and elements of type_λ, like o_λ and $(\mathsf{arrow}_\lambda\ \mathsf{o}_\lambda\ \mathsf{o}_\lambda)$, that are codes for types of the object language.

Then we define the set $\mathsf{operation}_\lambda$ of names of operations: abstraction and application are parameterized over types, so there are operations names $(\mathsf{abs}_\lambda\ A\ B)$ and $(\mathsf{app}_\lambda\ A\ B)$ for every pair of type codes A and B.

In type theory, the sets of codes for types and names of operations are defined by the following inductive declarations:

Inductive $\mathsf{type}_\lambda : \mathsf{Set} :=$
 $\mathsf{o}_\lambda : \mathsf{type}_\lambda$
 $\mathsf{arrow}_\lambda : \mathsf{type}_\lambda \to \mathsf{type}_\lambda \to \mathsf{type}_\lambda$

Inductive $\mathsf{operation}_\lambda : \mathsf{Set} :=$
 $\mathsf{abs}_\lambda : \mathsf{type}_\lambda \to \mathsf{type}_\lambda \to \mathsf{operation}_\lambda$
 $\mathsf{app}_\lambda : \mathsf{type}_\lambda \to \mathsf{type}_\lambda \to \mathsf{operation}_\lambda.$

It is necessary, in order to prove some of the results of the formal language, that the set of sorts, type_λ in the example, has a decidable equality. This requirement is expressed in type theory by:

$$\mathsf{dec}_\lambda : \forall t_1, t_2 : \mathsf{type}_\lambda, \{t_1 = t_2\} + \{t_1 \neq t_2\}.$$

Precisely, dec_λ is a computable function that maps every pair of codes for types t_1 and t_2 to either a proof that they are equal or a proof that they are distinct.

The last component of the signature is a specification of the typing rules for the operations. Every sort code A will be associated to a set of terms $\mathsf{Term}\,A$ and every operation name f will be associated to a (possibly higher-order) function $\mathsf{Opr}\,f$ on terms. Every operation name is associated to a rule specifying the sorts and bindings of its arguments and the sort of its result. In our case, our goal is to obtain the following typing rules for abstraction and application:

$$\frac{\begin{array}{c}[x : \mathsf{Term}\,A] \\ \vdots \\ b : \mathsf{Term}\,B\end{array}}{\mathsf{Opr}\,(\mathsf{abs}_\lambda\,A\,B)\,\{\mathsf{fun}\,x \mapsto b\} : \mathsf{Term}\,(\mathsf{arrow}_\lambda\,A\,B)};$$

$$\frac{f : \mathsf{Term}\,(\mathsf{arrow}_\lambda\,A\,B) \quad a : \mathsf{Term}\,A}{\mathsf{Opr}\,(\mathsf{app}_\lambda\,A\,B)\,\{f, a\} : \mathsf{Term}\,B}.$$

Technically, rules are themselves formal objects, that is, they are codes specifying the intended use of the operation names. We define the exact form of rules later; here we just show how the user specifies the two previous informal rules for the operations of the simply typed λ-calculus:

$$\begin{array}{l}\mathsf{rule}_\lambda : \mathsf{operation}_\lambda \to \mathsf{operation\_type}\,\mathsf{type}_\lambda \\ \mathsf{rule}_\lambda\,(\mathsf{abs}_\lambda\,A\,B) = [[A] \vdash B] \,/\!/\, (\mathsf{arrow}_\lambda\,A\,B); \\ \mathsf{rule}_\lambda\,(\mathsf{app}_\lambda\,A\,B) = [[] \vdash (\mathsf{arrow}_\lambda\,A\,B), [] \vdash A] \,/\!/\, B.\end{array}$$

Informally, rule_λ states that: $(\mathsf{abs}_\lambda\,A\,B)$ is the name of an operation taking as argument a term of type B, binding a variable of type A, giving a result of type $(\mathsf{arrow}_\lambda\,A\,B)$; $(\mathsf{app}_\lambda\,A\,B)$ is the name of an operation taking two arguments of type $(\mathsf{arrow}_\lambda\,A\,B)$ and A, without any binding, giving a result of type B.

Thus, signatures are defined as records:

$$\begin{array}{l}\textbf{\textit{Record}}\ \mathsf{Signature} := \mathsf{signature} \\ \quad \{\ \mathsf{sig\_sort} : \mathsf{Set}; \\ \qquad \mathsf{sig\_dec} : \forall s_1, s_2 : \mathsf{sig\_sort}, \{s_1 = s_2\} + \{s_1 \neq s_2\}; \\ \qquad \mathsf{sig\_operation} : \mathsf{Set}; \\ \qquad \mathsf{sig\_rule} : \mathsf{sig\_operation} \to \mathsf{operation\_type}\,\mathsf{sig\_sort}\ \}\end{array}$$

Let a signature σ : Signature be fixed from now on, and take $\mathsf{sort} := \mathsf{sig\_sort}\,\sigma$, $\mathsf{operation} := \mathsf{sig\_operation}\,\sigma$, and $\mathsf{rule} := \mathsf{sig\_rule}\,\sigma$. The function rule maps every operation f to an element of $(\mathsf{operation\_type}\,\mathsf{sort})$, whose general form is:

$$(\mathsf{rule}\,f) = [[A_{11}, \ldots, A_{1k_1}] \vdash B_1, \ldots, [A_{n1}, \ldots, A_{nk_n}] \vdash B_n] \,/\!/\, C$$

where the A_{ij}s, B_is, and C are sorts.

Let us explain informally the meaning of this rule. It is comprised of a list of argument specifications and a result sort; each argument has a list of bindings and a sort. This states that f is an operation that takes n arguments. The ith argument is a function from the sorts $A_{i1}, \ldots, A_{ik_i}$ to B_i; or, in more syntactical terms, it is a term of sort B_i with bindings of variables of sorts $A_{i1}, \ldots, A_{ik_i}$. Formally we define (arg_type sort) as the set of argument types: its elements are pairs $\langle [A_1, \ldots, A_k], B \rangle$ consisting of a list of sorts (the types of bindings) and sort (the type of the argument); we use the symbol $\vdash$ as a pairing operator. Then (rule f) is a pair $\langle [T_1, \ldots, T_n], C \rangle$ consisting of a list of argument types and a result sort; we use // as a pairing operator. We use the square bracket notation exclusively for lists of elements of sort and of (arg_type sort), to distinguish them from other kinds of lists. The various components of the rule can be extracted by the following functions:

$$\begin{aligned}
&\mathsf{op_arguments}\,(\mathsf{rule}\, f) = [T_1, \ldots, T_n] : \mathsf{list}\,(\mathsf{arg_type\,sort})\\
&\mathsf{op_result}\,(\mathsf{rule}\, f) = C : \mathsf{sort}\\
&\mathsf{arg_bindings}\, T_i = [A_{i1}, \ldots, A_{ik_i}] : \mathsf{list\,sort}\\
&\mathsf{arg_result}\, T_i = B_i : \mathsf{sort}.
\end{aligned}$$

Therefore, the specification of f is equivalent to the following higher-order introduction rule for the type (Term C) of terms of sort C, where Opr is the higher-order application operator:

$$\frac{\begin{array}{ccc} [x_{1j} : \mathsf{Term}\, A_{1j}]_{j=1}^{k_1} & & [x_{nj} : \mathsf{Term}\, A_{nj}]_{j=1}^{k_n} \\ \vdots & \cdots & \vdots \\ b_1 : \mathsf{Term}\, B_1 & & b_n : \mathsf{Term}\, B_n \end{array}}{\mathsf{Opr}\, f\, \{\mathsf{fun}\, \langle x_{11}, \ldots, x_{1k_1} \rangle \mapsto b_1, \cdots, \mathsf{fun}\, \langle x_{n1}, \ldots, x_{nk_n} \rangle \mapsto b_n\} : \mathsf{Term}\, C}.$$

We can exploit Coq's type inference mechanism to define, straightforwardly, a compact notation that reflects the informal way of writing (Church style [5]) λ-terms. In the case of the simply typed λ-calculus, we defined the following notations:

informal	*formal compact*	*formal long*
$(\lambda x : A.b)$	(Lambda x in A gives b)	(Opr (abs$_\lambda$ A B) (fun $x \mapsto b$))
$(f\ a)$	(f of a)	(Opr (app$_\lambda$ A B) f a).

Once the user declares the signature σ, the system automatically generates, for each sort A : sort, the set of terms Term A. The user-accessible notation for these terms is higher-order syntax, generated starting from free and bound variables using the binding operators of the form Opr f for f : operation. The user never needs to see the internal de Bruijn syntax representation of the term. It will be explained in the next section.

There are two distinct sets of variables, free and bound. Each is indexed on the sorts and the natural numbers. Free variables, in the form v_i^A for A a sort and i a natural number, are treated as parameters whose interpretation is arbitrary but fixed. Bound variables, in the form x_j^A, are internally represented as de Bruijn indices, and their interpretation may change.

The argument functions $(\mathsf{fun}\,\langle x_{i1},\ldots,x_{ik_i}\rangle \mapsto b_i)$ in the introduction rule for operations are not restricted. The user may exploit every construction allowed in the logical framework to define such functions, opening the door to exotic terms. We want to prevent this from happening: b_i should be uniform in the variables, that is, it should be a term constructed solely by the given rules, other operators and recursors from the logical framework should be banned. Two options are available. First, we could simply add to the rule for Opr some conditions requiring the uniformity of the arguments. This has the disadvantage that even in the case when the uniformity is trivial (always in practice) a proof must be provided, cluttering the syntax. Furthermore, in an intensional system like Coq, terms would depend on the proof of uniformity and we may have that two terms differing only by the proofs of uniformity cannot be proven equal. The second solution, which we adopt, is to perform an automatic normalization of the functions. As an example, let's consider the following non-uniform function allowable in our encoding of the simply typed λ-calculus:

$$h : \mathsf{Term}\,A \to \mathsf{Term}\,B$$
$$h\,x = \textbf{\textit{if}}\ x = \mathsf{x}_0^A\ \textbf{\textit{then}}\ \mathsf{v}_0^B\ \textbf{\textit{else}}\ \mathsf{v}_1^B.$$

The function h maps the term x_0^A to v_0^B and all other terms to v_1^B. In a naive implementation of higher-order syntax, the abstraction $\lambda(h)$ would be an exotic term: Inside the formal language of the simply typed λ-calculus, it is not possible to make a case distinction on the syntactic form of a term in this way. In our implementation, the higher-order term $\mathsf{Opr}\,(\mathsf{abs}_\lambda\,A\,B)\,h$ performs an automatic uniformization of h by choosing a fresh variable, v_0^A will do in this case, and forcing h to behave on every argument in the same way as it behaves on the fresh variable:

$$h \rightsquigarrow \underline{h} = \big(\mathsf{fun}\,x \mapsto (h\,\mathsf{v}_0^A)[\mathsf{v}_0^A := x]\big)$$
$$= \big(\mathsf{fun}\,x \mapsto \mathsf{v}_1^B[\mathsf{v}_0^A := x]\big) = \big(\mathsf{fun}\,x \mapsto \mathsf{v}_1^B\big).$$

It is important to keep in mind here that x is a *metavariable*, i.e. a Coq variable, of type $\mathsf{Term}\,A$, while x_0^A and v_0^A are object variables of sort A. The notation $[\mathsf{v} := x]$ expresses a defined substitution operation that replaces occurrences of free variable v with x; in its general form it allows simultaneous substitution of multiple variables. In our tool the function h is transformed into $\underline{h}$ behind the scene, so $\mathsf{Opr}\,(\mathsf{abs}_\lambda\,A\,B)\,h$ is actually equivalent to $\mathsf{Opr}\,(\mathsf{abs}_\lambda\,A\,B)\,\underline{h}$. Since $\underline{h}$ is uniform, the exotic term has disappeared.

Now for the general case: Given a function $h = \mathsf{fun}\,\langle x_{i1},\ldots,x_{ik_i}\rangle \mapsto b_i$, we define its *uniformization* $\underline{h}$ by simply taking arbitrary fresh variables $\mathsf{v}_{i1},\ldots,\mathsf{v}_{ik_i}$, applying h to them and then stating that $\underline{h}$ acts on every input in the same way that h acts on these variables. In symbols:

$$\underline{h}\,\langle x_{i1},\ldots,x_{ik_i}\rangle = (h\,\langle \mathsf{v}_{i1},\ldots,\mathsf{v}_{ik_i}\rangle)\big[\mathsf{v}_{i1} := x_{i1},\ldots,\mathsf{v}_{ik_i} := x_{ik_i}\big].$$

Given a list of functions $\vec{h} = \{\mathsf{fun}\,\vec{x_1} \mapsto b_1, \cdots, \mathsf{fun}\,\vec{x_n} \mapsto b_n\}$, before applying $\mathsf{Opr}\,f$ to it, we normalize it to $\underline{\vec{h}}$. See Subsection 3.3 for a rigorous definition of uniformization.

The system automatically generates recursion and induction principles on the higher-order syntax. The (non-dependent) recursion principle also provides lemmas validating the recursion equations.

When defining a recursive function on the object language, we want to map each term to a result in a certain type. In general, it is possible that terms of different sorts are mapped to results with different types. Therefore the general type of a recursive function is $\forall A : \mathsf{sort}, \mathsf{Term}\,A \to F\,A$, where F is a family of sets indexed on sorts, $F : \mathsf{sort} \to \mathsf{Set}$.

Let us give a couple of examples on the simply typed λ-calculus.

As a first example, suppose that we want to define a size function on terms that counts the number of operations occurring in them. In this case every term is mapped to a natural number, so $F = \mathsf{fun}\,s \mapsto \mathbb{N}$ and $\mathsf{size} : \forall A : \mathsf{type}_\lambda, \mathsf{Term}\,A \to \mathbb{N}$.

For the second example, we consider the definition of a set-theoretic model of the λ-calculus. In this case, every sort (that is, every simple type) is mapped to a set and every term is mapped to an element of the corresponding set. Given a fixed set X to interpret the base type o_λ, we define a family of sets Model_X by recursion on type_λ:

$$\begin{array}{l} \mathsf{Model}_X : \mathsf{type}_\lambda \to \mathsf{Set} \\ \mathsf{Model}_X\,\mathsf{o}_\lambda = X \\ \mathsf{Model}_X\,(\mathsf{arrow}_\lambda\,A\,B) = (\mathsf{Model}_X\,A) \to (\mathsf{Model}_X\,B). \end{array}$$

Then an interpretation of the typed λ-calculus can be defined by recursion with $F = \mathsf{Model}_X$, as a map $\mathsf{interpret}_X : \forall A : \mathsf{sort}, \mathsf{Term}\,A \to \mathsf{Model}_X\,A$. To be precise, we need also an extra argument giving the assignment of a value in the model for all dangling variables. See the formal definition at the end of this section.

Before formulating the recursion principle, we need to introduce some notation. Given a family of sets indexed on the sorts, $F : \mathsf{sort} \to \mathsf{Set}$, we define a type of dependent lists of elements of F indexed on lists of sorts:

$$\mathsf{sort\_list}\,F\,[A_1, \ldots, A_k] \cong (F\,A_1) \times \cdots \times (F\,A_k).$$

We use $\cong$ to denote provable equivalence; we omit the exact definition of sort_list. An element of this type has the form $\langle a_1, \ldots, a_k \rangle$, with $a_i : F\,A_i$. Functions within the family F can be specified by an argument type. To an argument type $[A_1, \ldots, A_k] \vdash B$ we associate the type of functions from $(F\,A_1) \times \cdots \times (F\,A_k)$ to $F\,B$:

$$\begin{array}{l} \mathsf{arg\_map}\,F : \mathsf{arg\_type} \to \mathsf{Set} \\ \mathsf{arg\_map}\,F\,T = \mathsf{sort\_list}\,F\,(\mathsf{arg\_bindings}\,T) \to F\,(\mathsf{arg\_result}\,T). \end{array}$$

In other words, the $\mathsf{arg\_map}\, F\ ([A_1, \ldots, A_k] \vdash B)$ is the type of functions $(F\ A_1) \times \cdots \times (F\ A_k) \to F\ B$. We extend it to lists of argument types:

$$\begin{aligned}&\mathsf{args\_map}\, F : \mathsf{list\ arg\_type} \to \mathsf{Set}\\&\mathsf{args\_map}\, F\, [T_1, \ldots, T_n] = \{\!\!\{\mathsf{arg\_map}\, F\ T_1, \ldots, \mathsf{arg\_map}\, F\ T_n\}\!\!\}.\end{aligned}$$

The notation means that an element of this type is a dependent list $\{g_1, \ldots, g_n\}$, where g_i has type $(\mathsf{sort\_list}\, F\ \vec{A_i} \to F\ B_i)$ if $T_i = \vec{A_i} \vdash B_i$, for $1 \leq i \leq n$.

If we instantiate the previous definitions using the family Term for F, we obtain the representation of higher-order arguments as functions on dependent lists of terms. We call $\mathsf{Term\_arg}$ and $\mathsf{Term\_args}$ the instantiations of $\mathsf{arg\_map}$ and $\mathsf{args\_map}$ for $F := \mathsf{Term}$. An argument with specification $[A_1, \ldots, A_k] \vdash B$ will have the type:

$$\begin{aligned}\mathsf{Term\_arg}\ ([A_1, \ldots, A_k] \vdash B) &:= \mathsf{arg\_map\ Term}\ ([A_1, \ldots, A_k] \vdash B)\\&= \mathsf{sort\_list\ Term}\, [A_1, \ldots, A_k] \to \mathsf{Term}\ B\\&\cong (\mathsf{Term}\ A_1) \times \cdots \times (\mathsf{Term}\ A_k) \to \mathsf{Term}\ B.\end{aligned}$$

And a list of arguments will have the following type:

$$\begin{aligned}&\mathsf{Term\_args} : \mathsf{list\ arg\_type} \to \mathsf{Set}\\&\mathsf{Term\_args}\, [\vec{A_1} \vdash B_1, \ldots, \vec{A_n} \vdash B_n]\\&\qquad := \mathsf{args\_map\ Term}\, [\vec{A_1} \vdash B_1, \ldots, \vec{A_n} \vdash B_n]\\&\qquad = \{\!\!\{\mathsf{sort\_list\ Term}\ \vec{A_1} \to \mathsf{Term}\ B_1, \ldots, \mathsf{sort\_list\ Term}\ \vec{A_n} \to \mathsf{Term}\ B_n\}\!\!\}.\end{aligned}$$

When defining a recursive function on terms, the results associated to bound variables are stored in an assignment. Formally assignments are defined using *streams*: a stream s : Stream X is an infinite sequence of elements of X. An assignment is a family of streams.

DEFINITION 1. An *assignment* is a family of streams of the family F indexed on the sorts: $\mathsf{Assignment}\, F = \forall A : \mathsf{sort}, \mathsf{Stream}\ (F\ A)$. We use the symbol α to denote a generic assignment.

This definition means that for every sort name A, we have an infinite sequence $\alpha_A = a_0, a_1, a_2, \ldots$ of elements of $F\ A$.

Assignments are used to give interpretations for the de Bruijn variables during the definition of a function by recursion. So the variable x_j^A will be interpreted as $(\alpha\ A)_i$, the ith element of the stream associated with the sort A. Representing assignments as streams harmonizes nicely with the use of de Bruijn indices: whenever we go under a binder, the interpretations of the new bound variables can be simply appended in front of the stream; the old variables will be shifted to the right, automatically performing the required index increment. If $\vec{a} : \mathsf{sort\_list}\, F\ \vec{A}$, then we denote by $[\vec{a}]\alpha$ the assignment obtained by appending the elements of $\vec{a}$ in front of the streams in α.

With these notions we are ready to formulate the recursion principle.

THEOREM 1. *Given the step functions* Fvar *for free variables and* FOpr *for operation application, of the following types*:

$$\begin{aligned}
&\mathsf{Fvar} : \forall A : \mathsf{sort}, \mathbb{N} \to F\ A \\
&\mathsf{FOpr} : \ \forall f : \mathsf{operation}, \\
&\qquad \mathsf{Term_args}\,(\mathsf{op_arguments}\,(\mathsf{rule}\,f)) \to \\
&\qquad \mathsf{args_map}\,F\ (\mathsf{op_arguments}\,(\mathsf{rule}\,f)) \to \\
&\qquad F\ (\mathsf{op_result}\,(\mathsf{rule}\,f)).
\end{aligned}$$

we can construct recursive functions ϕ *and* $\vec{\phi}$ *of the following types*:

$$\begin{aligned}
&\phi : \mathsf{Assignment} \to \forall A : \mathsf{sort}, \mathsf{Term}\ A \to F\ A \\
&\vec{\phi} : \mathsf{Assignment} \to \forall \vec{T} : \mathsf{list\ arg_type}, \mathsf{Term_args}\ \vec{T} \to \mathsf{args_map}\,F\ \vec{T}
\end{aligned}$$

satisfying the reduction behaviour given by the equations:

$$\begin{aligned}
&\phi\,\alpha\,A\,\mathsf{v}_i^A = \mathsf{Fvar}\,A\,i \\
&\phi\,\alpha\,A\,\mathsf{x}_j^A = (\alpha\,A)_j \\
&\phi\,\alpha\,C\,(\mathsf{Opr}\,f\,\vec{h}) = \mathsf{FOpr}\,f\,\underline{\vec{h}}\,(\vec{\phi}\,\alpha\,_\,\vec{h})
\end{aligned}$$

$$\begin{aligned}
&\vec{\phi}\,\alpha\,[]\,\{\} = \{\} \\
&\vec{\phi}\,\alpha\,(\vec{A} \vdash B :: \vec{T})\,\{h :: \vec{h}\} = \{(\mathsf{fun}\,\vec{a} \mapsto \phi\,[\vec{a}]\alpha\,(\mathsf{bind}\,_\,h)) :: \vec{\phi}\,\alpha\,\vec{T}\,\vec{h}\}.
\end{aligned}$$

PROOF. See Subsection 3.4. ⊣

The operation bind takes a higher-order (functional) argument h and *flattens* it by replacing the binding meta-variables with de Bruijn indices:

$$\begin{aligned}
&h : \mathsf{Term_arg}\,([A_1,\ldots,A_k] \vdash B) \cong (\mathsf{Term}\,A_1)\times\cdots\times(\mathsf{Term}\,A_k) \to (\mathsf{Term}\,B) \\
&(\mathsf{bind}\,_\,h) : \mathsf{Term}\,B := \ \text{compute } h(x_1,\ldots,x_k) \text{ and then replace} \\
&\qquad\qquad\qquad\qquad \text{metavariable } x_i \text{ with a de Bruijn variable } \mathsf{x}_{i'}^{A_i}.
\end{aligned}$$

This intuitive definition is enough to follow the discussion in this section. In Section 3, we give a formal definition.

At the higher level, the user does not need to know how de Bruijn indices work. To understand the last recursion equation, it is enough to know that the function $(\mathsf{fun}\,\vec{a} \mapsto \phi\,[\vec{a}]\alpha\,(\mathsf{bind}\,\_\,h))$ maps a list $\vec{a} = \{a_1,\ldots,a_k\}$: $\mathsf{sort\_list}\,F\,[A_1,\ldots,A_k]$ to the result of the recursive call of ϕ on h where the occurrences of the meta-variables of type $\mathsf{Term}\,A_1,\ldots,\mathsf{Term}\,A_k$ are mapped to $a_1,\ldots,a_k$, respectively.

We also get an induction principle on the higher-order abstract syntax. Let $P : \forall A : \mathsf{sort}, \mathsf{Term}\,A \to \mathsf{Prop}$ be a predicate on terms. We can extend it to lists of higher-order arguments in the following way:

If $\{g_1,\ldots,g_n\} : \mathsf{Term\_args}\,[[A_{11},\ldots,A_{1k_1}] \vdash B_1,\ldots,[A_{n1},\ldots,A_{nk_n}] \vdash B_n]$, then $\vec{P}\,\{g_1,\ldots,g_n\} = (P\,\_\,(\mathsf{bind}\,\_\,g_1)) \wedge \cdots \wedge (P\,\_\,(\mathsf{bind}\,\_\,g_n))$.

THEOREM 2. *If the following hypotheses are true*:

$$\forall(A : \mathsf{sort})(i : \mathbb{N}), P\, A\, \mathsf{v}_i^A$$
$$\forall(A : \mathsf{sort})(i : \mathbb{N}), P\, A\, \mathsf{x}_i^A$$
$$\forall(f : \mathsf{operation})(\vec{g} : \mathsf{Term\_args}\,(\mathsf{op\_arguments}\,(\mathsf{rule}\, f))),$$
$$\vec{P}\, \vec{g} \rightarrow P\, \_\, (\mathsf{Opr}\, f\, \vec{g});$$

then, for every sort A and term t : Term A, $P\, A\, t$ *is true.*

PROOF. See Subsection 3.5. ⊣

Theorem 2 instantiates directly to an induction principle for our running example. For use in practice, it is convenient if the inductive hypothesis for Opr is formulated as two separate hypotheses for abs_λ and app_λ. In the case of abs_λ we also define the *body of a function* F : Term $A \rightarrow$ Term B to be:

$$\mathsf{body}\; h = \mathsf{bind}\,([A] \vdash B)\,(\mathsf{fun}\,\langle x\rangle \mapsto h\, x) : \mathsf{Term}\, B.$$

Then we can easily convert the general induction principle to one for the typed λ-calculus.

THEOREM 3. *Let* $P : \forall A : \mathsf{type}_\lambda, \mathsf{Term}\, A \rightarrow \mathsf{Prop}$ *be a predicate on terms. If the following hypotheses are true*:

$$\forall(A : \mathsf{type}_\lambda)(i : \mathbb{N}), P\, A\, \mathsf{v}_i^A$$
$$\forall(A : \mathsf{type}_\lambda)(i : \mathbb{N}), P\, A\, \mathsf{x}_i^A$$
$$\forall(A, B : \mathsf{type}_\lambda)(h : \mathsf{Term}\, A \rightarrow \mathsf{Term} B),$$
$$P\, B\, (\mathsf{body}\, h) \rightarrow P\, (\mathsf{arrow}_\lambda\, A\, B)\, (\mathsf{Lambda}\; x\; \mathsf{in}\; A\; \mathsf{gives}\; (h\, x))$$
$$\forall(A, B : \mathsf{type}_\lambda)(f : \mathsf{Term}\, (\mathsf{arrow}_\lambda\, A\, B))(a : \mathsf{Term}\, A),$$
$$P\, (\mathsf{arrow}_\lambda\, A\, B)\, f \rightarrow P\, A\, a \rightarrow P\, B\, (f\; \mathsf{of}\; a)$$

Then, for every A : type_λ *and λ-term t* : Term A, $P\, A\, t$ *is true.*

In a similar way we can adapt the recursion principle to our specific object language. Here is its instantiation for the simply typed λ-calculus.

THEOREM 4. *Given the step functions* Fvar *for free variables*, Fabs *for abstractions, and* Fapp *for application, of the following types*:

$$\begin{aligned}
\mathsf{Fvar} :\;& \forall A : \mathsf{type}_\lambda, \mathbb{N} \rightarrow F\, A \\
\mathsf{Fabs} :\;& \forall A, B : \mathsf{type}_\lambda, (\mathsf{Term}\, A \rightarrow \mathsf{Term}\, B) \\
& \rightarrow (F\, A \rightarrow F\, B) \rightarrow F\, (\mathsf{arrow}_\lambda\, A\, B) \\
\mathsf{Fapp} :\;& \forall A, B : \mathsf{type}_\lambda, (\mathsf{Term}\, (\mathsf{arrow}_\lambda\, A\, B) \times \mathsf{Term}\, B) \\
& \rightarrow (F\, (\mathsf{arrow}_\lambda\, A\, B) \times F\, B) \rightarrow F\, B
\end{aligned}$$

we can construct a recursive function ϕ *of the following type*:

$$\phi : \mathsf{Assignment} \rightarrow \forall A : \mathsf{type}_\lambda, \mathsf{Term}\, A \rightarrow F\, A$$

satisfying the reduction behaviour given by the equations:

$$\begin{aligned}
&\phi\,\alpha\,A\,\mathsf{v}_i^A = \mathsf{Fvar}\,A\,i\\
&\phi\,\alpha\,A\,\mathsf{x}_j^A = (\alpha\,A)_j\\
&\phi\,\alpha\,(\mathsf{arrow}_\lambda\,A\,B)\,(\mathsf{Lambda}\,x\,\mathsf{in}\,A\,\mathsf{gives}\,(h\,x))\\
&\qquad\qquad = \mathsf{Fabs}\,A\,B\,\underline{h}\,(\mathsf{fun}\,a \mapsto \phi\,[a]\alpha\,(\mathsf{body}\,h))\\
&\phi\,\alpha\,B\,(f\,\mathsf{of}\,a) = \mathsf{Fapp}\,A\,B\,\langle f, a\rangle\,\langle \phi\,\alpha\,(\mathsf{arrow}_\lambda\,A\,B)\,f, \phi\,\alpha\,A\,a\rangle
\end{aligned}$$

We omitted the extension $\vec{\phi}$ of ϕ to higher-order arguments, and instead directly unfolded its occurrence in the reduction rule for

$$(\mathsf{Lambda}\,x\,\mathsf{in}\,A\,\mathsf{gives}\,(h\,x)).$$

Notice the true higher-order nature of this recursion principle: in the recursion step Fabs, the recursive argument is not just the result on the body of the function, but a mapping giving the result for all possible assignments to the abstracted variable.

For example, the function size mentioned earlier is defined by the following step functions:

$$\begin{aligned}
&\mathsf{Fvar}\,A\,i = 1\\
&\mathsf{Fabs}\,A\,B\,h\,g = (g\,1) + 1\\
&\mathsf{Fapp}\,A\,B\,\langle f, a\rangle\,\langle n_f, n_a\rangle = n_f + n_a + 1.
\end{aligned}$$

In this case we want to give to every bound variable the size 1, so we define size $= \phi\,\vec{1}$, where $\vec{1}$ is the constant stream where all elements are 1, resulting in the following reduction behaviour:

$$\begin{aligned}
&\mathsf{size}\,A\,\mathsf{v}_j^A = 1\\
&\mathsf{size}\,A\,\mathsf{x}_j^A = 1\\
&\mathsf{size}\,(\mathsf{arrow}_\lambda\,A\,B)\,(\mathsf{Lambda}\,x\,\mathsf{in}\,A\,\mathsf{gives}\,(h\,x)) = (\mathsf{size}\,(\mathsf{body}\,h)) + 1\\
&\mathsf{size}\,B\,(f\,\mathsf{of}\,a) = (\mathsf{size}\,(\mathsf{arrow}_\lambda\,A\,B)\,f) + (\mathsf{size}\,A\,a) + 1
\end{aligned}$$

as desired.

The size example is quite simple in that we didn't use the higher-order power of the recursion principle and the assignment argument. The second order recursive call g in the abstraction case was used only with constant argument 1, and the assignment is constantly equal to $\vec{1}$ during reduction.

To illustrate what a truly higher-order application of the recursion principle is, let us show the definition of the model construction $\mathsf{interpret}_X$ mentioned earlier. We assume as parameter a fixed interpretation of the free variables: $\mathsf{Fvar}\,A\,i : \forall A : \mathsf{type}_\lambda, \mathbb{N} \to \mathsf{Model}_X\,A$. The step functions are the following:

$$\begin{aligned}
&\mathsf{Fabs}\,A\,B\,h\,g = g\\
&\mathsf{Fapp}\,A\,B\,\langle f, a\rangle\,\langle g, x\rangle = g\,x
\end{aligned}$$

Setting $\mathsf{interpret}_X = \phi$, we obtain the following reduction rules:

$$\begin{aligned}
&\mathsf{interpret}_X\, \alpha\, A\, \mathsf{v}_i^A = \mathsf{Fvar}\, A\, i\\
&\mathsf{interpret}_X\, \alpha\, A\, \mathsf{x}_j^A = (\alpha\, A)_j\\
&\mathsf{interpret}_X\, \alpha\, (\mathsf{arrow}_\lambda\, A\, B)\, (\mathsf{Lambda}\, x\, \mathsf{in}\, A\, \mathsf{gives}\, (h\, x))\\
&\qquad\qquad = \mathsf{fun}\, a \mapsto \mathsf{interpret}_X\, [a]\alpha\, (\mathsf{body}\, h)\\
&\mathsf{interpret}_X \alpha\, B\, (f\ \mathsf{of}\ a) = (\mathsf{interpret}_X\, \alpha\, (\mathsf{arrow}_\lambda\, A\, B)\, f)\, (\mathsf{interpret}_X\, \alpha\, A\, a).
\end{aligned}$$

Notice how the usual interpretation of abstraction in a set-theoretic model, as the function mapping an argument to the interpretation of the body under a modified assignment, is automatically validated by the recursion principle from the simple definition of Fabs.

§3. **Technical details.** Let us explain some details of our Coq implementation. Under the higher-order syntax described in the previous section, we have a de Bruijn syntax defined as a standard Coq inductive type.

3.1. De Bruijn syntax. To define terms over the signature σ, we need to define the type of terms simultaneously with the type of term lists with bindings. A list of terms with bindings is an object of the form:

$$\langle\!\langle [\mathsf{x}_{11}^{A_{11}}, \ldots, \mathsf{x}_{1k_1}^{A_{1k_1}}]\, t_1, \ldots, [\mathsf{x}_{n1}^{A_{n1}}, \ldots, \mathsf{x}_{nk_n}^{A_{nk_n}}]\, t_n \rangle\!\rangle.$$

It is the list of the terms $t_1, \ldots, t_n$, each binding a list of variables: the term t_i binds the variables $\mathsf{x}_{i1}, \ldots, \mathsf{x}_{ik_i}$, where the variable x_{ij} has sort A_{ij}. In keeping with the de Bruijn convention, we don't actually need to specify the names of the abstracted variables, but they will automatically be determined as indices, starting with index 0 for the rightmost variable.

Informally, terms and term lists are defined by the following rules (in the operation rule, assume that $(\mathsf{rule}\, f) = [\vec{A}_1 \vdash B_1, \ldots, \vec{A}_n \vdash B_n] \mathbin{/\!\!/} C$):

$$\mathsf{Term} : \mathsf{sort} \to \mathsf{Set}$$
$$\mathsf{TermList} : \mathsf{list}\, (\mathsf{arg\_type}\, \mathsf{sort}) \to \mathsf{Set}$$

$$\frac{A : \mathsf{sort} \quad i : \mathbb{N}}{\mathsf{v}_i^A : \mathsf{Term}\, A}\ \text{(free variable)} \qquad \frac{A : \mathsf{sort} \quad i : \mathbb{N}}{\mathsf{x}_i^A : \mathsf{Term}\, A}\ \text{(bound variable)}$$

$$\frac{f : \mathsf{operation} \quad \langle\!\langle t_1, \ldots, t_n \rangle\!\rangle : \mathsf{TermList}\, [\vec{A}_1 \vdash B_1, \ldots, \vec{A}_n \vdash B_n]}{(\mathsf{opr}\, f\, \langle\!\langle t_1, \ldots, t_n \rangle\!\rangle) : \mathsf{Term}\, C}$$

$$\frac{t_1 : \mathsf{Term}\, B_1 \quad \cdots \quad t_n : \mathsf{Term}\, B_n}{\langle\!\langle [\mathsf{x}_{1j}^{A_{1j}}]_{j=1}^{k_1} t_1, \ldots, [\mathsf{x}_{nj}^{A_{nj}}]_{j=1}^{k_n} t_n \rangle\!\rangle : \mathsf{TermList}\, [\vec{A}_1 \vdash B_1, \ldots, \vec{A}_n \vdash B_n]}.$$

In the formal definition, as mentioned, we don't need to explicitly mention the names of the variables. Also, in the last rule the list of bound sorts for each term in the list is specified by the type of the list. Therefore, we need to put explicitly in the list just the terms. The formal definition of terms and term lists is the following:

$$
\begin{array}{rl}
\textbf{\textit{Inductive}}\ \mathsf{Term}: & \mathsf{sort} \to \mathsf{Set} \\
\mathsf{TermList}: & \mathsf{list}\,(\mathsf{arg\_type\,sort}) \to \mathsf{Set} \\
& \\
\mathsf{var}: & \forall A : \mathsf{sort}, \mathbb{N} \to \mathsf{Term}\ A \\
\mathsf{bnd}: & \forall A : \mathsf{sort}, \mathbb{N} \to \mathsf{Term}\ A \\
\mathsf{opr}: & \forall f : \mathsf{operation}, \\
& \mathsf{TermList}\,(\mathsf{op\_arguments}\,(\mathsf{rule}\ f)) \to \mathsf{Term}\,(\mathsf{op\_result}\,(\mathsf{rule}\ f)) \\
& \\
\mathsf{nil\_term}: & \mathsf{TermList}\,[] \\
\mathsf{cons\_term}: & \forall T : \mathsf{arg\_type\,sort}, \forall \vec{T} : \mathsf{list}\,(\mathsf{arg\_type\,sort}), \\
& \mathsf{Term}\,(\mathsf{arg\_result}\ T) \to \mathsf{TermList}\ \vec{T} \to \mathsf{TermList}\,(T :: \vec{T}).
\end{array}
$$

So cons_term doesn't explicitly mention the bound variables in the term $t : \mathsf{Term}\,(\mathsf{arg\_result}\ T)$. The binding list is implicitly given by the argument type T using the de Bruijn convention for indexing. For example, if $T = [A, A', A, A, A'] \vdash B$, then the bound variables are $\mathsf{x}_2^A, \mathsf{x}_1^{A'}, \mathsf{x}_1^A, \mathsf{x}_0^A, \mathsf{x}_0^{A'}$. Bound variables are numbered from right to left starting from 0; each sort has an independent indexing.

We must translate the higher-order notation of the previous section into this de Bruijn syntax. The fundamental point is to establish a correspondence between the two ways to apply an operation:

$$\mathsf{Opr}\ f\ \{g_1, \dots\ g_n\} \leftrightsquigarrow \mathsf{opr}\ f\ \langle\!\langle a_1, \dots, a_n \rangle\!\rangle.$$

This transformation is performed by the mentioned bind operation. Before defining it, we need some definitions and results about variables and substitution.

3.2. Variables and binding. The operator newvar finds a new free variable of a specific sort with respect to a term. If $t : \mathsf{Term}\ A$ is a term and A' is a sort, then $(\mathsf{newvar}\ A'\ t)$ gives an index i such that $\mathsf{v}_i^{A'}$ does not occur in t. More precisely, all free variables of sort A' occurring in t have indices smaller that i. It is defined by recursion on the structure of t.

The identity of bound variables is determined by the number of bindings of their sort above them. If we have a term with bindings of sorts $[A_1, \dots, A_k]$, we want to know what index the variable $\mathsf{x}_i^{A'}$ will have under the bindings:

$$\cdots\ \mathsf{x}_i^{A'}\ \cdots (\mathsf{opr}\ f\ \langle\!\langle \cdots, \dots \mathsf{x}_j^{A'} \dots, \cdots \rangle\!\rangle) \cdots$$

In this expression $\mathsf{x}_i^{A'}$ and $\mathsf{x}_j^{A'}$ are meant to represent occurrences of the same de Bruijn variable. The indices of the bound variables are determined by the de Bruijn convention. The value of j depends on the number of occurrences of A' in the binding list of the argument of f where $\mathsf{x}_j^{A'}$ occurs. Suppose there are h occurrences of A' in it; then they bind the variables $\mathsf{x}_{h-1}^{A'}, \dots, \mathsf{x}_0^{A'}$.

Therefore, the indices of all other variables of sort A' are shifted by h, so we must have $j = i + h$.

This shifting is performed by the function

$$\mathsf{bind\_inc}\colon (\mathsf{bind\_inc}\,[A_1, \ldots, A_k]\,A'\,i)$$

gives the value $j = i + h$. It is defined by recursion on $[A_1, \ldots, A_k]$, the binding list of the argument of f. It uses the decidability of equality of sorts to check whether A' is equal to the A_js. Here is where the assumption sig_dec is used.

Given a term $t : \mathsf{Term}\, B$, a free variable v_i^A and a de Bruijn variable x_j^A, we can define the operation of swapping the two variables in t: $t[\mathsf{v}_i^A \leftrightarrow \mathsf{x}_j^A]$. It is defined by recursion on the structure of t, taking care to increment the index of x_j^A with bind_inc every time we go under a binding operation.

Note: We could have chosen to have only one indexing for the variables, regardless of their sorts. This would have avoided the need for decidability. However, it would have required carrying around an assignment of sorts to the indices everywhere, so we opted for independent indexing of every sort.

Now we have the machinery to bind meta-variables to de Bruijn indices. We start with the simple example of a single variable. We keep track of variables already bound with an argument $\vec{A}$: list sort.

The index of the new bound variable of sort A must be $j = (\mathsf{bind\_inc}\,\vec{A}\,A\,0)$:

$$\begin{aligned}
&\mathsf{one_arg_bind}\; A\; B\; \vec{A} : (\mathsf{Term}\; A \to \mathsf{Term}\; B) \to \mathsf{Term}\; B\\
&\mathsf{one_arg_bind}\; A\; B\; \vec{A}\, g = (g\; \mathsf{v}_i^A)[\mathsf{v}_i^A \leftrightarrow \mathsf{x}_j^A]\\
&\qquad \text{where } i = \mathsf{newvar}\; A\; (g\; \mathsf{x}_0^A)\\
&\qquad\qquad\quad j = \mathsf{bind_inc}\; \vec{A}\; A\; 0
\end{aligned}$$

Note: We must require that v_i^A is a new variable for g. Since g is a function, it is not immediately clear what this means. It does *not* mean that v_i^A is new for every instance of g (for example, if g is the identity, no variable is new for all instances). We are mainly interested in the case where g is uniform, that is, g is of the form $\mathsf{fun}\; X : \mathsf{Term}\; A \mapsto C[X]$ where $C[\_]$ is a term build up using only variables and operations of the signature. We want that v_i^A doesn't occur in C. So it is enough to apply g to a dummy argument that doesn't add new free variables and find a new variable for that term. That is why we use $i = (\mathsf{newvar}\; A\; (g\; \mathsf{x}_0^A))$.

For bindings of several variables, we use the same process simultaneously on all the variables. Similarly to the definition of swapping for a single variable, we can swap several variables simultaneously. We use the notation $\mathbb{N}^{[A_1,\ldots,A_k]}$ (Coq notation: `(indices [A1,...,Ak])`) for the type $(\mathsf{sort\_list}\,(\mathsf{fun}\; A : \mathsf{sort} \mapsto \mathbb{N})\,[A_1, \ldots, A_k])$. Its elements are lists of indices $\langle i_1, \ldots, i_k\rangle$ denoting either free or de Bruijn variables; let us use the notation $\boldsymbol{v}_{\langle i_1,\ldots,i_k\rangle}$ (Coq: `vars As is`) for $\langle \mathsf{v}_{i_1}^{A_1}, \ldots, \mathsf{v}_{i_k}^{A_k} \rangle$. We use the similar notation $\boldsymbol{x}_{\langle i_1,\ldots,i_k \rangle}$ (Coq: `bnds As is`) for de Bruijn variables.

Let now $\langle i_1, \ldots, i_k \rangle$ and $\langle j_1, \ldots, j_k \rangle$ be two elements of $\mathbb{N}^{[A_1,\ldots,A_k]}$, denoting a list of indices for free variables and a list of indices for de Bruijn variables, respectively. We then define the operation of simultaneously swapping each free variable with the corresponding de Bruijn variable:

$$\mathsf{vars\_swap}\,[A_1, \ldots, A_k]\,\langle i_1, \ldots, i_k \rangle\,\langle j_1, \ldots, j_k \rangle$$

which we also denote by the notation

$$t[\mathsf{v}_{i_1}^{A_1} \leftrightarrow \mathsf{x}_{j_1}^{A_1}, \ldots, \mathsf{v}_{i_k}^{A_k} \leftrightarrow \mathsf{x}_{j_k}^{A_k}] \quad \text{or} \quad t[\boldsymbol{v}_{\langle i_1,\ldots,i_k \rangle} \leftrightarrow \boldsymbol{x}_{\langle j_1,\ldots,j_k \rangle}].$$

As mentioned earlier, given a list of sorts $[A_1, \ldots, A_k]$, the indices of the corresponding bound variables are determined by the de Bruijn convention. This is formalized by the following operator $\mathsf{bind\_vars}\,[A_1, \ldots, A_k] : \mathbb{N}^{[A_1,\ldots,A_k]}$ for which we use the notation $\boldsymbol{x}^{[A_1,\ldots,A_k]}$. For example $\boldsymbol{x}^{[A,A',A,A,A']} = \langle 2, 1, 1, 0, 0 \rangle$.

We also have an operator that defines a list of new variables of specified sorts with respect to a given term: $(\mathsf{newvars}\,[A_1, \ldots, A_k]\,t) = \langle i_1, \ldots, i_k \rangle$ such that the variables $\mathsf{v}_{i_1}^{A_1}, \ldots, \mathsf{v}_{i_k}^{A_k}$ do not occur in t.

We define the operation of binding meta-variables to de Bruijn variables in a similar way to what we have done for a single variable:

$$\begin{array}{l}
\mathsf{bind_args}\,[A_1, \ldots, A_k]\,B : (\mathsf{sort_list\ Term}\,[A_1, \ldots, A_k] \to \mathsf{Term}\,B) \to \mathsf{Term}\,B \\
\mathsf{bind_args}\,[A_1, \ldots, A_k]\,B\,g = (g\,\boldsymbol{v}_{\langle i_1,\ldots,i_k \rangle})[\boldsymbol{v}_{\langle i_1,\ldots,i_k \rangle} \leftrightarrow \boldsymbol{x}_{\langle j_1,\ldots,j_k \rangle}] \\
\qquad \text{where } \langle i_1, \ldots, i_k \rangle = \mathsf{newvars}\,[A_1, \ldots, A_k]\,(g\,\boldsymbol{x}_{\langle j_1,\ldots,j_k \rangle}) \\
\qquad\qquad \langle j_1, \ldots, j_k \rangle = \boldsymbol{x}^{[A_1,\ldots,A_k]}.
\end{array}$$

If we apply this binding operator directly to meta-arguments, we obtain the bind operation that we mentioned before:

$$\begin{array}{l}
\mathsf{bind} : \forall T : \mathsf{arg_type}, (\mathsf{Term_arg}\,T) \to \mathsf{Term}\,(\mathsf{arg_result}\,T) \\
\mathsf{bind}\,T = \mathsf{bind_args}\,(\mathsf{arg_bindings}\,T)\,(\mathsf{arg_result}\,T).
\end{array}$$

We extend it to lists of argument types, so we can simultaneously bind different variables in different arguments.

$$\begin{array}{l}
\mathsf{binds}\,\vec{T} : \mathsf{Term_args}\,\vec{T} \to \mathsf{TermList}\,\vec{T} \\
\mathsf{binds}\,[T_1, \ldots, T_n]\,\{g_1, \ldots, g_n\} = \langle\!\langle \mathsf{bind}\,T_1\,g_1, \ldots, \mathsf{bind}\,T_n\,g_n \rangle\!\rangle.
\end{array}$$

We have now all the tools needed to define the higher-order application operator:

$$\begin{array}{l}
\mathsf{Opr}\,f : \mathsf{Term_args}\,(\mathsf{op_arguments}\,(\mathsf{rule}\,f)) \to \mathsf{Term}\,(\mathsf{op_result}\,(\mathsf{rule}\,f)) \\
\mathsf{Opr}\,f\,\vec{g} = \mathsf{opr}\,f\,(\mathsf{binds}\,(\mathsf{op_arguments}\,(\mathsf{rule}\,f))\,\vec{g}).
\end{array}$$

Assume that the operation f has the rule $[\vec{A_1} \vdash B_1, \ldots, \vec{A_n} \vdash B_n] \,/\!\!/\, C$. Then:

$$\begin{array}{l}
\mathsf{Opr}\,f : \{\!\{\mathsf{sort_list\ Term}\,\vec{A_1} \to B_1, \ldots, \mathsf{sort_list\ Term}\,\vec{A_n} \to B_n\}\!\} \to \mathsf{Term}\,C \\
\mathsf{Opr}\,f\,\{g_1, \ldots, g_n\} = \mathsf{opr}\,f\,\langle\!\langle \mathsf{bind}\,(\vec{A_1} \vdash B_1)\,g_1, \ldots, \mathsf{bind}\,(\vec{A_n} \vdash B_n)\,g_n \rangle\!\rangle.
\end{array}$$

3.3. Application. The inverse operation of binding is application of a de Bruijn term with bindings to a list of arguments. This is the same as substitution of a de Bruijn variable with terms, keeping track of the increase of the variable index under abstraction. We want to do this simultaneously for several variables. We use the notation $t[\mathbf{x}_{\vec{i}}/\vec{a}]$ (Coq: `sub t As is al`) for the simultaneous substitution of the list of de Bruijn variables $\mathbf{x}_{\vec{i}}$ with the list of terms $\vec{a}$ of the correct sort. It is defined by recursion on t, mutually with its extension subs to lists of terms with bindings.

Application is just the substitution of the abstracted variables: $t[\mathbf{x}^{\vec{A}}/\bullet]$, where $\vec{A}$ is the list of the variables abstracted in t. The bullet $\bullet$ indicates that the substituted terms are meta-abstracted: $t[\mathbf{x}^{\vec{A}}/\bullet] := (\mathsf{fun}\,\vec{x} \mapsto t[\mathbf{x}^{\vec{A}}/\vec{x}])$. We define application directly on lists of terms with bindings:

$$\begin{aligned}
&\mathsf{aps}\,\vec{T} : \mathsf{TermList}\,\vec{T} \to \mathsf{Term_args}\,\vec{T}\\
&\mathsf{aps}\,[]\,\langle\!\langle\rangle\!\rangle = \{\}\\
&\mathsf{aps}\,(T :: \vec{T})\,\langle\!\langle t :: \vec{t}\rangle\!\rangle = \{t[\mathbf{x}^{\mathsf{arg_bindings}\,T}/\bullet] :: (\mathsf{aps}\,\vec{T}\,\vec{t})\}.
\end{aligned}$$

THEOREM 5. *Application is a right inverse of binding*:

$$\forall(\vec{T} : \mathsf{list\,arg\_type})(\vec{t} : \mathsf{TermList}\,\vec{T}), \vec{t} = \mathsf{binds}\,\vec{T}\,(\mathsf{aps}\,\vec{T}\,\vec{t}).$$

PROOF. We have that:

$$\begin{aligned}
&\mathsf{binds}\,[T_1\ \ldots, T_n]\,(\mathsf{aps}\,[T_1\ \ldots, T_n]\,\langle\!\langle t_1, \ldots\ t_n\rangle\!\rangle)\\
&= \langle\!\langle \mathsf{bind}\,T_1\,(t_1[\mathbf{x}^{\mathsf{arg_bindings}\,T_1}/\bullet]), \ldots, \mathsf{bind}\,T_n\,(t_n[\mathbf{x}^{\mathsf{arg_bindings}\,T_n}/\bullet])\rangle\!\rangle
\end{aligned}$$

So it is sufficient to prove that for every $T : \mathsf{arg\_type}$ and $t : \mathsf{Term}\,(\mathsf{op\_result}\,T)$ we have $t = \mathsf{bind}\,T\,(t[\mathbf{x}^{\mathsf{arg\_bindings}\,T}/\bullet])$. Assume that $T = \vec{A} \vdash B$, then:

$$\begin{aligned}
&\mathsf{bind}\,T\,(t[\mathbf{x}^{\mathsf{arg_bindings}\,T}/\bullet]) = \mathsf{bind_args}\,\vec{A}\,B\,(t[\mathbf{x}^{\vec{A}}/\bullet])\\
&= t[\mathbf{x}^{\vec{A}}/\mathbf{v}_{\vec{i}}][\mathbf{v}_{\vec{i}} \leftrightarrow \mathbf{x}_{\vec{j}}]
\end{aligned}$$

where $\mathbf{x}_{\vec{j}} = \mathbf{x}^{\vec{A}}$ and $\mathbf{v}_{\vec{i}} = \mathsf{newvars}\,\vec{A}\,t[\mathbf{x}^{\vec{A}}/\mathbf{x}_{\vec{j}}] = \mathsf{newvars}\,\vec{A}\,t$. Since the variables $\mathbf{v}_{\vec{i}}$ are fresh for t, we have $t[\mathbf{x}^{\vec{A}}/\mathbf{v}_{\vec{i}}][\mathbf{v}_{\vec{i}} \leftrightarrow \mathbf{x}_{\vec{j}}] = t$ as desired. ⊣

Note that in general aps is not the left inverse of binds. In fact, if we start with a list of functions, bind their meta-arguments, and then lift the result again to the meta-level; we won't in general get back the original functions. If some of them were not uniform, we will instead obtain their *uniformization*, that is, the system will choose one instance of the function and generalize it to all other arguments.

DEFINITION 2. The *uniformization* of a list of functions $\vec{h} : \mathsf{Term\_args}\,\vec{T}$ is defined as the list of functions: $\underline{\vec{h}} = \mathsf{aps}\,\vec{T}\,(\mathsf{binds}\,\vec{T}\,\vec{h})$.

We expect lists of functions used as arguments of operations in higher-order syntax to be extensionally equal to their uniformization. If they are not, then we have exotic terms.

3.4. The Recursion Principle. The higher-order recursion principle is translated internally into the structural recursion principle on de Bruijn notation. This is a standard principle that can easily be derived in Coq.

Let H : sort $\to$ Set be a family of types indexed on the sorts. The *recursion principle on the de Bruijn notation* is the standard structural recursion in Coq (but it has to be explicitly given by a Fixpoint definition, because the automatic recursor does not perform mutual recursion).

Let the following step functions be given:

$$\begin{aligned}
\mathsf{Hvar}: &\ \forall A : \mathsf{sort}, \mathbb{N} \to H\ A \\
\mathsf{Hbnd}: &\ \forall A : \mathsf{sort}, \mathbb{N} \to H\ A \\
\mathsf{Hopr}: &\ \forall (f : \mathsf{operation}), \\
&\qquad \mathsf{TermList}\,(\mathsf{op_arguments}\,(\mathsf{rule}\,f)) \to \\
&\qquad \mathsf{sort_list}\,H\,(\mathsf{op_arguments}\,(\mathsf{rule}\,f)) \to H\,(\mathsf{op_result}\,(\mathsf{rule}\,f)).
\end{aligned}$$

(The actual principle is a bit more general in that you can use any family on term lists in place of sort_list H and have recursion steps on it for lists; but we only need the case with sort_list H.) We obtain two recursive functions on Term and TermList:

$$\begin{aligned}
&\mathsf{debruijn_term_recursion}\ H\ \mathsf{Hvar}\ \mathsf{Hbnd}\ \mathsf{Hopr}: \\
&\qquad \forall A : \mathsf{sort}, \mathsf{Term}\ A \to H\ A \\
&\mathsf{debruijn_term_list_recursion}\ H\ \mathsf{Hvar}\ \mathsf{Hbnd}\ \mathsf{Hopr}: \\
&\qquad \forall \vec{T} : \mathsf{list\ arg_type}, \mathsf{TermList}\ \vec{T} \to \mathsf{sort_list}\ H\ \vec{T}.
\end{aligned}$$

For brevity, let us denote them by θ and $\vec{\theta}$, respectively. They satisfy the recursive equations:

$$\begin{aligned}
\theta\ A\ \mathsf{v}_i^A &= \mathsf{Hvar}\ A\ i \\
\theta\ A\ \mathsf{v}_j^A &= \mathsf{Hbnd}\ A\ j \\
\theta\ _\,(\mathsf{opr}\ f\ \vec{t}) &= \mathsf{Hopr}\ f\ \vec{t}\,(\vec{\theta}\ _\ \vec{t})
\end{aligned}$$

$$\begin{aligned}
&\vec{\theta}\,[]\,\langle\!\langle\rangle\!\rangle = \langle\rangle \\
&\vec{\theta}\,(\vec{A} \vdash B :: \vec{T})\,\langle\!\langle t :: \vec{t}\rangle\!\rangle = \langle(\theta\ B\ t) :: (\vec{\theta}\ \vec{T}\ \vec{t})\rangle
\end{aligned}$$

So in conclusion we have that:

$$\theta\ \_\,(\mathsf{opr}\ f\ \langle\!\langle t_1, \ldots, t_n\rangle\!\rangle) = \mathsf{Hopr}\ f\ \langle\!\langle t_1, \ldots, t_n\rangle\!\rangle\ \langle \theta\ \_\ t_1, \ldots, \theta\ \_\ t_n\rangle$$

Notice that this is a purely syntactic recursion principle; the bindings in the term lists are completely ignored. In particular, the results on de Bruijn variables are computed by Hbnd, which may give different outputs for different de Bruijn indices, even if they happen to correspond to the same bound variable.

We must show how the higher-order function ϕ can be defined in terms of θ. First of all, let us define an appropriate H in terms of F. We take: $H\ A = \mathsf{Assignment} \to F\ A$. Then we need to define Hvar, Hbnd, and Hopr in terms of Fvar and FOpr:

$$\begin{aligned}
&\mathsf{Hvar}\,A\,i = \mathsf{fun}\,\alpha \mapsto \mathsf{Fvar}\,A\,i\\
&\mathsf{Hbnd}\,A\,i = \mathsf{fun}\,\alpha \mapsto (\alpha\,A)_i\\
&\mathsf{Hopr}\,f\,\vec{t}\,\vec{r} = \mathsf{fun}\,\alpha \mapsto \mathsf{FOpr}\,f\,(\mathsf{aps}\,_\,\vec{t})\,(\vec{r} \bullet \alpha).
\end{aligned}$$

where $\vec{g} = \vec{r} \bullet \alpha$ is the list of functions defined in the following way, assuming $(\mathsf{rule}\,f) = [[A_{11},\dots,A_{1k_1}] \vdash B_1,\dots,[A_{n1},\dots,A_{nk_n}] \vdash B_n] \,/\!/\, C$ and $\vec{r} = \langle r_1,\dots,r_n\rangle$:

$$\begin{aligned}
&\vec{g} : \mathsf{args_map}\,F\,[[A_{11},\dots,A_{1k_1}] \vdash B_1,\dots,[A_{n1},\dots,A_{nk_n}] \vdash B_n]\\
&\vec{g} = \langle g_1,\dots,g_n\rangle\\
&g_i = \mathsf{fun}\,\vec{a}_i : \mathsf{sort_list}\,F\,[A_{i1},\dots,A_{ik_i}] \mapsto r_i\,[\vec{a}_i]\alpha
\end{aligned}$$

Having defined θ using these parameters, we define ϕ and $\vec{\phi}$ as:

$$\begin{aligned}
&\phi\,\alpha\,A\,t = \theta\,A\,t\,\alpha\\
&\vec{\phi}\,\alpha\,\vec{T}\,\vec{h} = (\vec{\theta}\,\vec{T}\,(\mathsf{binds}\,\vec{T}\,\vec{h})) \bullet \alpha
\end{aligned}$$

We have to prove that they satisfy the higher-order recursion relations of Theorem 1. The equations for free and bound variables are immediate. The proof of the equation for operation application follows from Theorem 5 in the following way:

$$\begin{aligned}
\phi\,\alpha\,C\,(\mathsf{Opr}\,f\,\vec{h}) &= \theta\,C\,(\mathsf{Opr}\,f\,\vec{h})\,\alpha\\
&= \theta\,C\,(\mathsf{opr}\,f\,(\mathsf{binds}\,_\,\vec{h}))\,\alpha\\
&= \mathsf{Hopr}\,f\,(\mathsf{binds}\,_\,\vec{h})\,(\vec{\theta}\,_\,(\mathsf{binds}\,_\,\vec{h}))\,\alpha\\
&= \mathsf{FOpr}\,f\,(\mathsf{aps}\,_\,(\mathsf{binds}\,_\,\vec{h}))\,((\vec{\theta}\,_\,(\mathsf{binds}\,_\,\vec{h})) \bullet \alpha)\\
&= \mathsf{FOpr}\,f\,\underline{\vec{h}}\,(\vec{\phi}\,\alpha\,_\,\vec{h}).
\end{aligned}$$

The equations for $\vec{\phi}$ can easily be proved by noticing that they are equivalent to the single equation $\vec{\phi}\,\alpha\,\vec{T}\,\vec{h} = \vec{r} \bullet \alpha$, with $r_i = \mathsf{fun}\,\alpha' \mapsto \phi\,\alpha'\,\_\,(\mathsf{bind}\,\_\,h_i)$.

3.5. The Induction Principle. The proof of the induction principle is very similar to that of the recursion principle. Therefore, we only point out two differences.

In one respect the proof is easier, because we don't need to use an assignment for de Bruijn variables: It is enough that the predicate P is true for the variables, we are not interested in changing those proofs when going under an abstraction. We also don't need to prove reduction equations.

In another respect the proof is slightly more difficult: The conclusion $(P\,A\,t)$ depends on the term t, while in the recursion principle the conclusion $(F\,A)$ only depended on the sort A. As a consequence, when we apply structural induction on the de Bruijn syntax, the inductive case for operation application requires a proof of $(P\,\_\,(\mathsf{opr}\,f\,\vec{t}))$ for any operation f and argument list $\vec{t}$. We cannot directly apply the induction hypothesis given by the statement of

Theorem 2, because its conclusion is $(P\ \_\ (\mathsf{Opr}\ f\ \vec{g}))$. We must first convert $(\mathsf{opr}\ f\ \vec{t})$ to the form $(\mathsf{Opr}\ f\ \vec{g})$. We do this by applying Theorem 5 and the definition of Opr:

$$\begin{aligned}\mathsf{opr}\ f\ \vec{t} &= \mathsf{opr}\ f\ (\mathsf{binds}\ (\mathsf{op\_arguments}\ (\mathsf{rule}\ f))\ (\mathsf{aps}\ \_\ \vec{t}))\\ &= \mathsf{Opr}\ f\ (\mathsf{aps}\ \_\ \vec{t})\end{aligned}$$

We can then apply the hypothesis with $\vec{g} = (\mathsf{aps}\ \_\ \vec{t})$ and the proof goes through without problems.

3.6. Adequacy. It is important to show that our encodings of de Bruijn terms and object languages are adequate, i.e., that there is a bijection between the language we are encoding and its encoded form such that substitution commutes with the encoding. The first adequacy result for the original Hybrid [2] appears in [11], and since we use a standard de Bruijn encoding, the adequacy of our representation should follow directly. Adequacy for object languages such as the simply-typed λ-calculus should also follow similarly from adequacy results for Hybrid. The main difference here is that when descending through binders via the bind operator, a meta-variable becomes a *dangling index*, i.e., an index representing a bound variable that does not have a corresponding binder in the term. Establishing adequacy in this case requires mapping free variables of the object language to two kinds of variables in the formalization: free variables *and* dangling indices.

§4. Related work.

The slides of a recent talk by Randy Pollack [30] give a good summary of the literature on approaches to the implementation of languages with binders. We discuss here some that are most closely related to ours.

One of the basic ideas used in this work is the notion of translation between a high-level notation and a lower-level de Bruijn representation. This idea appears in the work of Gordon [17], where bound variables are presented to the user as strings. The idea of replacing strings with a binding operator was introduced by Ambler et al. [2], and adopted directly here. Gordon and Melham [18] used the name-as-strings syntax approach and developed a general theory of untyped λ-terms up to α-conversion, including induction and recursion principles. Norrish building on this work, improves the recursion principles [25], allowing greater flexibility in defining recursive functions on this syntax. Schürmann et. al. have also worked on designing a new calculus for defining recursive functions directly on higher-order syntax [33]. Built-in primitives are provided for the reduction equations for the higher-order case, in contrast to our approach where we define the recursion principle on top of the base level de Bruijn encoding, and prove the reduction equations as lemmas.

Multi-level approaches [14, 23] in Coq and Isabelle, respectively, have been adopted to facilitate reasoning by induction on object-level judgments. We should be able to directly adopt these ideas to create a multi-level version of our system. This kind of approach is inspired by logics such as $FO\lambda^{\Delta\mathbb{N}}$ [22], which was developed specifically to reason using higher-order syntax. In such logics, one can encode an intermediate "specification logic" between the meta-logic and the object language. Negative occurrences in inductive types representing object-level judgments are avoided by using the specification logic instead.

The Twelf system [26], which implements the Logical Framework (LF) has also been used as a framework for reasoning using higher-order syntax. In particular, support for meta-reasoning about object logics expressed in LF has been added [27, 32]. The design of the component for reasoning by induction does not include induction principles for higher-order encodings. Instead, it is based on a realizability interpretation of proof terms. The Twelf implementation of this approach includes powerful automated support for inductive proofs such as termination and coverage checkers.

Weak Higher-Order Abstract Syntax (WHOAS) tries to solve the problems of HOAS by turning the negative occurrences of the type of terms in the definition of a data-type into a parameter. In the case of the λ-calculus, the abstraction operator has type:

$$\mathsf{abs} : (\mathsf{Var} \to \Lambda) \to \Lambda.$$

where Var is a type parameter. This approach was introduced by Despeyroux, Felty, and Hirschowitz [13]. Exotic terms were discussed and a predicate was defined to factor them out. Honsell, Miculan, and Scagnetto [20] use a WHOAS approach; they have considered a variety of examples and developed a Theory of Contexts to aid reasoning about variables. A drawback of this approach is that it needs to assume axiomatically several properties of Var.

Gabbay and Pitts [16, 28] define a variant of classical set theory that includes primitives for variable renaming and variable freshness, and a new "freshness quantifier." Using this set theory, it is possible to prove properties by structural induction and also to define functions by recursion over syntax. This approach has been used by Urban and others to solve unification problems [36] and to formalize results on the λ-calculus in Isabelle/HOL [37].

Every object of a nominal set is associated with a *support*, a set of *atoms* (variable names) which generalizes the notion of the set of free variables of a term. There is a notion of swapping of atoms: $(a\,b)\cdot t$ intuitively interchanges the free occurrences of the atoms a and b in t. This notion can be extended to standard type constructors like pairs and functions. In particular, if we have a notion of swapping for nominal sets A and B, we can define swapping for the type of functions $A \to B$: $(a\,b)\cdot f = \lambda x.(a\,b)\cdot(f\ (a\,b)\cdot x)$. Once swapping is defined, the support of an object (its free variables) can be defined as:

$$\mathsf{support}\, t = \{a \mid \{b \mid (a\,b)\cdot t \neq t\} \text{ is infinite}\}.$$

Therefore it is possible to define a notion of free variable also for functions. The freshness relation $a \# t$ expresses the fact that $a \notin \mathsf{support}\, t$ (a is a fresh variable for t). The fact that these definitions generalize to function and product types allows the authors to impose freshness conditions on the recursive definition of functions on the syntax, thus guaranteeing the preservation of α-equality. The possibility of defining a set of free variables for a function is of interest for our work. However, this method is not constructive and cannot be used to generate effectively a fresh variable for a function. For this reason we are forced, in our formalization, to adopt a different, less elegant but computable solution. The nominal approach has been implemented to give representations of object languages with easy recursion principles on α-equivalence classes in Isabelle/HOL [35] and in Coq [4].

Schürmann, Despeyroux, and Pfenning [31] develop a modal meta-theory that allows the formalization of higher-order abstract syntax with a primitive recursive principle. They introduce a modal operator $\Box$. Intuitively, for every type A there is a type $\Box A$ of *closed* objects of type A. Besides the regular function type $A \to B$, there is a more restricted type $A \Rightarrow B = \Box A \to B$ of uniform functions. Functions used as arguments for higher-order constructors are of this kind. For example, in formalizing the pure λ-calculus, the abstraction operator has type

$$\mathsf{abs} : (\Lambda \Rightarrow \Lambda) \to \Lambda.$$

A structural recursion principle is provided. It can be used to define functions of the regular type $\Lambda \to B$. On the other hand, we are not allowed to use structural recursion to define a function of type $\Box\Lambda \to B$. This avoids the usual paradoxes associated with recursion for types with non-positive occurrences in their definition. Intuitively, we can explain the method as follows: While defining a type A, we cannot assume knowledge of the type as a whole. Think of $\Box A$ as a non-completed version of A, that is, a type that contains some elements of A but may still be extended in the future. Since $\Box A$ is not complete, we are not allowed to do recursion on it. If a constructor requires an argument that is a function of A, we must use $\Box A$, because the function should be compatible with future extensions.

§5. **Conclusion.** We have developed an approach to reasoning using higher-order abstract syntax which is built on a formalization of a higher-order universal algebra with bindings. This approach generalizes the Hybrid approach where an underlying de Bruijn notation is used. Higher-order syntax encodings are defined in such a way that expanding definitions results in the low-level de Bruijn representation of terms. Reasoning, however, is carried out at the level of higher-order syntax, allowing details of the lower-level implementation to be hidden. In our generalized version, an object language is defined by simply giving its signature, and the resulting tools for reasoning about the object language, such as a higher-order notation, induction principles, and recursion principles are directly available by simply instantiating the general theorems.

Future work includes considering a variety of object languages and completing more extensive proofs. In our earlier work [8], we expressed induction and recursion principles more directly for each object language, but proving them was not just simple instantiation and instead required some proof effort. In that setting, we illustrated the approach with examples showing that reasoning about object languages was direct and simple. In the new setting, after instantiating our general induction and recursion theorems, we expect the reasoning to be equally direct and simple. We also plan to apply our approach to more complex examples such as the POPLMARK challenge [3].

Acknowledgment. The authors would like to thank Alberto Momigliano for useful discussions on Hybrid. The work described here is supported in part by the Natural Sciences and Engineering Research Council of Canada.

REFERENCES

[1] *The LEGO proof assistant*, `www.dcs.ed.ac.uk/home/lego/`, 2001.

[2] SIMON J. AMBLER, ROY L. CROLE, and ALBERTO MOMIGLIANO, *Combining higher order abstract syntax with tactical theorem proving and (co)induction*, ***Fifteenth International Conference on Theorem Proving in Higher-Order Logics***, Lecture Notes in Computer Science, vol. 2410, Springer-Verlag, 2002, pp. 13–30.

[3] BRIAN E. AYDEMIR, AARON BOHANNON, MATTHEW FAIRBAIRN, J. NATHAN FOSTER, BENJAMIN C. PIERCE, PETER SEWELL, DIMITRIOS VYTINIOTIS, GEOFFREY WASHBURN, STEPHANIE WEIRICH, and STEVE ZDANCEWIC, *Mechanized metatheory for the masses: The* POPLMARK *challenge*, ***Eighteenth International Conference on Theorem Proving in Higher-Order Logics***, Lecture Notes in Computer Science, vol. 3605, Springer-Verlag, 2005, `http://fling-l.seas.upenn.edu/~plclub/cgi-bin/poplmark/index.php?title=The_POPLmark_Challenge`, pp. 50–65.

[4] BRIAN E. AYDEMIR, AARON BOHANNON, and STEPHANIE WEIRICH, *Nominal reasoning techniques in Coq*, ***International Workshop on Logical Frameworks and Meta-Languages: Theory and Practice***, 2006, to appear in ***Electronic Notes in Theoretical Computer Science***.

[5] HENK BARENDREGT, *Lambda calculi with types*, ***Handbook of Logic in Computer Science, Volume 2*** (S. Abramsky, Dov M. Gabbay, and T. S. E. Maibaum, editors), Oxford University Press, 1992, pp. 117–309.

[6] YVES BERTOT and PIERRE CASTÉRAN, ***Interactive Theorem Proving and Program Development. Coq'art: The Calculus of Inductive Constructions***, Springer, 2004.

[7] VENANZIO CAPRETTA, ***Abstraction and Computation***, Ph.D. thesis, Computing Science Institute, University of Nijmegen, 2002.

[8] VENANZIO CAPRETTA and AMY P. FELTY, *Combining de Bruijn indices and higher-order abstract syntax in Coq*, ***Proceedings of Types 2006***, Lecture Notes in Computer Science, Springer-Verlag, 2007, to appear.

[9] COQ DEVELOPMENT TEAM, LOGICAL PROJECT, ***The Coq Proof Assistant Reference Manual: Version 8.0***, Technical Report, 2006.

[10] THIERRY COQUAND and GÉRARD HUET, *The Calculus of Constructions*, ***Information and Computation***, vol. 76 (1988), pp. 95–120.

[11] ROY L. CROLE, *A Representational Adequacy Result for Hybrid*, submitted for journal publication, April 2006.

[12] N. G. DE BRUIJN, *Lambda-calculus notation with nameless dummies: a tool for automatic formula manipulation with application to the Church-Rosser theorem*, ***Indagationes Mathematicae***, vol. 34 (1972), pp. 381–392.

[13] JOËLLE DESPEYROUX, AMY FELTY, and ANDRÉ HIRSCHOWITZ, *Higher-order abstract syntax in Coq*, ***Second International Conference on Typed Lambda Calculi and Applications***, Lecture Notes in Computer Science, vol. 902, Springer-Verlag, April 1995, pp. 124–138.

[14] AMY P. FELTY, *Two-level meta-reasoning in Coq*, ***Fifteenth International Conference on Theorem Proving in Higher-Order Logics***, Lecture Notes in Computer Science, Springer-Verlag, August 2002, pp. 198–213.

[15] MARCELO P. FIORE, GORDON D. PLOTKIN, and DANIELE TURI, *Abstract syntax and variable binding.*, ***Fourteenth Annual IEEE Symposium on Logic in Computer Science***, IEEE-Computer Society, 1999, pp. 193–202.

[16] MURDOCH J. GABBAY and ANDREW M. PITTS, *A new approach to abstract syntax with variable binding*, ***Formal Aspects of Computing***, vol. 13 (2001), pp. 341–363.

[17] ANDREW D. GORDON, *A mechanisation of name-carrying syntax up to alpha-conversion*, ***Higher-Order Logic Theorem Proving and its Applications***, Lecture Notes in Computer Science, vol. 780, Springer-Verlag, 1993, pp. 414–426.

[18] ANDREW D. GORDON and TOM MELHAM, *Five axioms of alpha-conversion*, ***9th International Conference on Higher-Order Logic Theorem Proving and its Applications***, Lecture Notes in Computer Science, vol. 1125, Springer-Verlag, 1996, pp. 173–191.

[19] MICHAEL J. C. GORDON and TOM F. MELHAM, ***Introduction to HOL: A Theorem Proving Environment for Higher-Order Logic***, Cambridge University Press, 1993.

[20] FURIO HONSELL, MARINO MICULAN, and IVAN SCAGNETTO, *An axiomatic approach to metareasoning on nominal algebras in HOAS*, ***28th International Colloquium on Automata, Languages and Programming***, Lecture Notes in Computer Science, vol. 2076, Springer Verlag, 2001, pp. 963–978.

[21] ZHAOHUI LUO, ***Computation and Reasoning: A Type Theory for Computer Science***, International Series of Monographs on Computer Science, vol. 11, Oxford University Press, 1994.

[22] RAYMOND MCDOWELL and DALE MILLER, *Reasoning with higher-order abstract syntax in a logical framework*, ***ACM Transactions on Computational Logic***, vol. 3 (2002), no. 1, pp. 80–136.

[23] ALBERTO MOMIGLIANO and SIMON J. AMBLER, *Multi-level meta-reasoning with higher-order abstract syntax*, ***Sixth International Conference on Foundations of Software Science and Computational Structures***, Lecture Notes in Artificial Intelligence, vol. 2620, Springer-Verlag, 2003, pp. 375–391.

[24] TOBIAS NIPKOW, LAWRENCE C. PAULSON, and MARKUS WENZEL, ***Isabelle/HOL: A Proof Assistant for Higher-Order Logic***, Lecture Notes in Computer Science, vol. 2283, Springer Verlag, 2002.

[25] MICHAEL NORRISH, *Recursive function definition for types with binders*, ***Seventeenth International Conference on Theorem Proving in Higher Order Logics***, Lecture Notes in Computer Science, vol. 3223, Springer-Verlag, 2004, pp. 241–256.

[26] FRANK PFENNING and CARSTEN SCHÜRMANN, *System description: Twelf — a meta-logical framework for deductive systems*, ***Sixteenth International Conference on Automated Deduction***, Lecture Notes in Computer Science, vol. 1632, Springer-Verlag, 1999, pp. 202–206.

[27] BRIGITTE PIENTKA, *Verifying termination and reduction properties about higher-order logic programs*, ***Journal of Automated Reasoning***, vol. 34 (2005), no. 2, pp. 179–207.

[28] ANDREW M. PITTS, *Alpha-structural recursion and induction*, ***Journal of the Association of Computing Machinery***, vol. 53 (2006), no. 3, pp. 459–506.

[29] GORDON D. PLOTKIN, *An illative theory of relations*, ***Situation Theory and its Applications, Volume 1*** (Robin Cooper, Kuniaki Mukai, and John Perry, editors), CSLI, 1990, pp. 133–146.

[30] RANDY POLLACK, *Reasoning about languages with binding*, Presentation available at `http://homepages.inf.ed.ac.uk/rap/export/bindingChallenge_slides.pdf`, 2006.

[31] Carsten Schürmann, Joëlle Despeyroux, and Frank Pfenning, *Primitive recursion for higher-order abstract syntax*, ***Theoretical Computer Science***, vol. 266 (2001), pp. 1–57.

[32] Carsten Schürmann and Frank Pfenning, *A coverage checking algorithm for LF*, ***Sixteenth International Conference on Theorem Proving in Higher Order Logics***, Lecture Notes in Computer Science, vol. 2758, Springer-Verlag, 2003, pp. 120–135.

[33] Carsten Schürmann, Adam Poswolsky, and Jeffrey Sarnat, *The ∇-calculus. functional programming with higher-order encodings*, ***Seventh International Conference on Typed Lambda Calculi and Applications***, Lecture Notes in Computer Science, vol. 3461, Springer-Verlag, 2005, pp. 339–353.

[34] Morten Heine B. Sørensen and P. Urzyczyn, ***Lectures on the Curry-Howard Isomorphism***, Elsevier Science, 2006.

[35] Christian Urban and Stefan Berghofer, *A recursion combinator for nominal datatypes implemented in Isabelle/HOL*, ***Third International Joint Conference on Automated Reasoning***, Lecture Notes in Computer Science, vol. 4130, Springer-Verlag, 2006, pp. 498–512.

[36] Christian Urban, Andrew M. Pitts, and Murdoch J. Gabbay, *Nominal unification*, ***Theoretical Computer Science***, vol. 323 (2004), pp. 473–497.

[37] Christian Urban and Christine Tasson, *Nominal techniques in Isabelle/HOL*, ***Twentieth International Conference on Automated Deduction***, Lecture Notes in Computer Science, vol. 3632, Springer-Verlag, 2005, pp. 38–53.

[38] Benjamin Werner, ***Méta-théorie du Calcul des Constructions Inductives***, Ph.D. thesis, Université Paris 7, 1994.

SCHOOL OF INFORMATION TECHNOLOGY AND ENGINEERING
AND DEPARTMENT OF MATHEMATICS AND STATISTICS
UNIVERSITY OF OTTAWA, CANADA
Current address: School of Computer Science, University of Nottingham, UK
E-mail: vxc@cs.nott.ac.uk

SCHOOL OF INFORMATION TECHNOLOGY AND ENGINEERING
AND DEPARTMENT OF MATHEMATICS AND STATISTICS
UNIVERSITY OF OTTAWA, CANADA
E-mail: afelty@site.uottawa.ca

AN INTRODUCTION TO b-MINIMALITY

RAF CLUCKERS

Abstract. We give a survey with some explanations but no proofs of the new notion of b-minimality by the author and F. Loeser [b-minimality, Journal of Mathematical Logic, vol. 7 (2007), no. 2, pp. 195–227, `math.LO/0610183`]. We compare this notion with other notions like o-minimality, C-minimality, p-minimality, and so on.

§1. Introduction. As van den Dries notes in his book [17], Grothendieck's dream of tame geometries found a certain realization in model theory, at first by the study of the geometric properties of definable sets for some nice structure like the field of real numbers, and then by axiomatizing these properties by notions of o-minimality, minimality, C-minimality, p-minimality, v-minimality, t-minimality, b-minimality, and so on. Although there is a joke speaking of x-minimality with $x = a, b, c, d, \ldots$, these notions are useful and needed in different contexts for different kinds of structures, for example, o-minimality is for ordered structures, and v-minimality is for algebraically closed valued fields.

In recent work with F. Loeser [4], we tried to unify some of the notions of x-minimality for different x, for certain x only under extra conditions, to a very basic notion of b-minimality. At the same time, we tried to keep this notion very flexible, very tame with many nice properties, and able to describe complicated behavior.

An observation of Grothendieck's is that instead of looking at objects, it is often better to look at morphisms and study the fibers of the morphisms. In one word, that is what b-minimality does: while most notions of x-minimality focus on sets and axiomatize subsets of the line to be simple (or tame), b-minimality focuses on definable functions and gives axioms on the existence of definable functions with nice fibers.

We give a survey on the new notion of b-minimality and put it in context, without giving proofs, and refer to [4] for the proofs.

§2. A context. There is a plentitude of notions of tame geometries, even just looking at variants of o-minimality like quasi-o-minimality, d-minimality, and

Logic Colloquium '06
Edited by Barry Cooper, Herman Geuvers, Anand Pillay, and Jouko Väänänen
Lecture Notes in Logic, 32

so on. Hence there is a need for unifying notions. Very recently, A. Wilkie [18] expanded the real field with entire analytic functions other than exp, where the zeros are like the set of integer powers of 2, and he shows this structure still has a very tame geometry. Since an o-minimal structure only allows finite discrete subsets of the line, Wilkie's structure is not o-minimal, but it still probably is d-minimal [12], where an expansion of the field $\mathbb{R}$ is called d-minimal if for every m and definable $A \subseteq \mathbb{R}^{m+1}$ there is some N such that for all $x \in \mathbb{R}^m$, the fiber $A_x := \{y \in \mathbb{R} \mid (x, y) \in A\}$ of A above x either has nonempty interior or is a union of N discrete sets. A similar problem exists on algebraically closed valued fields: if one expands them with a nontrivial entire analytic function on the line, one gets infinitely many zeros, and thus such a structure can not be C-minimal nor v-minimal. This shows there is a need for flexible notions of tame geometry.

In [5], a general theory of motivic integration is developed, where dependence on parameters is made possible. It is only developed for the Denef-Pas language for Henselian valued fields (a semi-algebraic language), although only a limited number of properties of this language are used. Hence, one needs a notion of tame geometry for Henselian valued fields that is suitable for motivic integration. In this paper, we give an introduction to a notion satisfying to some extend these requirements, named b-minimality, developed in detail in [4].

§**3.** ***b*-minimality.** This section is intended to sharpen the reader's intuition before we give the formal definitions, by giving some informal explanations on b-minimality. The reader who wants to see formal definitions first, can proceed directly with Section 4, or go back and forth between this and the subsequent section.

In a b-minimal set-up, there are two basic kinds of sets: balls and points. The balls are subsets of the main sort and are given by the fibers of a single predicate B in many variables, under some coordinate projection. There are also two kinds of sorts: there is a unique main sort and all other sorts are called auxiliary sorts (hence there is a partitioning of the sorts in one main sort and some auxiliary sorts). The points are just singletons. The b from b-minimality refers to the word balls.

One sees that in any of the notions x-minimal with $x = o, v, C, p$, a ball makes sense (for example, open intervals in o-minimal structures), and thus b-minimality a priori can make sense.

The formal definition of b-minimality will be given in Section 4, but here we describe some reasonable and desirable properties, of which the axioms will be an abstraction.

To be in a b-minimal setup, a definable subset X of the line M in the main sort should be a disjoint union of balls and points. Such unions might be finite, but can as well be infinite, as long as they are "tame" in some sense.

Namely, by a tame union, we mean that this union is the union of the fibers of a definable family, parameterized by auxiliary parameters. Hence, infinite unions are "allowed" as long as they are tame in this sense. To force cell decomposition to hold, such family should be A-definable as soon as X is A-definable, with A some parameters. This is the content of the first axiom for b-minimality. Thus so to speak, there is a notion of "allowed" infinite (disjoint) union of balls points, and any subset of the line should be such a union.

Secondly, we really want balls to be different from points, and the auxiliary sorts to be really different from the main sort. A ball should not be a union of points (that is to say, an "allowed" union of points). This is captured in the second axiom for b-minimality.

For the third axiom, the idea of a "tame" disjoint union in a b-minimal structure is needed to formulate piecewise properties. In the third axiom, we assume a tameness property on definable functions from the line M in the main sort to M. Roughly, a definable function $f : M \to M$ should be piecewise constant or injective, where the pieces are forming a tame disjoint union, that is, there exists a definable family whose fibers form a disjoint union of M, and whose parameters are auxiliary, and on the fibers of this family the function f is constant or injective.

One more word on tame disjoint unions partitioning a set X. Instead of speaking of "a" definable family whose fibers form a partition of X and whose parameters are auxiliary, we will just speak of a definable function

$$f : X \to S$$

with S auxiliary, and the fibers of f then form such a tame union.

§4. b-minimality: the definition.

4.1. Some conventions. All languages will have a unique main sort, the other sorts are auxiliary sorts. An expansion of a language may introduce new auxiliary sorts. If a model is named $\mathcal{M}$, then the main sort of $\mathcal{M}$ is denoted by M.

By *definable* we shall always mean definable with parameters, as opposed to $\mathcal{L}(A)$-definable or A-definable, which means definable with parameters in A. By a *point* we mean a singleton. A definable set is called *auxiliary* if it is a subset of a finite Cartesian product of (the universes of) auxiliary sorts.

If S is a sort, then its Cartesian power S^0 is considered to be a point and to be $\emptyset$-definable.

Recall that o-minimality is about expansions of the language $\mathcal{L}_<$ with one predicate $<$, with the requirement that the predicate $<$ defines a dense linear order without endpoints. In the present setting we shall study expansions of a language $\mathcal{L}_B$ consisting of one predicate B, which is nonempty and which has fibers in the M-sort (by definition called balls). In both instances of tame geometry, the expansion has to satisfy extra properties. A priori, it is not determined to which product of sorts the predicate B corresponds; this will always be fixed by the context, or it will be supposed to be fixed later on by some context, when it needs to be fixed.

4.2. Let $\mathcal{L}_B$ be the language with one predicate B. We require that B is interpreted in any $\mathcal{L}_B$-model $\mathcal{M}$ with main sort M as a nonempty set $B(\mathcal{M})$ with

$$B(\mathcal{M}) \subset A_B \times M$$

where A_B is a finite Cartesian product of (the universes of) some of the sorts of $\mathcal{M}$.

When $a \in A_B$ we write $B(a)$ for

$$B(a) := \{m \in M \mid (a, m) \in B(\mathcal{M})\},$$

and if $B(a)$ is nonempty, we call it a *ball* (in the structure $\mathcal{M}$), or B-ball when useful.

DEFINITION 4.2.1 (b-minimality). Let $\mathcal{L}$ be any expansion of $\mathcal{L}_B$. We call an $\mathcal{L}$-model $\mathcal{M}$ *b-minimal* when the following three conditions are satisfied for every set of parameters A (the elements of A can belong to any of the sorts), for every A-definable subset X of M, and for every A-definable function $F : X \to M$.

(b1) There exists a A-definable function $f : X \to S$ with S an auxiliary set such that for each $s \in f(X)$ the fiber $f^{-1}(s)$ is a point or a ball.
(b2) If g is a definable function from an auxiliary set to a ball, then g is not surjective.
(b3) There exists a A-definable function $f : X \to S$ with S an auxiliary set such that for each $s \in f(X)$ the restriction $F_{|f^{-1}(s)}$ is either injective or constant.

We call an $\mathcal{L}$-theory *b-minimal* if all its models are b-minimal.

§5. Cell decomposition. In his paper on decision procedures, Cohen [6] develops cell decomposition techniques for real and p-adic fields, by a kind of Taylor approximation of roots of polynomials. At that time, the writing was rather complicated and it was only through the work by Denef [7, 8] that some concrete notion of p-adic cells became apparent. One should keep in mind that there was no ideological framework of o-minimality which later on formed intuition of what cells should be and what they should do. An example of a fracture with actual o-minimal intuition about cells was that these original p-adic cells were not literally designed to partition definable sets into cells, they merely helped to partion into some nice pieces. On these nice pieces, one could get good properties of functions defined on them, which helped to calculate p-adic integrals [7, 8, 13, 14].

Another aspect of o-minimal intuition is that cells in one variable should be simple and defined by induction on the variables, both aspects were not so clear for the original p-adic cells and became even more complicated in the Pas-framework. Also cell decomposition for C-minimal structures [9] is somehow complicated. In v-minimality [11], cell decomposition appears mainly implicitely.

The notion of b-minimality is intended to give a blueprint for a versatile kind of cell decomposition for tame geometries that is simple in one variable and defined by induction on the variables. A (1)-cell is a tame union of balls, and a (0)-cell is a tame union of points. Then one builds further with more variables.

§6. Cell decomposition: the definitions. Let $\mathcal{L}$ be any expansion of $\mathcal{L}_B$, as before, and let $\mathcal{M}$ be an $\mathcal{L}$-model.

Definition 6.0.2 (Cells). If all fibers of some $f : X \to S$ as in (b1) are balls, then call (X, f) a (1)-cell with presentation f. If all fibers of f as in (b1) are points, then call (X, f) a (0)-cell with presentation f. For short, call such X a cell.

Let $X \subset M^n$ be definable and let $(j_1, \dots, j_n)$ be in $\{0, 1\}^n$. Let $p_n : X \to M^{n-1}$ be the projection. Call X a $(j_1, \dots, j_n)$-cell with presentation

$$f : X \to S$$

for some auxiliary S, when for each $\hat{x} := (x_1, \dots, x_{n-1}) \in p_n(X)$, the set $p_n^{-1}(\hat{x}) \subset M$ is a (j_n)-cell with presentation

$$p_n^{-1}(\hat{x}) \to S : x_n \mapsto f(\hat{x}, x_n)$$

and $p_n(X)$ is a $(j_1, \dots, j_{n-1})$-cell with presentation

$$f' : p_n(X) \mapsto S'$$

for some f' satisfying $f' \circ p_n = p \circ f$ for some $p : S \to S'$.

One proves that if X is a $(i_1, \dots, i_n)$-cell, then X is not a $(i'_1, \dots, i'_n)$-cell, for the same ordering of the factors of M^n, for any tuple $(i'_1, \dots, i'_n)$ different from $(i_1, \dots, i_n)$. Thus $(i_1, \dots, i_n)$ can be called the *type* of the $(i_1, \dots, i_n)$-cell X.

One proves the cell decomposition theorem by compactness.

Theorem 6.1 (Cell decomposition). *Let $\mathcal{M}$ be a model of a b-minimal theory. Let $X \subset M^n$ be a definable set. Then there exists a finite partition of X into cells.*

§7. **Refinements.** Often, one has a cell decomposition of X, but one needs a finer cell decomposition, such that more properties hold on the parts. Here, it is not only the cells X_i that should be partitioned further into cells, but each X_i is already written as a union of fibers which resemble products of balls and points, and all these fibers should be partitioned into finer parts to speak of a genuine refinement.

DEFINITION 7.0.1. Let $\mathcal{P}$ and $\mathcal{P}'$ be two finite partitions of X into cells (X_i, f_i), respectively (Y_j, g_j). Call $\mathcal{P}'$ a *refinement of* $\mathcal{P}$ when for each i there exists j such that $Y_j \subset X_i$ and such that g_j is a refinement of $f_{ij} := f_{i|Y_j}$, that is, for each $a \in g_j(Y_j)$, there exists a (necessarily unique) $b \in f_{ij}(Y_j)$ such that

$$g_j^{-1}(a) \subset f_{ij}^{-1}(b).$$

One proves by compactness that refinements exist.

§8. **Relative cells.** Cells use an order of the variables, so they are very well suited to work relatively over some of the variables.

Since in a b-minimal set-up there are many sorts, not all definable sets are subsets of M^n, with M the main sort. Still, we want most notions to make sense for the main sort, and not to bother about the auxiliary sorts, as long as (b1), (b2) and (b3) are not violated. So there is a need to define all the concepts for definable subsets of $S \times M^n$ with S auxiliary, or more generally, for definable subsets X of $Y \times M^n$ for any definable Y. That way, one defines relative dimension over Y, cells over Y, a presentation over Y, and so on.

We just define a (i)-cell over Y with $i = 0, 1$. A definable set $X \subset Y \times M$ is called a i-cell with presentation

$$f : X \to Y \times S$$

with S auxiliary if f commutes with the projections $Y \times S \to Y$ and $p : X \to Y$ to Y, and for each y, the set $p^{-1}(y)$ is a (i)-cell with presentation

$$p^{-1}(y) \to S : m \mapsto f(y, m),$$

where we have identified $\{y\} \times S$ with S and $p^{-1}(y)$ with a subset of M.

§9. **Dimension theory.** Very similar to the o-minimal dimension as in [17], a dimension theory for b-minimal structures unfolds.

There are many sorts, but we want the dimension to live in the main sort.

DEFINITION 9.0.2. The dimension of a nonempty definable set $X \subset M^n$ is defined as the maximum of all sums

$$i_1 + \cdots + i_n$$

where $(i_1, \ldots, i_n)$ runs over the types of all cells contained in X, for all orderings of the n factors of M^n. To the empty set we assign the dimension $-\infty$.

If $X \subset S \times M^n$ is definable with S auxiliary, the dimension of X is defined as the dimension of $p(X)$ with $p : S \times M^n \to M^n$ the projection.

Many properties as in [17] follows, for example, a $(i_1, \dots, i_n)$-cell has dimension $\sum_j i_j$, and if $f : X \to Y$ is a definable functions, then $\dim(X) \geq \dim(f(X))$.

§**10. Preservation of balls.** For o-minimal structures, piecewise monotonicity of definable functions plays a key role. On a general b-minimal structure, there is no order $<$, so functions cannot be called monotone. Nevertheless, the Monotonicity Theorem for o-minimal structures does have an analogue for b-minimal structures. It is not a consequence of b-minimality but has to be required as an extra property, named preservation of (all) balls. When we look at an o-minimal structure as a b-minimal structure as we do below, preservation of all balls is a consequence of the Monotonicity Theorem. The notion is especially useful for Henselian valued fields in the context of motivic integration [5], where it is used for the change of variables in one variable, see below.

Definition 10.1 (Preservation of balls). Let $\mathcal{M}$ be a b-minimal $\mathcal{L}$-model. We say that $\mathcal{M}$ *preserves balls* if for every set of parameters A and A-definable function

$$F : X \subset M \to M$$

there is a A-definable function

$$f : X \to S$$

as in (b1) such that for each $s \in S$

$$F(f^{-1}(s))$$

is either a ball or a point.

If moreover there exists such f such that for every map $f_1 : X \to S_1$ as in (b1) refining f (in the sense that the fibers of f_1 partition the fibers of f) the set

$$F(f_1^{-1}(s_1))$$

is also either a ball or a point for each $s_1 \in S_1$, then say that $\mathcal{M}$ *preserves all balls*.

We say that a b-minimal theory *preserves balls* (resp. *preserves all balls*) when all its models do.

10.2. Let's give an example of p-adic integration and its change of variables formula in one variable, using preservation of balls.

If one integrates $|f(x)|$ over $\mathbb{Z}_p$ with f, say, a semi-algebraic function $\mathbb{Z}_p \to \mathbb{Z}_p$, and $|\cdot|$ the p-adic norm, it is useful to know a b-minimal cell decomposition of $\mathbb{Z}_p$ relative to $\operatorname{ord}(f)$. That is, one takes for X the definable subset of $\mathbb{Z}_p \times (\mathbb{Z} \cup \{+\infty\})$ given by $\mathrm{ord}(f(x)) = a$ for x in $\mathbb{Z}_p$ and a in $\mathbb{Z}\cup\{+\infty\}$ and one takes a b-minimal cell decomposition of X over $\mathbb{Z}\cup\{+\infty\}$ to find cells X_j over $\mathbb{Z}\cup\{+\infty\}$ with presentation $f_j : X_j \to (\mathbb{Z}\cup\{+\infty\})\times\mathbb{Z}^m$ for some m. The fibers of f_j are either balls or points, depending on j only, and since points have zero measure we can focus on 1-cells. Then

$$\int_{\mathbb{Z}_p} |f(x)||dx|, \tag{10.2.1}$$

with $|dx|$ the Haar measure, is easily integrated, since the volume of a ball is an easy function of its size, and since $\mathrm{ord}(f(x))$ by construction is constant on the fibers of the f_j. Since the measure of a ball is of the form p^b for some $b \in \mathbb{Z}$, and since $|f(x)|$ for any x is of the form $p^{-b'}$ for some $b' \in \mathbb{Z}\cup\{+\infty\}$, the integral (10.2.1) equals a converging sum

$$\sum_{a\in S} p^{-b(a)}, \tag{10.2.2}$$

with S a Presburger set, and $b : S \to \mathbb{Z} \cup \{+\infty\}$ a Presburger function. Indeed, one rewrites $\mathbb{Z}_p$ as the "tame" disjoint union of the balls occurring in the 1-cells (these balls are parameterized by a single Presburger set S), and on each such ball, say parameterized by $a \in S$, one multiplies the volume of the ball, $p^{v(a)}$, with the value of $|f(x)| = p^{-w(a)}$, where v and w are Presburger functions, to obtain $p^{-b(a)} = p^{v(a)-w(a)}$, and one then sums $p^{-b(a)}$ over S.

In a semi-algebraic setup, preservation of balls holds such that moreover the size of the balls is changed in a way compatible with the Jacobian. If $g : A \subset \mathbb{Z}_p \to \mathbb{Z}_p$ is a semi-algebraic bijection, then

$$\int_{\mathbb{Z}_p} |f(x)||dx| = \int_A |f\circ g(y)||\mathrm{Jac}(g)(y)||dy|,$$

by the change of variables formula. This change of variables formula holds here by general theory of the Haar measure on p-adic fields, but such arguments fail for motivic integrals because they involve much more general valued fields, like $k((t))$ with k of characteristic zero. However, if one takes the above cell decomposition such that balls are preserved through g^{-1} and such that their sizes change as predicted by the Jacobian, then we can translate both integrals into Presburger sums as (10.2.2) which one sees are exactly the same Presburger sums. Indeed, the norm of the Jacobian makes up for the difference in size of a ball B_a and its inverse image $g^{-1}(B_a)$. Thus one finds an alternative proof of the change of variables formula in one variable. In [4] the motivic case is worked out. That the preservation of balls also changes the size of the balls w.r.t. the Jacobian, is a corollary of Weierstrass division and thus also holds in a motivic setting and even in subanalytic motivic settings, as long as the Henselian valued field has characteristic zero.

§11. Some examples of b-minimal structures.

11.1. o-minimal structures and non o-minimal expansions. Any o-minimal structure R admits a natural b-minimal expansion by taking as main sort R with the induced structure, the two point set $\{0, 1\}$ as auxiliary sort and two constant symbols to denote these auxiliary points. A possible interpretation for B is clear, for example,

$$B = \{(x, y, m) \in R^2 \times R \mid x < m < y \text{ when } x < y,\\ x < m \text{ when } x = y,\\ \text{and } m < y \text{ when } x > y\},$$

so that in the m variable one gets all open intervals as fibers of B above R^2. Property (b3) and preservation of all balls is in this case a corollary of the Monotonicity Theorem for o-minimal structures.

The notion of b-minimality leaves much more room for expansions than the notion of o-minimality: some structures on the real numbers are not o-minimal but are naturally b-minimal, for example, the field of real numbers with a predicate for the integer powers of 2 are b-minimal by [16] when adding to the above language the set of integer powers of 2 as auxiliary sort and the natural inclusion of it into $\mathbb{R}$ as function symbol.

Recently [18], Wilkie extended van den Dries's construction to polynomially bounded structures, hence finding new entire analytic functions on the reals (other than exp) with tame geometry. These structures seem to be b-minimal as well, w.r.t. similar auxiliary sorts as for van den Dries's structure $\mathbb{R}, 2^{\mathbb{Z}}$.

11.2. Henselian valued fields of characteristic zero. In [4] is proved that the theory of Henselian valued fields of characteristic zero is b-minimal, in a natural definitial expansion of the valued field language, by adapting the Cohen - Denef proof. As far as we know, this is the first written instance of cell decomposition in mixed characteristic for unbounded ramification.

Let Hen denote the collection of all Henselian valued fields of characteristic zero (hence mixed characteristic is allowed).

For K in Hen, write K° for the valuation ring and M_K for the maximal ideal of K°.

For $n > 0$ an integer, set $nM_K = \{nm \mid m \in M_K\}$ and consider the natural group morphism

$$rv_n : K^\times \to K^\times/1 + nM_K$$

which we extend to $rv_n : K \to (K^\times/1 + nM_K) \cup \{0\}$ by sending 0 to 0.

For every $n > 0$ we write $RV_n(K)$ for

$$RV_n(K) := (K^\times/1 + nM_K) \cup \{0\},$$

rv for rv_1 and RV for RV_1.

We define the family $B(K)$ of balls by

$$B(K) = \{(a, b, x) \in K^\times \times K^2 \mid |x - b| < |a|\}.$$

Hence, a ball is by definition any set of the form $B(a, b) = \{x \in K \mid |x - b| < |a|\}$ with a nonzero.

It is known that T_{Hen} allows elimination of valued field quantifiers in the language L_{Hen} by results by Scanlon [15], F.V. Kuhlmann and Basarab [1].

THEOREM 11.2.1. *The theory* T_{Hen} *is b-minimal. Moreover,* T_{Hen} *preserves all balls.*

REMARK 11.2.2. In fact, a slightly stronger cell decomposition theorem than Theorem 6.1 holds for T_{Hen}, namely a cell decomposition with centers. We refer to [4] to find back the full statement of cell decomposition with centers and the definition of a center of a cell in a b-minimal context.

The search for an expansion of T_{Hen} with a nontrivial entire analytic function is open and challenging. Nevertheless, in [2] b-minimality for a broad class of analytic expansions of T_{Hen} is proved. This class of analytic expansions is an axiomatization of previous work [3].

§12. A further study and context. Among other things, b-minimality is an attempt to lay the fundamentals of a tame geometry on Henselian valued fields that is suitable for motivic integration, as in [5]. We hope to develop this theory in future work. One goal is to generalize the study in [11] by Hrushovski and Kazhdan on Grothendieck rings in a v-minimal context to a b-minimal context.

Theories which are v-minimal [11], or p-minimal [10] plus an extra condition, are b-minimal, namely, for the p-minimal case, under the extra condition of existence of definable Skolem functions. Also for C-minimality, some extra conditions are needed to imply b-minimality. For p-minimality, for example, cell decomposition lacks exactly when there are no definable Skolem functions. A possible connection with d-minimality needs to be investigated further.

For notions of x-minimality with $x = p, C, v, o$, an expansion of a field with an entire analytic function (other than exp on the real field) is probably impossible, intuitively since such functions have infinitely many zeros. In a b-minimal context, an infinite discrete set does not pose any problem, see Section 11.1 for an example, as long as it is a "tame" union of points. So, one can hope for nontrivial expansions of b-minimal fields by entire analytic functions, as done by Wilkie with exp and other entire functions on the reals, see Section 11.1.

We give some open questions to end with:

Does a b-minimal L_{Hen}-theory of valued fields imply that the valued fields are Henselian?

As soon as the main sort M is a normed field, is a definable function $f : M^n \to M$ then automatically C^1, that is, continuously differentiable?

Is there a weaker condition for expansions of $\mathcal{L}_B$ than preservation of (all) balls that together with (b1), (b2) and (b3) implies preservation of (all) balls?

Acknowledgments. Many thanks to Anand Pillay for the invitation to write this article and to Denef, Hrushovski, and Pillay for useful advice on b-minimality.

During the writing of this paper, the author was a postdoctoral fellow of the Fund for Scientific Research - Flanders (Belgium) (F.W.O.) and was supported by The European Commission - Marie Curie European Individual Fellowship with contract number HPMF CT 2005-007121.

Added in proof. Since the reference to [18] is only a reference to a lecture, I suggest that the reader compares any claims made in this paper about [18] with the results by C. Miller in [12] and that one uses the results of [12] for future reference. The author only recently learned that C. Miller in [12] has proved results which are very close to results of [18].

REFERENCES

[1] Ş. Basarab and F.-V. Kuhlmann, *An isomorphism theorem for Henselian algebraic extensions of valued fields*, ***Manuscripta Mathematica***, vol. 77 (1992), no. 2-3, pp. 113–126.

[2] R. Cluckers and L. Lipshitz, *Fields with Analytic Structure*, submitted, preprint available at `http://www.dma.ens.fr/~cluckers/`.

[3] R. Cluckers, L. Lipshitz, and Z. Robinson, *Analytic cell decomposition and analytic motivic integration*, ***Annales Scientifiques de l'École Normale Supérieure. Quatrième Série***, vol. 39 (2006), no. 4, pp. 535–568.

[4] R. Cluckers and F. Loeser, *b-minimality*, ***Journal of Mathematical Logic***, vol. 7 (2007), no. 2, pp. 195–227.

[5] ———, *Constructible motivic functions and motivic integration*, ***Inventiones Mathematicae***, vol. 173 (2008), no. 1, pp. 23–121.

[6] P. J. Cohen, *Decision procedures for real and p-adic fields*, ***Communications on Pure and Applied Mathematics***, vol. 22 (1969), pp. 131–151.

[7] J. Denef, *On the evaluation of certain p-adic integrals*, ***Séminaire de Théorie des Nombres, Paris 1983–84***, Progress in Mathematics, vol. 59, Birkhäuser, Boston, MA, 1985, pp. 25–47.

[8] ———, *p-adic semi-algebraic sets and cell decomposition*, ***Journal für die Reine und Angewandte Mathematik***, vol. 369 (1986), pp. 154–166.

[9] D. Haskell and D. Macpherson, *Cell decompositions of C-minimal structures*, ***Annals of Pure and Applied Logic***, vol. 66 (1994), no. 2, pp. 113–162.

[10] ———, *A version of o-minimality for the p-adics*, ***The Journal of Symbolic Logic***, vol. 62 (1997), no. 4, pp. 1075–1092.

[11] E. Hrushovski and D. Kazhdan, *Integration in valued fields*, ***Algebraic Geometry and Number Theory***, Progress in Mathematics, vol. 253, Birkhäuser Boston, Boston, MA, 2006, pp. 261–405.

[12] C. Miller, *Tameness in expansions of the real field*, ***Logic Colloquium '01***, Lecture Notes in Logic, vol. 20, ASL, Urbana, IL, 2005, pp. 281–316.

[13] J. Pas, *Uniform p-adic cell decomposition and local zeta functions*, ***Journal für die Reine und Angewandte Mathematik***, vol. 399 (1989), pp. 137–172.

[14] ———, *Cell decomposition and local zeta functions in a tower of unramified extensions of a p-adic field*, ***Proceedings of the London Mathematical Society. Third Series***, vol. 60 (1990), no. 1, pp. 37–67.

[15] T. Scanlon, *Quantifier elimination for the relative Frobenius*, ***Valuation Theory and Its Applications, Vol. II (Saskatoon, SK, 1999)***, Fields Institute Communications, vol. 33, AMS, Providence, RI, 2003, pp. 323–352.

[16] L. van den Dries, *The field of reals with a predicate for the powers of two*, ***Manuscripta Mathematica***, vol. 54 (1985), no. 1-2, pp. 187–195.

[17] ———, ***Tame Topology and o-Minimal Structures***, London Mathematical Society Lecture Note Series, vol. 248, Cambridge University Press, Cambridge, 1998.

[18] A. J. Wilkie, *Adding a multiplicative group to a polynomially bounded structure*, (October 13, 2006), Lecture at the ENS, Paris.

KATHOLIEKE UNIVERSITEIT LEUVEN
DEPARTEMENT WISKUNDE, CELESTIJNENLAAN 200B
B-3001 LEUVEN, BELGIUM

Current address: École Normale Supérieure, Département de mathématiques et applications, 45 rue d'Ulm, 75230 Paris Cedex 05, France

E-mail: cluckers@ens.fr

URL: `www.wis.kuleuven.be/algebra/Raf/`

THE SIXTH LECTURE ON ALGORITHMIC RANDOMNESS

ROD DOWNEY

Abstract. This paper follows on from the author's *Five Lectures on Algorithmic Randomness*. It is concerned with material not found in that long paper, concentrating on Martin-Löf lowness and triviality. We present a hopefully user-friendly account of the decanter method, and discuss recent results of the author with Peter Cholak and Noam Greenberg concerning the class of strongly jump traceable reals introduced by Figueira, Nies and Stephan.

§1. Introduction. This paper is a follow-up to the author's paper *Five Lectures on Algorithmic Randomness*, Downey [6], and covers material not covered in those lectures. In particular, I plan to look at lowness which was the basis of one of my lectures in Nijmegen, and to try to make accessible the *decanter* method whose roots come from the paper Downey, Hirschfeldt, Nies and Stephan [10], and whose full development was in Nies [23]. The class of K-trivial reals has turned out to be a remarkable "natural" class with really fascinating properties and connections with other areas of mathematics. For instance, K-trivial reals allow us to solve Post's problem "naturally" (well, reasonably naturally), without the use of a priority argument, or indeed without the use of requirements. Also, Kučera and Slaman [17] have used this class to solve a longstanding question in computable model theory, namely given a noncomputable Y in a Scott set S, there exists in S an element X Turing incomparable with Y.

As well I will report on recent work of the author with Peter Cholak and Noam Greenberg [4] on the computably enumerable *strongly jump traceable* reals which form a *proper* subclass of the K-trivials.

§2. Notation. I will keep the notation identical to that of Downey [6] and Downey and Hirschfeldt [9]. As a brief reminder, I will use C for plain complexity and K for prefix free complexity. We will be working in Cantor space 2^ω with uniform measure $\mu([\sigma]) = 2^{-|\sigma|}$, where the subbasic clopen sets

Research supported by the Marsden Fund of New Zealand. Special thanks to Noam Greenberg for providing extensive corrections.

Logic Colloquium '06
Edited by Barry Cooper, Herman Geuvers, Anand Pillay, and Jouko Väänänen
Lecture Notes in Logic, 32

are $[\sigma] =_{\text{def}} \{\sigma\alpha : \alpha \in 2^\omega \wedge \sigma \in 2^{<\omega}\}$. Elements of 2^ω will be referred to as *reals*. For any other notations and the like, we refer the reader to either [6] or [9].

§3. **Plain complexity characterizations of computable sets.** The first result establishing the fact that the complexity of initial segment of a real can have significant impact on its algorithmic complexity is due to Loveland. It concerned conditional complexity. It states that if all initial segments of a real are as compressible as they can be then the real must be computable. Notice that if α is a computable real then since there is a computable function g such that $g(n) = \alpha \upharpoonright n$ for all n, $C(\alpha \upharpoonright n|n)$ is a constant: we read n then apply g.

THEOREM 3.1 (Loveland [20]). *Suppose that there is e such that for all x, $C(\alpha \upharpoonright x|x) \leq e$. Then α is computable. Moreover for each e there are only finitely many α with $C(\alpha \upharpoonright x|x) \leq e$ for all x. (Indeed the same is true even if we replace "for all x" by "for all x in some infinite computable set.".)*

PROOF (SKETCH). The main idea in the proof below is that the possibilities for α reduce to a Π^0_1 class with a finite number of paths, and hence all such α all will be computable. If $C(\alpha \upharpoonright x|x) \leq e$ for all x, then there are at most $f = \mathcal{O}(2^e)$ many programs of size e or less. There is a maximum collection of such programs $p_1, \dots, p_m$ which are hit infinitely often.

Working above some length (taken to be 0 for simplicity), inductively form the Π^0_1 class of strings σ as follows. For the first step, find a set of m strings $\sigma_1, \dots, \sigma_m$ of the same length ℓ such that all of them appear to have $C(\sigma_i \upharpoonright p|p) \leq e$ for all $p \leq \ell$, and $C(\sigma_i|\ell) < e$ via p_i. For the second step, we simply use extensions of these σ_i in the same way. Then this will generate a Π^0_1 class of width at most m which must contain α. Therefore α is computable. ⊣

The same proof will also work for any usual complexity measure such as K or KM or Km etc. in place of C *for the conditional case.*

A mild generalization of Loveland's Theorem is the result of Chaitin, which gives another information-theoretical characterization of computable sets.

THEOREM 3.2 (Chaitin [3]). *Suppose that $C(\alpha \upharpoonright n) \leq C(n) + \mathcal{O}(1)$ for all n (or for an infinite computable set of n), or $C(\alpha \upharpoonright n) \leq \log n + \mathcal{O}(1)$, for all n. Then α is computable (and conversely). Furthermore for a given constant $\mathcal{O}(1) = d$, there are only finitely many $(\mathcal{O}(2^d))$ such α.*

In the same way that Loveland's Theorem needed a basic finiteness condition to work, the same holds for Chaitin's theorem. The following is the main lemma. Let $D : \Sigma^* \mapsto \Sigma^*$ be partial computable. Then we define *D-description* of σ to be a pre-image of σ.

LEMMA 3.3 (Chaitin [3]). *For all d and D, there is a number $f(d)$ such that for each $\sigma \in \Sigma^*$,*

$$|\{q : D(q) = \sigma \wedge |q| \leq C(\sigma) + d\}| \leq f_D(d).$$

That is, the number of D-descriptions of length $\leq C(\sigma) + d$, is bounded by an absolute constant depending upon d, D alone (and not on σ)

PROOF. Let σ be given, and $k = C(\sigma) + d$. For each m there are at most $2^{k-m} - 1$ strings with at least 2^m D-descriptions of length at most k, since there are $2^k - 1$ strings in total. Given k and m we can effectively list strings σ with at least 2^m D-descriptions of length below k, uniformly in k, m. (Wait till you see 2^m q's of length $\leq k$ with $D(q) = v$ and and then put v on the list $L_{k,m}$.) The list $L_{k,m}$ has length $\leq 2^{k-m}$.

If σ has at least 2^m D-descriptions of length at most k, then σ can be specified by

- m,
- a string q of length 2^{k-m},

the latter indicating the position of σ in $L_{k,m}$. This description has length bounded by $2\log m + k - m + c$ where c depends only upon D. If we choose m large enough so that $2\log m + k - m + c < k - d$, we can then get a description of σ of length $< k - d = C(\sigma)$. If we let $f(d)$ be 2^n where n is the least m with $2\log m + c + d < m$ then we are done. ⊣

In the next lemma, we will apply Lemma 3.3 with D being the universal prefix-free machine. The next lemma tells us that there are relatively few string with short descriptions, and the number depends on d alone.

LEMMA 3.4 (Chaitin [3]). *There is a computable h depending only on d* ($h(d) = \mathcal{O}(2^d)$) *such that, for all n,*

$$|\{\sigma : |\sigma| = n \wedge C(\sigma) \leq C(n) + d\}| \leq h(d).$$

PROOF. Consider the partial computable function D defined via $D(p)$ is $1^{|U(p)|}$. Then let $h(d) = f_D(d)$, with f given by the previous lemma. Suppose that $C(\sigma) \leq C(n) + d$, and pick the shortest p with $U(p) = \sigma$. Then p is a D-description of n and $|p| \leq C(n) + d$. Thus there at most $f(d)$ many $p's$, and hence σ's. ⊣

Now we can use the same Π^0_1 class argument to establish Chaitin's theorem, Theorem 3.2. We will have the width determined by Lemma 3.4, and hence if $C(\alpha \restriction n) \leq \log n + c$ for all n, then α will be computable. This is because there will always be a length between k and 2^k where $C(n) = \log n$.

The same argument will also show that for monotone complexity (or any other where C is a special case) if $Km(\alpha \restriction n) \leq \log n + c$ for all n, then α is computable.

Frank Stephan (see [9]) has shown that for left c.e. reals, α and β, if for all n,

$$C(\alpha \restriction n) \leq C(\beta \restriction n) + O(1),$$

then $\alpha \leq_T \beta$.

§4. K-trivials are Δ_2^0, and there are few of them. Chaitin's Theorem 3.2 shows that if, for all n, $C(\alpha \upharpoonright n) \leq C(n) + O(1)$, then α is computable. A compact way of expressing this is to use $\alpha \leq_Q \beta$, for a measure of complexity Q, iff $Q(\alpha \upharpoonright n) \leq Q(\beta \upharpoonright n) + O(1)$. Then Chaitin's Theorem says that $\alpha \leq_C 1^\omega$ iff α is computable. Chaitin asked if the same result held for K in place of C. We define the following for K, for other measures of relative complexity, there are similar notions (most of which coincide with being computable!).

DEFINITION 4.1 (Downey, Hirschfeldt, Nies and Stephan [10]). We will call a real α *K-trivial* if $\alpha \leq_K 1^\omega$.

The first limitation on the complexity of a K-trivial real was given by Chaitin. It relies on a analog of Lemma 3.4 above for prefix-free Kolmogorov complexity.

THEOREM 4.2 (Chaitin [3], Zambella [34]). *For any prefix-free machine D, there is a constant d such that for all c and all σ,*

$$|\{\nu : D(\nu) = \sigma \wedge |\nu| \leq K(\sigma) + c\}| \leq d2^c.$$

The proof of Theorem 4.2 uses the Coding Theorem (Levin [19]) (see Downey [6], Downey and Hirschfeldt [9]). Essentially, if there were too many elements in $\{\nu : D(\nu) = \sigma \wedge |\nu| \leq K(\sigma) + c\}$, then we could use the total measure they provide to make an even shorter description of σ, that being a contradiction. Here is our analog of Lemma 3.4.

THEOREM 4.3. *The set $KT(d, n) = \{\sigma : K(\sigma) < K(|\sigma|) + d\}$ has at most $O(2^d)$ many strings of length n.*

PROOF. We build a prefix-free machine V. Suppose that $U_s(\nu) = \sigma \wedge |\nu| \leq K_s(|\sigma|) + d$. Then define $V(\nu) = |\sigma|$. Then V is prefix-free as U is. Moreover, by Theorem 4.2,

$$|\{\nu : V(\nu) = |\sigma| \wedge |\nu| \leq K(|\sigma|) + d\}| \leq c2^d.$$

But then by construction, there is a constant p such that $KT(d, n)$ has at most $pc2^d$ many members. ⊣

Now we can run the proof that all C-trivial reals are computable using K in place of C. However, we do not know any value of K[1]. The key thing we use in the C case is that between k and 2^k there is a C-random length which will have complexity $\log p$. Using $\emptyset'$ as an oracle we *can* know K. The same argument will show the following.

THEOREM 4.4 (Chaitin [3]). *If α is K-trivial, then $\alpha \leq_T \emptyset'$.*

---

[1]It is also true that we don't know the value of $C(\sigma)$ for any *given* σ. *However,* we do know that there is some random length between k and 2^k for each k, and such a length m will have maximum complexity which in the C case *is* m. In the K case, we also know that there is some random length, but we don't know what its complexity is, as its complexity would be $m + K(m)$, *bounded* by $m + 2\log m$.

This proof has the following corollary, first observed by Zambella. To state Zambella's result, we let $KT(d)$ denote the class of reals which are K-trivial with constant d, meaning that for all n,

$$K(\alpha \upharpoonright n) \leq K(n) + d.$$

COROLLARY 4.5 (Zambella [34]). *The number of elements in $KT(d)$ is $\leq b2^d$ for some constant b independent of d.*

The reader should note that to be K-trivial, it is enough that the real be K-trivial on an infinite computable set. That is, the following piece of folklore is true.

PROPOSITION 4.6. *Suppose that we have a computable set $A = \{a_1, a_2, \dots\}$ listed in increasing order of magnitude, and for all i, $K(A \upharpoonright a_i) \leq K(a_i) + O(1)$. Then A is K-trivial.*

PROOF. Let $h(n) = a_n$, be computable. Then $K(n) = K(h(n)) + O(1) = K(a_n) + O(1)$. Notice that $K(A \upharpoonright n) \leq K(a_n) + O(1)$, since to compute $A \upharpoonright n$, take the program for $A \upharpoonright h(n)$, then reconstruct n from $h(n)$ and truncate $A \upharpoonright h(n)$ to get $A \upharpoonright n$. Then $K(A \upharpoonright n) \leq K(A \upharpoonright h(n)) + O(1) \leq K(h(n)) + O(1) \leq K(n) + O(1)$. ⊣

Zambella's Theorem, Theorem 4.5, leads one to speculate as to how many K-trivials there are. This has been investigated in unpublished work of Downey, Miller and Yu.

DEFINITION 4.7 (Downey, Miller, Yu [12]). Let

$$G(d) = |KT(d)|.$$

G seems a strangely complicated object. We calculate some arithmetical bounds on G. To do this we will need the following combinatorial result.

THEOREM 4.8 (First Counting Theorem, [12]). (i) *$lim_c \frac{G(c)}{2^c} = 0$.*
(ii) *Indeed,*

$$\sum_{c \in \mathbb{N}} \frac{G(c)}{2^c} \text{ is finite.}$$

PROOF. We define $G(c, n) = |KT(c, n)|$.

LEMMA 4.9. $\sum_{c \in \mathbb{N}} \frac{G(c)}{2^c} \leq \liminf_n \sum_c \frac{G(c,n)}{2^c}$.

The proof of Lemma 4.9 is almost immediate. Any finite partial sum on the left represents a finite number of K-trivials. For sufficiently large n each of these reals is isolated, so that the sum on the right exceeds that of the left.

LEMMA 4.10. *There is a finite q such that, for all n,*

$$\sum_{c \in \mathbb{N}} \frac{G(c, n)}{2^c} \leq q.$$

PROOF OF LEMMA 4.10. By definition of $G(\cdot,\cdot)$ we have that for some constant q, for all n,

$$\sum_{c\in\mathbb{N}} G(c,n)2^{-K(n)-c} \leq 2\sum_{\sigma\in 2^n} 2^{-K(\sigma)}.$$

The first inequality follows from definition of $G(\cdot,\cdot)$. On the other hand, K is a minimal universal computably enumerable semi-measure (by the Coding Theorem, see [6, 9]), and hence there is a constant q such that for every n,

$$2\sum_{\sigma\in 2^n} 2^{-K(\sigma)} \leq q2^{-K(n)},$$

since the quantity on the left is a computably enumerable semi-measure. ⊣

The proof is now finished by putting together Lemmas 4.9 and 4.10. ⊣

Before we turn to analyzing the Turing complexity of G, we point out that the result above is relatively sharp in in the following sense.

THEOREM 4.11 (Second Counting Theorem, [12]).

$$G(b) = \Omega\left(\frac{2^b}{b^2}\right).$$

PROOF. For any string σ, we know that

$$K(\sigma 0^r) \leq^+ K(\sigma) + K(0^r) \leq^+ K(\sigma) + K(0^{|\sigma|+r}).$$

Now if we choose σ of length below $b - 2\log b$, we know that $K(\sigma) \leq b$, and hence $K(\sigma 0^r) \leq K(|\sigma 0^r|) + b$. There are $\Omega(\frac{2^b}{b^2})$ such strings σ. ⊣

THEOREM 4.12 (Downey, Miller, Yu [12]). $G \not\leq_T \emptyset'$.

PROOF. [2] Assume that G is Δ^0_2, and hence by the Limit Lemma, $G(n) = \lim_s G(n,s)$, where this time $G(n,s)$ denotes a computable approximation to $G(n)$. Also we assume that we know k such that that $0^\omega \in KT(k)$.

Using Kraft-Chaitin, we build a machine M with coding constant d, which we know in advance. M looks for the first $c \geq k, 2^d$ such that

$$\frac{G(c)}{2^c} < 2^{-d}.$$

We can approximate the c in a Δ^0_2 manner, so that $c = \lim_s c_s$, where c_s is the stage s approximation of c.

The key idea behind the definition of M is that M wants to ensure that there are at least 2^{c-d} $KT(c)$ reals. If it does this, then we have a contradiction.

At stage s, M will pick (at most) 2^{c_s-d} strings of length s extending those of length $s-1$ already defined, and promise to give them each M-descriptions

[2] If the reader is unfamiliar with the use of Kraft-Chaitin to build prefix free Turing machines, then they might delay the reading of this proof until looking at the first construction of the next section.

of length $K(s)+c_s-d$ (hence they are $KT(c_s)$ *strings*). It clearly has enough room to do this, since

$$2^{c_s-d}2^{-(K(s)+c_s-d)} = 2^{-K(s)}.$$

At stage s, if c_s has a new value then all of M's old work is abandoned and M starts building 2^{c_s-d} $KT(c_s)$ reals which branch off of 0^ω starting at length s. If c_s has the same old value, then M continues building the 2^{c_s-d} $KT(c_s)$ reals. Eventually c_s stabilizes and M gets to build his 2^{c-d} $KT(c)$ reals, contradicting the definition of c. ⊣

We remark that a crude upper bound on G is that it is computable in $\mathbf{0}'''$. It is unknown if $G \leq_T \emptyset''$, or even if this question is machine dependent.

§5. The basic construction: limiting damage and quanta recycling. Solovay was the first to construct a (Δ^0_2) K-trivial real. This method was adapted by Zambella [34] and later Calude and Coles [2] to construct a c.e. K-trivial real. In [10], Downey, Hirschfeldt, Nies, and Stephan gave a new construction of a K-trivial real, and this time the real was strongly c.e., that is, the characteristic function of a c.e. set. (Independently, this had been found by Kummer in an unpublished manuscript.) As we will later see this is a priority-free and later requirement free solution to Post's problem.

THEOREM 5.1. (Downey, Hirschfeldt, Nies, and Stephan [10], Calude and Coles [2], after Solovay [30]) *There is a noncomputable c.e. set B such that $B \leq_K 1^\omega$.*

PROOF. The proof below is surely becoming pretty well known. We only give it for completeness. As we will later see, the only way to to construct a K-trivial real is to build a machine to demonstrate the fact that it is K-trivial. By that, if B is the relevant target set, we must show that $K(B \upharpoonright n) \leq K(n) + O(1)$. In most constructions, the $O(1)$ term is the overhead of the Recursion, or at least the s-m-n theorem.

We view this as a game: The opponent will give descriptions of n's in the construction. That is, at each stage s, he will say $K_s(n) = p$, say, by giving some string v of length p, (and $n \leq s$) *shorter than* $K_{s-1}(n)$ and have $U(v) = n[s]$. It is *our* job to build a machine M, by using Kraft-Chaitin, saying saying there is a description of $B \upharpoonright n$ of length $p+1$ (the "+1" option here being for technical reasons). We enumerate an axiom $\langle p, B \upharpoonright n\rangle$.

Notice that to be a Kraft-Chaitin set we only need that the sum of 2^{-p} (indeed $2^{-(p+1)}$) for such requests is below 1. In the case that B was, say, computable, no problem, for we are only making request of *exactly the same size* as the opponent has used. So our requests (to make M) are bounded by $\frac{1}{2}$ the amount he spends, namely $\frac{\Omega}{2}$.

Unfortunately, B being noncomputable, is not actually given to us. We will only know B_s, where $B = \cup_s B_s$; that is, approximations to the real B. Here is the main asymmetry:

We might describe $\langle p, B_s \upharpoonright n\rangle$ *but* it might be that to make B noncomputable, we will need $B_{s+1} \upharpoonright n \neq B_s \upharpoonright n$, since we enumerate n into $B_{s+1} - B_s$. *This will then entail issuing new descriptions for parts of B which have changed since the last time. In particular, we will need to describe* $B_{s+1} \upharpoonright m$ *for* $m \in [n, s]$.

Thus we have a *cost* of

$$c(n, s) = \sum_{n \leq m \leq s} 2^{-(K_s(m)+1)}.$$

This cost is *not* chargeable to the opponent, and hence we will have to make sure that this is limited. Thus, if we are meeting a requirement R_e saying $\overline{W_e} \neq B$, we would be prepared to do this if the cost was less than, say, $2^{-(e+1)}$.

Thus we can define

$$B_{s+1} = B_s \cup \Big\{ n : n \in W_{e,s} \wedge W_{e,s} \cap B_s = \emptyset \wedge \\ n \geq 2e \wedge \sum_{n \leq m \leq s} 2^{-K_s(m)} \leq 2^{-(e+1)} \Big\}.$$

In that case B will be noncomputable and we can make it K-trivial since the overall cost of the injuries is bounded by the amount we can charge against the opponent $\frac{\Omega}{2}$, plus the amount we are not compensated for, namely, $\sum_{e \in \omega} 2^{-(e+1)} = \frac{1}{2}$. The reader should realize that the $\overline{B} \neq W_e$ since $\lim_n \sup_s c(n, s) \to 0$. ⊣

Actually, this construction can be made to have *no* visible requirements: In particular, if t is any computable function which dominated the overheads of the Recursion Theorem, then if we define, akin to the Dekker deficiency set,

$$B_{t(s+1)} = B_{t(s)} \cup \{b_n^{t(s)}, \dots, b_{t(s)}^{t(s)}\},$$

if $K_{t(s+1)}(m) < K_{t(s)}(m)$ for all $n \leq m \leq t(s)$, where $\overline{B_s} = \{b_j^s : s \in \omega\}$, then B is automatically K-trivial and noncomputable. Space limitations mean that we will not discuss this variation here. hence, we refer the reader to Downey, Hirschfeldt, Nies, Terwijn [11], or Downey and Hirschfeldt [9] for more details.

§6. **Lowness.** A very similar construction to that of the last section is due to Kučera and Terwijn [18], involving *lowness*. Let R be the collection of random sets for some randomness concept. Then R^X is the class obtained for the same concept using X as an oracle.

Definition 6.1. We say that a set X is R-*low* if $R = R^X$. We say that X is low for R-tests, R_0-low if, for every R^X-test $\{U_n^X : n \in \mathbb{N}\}$, there is a R-test $\{V_n : n \in \mathbb{N}\}$ such that $\cap_n V_n \supseteq \cap_n U_n$.

For instance, a set X is Martin-Löf low if the collection of reals Martin-Löf random relative to X is the same as the collection of Martin-Löf random reals. Hence X is no help in making reals non-random. Notice also that if X is R_0-low then it is automatically R-low, but the converse is not clear. However, since there is a universal Martin-Löf test, the collections Martin-Löf low and Martin-Löf low for tests coincide[3].

Martin-Löf low sets were first studied by Kučera and Terwijn [18], answering a question of van Lambalgen [33].

THEOREM 6.2 (Kučera and Terwijn [18]). *There is a noncomputable c.e. set A that is Martin-Löf low.*

PROOF. We give an alternative proof to that in [18], taken from Downey [7]. It is clear that there is a primitive recursive function f, so that $\{U^A_{f(n)} : n \in \omega\}$ is the universal Martin-Löf test relative to A. Let I^A_n denote the corresponding Solovay test. Then X is A-random iff X is in at most finitely many I^A_n. We show how to build a $\{J_n : n \in \omega\}$, a Solovay test, so that for each $[p] \in I^A_n$ is also in J_n. This is done by simple copying: if $p < s$ (as a number) is in $\cup_{j \leq s} I^{A_s}_j$ is not in $J_i : i \in s$, add it. Clearly this "test" has the desired property of covering I^A_n. We need to make A so that the "mistakes" are not too big. This is done in the *same* way as the construction of a K-trivial.

The crucial concept comes from Kučera and Terwijn: Let $M_s(y)$ denote the collection of intervals $\{I^{A_s}_n : n \leq s\}$ which have $A_s(y) = 0$ in their use function. Then we put $y > 2e$ into $A_{s+1} - A_s$ provided that e is least with $A_s \cap W_{e,s} = \emptyset$, and

$$\mu(M_s(y)) < 2^{-e}.$$

It is easy to see that this can happen at most once for e and hence the measure of the total mistakes is bounded by $\Sigma 2^{-n}$ and hence the resulting test is a Solovay test. The only thing we need to prove is that A is noncomputable. This follows since, for each e, whenever we see the some y with $\mu(M_s(y)) \geq 2^{-e}$, such y will *not* be added and hence this amount of the A-Solovay test will be protected. But since the total measure is bounded by 1, this cannot happen forever. ⊣

A related concept is the following.

DEFINITION 6.3. We call a set X *low for K* if there is a constant d such that for all σ, $K^X(\sigma) \geq K(\sigma) - d$.

This notion is due to An. A. Muchnik who, in unpublished work, constructed such a real. It is evident that the same method of proof (keeping the measure of the injury of the uses down) will establish the following.

[3]Not treated in these notes are lowness notions for other kinds of randomness such as Schnorr and Kurtz randomness. To the author's knowledge there is no notion of randomness where the potentially two lowness notions have not turned out to coincide.

THEOREM 6.4 (Muchnik, unpubl.). *There is a noncomputable c.e. set X that is low for K.*

The reader cannot miss the similarities in the proofs of the existence of K-trivials and K-lows and even low for K reals. These are all notions of K-antirandomness, and should somehow be related.

We will soon see, in some deep work of Nies, that these classes all *coincide*!

6.1. K-trivials solve Post's problem. The basic method introduced in this section, the *quanta pushing* or more colorfully, the *decanter*[4] method, is the basis for almost all results on the K-antirandom reals[5]. In this section we will look only at the basic method to aid the reader with the more difficult nonuniform applications in subsequent sections. The following result proves that they are a more-or-less natural solution to Post's problem.

THEOREM 6.5 (Downey, Hirschfeldt, Nies, Stephan [10]). *If a real α is K-trivial then α is Turing incomplete.*

6.2. The Decanter Method. In this section we will motivate a very important technique for dealing with K-trivial reals now called the decanter technique, or the golden run machinery. It evolved from attempted proofs that there were Turing complete K-trivial reals, the blockage being turned around into a proof that no K-trivial real was Turing complete in the time-honoured symmetry of computability theory. Subsequently many of the artifacts of the original proof were removed and streamlined particularly by Nies and we have now what appears to be a generic technique for dealing with this area. The account below models the one from Downey, Hirschfeldt, Nies and Terwijn [11].

The following result shows that K-trivials solve Post's Problem. This will be greatly improved in the later sections.

The proof of Theorem 6.5 below runs the same way whether A is Δ^0_2 or computably enumerable. We only need the relevant approximation being $A = \cup_s A_s$ or $A = \lim_s A_s$.

6.3. The first approximation, wtt-incompleteness. The fundamental tool used in all of these proofs is what can be described as *amplification*. Suppose that A is K-trivial with constant of triviality b, and we are building a machine M whose coding constant within the universal machine U is known to be d.

Now the import of these constants is that if we describe n by some KC-axiom $\langle p, n\rangle$ meaning that we describe n by something of length p, and hence has *weight* 2^{-p}, then in U we describe n by something of length $p + d$ and hence the opponent at some stage s must eventually give a description of $A_s \upharpoonright n$ of length $p + b + d$. The reader should think of this as meaning that the opponent has to play *less quanta* than we do for the same effect.

[4]The *decanter* method was introduced in [10] and is a linear one. It is used to prove, for example, that K-trivials solve Post's problem. The later extension, discussed later, where trees of strategies are used is due to Nies [23] and is refereed to as the *golden run* method.

[5]At least in this author's opinion, and at the present time.

What has this to do with us? Suppose that we are trying to claim that A is *not* K-trivial. Then we want to *force* U to issue too many descriptions of A, by using up all of its quanta.

The first idea is to make the opponent play many *times on the same length and hence amount of quanta.*

The easiest illustration of this method is to show that no K-trivial is *wtt*-complete.

PROPOSITION 6.6. *Suppose that A is K-trivial then A is wtt-incomplete.*

PROOF. We assume that we are given $A = \lim_s A_s$, a computable approximation to A. Using the Recursion Theorem, we build a c.e. set B, and a prefix free machine M. We suppose that $\Gamma^A = B$ is a weak truth table reduction with computable use γ. Again by the Recursion Theorem, we can know Γ, γ and we can suppose that the coding constant is d and the constant of triviality is b as above.

Now, we pick $k = 2^{b+d+1}$ many followers $m_k < \cdots < m_1$ targeted for B and wait for a stage where $\ell(s) > m_1$, $\ell(s)$ denoting the length of agreement of $\Gamma^A = B[s]$.

At this stage we will *load* an M-description of some *fresh, unseen* $n > \gamma(m_1)$ (and hence bigger than $\gamma(m_i)$ for all i) of size 1, enumerating an axiom $\langle 1, n\rangle$. The translation of course is that at some stage s_0 we must get a description of $A_{s_0} \upharpoonright n$ in U of length $b + d$ or less. That is, at least $2^{-(b+d)}$ must enter the domain of U devoted to describing this part of A.

At the first such stage s_0, we can put m_1 into $B_{s_0+1} - B_{s_0}$ causing a change in $A \upharpoonright n - A_{s_0} \upharpoonright n$. (We remark that in this case we could use any of the m_i's but later it will be important that the m_i's enter in reverse order.) Then there must be at some stage $s_1 > s_0$, with $\ell(s_1) > m_1$, a new $A_{s_1} \upharpoonright n \neq A_{s_0} \upharpoonright n$ *also* described by something of length $b + d$. Thus U must have at least $2^{-(b+d)}$ more in its domain. If we repeat this process one time for each m_i then eventually U runs out of quanta since $2^{-(b+d)}k > 1$. ⊣

6.4. The second approximation: impossible constants. The argument above is fine for weak truth table reducibility. There are clearly problems in the case that Γ is a *Turing* reduction, for example.

That is, suppose that our new goal is to show that no K-trivial is Turing complete. The problem with the above construction is the following.

When we play the M-description of n, we have used all of our quanta available for M to describe a single number. *Now it is in the opponents power to move the use $\gamma(m_1, s)$ (or $\gamma(m_k, s)$ even) to some value bigger than n* before *even it decides to match our description of n.* Thus it costs him very little to match our M-description of n:

He moves $\gamma(m_k, s)$ then describes $A_s \upharpoonright n$ and we can no longer cause *any* changes of A below n, as all the Γ-uses are too big.

It is at this point that we realize it is pretty dumb of us to try to describe n in one hit. All that really matters is that we load lots of quanta beyond some point were it is measured many times. For instance, in the *wtt* case, we certainly could have used many n's beyond $\gamma(m_1)$ loading each with, say, 2^{-e} for some small e, and only attacking once we have amassed the requisite amount beyond $\gamma(m_1)$.

This is the idea behind our second step.

Impossible assumption: We will assume that we are given a Turing reduction $\Gamma^A = B$ and the overheads of the coding and Recursion Theorem result in a constant of 0 for the coding, and the constant of triviality is 0.

Hence we have $\Gamma^A = B[s]$ in some stage by stage manner, and moreover when we enumerate $\langle q, n\rangle$ into M then the opponent will eventually enumerate something of length q into U describing $A_s \upharpoonright n$. Notice that with these assumptions, in the wtt case we'd only need one follower m. Namely in the wtt case, we could load (e.g.) $\frac{7}{8}$ onto n, beyond $\gamma(m)$ and then put m into B causing the domain of U to need $\frac{7}{4}$ since we count $A_s \upharpoonright n$ for two different $A_s \upharpoonright n$-configurations.

In the case that γ is a Turing reduction, the key thing to note is that we still have the problem outlined above. Namely if we use the dumb strategy, then the opponent will change $A_s \upharpoonright \gamma(m, s)$ moving some γ-use *before* he describes $A_s \upharpoonright n$. Thus he only needs to describe $A_s \upharpoonright n$ *once*.

Here is where we use the drip feed strategy for loading. What is happening is that we really have called a procedure $P(\frac{7}{8})$ asking us to load $\frac{7}{8}$ beyond $\gamma(m)$ and then use m to count it twice. It might be that whilst we are trying to load some quanta, the "change of use" problem might happen, a certain amount of "trash"will occur. This trash corresponds to axioms enumerated into M that do not cause the appropriate number of short descriptions to appear in U. We will need to show that this trash is small enough that it will not cause us problems.

Specifically, we would use a procedure $P(\frac{7}{8}, \frac{1}{8})$ asking for twice counted quanta (we call this a 2-*set*) of size $\frac{7}{8}$ but only having trash bounded by $\frac{1}{8}$.

Now, $\frac{1}{8} = \sum_j 2^{-(j+4)}$. Initially we might try loading quanta beyond the current use $\gamma(m, s_0)$ in lots of 2^{-4}. If we are successful in reaching our target of $\frac{7}{8}$ before A changes, then we are in the *wtt*-case and can simply change B to get the quanta counted twice.

Now suppose we load the quantum 2^{-4} on some $n_0 > \gamma(m, s_0)$. The opponent might, at this very stage, move $\gamma(m, s)$ to some new $\gamma(m, s_1) > n_0$, at essentially no cost to him. We would have played 2^{-4} for no gain, and would throw the 2^{-4} into the trash. Now we would begin to try to load anew $\frac{7}{8}$ beyond $\gamma(m, s_1)$ *but this time we would use chunks of size* 2^{-5}. Again if he moved immediately, then we would trash that quanta and next time use 2^{-6}.

Notice that if we assume that $\Gamma^A = B$ this movement can't happen forever, lest $\gamma(m, s) \to \infty$.

On the other hand, in the first instance, perhaps we loaded 2^{-4} beyond $\gamma(m, s_0)$ and he did not move $\gamma(m, s_0)$ at that stage, but simply described $A \upharpoonright n_0$ by some description of size 4. At the next step, we would pick another n beyond $\gamma_(m, s_0) = \gamma(m, s_1)$ and try again to load 2^{-4}. If the opponent *now* changes, then we lose the *second* 2^{-4} but he *must* count the *first* one (on n_0) twice. That is, whenever he actually does not move $\gamma(m, s)$ then he must match our description of the current n, and this *will* later be counted *twice* since either *he* moves $\gamma(m, s)$ over it (causing it to be counted twice) *or* we put m into B making $\gamma(m, s)$ change.

Thus, for this simplified construction, each time we try to load, he either matches us (in which case the amount will contribute to the 2-set, and we can return $2^{-\text{current } \beta}$ where β is the current number being used for the loading to the target) *or* we lose β, but gain in that $\gamma(m, s)$ moves again, and we put β in the trash, but make the next $\beta = \frac{\beta}{2}$.

If $\Gamma^A = B$ then at some stage $\gamma(m, s)$ must stop moving and we will succeed in loading our target $\alpha = \frac{7}{8}$ into the 2-set. Our cost will be bounded above by $\frac{7}{8} + \frac{1}{8} = 1$.

6.5. The less impossible case. Now we will remove the simplifying assumptions. The key idea from the *wtt*-case where the use is fixed but the coding constants are nontrivial, is that we must make the changes beyond $\gamma(m_k)$ a k-set. Our idea is to combine the two methods to achieve this goal. For simplicity, suppose we pretend that the constant of triviality is 0, but now the coding constant is 1. Thus when we play 2^{-q} to describe some n, the opponent will only use $2^{-(q+1)}$.

Emulating the wtt-case, we would be working with $k = 2^{1+1} = 4$ and would try to construct a 4-set of changes. What we will do is break the task into the construction of a 2-set of a certain weight, a 3-set and a 4-set of a related weight in a coherent way.

We view these as procedures P_j for $2 \le j \le 4$ which are called in in reverse order in the following manner.

Our overall goal begins with, say $P_4(\frac{7}{8}, \frac{1}{8})$ asking us to load $\frac{7}{8}$ beyond $\gamma(m_4, s_0)$ initially in chunks of $\frac{1}{8}$, this being a 4-set.

To do this we will invoke the lower procedures. The procedure P_j $(2 \le j \le 4)$ enumerates a j-set C_j. The construction begins by calling P_4, which calls P_3 several times, and so on down to P_2, which enumerates the 2-set C_2 and KC set L of axioms $\langle q, n\rangle$.

Each procedure P_j has rational parameters $q, \beta \in [0, 1]$. The *goal* q is the weight it wants C_j to reach, and the *garbage quota* β is how much it is allowed to waste.

In the simplified construction, where there was only one m, the goal was $\frac{7}{8}$ and the β evolved with time. The same thing happens here. P_4's goal *never* changes, and hence can never be met lest U use too much quanta. Thus A cannot compute B.

The main idea is that procedures P_j will ask that procedures P_i for $i < j$ do the work for them, with eventually P_2 "really" doing the work, but the goals of the P_i are determined inductively by the garbage quotas of the P_j above. Then if the procedures are canceled before completing their tasks then thc amount of quanta wasted is acceptably small.

We begin the construction by starting $P_4(\frac{7}{8}, \frac{1}{8})$. Its action will be to

1. Choose m_4 large.
2. Wait until $\Gamma^A(m_4)\downarrow$.

When this happens, P_4 will call $P_3(2^{-4}, 2^{-5})$. Note that here the idea is that P_4 is asking P_3 to enumerate the 2^{-4}'s which are the current quanta bits that P_4 would like to load beyond m_4's current Γ-use. (The actual numbers being used here are immaterial except that we need to make them converge, so that the total garbage will be bounded above.

Now if, while we are waiting, the Γ-use of m_4 changes, then we will go back to the beginning. But let's consider what happens on the assumption that this has not yet occurred.

What will happen is that $P_3(2^{-4}, 2^{-5})$ will pick some m_3 large, wait for $\Gamma(m_3)$ convergence, and then it will invoke $P_2(2^{-5}, 2^{-6})$, say. This will pick its own number m_2 again large, wait for $\Gamma^A(m_2)\downarrow$ and finally now we will get to enumerate something into L. Thus, at this very stage we would try to load 2^{-5} beyond $\gamma(m_2, s)$ in chunks of 2^{-6}.

Now whilst we are doing this, many things can happen. The simplest case is that nothing happens to the uses, and hence, as with the wtt case, we would successfully load this amount beyond $\gamma(m_2, s)$. Should we do this then we can enumerate m_2 into B and hence cause this amount to be a 2-set C_2 of weight 2^{-5} and we have reached our target.

This would return to P_3 which would realize that it now has 2^{-5} loaded beyond $\gamma(m_3, s)$, and it would like another such 2^{-5}. Thus it would again invoke $P_2(2^{-5}, 2^{-6})$. If it did this successfully, then we would have seen a 2-set of size 2^{-5} loaded beyond $\gamma(m_3, s)$ (which is unchanged) and hence if we enumerate m_3 into B we could make this a 3-set of size 2^{-5}, which would help P_4 towards its goals.

Then of course P_4 would need to invoke P_3 again and then down to P_2. The reader should think of this as "wheels within wheels within wheels" spinning ever faster.

Of course, the problems all come about because uses can change. The impossible case gave us a technique to deal with that. For example, if only the outer layer P_2 has its use $\gamma(m_2, s)$ change, then as we have seen, the amount already matched would still be a 2-set, but the latest attempt would be wasted. We would reset its garbage quota to be half of what it was, and then repeat. Then we could rely on the fact that (assuming that all the other procedures have m_i's with stable uses) $\lim_s \gamma(m_2, s) = \gamma(m_2)$ exists, eventually we get to build the 2-set of the desired target with acceptable garbage, build ever more slowly with ever lower quanta.

In general, the inductive procedures work the same way. Whilst waiting, if uses change, then we will initialize the lower procedures, reset their garbages to be ever smaller, but not throw away any work that has been successfully completed. Then in the end we can argue by induction that all tasks are completed.

Proof of Theorem 6.5. We are now ready to describe the construction. Let $k = 2^{b+d+1}$ The method below is basically the same for all the constructions with one difference as we later see.

As in the wtt case, our construction will build a k-set C_k of weight $> 1/2$ to reach a contradiction.

The procedure P_j $(2 \leq j \leq k)$ enumerates a j-set C_j. The construction begins by calling P_k, which calls P_{k-1} several times, and so on down to P_2, which enumerates L (and C_2).

Each procedure P_j has rational parameters $q, \beta \in [0, 1]$. The *goal* q is the weight it wants C_j to reach, and the *garbage quota* β is how much it is allowed to waste.

We now describe the procedure $P_j(q, \beta)$, where $1 < j \leq k$, and the parameters $q = 2^{-x}$ and $\beta = 2^{-y}$ are such that $x \leq y$.

1. Choose m large.
2. Wait until $\Gamma^A(m)\downarrow$.
3. Let $v \geq 1$ be the number of times P_j has gone through step 2.
   $j = 2$: Pick a large number n. Put $\langle r_n, n\rangle$ into L, where $2^{-r_n} = 2^{-v}\beta$. Wait for a *stage* t such that $K_t(n) \leq r_n + d$, and put n into C_1. (If M_d is a prefix-free machine corresponding to L, then t exists.)
   $j > 2$: Call $P_{j-1}(2^{-v}\beta, \beta')$, where $\beta' = \beta 2^{j-k-w-1}$, and w is the number of P_{j-1} procedures started so far.
   In any case, if weight$(C_{j-1}) < q$ then repeat step 3, and otherwise return.
4. Put m into B. This forces A to change below $\gamma(m) < \min(C_{j-1})$, and hence makes C_{j-1} a j-set (if we assume inductively that C_{j-1} is a $(j-1)$-set). So put C_{j-1} into C_j, and declare $C_{j-1} = \emptyset$.

If $\gamma^A(m)$ changes during the execution of the loop at step 3, then cancel the run of all subprocedures, and go to step 2. Despite the cancellations, C_{j-1} is now a j-set because of this very change. (This is an important point, as it ensures that the measure associated with numbers already in C_{j-1} is not wasted.) So put C_{j-1} into C_j, and declare $C_{j-1} = \emptyset$.

This completes the description of the procedures. The construction consists of calling $P_k(\frac{7}{8}, \frac{1}{8})$ (say). It is easy to argue that since quotas are inductively halved each time they are injured by a use change, they are bounded by, say $\frac{1}{4}$. Thus L is a KC set. Furthermore C_k is a k-set, and this is a contradiction since then the total quanta put into U exceeds 1. ⊣

The following elegant description of Nies' is taken from [11]:

We can visualize this construction by thinking of a machine similar to Lerman's pinball machine (see [29, Chapter VIII.5]). However, since we enumerate rational quantities instead of single objects, we replace the balls in Lerman's machine by amounts of a precious liquid, say 1955 Biondi-Santi Brunello wine. Our machine consists of decanters $C_k, C_{k-1}, \ldots, C_0$. At any stage C_j is a j-set. We put C_{j-1} above C_j so that C_{j-1} can be emptied into C_j. The height of a decanter is changeable. The procedure $P_j(q, \beta)$ wants to add weight q to C_j, by filling C_{j-1} up to q and then emptying it into C_j. The emptying corresponds to adding one more A-change.

The emptying device is a hook (the $\gamma^A(m)$-marker), which besides being used on purpose may go off finitely often by itself. When C_{j-1} is emptied into C_j then $C_{j-2}, \ldots, C_0$ are spilled on the floor, since the new hooks emptying $C_{j-1}, \ldots, C_0$ may be much longer (the $\gamma^A(m)$-marker may move to a much bigger position), and so we cannot use them any more to empty those decanters in their old positions.

We first pour wine into the highest decanter C_0, representing the left domain of L, in portions corresponding to the weight of requests entering L. We want to ensure that at least half the wine we put into C_0 reaches C_k. Recall that the parameter β is the amount of garbage $P_j(q, \beta)$ allows. If v is the number of times the emptying device has gone off by itself, then P_j lets P_{j-1} fill C_{j-1} in portions of size $2^{-v}\beta$. Then when C_{j-1} is emptied into C_j, at most $2^{-v}\beta$ much liquid can be lost because of being in higher decanters $C_{j-2}, \ldots, C_0$. The procedure $P_2(q, \beta)$ is special but limits the garbage in the same way: it puts requests $\langle r_n, n\rangle$ into L where $2^{-r_n} = 2^{-v}\beta$. Once it sees the corresponding $A \upharpoonright n$ description, it empties C_0 into C_1 (but C_0 may be spilled on the floor before that because of a lower decanter being emptied).

6.6. K-trivials form a robust class. It turns out that the K-trivials are a remarkably robust class, and coincide with a host of reals defined in other ways. This coincidence also has significant degree-theoretical implications. For example, as we see, not only are the K-trivials Turing incomplete, but are closed downwards under Turing reducibility and form a natural Σ^0_3 ideal in the Turing degrees.

What we need is another view of the decanter method. In the previous section it was shown that K-trivials solve Post's Problem. Suppose however, we actually applied the method above to a K-trivial and a partial functional Γ. Then what would happen would be that for some i the procedure P_i *would not return.* This idea forms the basis for most applications of the decanter method, and the run that does not return would be called a *golden run.* This idea is from Nies [23]

For instance, suppose that we wanted to show Nies' result that all K-trivials are superlow.

Let A be K-trivial. Our task is to build a functional $\Gamma^K(e)$ computing whether $\Phi_e^A(e)\downarrow$. For ease of notation, let us denote $J^A(e)$ to be the partial function that computes $\Phi_e^A(e)$. Now the obvious approach to this task is to monitor $J^A(e)[s]$. Surely if $J^A(e)[s]$ never halts then we will never believe that $J^A(e)\downarrow$. However, we are in a more dangerous situation when we see some stage s where $J^A(e)[s]\downarrow$. If we define $\Gamma^K(e) = J(e)[s]$ then we risk the possibility that this situation could repeat itself many times since it is in the opponent's power the changes $A_s \upharpoonright \varphi_e(e, s)$.

Now if *we* were building A, then we would know what to do. We should *restrain* A in a familiar way and hence with finite injury A is low. However, the *opponent* is building A, and *all we know is that A is K-trivial.*

The main idea is to load up quanta beyond the use of the e-computation, before we change the value of $\Gamma^A(e)[s]$, that is changing our belief from divergence to convergence. Then, if A were to change on that use after we had successfully loaded, it would negate our belief and causing us to reset $\Gamma^K(e)$. Our plan is to make sure that *it would cost the opponent, dearly.*

As with the proof above, this cannot happen too many times for any particular argument, and in the construction to be described, there will be a golden run which does not return. The interpretation of this non-returning is that the $\Gamma_R^K(e)$ *that this run R builds* will actually work. Thus, the construction of the lowness index is *non-uniform.*

Thus, the idea would be to have *tree* of possibilities. The height of the tree is $k = 2^{b+d+1}$, and the tree is ω branching. A node $\sigma\widehat{\ }i$ denotes the action to be performed for $J^A(i)$ more or less assuming that now σ is the highest priority node that does not return.

At the top level we will be working at a procedure $P_k(\frac{7}{8}, \frac{1}{8})$ yet again, and we know in advance that this won't return with a k-set of that size.

What we do is distribute the tasks out to the successors of λ, the empty node. Thus outcome e would be devoted to solving $J^A(e)$ via a k-set. It will be given quanta, say, $2^{-(e+1)}\alpha_k$ where $\alpha_k = \frac{7}{8}$. (Here this choice is arbitrary, save that it is suitably convergent. For instance, we could use the series $\sum_{n\in\mathbb{N}} \frac{1}{n^2}$ which would sharpen the norms of the wtt-reductions to n^2.) To achieve its goals, when it sees some apparent $J^(A)(e)\downarrow[s]$, It will invoke $P_k(2^{-(e+1)}\alpha_k, 2^{-e}\beta_k)$ where $\beta_k = \frac{1}{8}$. We denote this procedure by $P_k(e, 2^{-(e+1)}\alpha_k, 2^{-(e+1)}\beta_k)$. This procedure will look for a k-set of the appropriate size, which, when it achieves its goal, the version of Γ at the empty string says it believes.

Again, notice that if this P_k returns (and this is the idea below) 2^{e+2} many times then $P_k(\alpha_k, \beta_k)$ would return, which is impossible.

As with the case of Theorem 6.5, to achieve its goals, *before it believes that* $J^A(e)\downarrow [s]$, it needs to get its quanta by invoking the team via nodes below e. These are, of course, of the form $e\widehat{\ }j$ for $j \in \omega$. They will be asked to try to achieve a $k-1$ set and $P_k(e, 2^{-(e+1)}\alpha_k, 2^{-(e+1)}\beta_k)$ by using $P_{k-1}(e\widehat{\ }j, 2^{(e+1)}2^{-(j+1)}\beta_k, 2^{-(j+1)}2^{-(e+1)}\beta_{k-1})$, with $\beta_{k-1} \ll \beta_k$ and j chosen appropriately. Namely we will choose those j, say, with $j > \varphi_e^A(e)[s]$. That is, these j will, by convention, have their uses beyond $\varphi_e^A(e)[s]$ and hence will be working similarly to the "next" m_i "down" in the method of the incompleteness proof of Theorem 6.5. Of course such procedures would await $\varphi_e^A(j)[s]$ and try to load quanta in the form of a $k-1$ set beyond the $\varphi_j^A(j,s)$ ($> \varphi_e^A(e)$), the relevant j-use.

The argument procedures working in parallel work their way down the tree. As above when procedure $\sigma\widehat{\ }i$ is injured because the $J^A(t)[s]$ is unchanged for all the uses of $t \in \sigma$, yet $J^A(i)[s]$ changes before the procedure returns then we reset all the garbage quotas in a systematic way, so as to make the garbage quota be bounded.

Now the argument is the same. There is some m least σ of length m, and some final α, β for which $P_m(e, \alpha, \beta)$ is invoked and never returns. Then the procedure built at σ will be correct on all $j > \varphi_e^A(e)$. Moreover, we can always calculate how many times it would be that some called procedure would be invoked to fulfill $P_m(\alpha, \beta)$. Thus the can bound the number of times that we would change our mind on $\Gamma^K_{P_m(\alpha,\beta)}(i)$ for any argument i. That is, A is superlow. In fact, as Nies pointed out, this gives a little more. Recall that an *order* is a computable nondecreasing function with infinite limit.

Definition 6.7 (Nies [24, 23]). We say that a set B is jump traceable iff there is a computable order h and a weak array (or not necessarily disjoint c.e. sets) $\{W_{g(j)} : j \in \mathbb{N}\}$, such that $|W_{g(j)}| \leq h(j)$, and $J^B(e) \in W_{g(e)}$.

Theorem 6.8 (Nies [24, 23]). *Suppose that A is K-trivial. Then A is jump traceable.*

The proof is to observe that we are actually constructing a trace. A mild variation of the proof above also shows the following.

Theorem 6.9 (Nies [24]). *Suppose that A is K-trivial. Then there exists a K-trivial computably enumerable B with $A \leq_{tt} B$.*

Proof (Sketch). Again the golden run proof is more or less the same, our task being to build B. This is farmed out to outcomes e in the ω-branching tree, where we try to build $\Gamma^B(e) = A(e)$. Again at level j, the size of the use will be determined by the number of times the module can act before it returns enough quanta to give the node above the necessary $j-1$ set. This is a computable calculation. Now when the opponent seeks to load quanta beyond $\gamma^j(e,s)$ before we believe this, we will load matching quanta beyond e for A. The details are then more or less the same. ⊣

Other similar arguments show that K-triviality is basically a *computably enumerable* phenomenon[6]. That is, the following is true.

THEOREM 6.10 (Nies [23]). *The following are equivalent*

(i) *A is K-trivial.*

(ii) *A has a Δ_2^0 approximation $A = \lim_s A_s$ which reflects the cost function construction. That is,*

$$\left\{ \sum_{x \leq y \leq s} \frac{1}{2} c(y, s) : x \text{ minimal with } A_s(x) \neq A_{s-1}(x) \right\} < \frac{1}{2}.$$

6.7. More characterizations of the K-trivials. We have seen that the K-trivials are all jump traceable. In this section, we sketch the proofs that the class is characterized by other "antirandomness" properties.

THEOREM 6.11 (Nies and Hirschfeldt [23]). *Suppose that A is K-trivial. Then A is low for K.*

COROLLARY 6.12. *The following are equivalent*:

(i) *A is K-trivial.*

(ii) *A is low for K.*

PROOF. The corollary is immediate by the implication (ii) $\rightarrow$ (i). We prove Theorem 6.11. Again this is another golden run construction. This proof proceeds in a similar way to that showing that K-trivials are low, except that $P_{j,\tau}$ calls procedures $P_{j-1,\sigma}$ based on computations $U^A(\sigma) = y[s]$ (since we now want to enumerate requests $\langle |\sigma| + d, y \rangle$), and the marker $\gamma(m, s)$ is replaced by the use of this computation. That is, we wish to believe a computation, $U^A(\sigma) = y[s]$ and to do so we want to load quanta beyond the use $u(\sigma, s)$. This is done more or less exactly the same way, beginning at P_k and descending down the nodes of the tree. Each node v will this time build a machine $\hat{U}_v$, which will copy v-believed computations; namely those for which we have successfully loaded the requisite $|v|$-set. We need to argue that for the golden v, the machine is real. The garbage is bounded by, say, $\frac{1}{8}$ by the way we reset it. The machine otherwise is bounded by U itself. ⊣

It is also possible to use similar methods to establish the following.

THEOREM 6.13 (Nies [23]). *A is K-trivial iff A is low for Martin-Löf randomness.*

COROLLARY 6.14 (Nies [23]). *The K-trivials are closed downward under $\leq_T$.*

From Downey [6], we know that real addition $+$ induces a join on the Solovay (and hence K- and C-) degrees of left c.e. reals. It is not hard to show that the following also holds.

[6]To the author's knowledge, K-triviality is the only example of a fact in computability theory that relies purely on enumerations. It would appear, for instance, that forcing the existence of a K-trivial is impossible.

THEOREM 6.15 (Downey, Hirschfeldt, Nies and Stephan [10]). *Suppose that α and β are K-trivial. Then so is $\alpha+\beta$, and hence $\alpha\oplus\beta$.*

PROOF. Assume that α, β are two K-trivial reals. Then there is a constant c such that $K(\alpha\upharpoonright n)$ and $K(\beta\upharpoonright n)$ are both below $K(n)+c$ for every n. By Theorem 4.3 there is a constant d such that for each n there are at most d strings $\tau\in\{0,1\}^n$ satisfying $K(\tau)\le K(n)+c$. Let $e=n^*$ be the shortest program for n. One can assign to $\alpha\upharpoonright n$ and $\beta\upharpoonright n$ numbers $i, j\le d$ such that they are the i-th and the j-th string of length n enumerated by a program of length up to $|e|+c$.

Let U be a universal prefix-free machine. We build a prefix-free machine V witnessing the K-triviality of $\alpha+\beta$. Representing i, j by strings of the fixed length d and taking $b\in\{0,1\}$, $V(eijb)$ is defined by first simulating $U(e)$ until an output n is produced and then continuing the simulation in order to find the i-th and j-th string α and β of length n such that both are generated by a program of size up to $n+c$. Then one can compute $2^{-n}(\alpha+\beta+b)$ and derive from this string the first n binary digits of the real $\alpha+\beta$. These digits are correct provided that e, i, j are correct and b is the carry bit from bit $n+1$ to bit n when adding α and β – this bit is well-defined unless $\alpha+\beta=z\cdot 2^{-m}$ for some integers m, z, but in that case $\alpha+\beta$ is computable and one can get the first n bits of $\alpha+\beta$ directly without having to do the more involved construction given here. Notice that $\alpha\oplus\emptyset\le_K\alpha$ trivially, and hence if α and β are K-trivial, $\alpha\oplus\beta$ will also be K-trivial. ⊣

Since the K-trivials are closed downwards in $\le_T$, we have the following.

COROLLARY 6.16 (Nies [23]). *The K-trivials form a nonprincipal Σ^0_3 ideal in the Turing degrees.*

The K-trivial form the only known such ideal. The proof that the ideal is nonprincipal is in [10]. It is also known that there is a low_2 computably enumerable degree above this ideal (Nies, see Downey and Hirschfeldt [9] for a proof). (It is known that no low c.e. degree can be above the K-trivials, but it is quite possible that there is a Δ^0_2 low degree above them all, perhaps even a Martin-Löf random one. This is an apparently difficult open question[7].)

In passing we mention two further characterizations of the K-trivials. First we can define B be a *base of a cone of Martin-Löf randomness* if $B\le_T A$ where A is B-random. Kučera was the first to construct such a noncomputable set B.

THEOREM 6.17 (Hirschfeldt, Nies and Stephan [15]). *A is K-trivial iff A is a base of a cone of Martin-Löf randomness.*

Finally, in recent work there has been a very surprising new characterization of K-triviality. We will say that A is *low for weak 2-randomness tests* iff for all Π^A_2 nullsets $\mathcal{N}$, there is a Π^0_2 nullset $\mathcal{M}\supseteq\mathcal{N}$.

[7]Since the writing of this article, Kučera and Slaman have shown that there is a low Δ^0_2 degree above all the K-trivials. It is still unknown if such a low upper bound can be random.

THEOREM 6.18 (Downey, Nies, Weber, Yu [13]+Nies [26]+Miller [21]). *A is K-trivial iff A is low for weak 2-randomness.*

§7. A proper subclass.

7.1. Jump traceability and strong jump traceability. Again we return to the theme of jump traceability. Recall that that a set B is jump traceable iff there is a computable order h and a weak array $\{W_{g(j)} : j \in \mathbb{N}\}$, such that $|W_{g(j)}| \leq h(j)$, and $J^B(e) \in W_{g(e)}$. We would say that A is *jump traceable via the order h*. Recall that Nies [23] showed that if A is K-trivial, then A is jump traceable.

Such considerations lead Figueira, Nies and Stephan to investigate a new class of reals.

DEFINITION 7.1 (Figueira, Nies and Stephan [14]). We say that A is *strongly jump traceable* iff for *all* (computable) orders h, A is jump traceable via h.

Nies' Theorem shows that if A is K-trivial then it is jump traceable via $h(e)$ about $e \log e$. Interestingly, Figueira, Nies and Stephan [14] showed that there are $2^{\aleph_0}$ many reals which are jump traceable at order $h(e) = 2^{2^e}$. Using a rather difficult argument, Downey and Greenberg proved the following.

THEOREM 7.2 (Downey and Greenberg, unpubl.). *For $h(e) = \log \log e$ all reals jump traceable with order $h(e)$ are Δ^0_2 and hence there are only countable many.*

It is not altogether clear that *strongly* jump traceable reals should exist.

THEOREM 7.3 (Figueira, Nies and Stephan [14]). *There exist c.e. promptly simple strongly jump traceable sets.*

PROOF. The following proof is due to Keng Meng Ng. It is really a Π^0_2 argument since the guess that a particular partial computable function is an order is a Π^0_2 fact. There is a promptly simple c.e. set A, which is strongly jump traceable.

Requirements

We build an c.e. set A satisfying the following requirements:

$$\mathcal{P}_e : W_e \text{ is infinite } \Rightarrow \exists x, s(x \in W_{e,\text{ats}} \wedge x \in A_{s+1}),$$

$$\mathcal{N}_e : h_e \text{ is an order } \Rightarrow A \text{ is jump traceable via } h_e.$$

Here, h_e is the eth partial computable function of a single variable. The negative requirement $\mathcal{N}_e$ will build the sequence $V_{e,0}, V_{e,1}, \dots$ of c.e. sets such that $|V_{e,i}| \leq h_e(i)$ and $J^A(i) \in V_{e,i}$ for all i, if h_e is an order.

Strategy

We will describe the strategy used to satisfy $\mathcal{N}_0$. The general strategy for $\mathcal{N}_e$ is similar. Suppose that h_0 is an order function, our aim is to build the uniformly c.e. sequence $V_{0,0}, V_{0,1}, \dots$. Consider the sequence of intervals $I_1, I_2, \dots$, initial segments of $\mathbb{N}$ such that $h_0(x) = n$ for all $x \in I_n$.

For $i \in I_1$, whenever $J^A(i)[s]\downarrow$ with use $u(i)$, we would enumerate the value $J^A(i)[s]$ into $V_{0,i}$ and preserve the value $J^A(i)[s]$ by preventing any positive requirement from enumerating an $x \in A\restriction_{u(i)}$. If $i' \in I_2$ and $J^A(i')[s']\downarrow$ with use $u(i')$ at stage s', we will also enumerate $J^A(i')[s']$ into $V_{0,i'}$. Since $V_{0,i'}$ can take two values, it is therefore not essential that the computation $\{i'\}^A(i')[s']$ at stage s' be preserved forever. We could allow $\mathcal{P}_0$ to make an enumeration below $u(i')$ (but above $\max\{u(k) \mid k \in I_0\}$), and block all other positive requirements $\mathcal{P}_1, \mathcal{P}_2, \ldots$ from enumerating below $u(i')$. When a new value $J^A(i')[s'']$ appears after $\mathcal{P}_0$ acts, it will be put into $V_{0,i'}$ and the computation $\{i'\}^A(i')[s'']$ preserved forever.

In general, we would allow the requirements $\mathcal{P}_0, \ldots, \mathcal{P}_{n-2}$ to enumerate below the use of $J^A(x)[t]$ at any stage t and $x \in I_n$. This ensures that for $x \in I_n$ there will be at most n values placed in $V_{0,x}$. Therefore, $\mathcal{N}_0$ will impose different restraint on each positive requirement $\mathcal{P}_0, \mathcal{P}_1, \ldots$. In particular, the restraint imposed by $\mathcal{N}_0$ on $\mathcal{P}_e$ at a stage s is $r_0(e, s) >$ the use of any computation $J^A(x)[s]$, where $x \in I_1 \cup \cdots \cup I_{e+1}$.

The above strategy is designed to work in the case where h_0 is an order. At a stage s we could compute the values $h_0(0), \ldots, h_0(s)$ up to s steps, and use the values computed to see if h_0 looks like an order. If the first $l(s)$ (the length of convergence) many convergent values do not form a non-decreasing sequence, then we could cease all action for $\mathcal{N}_0$. On the other hand, if h_0 is a non-decreasing function such that $\lim_{n\to\infty} h_0(n) = m$, then at every stage s, the first $l(s)$ many convergent values of h_0 will always form a non-decreasing sequence. This would result in the positive requirements $\mathcal{P}_{m-1}, \mathcal{P}_m, \ldots$ having restraint $\to \infty$. To prevent this situation, we declare a number x to be active (at some stage s), if the length of convergence $l(s) > x$, and $h_0(x) < h_0(l(s))$. The requirement $\mathcal{N}_0$ would only act on those numbers i which are active, for if h_0 is indeed an order, it does no harm for us to wait until i becomes active before making enumerations into $V_{0,i}$.

Construction of A

We arrange the requirements in the order $P_0 < N_0 < P_1 < N_1 < \cdots$. Let $J^A(i)[s]$ denote the value of $\{i\}_s^{A_s}(i)$ if it is convergent, with the use $u(i, s)$, and $V_{e,i}[s]$ denote $V_{e,i}$ in the sth stage of formation. For each e, let the length of convergence of h_e at stage s be defined as

$$l(e, s) = \max\left\{y \leq s \mid (\forall x \leq y)\ \ (h_{e,s}(x)\downarrow\ \ \wedge\ h_e(x) \geq h_e(x-1))\right\}.$$

A number i is said to be e-active at stage s, if $i < l(e, s)$ and $h_e(i) < h_e(l(e, s))$. That is, a number i will become e-active when the length of convergence of h_e exceeds i, and we have received further confirmation that $h_e(x)$ is not an eventually constant function. For each $k < e$, we let $r_k(e, s)$ be the restraint imposed by $\mathcal{N}_k$ on the eth positive requirement $\mathcal{P}_e$ at stage s, defined as follows

$$r_k(e, s) = \max\left\{u(i, s) \mid i \text{ is } k\text{-active at stage } s\ \ \wedge\ \ h_k(i) \leq e+1\right\}.$$

We will let $r(e,s) := \max\{r_k(e,s) \mid k < e\}$. At stage s, we say that $\mathcal{P}_e$ requires attention if $A \cap W_{e,s} = \emptyset$, and $\exists x \in W_{e,s} - W_{e,s-1}$ such that $x > 2e$ and $x > r(e,s)$.

The construction at stage s involves the following actions:

(i) Pick the least $e < s$ such that $\mathcal{P}_e$ requires attention, and enumerate the least $x > \max\{2e, r(e,s)\}$ and $x \in W_{e,s} - W_{e,s-1}$ into A. For each $k \geq e$, we set $V_{k,i}[s+1] = \emptyset$ for all i.

(ii) For each $e < s$ we do the following for the sake of $\mathcal{N}_e$: For every currently e-active number i, we enumerate $J^A(i)[s]$ (if convergent) into $V_{e,i}[s]$.

Verification

Firstly note that for each e,

$$\exists^\infty s \;\; J^A(e)[s]\downarrow \;\Rightarrow J^A(e)\downarrow. \tag{1}$$

To see this for each e, pick an index $i > e$ such that $\forall n(h_i(n) = n)$. Since each positive requirement only enumerates at most one element into A, let s be a stage such that no $\mathcal{P}_k$ for any $k \leq i$ ever receives attention after stage s, and e is i-active after stage s. If $J^A(e)[t]\downarrow$ infinitely often, let $t > s$ be such that $J^A(e)[t]\downarrow$. Then, $A_t \upharpoonright_{u(e,t)} = A \upharpoonright_{u(e,t)}$ and so $J^A(e)\downarrow$. Note that this implies A is low, but A will actually be jump traceable and hence superlow.

It is not difficult to see that (1) implies $\lim_{s\to\infty} r(e,s) < \infty$ for every e: Fix a $k < e$, and we will argue that $\lim_{s\to\infty} r_k(e,s) < \infty$. There can only be finitely many numbers i such that i eventually becomes k-active with $h_k(i) \leq e+1$, and for each such i, $u(i,s)$ has to be bounded (or undefined) by (1).

Hence every $\mathcal{P}_e$ will be satisfied, and A is coinfinite and promptly simple. Next, we fix an e where h_e is an order. Fix an i, and we shall show that $|V_{e,i}| \leq h_e(i)$ and $J^A(i) \in V_{e,i}$. Let s be the last stage at which $V_{e,i}[s]$ is reset to $\emptyset$, and $t \geq s$ be the smallest stage such that i becomes e-active at stage t. Enumerations into $V_{e,i}$ start only after stage t (where i becomes active), and furthermore no requirement $\mathcal{P}_k$ for any $k > h_e(i) - 2$ can ever enumerate a number $x < u(i,t')$ at any stage $t' > t$. This means there are at most $h_e(i)$ many different values in $V_{e,i}$. Lastly if $J^A(i)\downarrow$ then $J^A(i) = J^A(i)[t'']$ for some $t'' > t$, and $J^A(i)[t'']$ is enumerated into $V_{e,i}[t'']$ at that stage. ⊣

Notice that the proof is yet another cost function construction. (Here the cost is how many things can potentially be put into some $V_{e,i}$.) Indeed, at first blush, it would seem that we are only getting the same class as the K-trivials.

THEOREM 7.4 (Cholak, Downey, Greenberg [4]). *Suppose that A is c.e. and strongly jump traceable. Then A is K-trivial. Indeed if A is c.e. and jump traceable via an order of size $\sqrt{\log e}$, then A is K-trivial.*

PROOF. Let A be strongly jump-traceable. As with the proof that K-trivials are low for K, we will need to cover U^A by an oracle-free machine, obtained via the Kraft-Chaitin theorem. When a string σ enters the domain of $U^A[s]$ we need to decide whether we believe the A-computation that put σ in dom U^A.

In the "K-trivial implies low for K" proof we put weight beyond the use of the relevant computation to enable us to certify it. In this construction, we will use another technique.

To test the $\sigma \in \text{dom } U^A[s]$ computation, let the use be u and let $\rho = A \upharpoonright u$ (at that stage). The naive idea is to pick some input x and define a function Ψ^A (which is partial computable in A) on the input x, with A-use u and value ρ. This function is traced by a trace $\langle T_x \rangle$; only if ρ is traced do we believe it is indeed an initial segment of A and so believe that $U^A(\sigma)$ is a correct computation. We can then enumerate $(|\sigma|, U^A(\sigma))$ into a Kraft-Chaitin set we build and so ensure that $K(U^A(\sigma)) \leq^+ |\sigma|$.

However, we need to make sure that the issued commands are a KC set. This would of course be ensured if we only believed correct computations, as $\mu(\text{dom } U^A)$ is finite. However, the size of most T_x is greater than 1, and so an incorrect ρ may be believed. We need to limit the mass of the errors.

A key new idea from Cholak, Downey and Greenberg [4] is that, rather than treat each string σ individually, we batch strings up in pieces of mass. When we have a collection of strings in dom U^A whose total mass is 2^{-k} we verify A up to a use that puts them all in dom U^A. The greater 2^{-k} is, the more stringent the test will be (ideally, in the sense that the size of T_x is smaller). We will put a limit m_k on the amount of times that a piece of size 2^{-k} can be believed and yet be incorrect. The argument will succeed if

$$\sum_{k<\omega} m_k 2^{-k}$$

is finite.

The reader should realize that once we use an input x to verify an A-correct piece, it cannot be used again for any testing, as $\Psi^A(x)$ becomes defined permanently. Following the naive strategy, we would need at least 2^k many inputs for testing pieces of size 2^{-k}. Even a single error on each x (and there will be more, as the size of T_x has to go to infinity) means that $m_k \geq 2^k$ is too large. The rest of the construction is a combinatorial strategy: which inputs are assigned to which pieces in such a way as to ensure that the number of possible errors m_k is sufficiently small. The strategy has two ingredients.

First, we note that two pieces of size 2^{-k} can be combined into a single piece of size $2^{-(k-1)}$. So if we are testing one such piece, and another piece, with comparable use, appears, then we can let the testing machinery for $2^{-(k-1)}$ take over. Thus, even though we need several testing locations for 2^{-k} (for example if a third comparable piece appears), at any stage, the testing at 2^{-k} is really responsible for at most one such piece.

The naive reader would imagine that it is now sufficient to let the size of T_x (for x testing 2^{-k}-pieces) be something like k and be done. However, the opponent's spoiling strategy would be to "drip-feed" small mass that aggregates to larger pieces only slowly (this is similar to the situation in decanter constructions.) In particular, fixing some small 2^{-k}, the opponent will first give us k pieces (of incomparable use) one after the other (so as to change A and remove one before giving us a new one.) At each such occurrence we would need to use the input x devoted to the first 2^{-k} piece, because at each such stage we only see one. Once the amount of errors we get from using x for testing is filled (T_x fills up to the maximum allowed size) the opponent gives us one correct piece of size $2^{-(k-1)}$ and then moves on to gives us k more incorrect pieces which we test on the next x. Overall, we get k errors on *each* x used for 2^{-k}-pieces. As we already agreed that we need something like 2^k many such x's, we are back in trouble.

Every error helps us make progress as the opponent has to give up one possible value in some T_x; fewer possible mistakes on x are allowed in the future. The solution is to make every single error count in our favour in all future testings of pieces of size 2^{-k}. In other words, what we need to do is to maximize the benefit that is given by a single mistake; we make sure that a single mistake on *some* piece will mean one less possible mistake on *every* other piece.

In the beginning, rather than just testing a piece on a single input x, we test it simultaneously on a large set of inputs and only believe it is correct if the use shows up in the trace of every input tested. If this is believed and more pieces show up then we use them on other large sets of inputs. If, however, one of these is incorrect, then we later have a large collection of inputs x for which the number of possible errors is reduced. We can then break up this collection into 2^k many smaller collections and keep working only with such x's.

This can be geometrically visualized as follows. If the naive strategy was played on a sequence of inputs x, we now have an m_k-dimensional cube of inputs, each side of which has length 2^k. In the beginning we test each piece on one hyperplane. If the testing on some hyperplane is believed and later found to be incorrect then from then on we work in that hyperplane, which becomes the new cube for testing pieces of size 2^{-k}; we test on hyperplanes of the new cube. If the size of T_x for each x in the cube is at most m_k then we never "run out of dimensions".

Further details can be found in Cholak, Downey and Greenberg [4], and Downey and Hirschfeldt [9]. ⊣

We remark *en passant* that it is unknown if this result is true with the hypothesis that A is c.e. is removed. Downey and Greenberg have conjectured that the answer is yes.

But finally we have an example of a class of reals, defined by cost functions, where we actually get a *proper* subclass of the K-trivials.

THEOREM 7.5 (Cholak, Downey, Greenberg [4]). *The c.e. strongly jump traceables form a proper subclass of the K-trivials. Again this is true at tracing order* $\log\log e$.

Proof. The easiest way to understand this proof is that the reader realize the following: since the K-trivials are closed under $\leq_T$, there must be ones of minimal degree, and hence not n-c.e. for any n. How would we make such a real directly? Now we have already seen that the only way to make K-trivials is to use a cost function construction. Thus in the basic construction, we will pick some n and monitor $c(n, s)$, the weight of the tail at s. If this was simply making A noncomputable, should that weight be too large, we would abandon this n and choose some $n' > s$.

Now if we are to make A not k-c.e. for any k, then we would need to perhaps put n into and take it out of A many times, perhaps $k + 1$ times.

The *problem* is that each time we change $A_s \upharpoonright n$, we must pay some *uncompensated* price $c(n, s)$. The basic idea, the reader will recall, is to keep this price bounded. The plan is to use a decanter kind of idea. Suppose, for instance, $k = 2$ so we need 3 attacks on some n. We would have an overall cost we are willing to pay of, say, $2^{-(e+1)}$, for this requirement R_e. Then the idea is to think of this cost as, initially, $[2^{-(e+2)}, 2^{-(e+4)}, 2^{-(e+6)}]$ where the *first* attack via some n will only cost *at most* $2^{-(e+6)}$. If we see that this cost exceeded at some stage s, *before* the first attack, we could abandon this attack at no cost choosing a new $n' > s$.

However, should we have done the first attack, we would choose a new $n' > s$ *but* give n' the quotas $[2^{-(e+2)}, 2^{-(e+4)}, 2^{-(e+9)}]$ (if the attack was abandoned *before* the second attack occurred, *or* give n' the quotas $[2^{-(e+2)}, 2^{-(e+6)}, 2^{-(e+9)}]$ should this happen *after* we did the second attack. (In general the third attack gets numbers of the form $2^{-(e+3p)}$ and the second $2^{-(e+2t)}$ (or any suitably convergent series). This is very similar to the decanter method.

Now in our argument, we need to make A K-trivial, yet not strongly jump traceable. Again we will have a suitable slowly growing order h about $\log\log e$ is enough. To kill some possible trace $W_{g(e)}$ we will control parts of the jump. Suppose that we are dealing with part where he is supposed to be able to jump trace with at most k members of $W_{g(e)}$. (This is some e we can put things into the jump.)

Then the idea is simple. We would pick some $a_s > s$ and put e into the jump with axiom s, e, a_s, saying $J^A(e) = s$ if $a_s \notin A$. Once this appears in $W_{g(e)}[s']$ for some $s' > s$, we can then remove this from $J^A(e)[t]$ defining it to be t instead of s by simply putting a_s into $A_{t+1} - A_t$ and this new value having a new use a_{t+1}. This would need to repeat itself at most $k + 1$ times. Each stage would cost us $c(a_t, t)$. Then, the argument is the same as the one above. We simply need a combinatorial counting argument to calculate how long the interval where $h(e) = k$ needs to be that we *must* succeed on some follower.

Details again can be found in Cholak, Downey, Greenberg [4]. ⊣

Notice the property of being strongly jump traceable is something closed downwards under $\leq_T$. Until very recently, it is an interesting open question as to whether they form an ideal. Keng Meng Ng constructed a strongly jump traceable c.e. set whose join with any strongly jump traceable set is strongly jump traceable. That construction was a careful analog of the construction of an *almost deep* c.e. degree by Cholak, Groszek and Slaman [5], which was a c.e. degree $\boldsymbol{a}$ such that for all low c.e. degree $\boldsymbol{b}$, $\boldsymbol{a} \cup$ was also low. The proof relied on a characterization of a c.e. set X being strongly jump traceable due to Nies, Figueira and Stephan: X is strongly jump traceable iff X' is *well-approximable* meaning that for all computable orders h, there is a computable enumeration $X'(z) = \lim_s f(z, s)$ with $|\{s : f(z, s+1) \neq f(z, s)\}| \leq h(z)$.

THEOREM 7.6 (Ng [22]). *There is a promptly simple c.e. set A, such that if W is a strongly jump traceable c.e. set, then $A \oplus W$ is strongly jump traceable.*

In an earlier version of the present paper we sketched a proof of Ng's Theorem. Recently, Cholak, Downey and Greenberg indeed verified that the c.e. strongly jump traceable degrees form an ideal. It is this last result whose proof we will sketch.

THEOREM 7.7 (Cholak, Downey and Greenberg [4]). *Suppose that A and B are c.e. and strongly jump traceable. Then so is $A \oplus B$.*

PROOF (SKETCH). Actually something more is proven. It is shown that given an order h we can construct a slower order k such that if A and B are jump traceable via k then $A \oplus B$ is jump traceable via h. The opponent must give us $W_{p(x)}$ jump tracing A and $W_{q(x)}$ jump tracing B, such that

$$|W_{p(x)}|, |W_{p(x)}| < k(x),$$

for all x. It is *our* task to construct a trace V_z tracing $J^{A\oplus B}(z)$ with $|V_z| < h(z)$. There are two obstacles to this task. We will treat them in turn.

Fundamentally, what happens is that we see some *apparent* jump computation $J^{A\oplus B}(x) \downarrow [s]$. The question is, *should we believe this computation?* The point is that we only have at most $h(x)$ many slots in the trace V_x to put possible values. (We will think of the V_x as a *box* of height x.) The opponent can change the computation by changing either A or B after stage s on the use.

Our solution is to build another part of the jump to test the A and B parts. Of course these locations are given by the Recursion Theorem. There will be many parts devoted to a single x. For each x the strategies will operate separately. We will denote parts of the jump for testing the A use by $a(x, i)$ and for the B side $b(x, i)$ for $i \in \omega$. (The reason for the large number will be seen later. It is kind of like a decanter of *infinite depth*, and is because of the *noncompletion* problem we need to solve.)

The basic idea is this: when stage s occurs for some a and b we will define

$$J^B[s](b) = j_B(x, s) \text{ and } J^A[s](a) = j_A(x, s),$$

where $j_C(x, s)$ denotes the C-use of the $J^{A\oplus B}(x)[s]$ computation.

Now of course this is not quite correct. We can do this, but *before* the real jump returns either of these computations, they can go away since A or B ranges on the relevant use. In particular, although we know that real jump values must occur in $W_{p(x)}$ and $W_{q(x)}$, the jump computations we have *purported* to define can become divergent on account of the relevant oracle changing. We will call this the *noncompletion problem*, and discuss its solution later.

To demonstrate the first idea, we will assume that this problem will not arise, so that the procedures return. That is, we see $j_A(x, s)$ occur in it trace: $W_{p(a)}$ and similarly for B. This would happen at some sage $t > s$ where without loss of generality, we can assume the $J^{A \oplus B}(x)$ computation is still around. Then at such a stage we would be prepared to believe it, by putting its present value v_1 into the first slot of V_x.

The simplest case is that we actually were working in the situation where the $W_{p(a)}$ and $W_{q(b)}$ were of size 1 (1-boxes) then we would be done. The computation for $A \oplus B$ *is* correct.

In the more general case we would have, say, the A and B boxes of, say, size 2, and the $A \oplus B$ one of size, say, 3. We will, as seen below, *manufacture* 1-boxes, if necessary.

Now, if the $A \oplus B$ computation is wrong, at least one of the A or B ones are too. We have arranged that the size of the V_x will be much bigger than that of the A and B. If both sides are wrong, or are shown wrong then there is are false jump computations in both of the $W_{p(a)}$ and $W_{q(b)}$. In that case then the boxes are now, in effect, 1-boxes as the top slot is filled with a false jump computation. Then the next time we get a $A \oplus B$ computation, we would be safe, assuming that we get a return.

If only one side, say the A side, is incorrect, then we have come to the first problem. The B-box $W_{q(b)}$ actually is returning correctly a jump computation and is thus useless for testing more computations. The next test would involve possibly the same a but would need a different b, and this could alternate.

The idea is to use more than one a, b as we now see. At the beginning we could use *two* A boxes and *two* B boxes of size 2. Suppose that, as above, we get a return on all of them and the A side was wrong. Then now we have *two promoted* 1-boxes. Then the next time we test a $A \oplus B$ computation, we could use only one of them and another B-box of size 2. Since the A-computation now must be correct, if the believed computation is wrong, it must be the B side which wrong the next time, now creating a new B-1-box. Finally the third time we test, we would have two 1-boxes.

This is all very fine, but the fact that we might not get a return into the boxes causes really deep problems. The problem is that we might enumerate into the boxes two A and B configurations and the computation might occur in only one side, say the A side, before the $A \oplus B$ computation vanishes. We have not used up any V_x slot, *but the probe is that the A side might be correct and hence that box is now useless.* The reader should not that this is even a problem if we were dealing with 1-boxes. What could happen is that the change side (before return) could alternate rendering the boxes corresponding to the (correct) other side useless, and we would run out of small boxes before the oracle decided to return correctly.

The idea is to use a decanter-like strategy to get rid of this problem. Initially, to test the 2 boxes for A and B we begin at 3-boxes. These will be *metaboxes* in the sense that they are amalgams of some large number of 3 boxes. Say, $W_{p(a_1)}, \ldots, W_{p(a_n)}$ and $W_{q(b_1)}, \ldots, W_{q(b_n)}$. Before we believe the $A \oplus B$ computation at x, we begin by testing the A side, then the B side one box at a time, alternating. Thus $W_{p(a_1)}$, $W_{q(b_1)}$, $W_{p(a_2)}$, and so forth, only moving on to the next box if the previous one returns, and hence the $A \oplus B$ computation remains unchanged. Now two things can happen.

The first possibility is that we get to the end of this process. It is only then that we would move to the 2-boxes and try to test as above. If we actually get to the end of the of this final procedure with no $A \oplus B$ change, then we would then return and use a slot of V_x.

The other possibility is that at some stage of this process one of A or B changes on the x-use. Suppose that this is A. The key new idea is then if this is in the last 2-box testing phase, then we have *created* many new 2-boxes, since the top slot of all of the $W_{p(a_i)}$ are filled with false jump computations of A. (In this case we have also created new A 1-boxes.) Box promotion is to a lesser extent true for the case where this fails in the first phase. We have also created A 1-boxes

Notice that all of the B boxes used in this processes are likely now useless. The function k will be slow enough growing that there will be plenty more 3 boxes for later work. Then the idea is that at the next try we would begin at the 4 boxes and recursively travel down to the 2-boxes only when we get returns from the higher boxes. (But at the 3 box stage, we would be using half of the now promoted original A metabox.) The key thing to observe is that for a correct computation, we must eventually make progress as the killed side always promotes. This can only happen finitely often since the height of the relevant metabox is fixed and the killed side will promote that box.

The details are a little messy but this is the general idea. Full details can be found in the paper [4]. $\dashv$

Several questions suggest themselves. First is it true that each strongly jump traceable is bounded by a c.e. strongly jump traceable? We have seen that they form an ideal in the Turing degrees? How complicated is the ideal in the c.e. case? It would seem likely that it could be Π^0_4 complete[8].

Finally we remark that there are several other examples of cost function constructions in the literature, whose relationship with the strongly jump

[8] Keng Meng Ng has recently showed that the ideal of c.e. strongly jump traceable sets is Π^0_4 complete.

traceables and the K-trivials is not yet clear. For example Nies has proven the following.

THEOREM 7.8 (Nies [25]). *There exists a c.e. set A such that for all B, if B is random then $A \oplus B$ does not compute $\emptyset'$.*

We know that such sets must be K-trivial. The question is whether this can be reversed. Barmpalias [1] has related material here, extending Nies' Theorem above. Finally, we will call a set A *almost complete* if $\emptyset'$ is K-trivial relative to A, and $A \leq_T \emptyset'$. Such sets can be constructed from the K-trivial construction and the pseudo-jump theorem. Hirschfeldt[9] (unpubl) has shown that there are c.e. (necessarily K-trivial) reals below all such *random* reals. The question is whether they coincide with the K-trivials[10].

§8. What have I left out this time? While I hope that this is the last lecture in the series, I should point out that there is a lovely series of results concerning lowness for other randomness notions. Terwijn and Zambella [32] characterized lowness for Schnorr randomness tests in terms of *computable traceability*, and this was extended by Kjos-Hanssen, Nies and Stephan [16] to the class of Schnorr randoms. Nies [23] proved that only the computable reals are low for computable randomness, and Downey and Griffiths [8], and later Stephan and Yu [31] investigated lowness for Kurtz randomness. To treat these and other related results properly would take another paper, and hence I will simply refer the reader to the source papers, or to Downey-Hirschfeldt [9], or Nies [28], and the recent survey Nies [27] for details.

REFERENCES

[1] G. BARMPALIAS, *Random non-cupping revisited*, ***Journal of Complexity***, vol. 22 (2006), no. 6, pp. 850–857.

[2] C. CALUDE and R. COLES, *Program-size complexity of initial segments and domination reducibility*, ***Jewels Are Forever*** (J. Karhümaki, H. Mauer, G. Paŭn, and G. Rozenberg, editors), Springer, Berlin, 1999, pp. 225–237.

---

[9] A simpler proof was found by Hirschfeldt and Miller and appears in Nies [27].

[10] In very recent work, Nies has shown that the strongly jump traceable c.e. sets are exactly the c.e. sets Turing below the superhigh randoms, and Greenberg and Nies also showed that they are also the c.e. sets below all superlow random reals. In his PhD thesis, David Diamondstone showed that if A is c.e. and strongly jump traceable and X is c.e. and superlow then $A \oplus X$ is incomplete, and Nies extended this to show that $A \oplus X$ is always superlow. Finally Ng has extended the notion of strong jump traceability to construct the strange class of hyper-jump traceable reals. These are (c.e.) sets A such that for all c.e. sets W and orders h^W, there is a trace T^W such that $|T^W(e)| < h^W(e)$ and $J^A(e) \in T^W(e)$ for all e. Noncomputable such c.e. sets exist and this class forms a Π^0_5 complete proper subclass of the strongly jump traceable sets. No hyper-jump traceable c.e. set is promptly simple, but they can be cuppable, and there are noncomputable c.e. sets A such that for all noncomputable $B \leq_T A$, B is not hyper-jump traceable. Details will appear in the forthcoming paper of Ng "Beyond Strong Jump Traceability".

[3] G. Chaitin, *Information-theoretic characterizations of recursive infinite strings*, ***Theoretical Computer Science***, vol. 2 (1976), no. 1, pp. 45–48.

[4] P. Cholak, R. Downey, and N. Greenberg, *Strong jump traceablilty I, the computably enumerable case*, ***Advances in Mathematics***, vol. 217 (2008), pp. 2045–2074.

[5] P. Cholak, M. Groszek, and T. Slaman, *An almost deep degree*, ***The Journal of Symbolic Logic***, vol. 66 (2001), no. 2, pp. 881–901.

[6] R. Downey, *Five lectures on algorithmic randomness*, to appear in ***Proceedings of Computational Prospects of Infinity*** World Scientific.

[7] ———, *Some computability-theoretic aspects of reals and randomness*, ***The Notre Dame Lectures***, Lecture Notes in Logic, vol. 18, ASL, Urbana, IL, 2005, pp. 97–147.

[8] R. Downey and E. Griffiths, *Schnorr randomness*, ***The Journal of Symbolic Logic***, vol. 69 (2004), no. 2, pp. 533–554.

[9] R. Downey and D. Hirschfeldt, *Algorithmic Randomness and Complexity*, Springer-Verlag, in preparation.

[10] R. Downey, D. Hirschfeldt, A. Nies, and F. Stephan, *Trivial reals*, ***Proceedings of the 7th and 8th Asian Logic Conferences***, Singapore University Press, Singapore, 2003, pp. 103–131.

[11] R. Downey, D. Hirschfeldt, A. Nies, and S. Terwijn, *Calibrating randomness*, ***The Bulletin of Symbolic Logic***, vol. 12 (2006), no. 3, pp. 411–491.

[12] R. Downey, J. Miller, and L. Yu, *On the quantity of K-trivial reals*, in preparation.

[13] R. Downey, A. Nies, R. Weber, and L. Yu, *Lowness and Π^0_2 nullsets*, ***The Journal of Symbolic Logic***, vol. 71 (2006), no. 3, pp. 1044–1052.

[14] S. Figueira, A. Nies, and F. Stephan, *Lowness properties and approximations of the jump*, ***Proceedings of the 12th Workshop on Logic, Language, Information and Computation (WoLLIC 2005)***, Electronic Lecture Notes in Theoretical Computer Science, vol. 143, Elsevier, Amsterdam, 2006, pp. 45–57.

[15] D. Hirschfeldt, A. Nies, and F. Stephan, *Using random sets as oracles*, ***Journal of the London Mathematical Society***, vol. 75 (2007), pp. 610–622.

[16] B. Kjos-Hanssen, A. Nies, and F. Stephan, *Lowness for the class of Schnorr random reals*, ***SIAM Journal on Computing***, vol. 35 (2005), no. 3, pp. 647–657.

[17] A. Kučera and T. Slaman, *Turing incomparability in Scott sets*, ***Proceedings of the American Mathematical Society***, vol. 135 (2007), no. 11, pp. 3723–3731.

[18] A. Kučera and S. Terwijn, *Lowness for the class of random sets*, ***The Journal of Symbolic Logic***, vol. 64 (1999), no. 4, pp. 1396–1402.

[19] L. Levin, ***Some Theorems on the Algorithmic Approach to Probability Theory and Information Theory***, Dissertation in Mathematics, Moscow, 1971.

[20] D. Loveland, *A variant of the Kolmogorov concept of complexity*, ***Information and Control***, vol. 15 (1969), pp. 510–526.

[21] J. Miller, personal communcation.

[22] Keng Meng Ng, *On strongly jump traceable reals*, ***Annals of Pure and Applied Logic***, vol. 154 (2008), no. 1, pp. 51–69.

[23] A. Nies, *Lowness properties and randomness*, ***Advances in Mathematics***, vol. 197 (2005), no. 1, pp. 274–305.

[24] ———, *Reals which compute little*, ***Logic Colloquium '02*** (Z. Chatzidakis, P. Koepke, and W. Pohlers, editors), Lecture Notes in Logic, vol. 27, ASL, La Jolla, CA, 2006, pp. 261–275.

[25] ———, *Non-cupping and randomness*, ***Proceedings of the American Mathematical Society***, vol. 135 (2007), no. 3, pp. 837–844.

[26] ———, personal communication.

[27] ———, *Eliminating concepts*, to appear in ***Proceedings of Computational Prospects of Infinity***, World Scientific.

[28] ———, ***Computability and Randomness***, Cambridge University Press, 2009.

[29] R. Soare, ***Recursively Enumerable Sets and Degrees***, Perspectives in Mathematical Logic, Springer-Verlag, Berlin, 1987, A study of computable functions and computably generated sets.

[30] R. Solovay, *Draft of paper (or series of papers) on Chaitin's work*, unpublished notes, May, 1975, 215 pages.

[31] F. Stephan and L. Yu, *Lowness for weakly* 1*-generic and Kurtz-random*, ***Theory and Applications of Models of Computation***, Lecture Notes in Computer Science, vol. 3959, Springer, Berlin, 2006, pp. 756–764.

[32] S. Terwijn and D. Zambella, *Computational randomness and lowness*, ***The Journal of Symbolic Logic***, vol. 66 (2001), no. 3, pp. 1199–1205.

[33] M. van Lambalgen, ***Random Sequences***, Ph.D. Diss., University of Amsterdam, 1987.

[34] D. Zambella, ***On Sequences with Simple Initial Segments***, ILLC technical report, ML-1990-05, University of Amsterdam, 1990.

SCHOOL OF MATHEMATICS, STATISTICS, AND COMPUTER SCIENCE
VICTORIA UNIVERSITY
PO BOX 600 WELLINGTON, NEW ZEALAND
E-mail: Rod.Downey@ecs.vuw.ac.nz

THE INEVITABILITY OF LOGICAL STRENGTH: STRICT REVERSE MATHEMATICS

HARVEY M. FRIEDMAN

Abstract. An extreme kind of logic skeptic claims that "the present formal systems used for the foundations of mathematics are artificially strong, thereby causing unnecessary headaches such as the Gödel incompleteness phenomena". The skeptic continues by claiming that "logician's systems always contain overly general assertions, and/or assertions about overly general notions, that are not used in any significant way in normal mathematics. For example, induction for all statements, or even all statements of certain restricted forms, is far too general - mathematicians only use induction for natural statements that actually arise. If logicians would tailor their formal systems to conform to the naturalness of normal mathematics, then various logical difficulties would disappear, and the story of the foundations of mathematics would look radically different than it does today. In particular, it should be possible to give a convincing model of actual mathematical practice that can be proved to be free of contradiction using methods that lie within what Hilbert had in mind in connection with his program". Here we present some specific results in the direction of refuting this point of view, and introduce the Strict Reverse Mathematics (SRM) program. See Corollary 11.12.

§1. Many sorted free logic, completeness. We present a flexible form of many sorted free logic, which is essentially the same as the one we found presented in [Fe95], Section 3. In [Fe95], Feferman credits this form of many sorted free logic to [Pl68, Fe75, Fe79], and [Be85, page 97–99].

We prefer to use many sorted free logic rather than ordinary logic, because we are particularly interested in the naturalness of our axioms, and want to avoid any cumbersome or ad hoc features.

We will not allow empty domains. We allow undefined terms. In fact, the proper use of undefined terms is the main point of free logic.

A signature σ (in many sorted free logic) consists of

i. A nonempty set $\mathrm{SRT}(\sigma)$ called the sorts.
ii. A set $\mathrm{CS}(\sigma)$ called constant symbols.
iii. A set $\mathrm{RS}(\sigma)$ called relation symbols.
iv. A set $\mathrm{FS}(\sigma)$ called function symbols.

This research was partially supported by NSF Grant DMS-0245349.

Logic Colloquium '06
Edited by Barry Cooper, Herman Geuvers, Anand Pillay, and Jouko Väänänen
Lecture Notes in Logic, 32

v. We require that CS(σ), RS(σ), FS(σ) be pairwise disjoint, and not contain $=$.
vi. A function ρ with domain CS(σ)$\cup$RS(σ)$\cup$FS(σ), and with the following properties.
vii. For $c \in$ CS(σ), $\rho(c) \in$ SRT(σ). This is the sort of c.
viii. For $R \in$ CS(σ), $\rho(R)$ is a nonempty finite sequence from SRT(σ). This is the sort of R.
ix. For $F \in$ CS(σ), $\rho(F)$ is a finite sequence from SRT(σ) of length ≥ 2. This is the sort of F.

We make the simplifying assumption that equality is present in each sort.
The σ variables are of the form v_n^α, n ≥ 1, where $\alpha \in$ SRT(σ).
The σ terms of σ, and their sorts, are defined inductively as follows.

i. The σ variable v_n^α is a σ term of sort α.
ii. If $c \in$ CS(σ) then c is a σ term of sort $\rho(c)$.
iii. If $t_1, \ldots, t_k$ are σ terms of sorts $\alpha_1, \ldots, \alpha_k$, $k \geq 1$, and $F \in$ FS(σ), F has sort $(\alpha_1, \ldots, \alpha_{k+1})$, then $F(t_1, \ldots, t_k)$ is a σ term of sort α_{k+1}.

The atomic formulas of σ are defined inductively as follows.

i. If s is a σ term then $s\uparrow$, $s\downarrow$ are atomic formulas of σ.
ii. If s, t are terms of the same sort, then $s = t$, $s \cong t$ are atomic formulas of σ.
iii. If $s_1, \ldots, s_k$ are terms of respective sorts $\alpha_1, \ldots, \alpha_k$, $k \geq 1$, and $R \in$ RS(σ) of sort $(\alpha_1, \ldots, \alpha_k)$, then $R(s_1, \ldots, s_k)$ is an atomic formula of σ.

The σ formulas are defined inductively as follows.

i. Every atomic formula of σ is a σ formula.
ii. If φ, ψ are σ formulas, then $(\neg\varphi)$, $(\varphi \wedge \psi)$, $(\varphi \vee \psi)$, $(\varphi \rightarrow \psi)$, $(\varphi \leftrightarrow \psi)$ are σ formulas.
iii. If v is a σ variable and φ is a σ formula, then $(\forall v)(\varphi)$, $(\exists v)(\varphi)$ are σ formulas.

The free logic aspect is associated with the use of $\uparrow$, $\downarrow$, $=$, $\cong$. As will be clear from the semantics, $\uparrow$ indicates "undefined", $\downarrow$ indicates "defined", $=$ indicates "defined and equal", $\cong$ indicates "either defined and equal, or both undefined". Also variables and constants always denote, and a term is automatically undefined if any subterm is undefined.

We now present the semantics for many sorted free logic.

A σ structure M consists of the following.

i. A nonempty set DOM(α) associated with every sort $\alpha \in$ SRT(σ).
ii. For each $c \in$ CS(σ), an element $c^* \in$ DOM($\rho(c)$). This is the interpretation of the constant symbol c.
iii. For each $R \in$ RS(σ), a relation $R^* \subseteq$ DOM(α_1) $\times \cdots \times$ DOM(α_k), where R has sort $(\alpha_1, \ldots, \alpha_k)$. This is the interpretation of the relation symbol R.

iv. For each $F \in \mathrm{FS}(\sigma)$, a partial function F^* from $\mathrm{DOM}(\alpha_1) \times \cdots \times \mathrm{DOM}(\alpha_k)$ into $\mathrm{DOM}(\alpha_{k+1})$, where $\rho(F) = (\alpha_1, \ldots, \alpha_{k+1})$. This is the interpretation of the function symbol F.

A σ assignment is a function γ which assigns to each σ variable v_n^α, $\alpha \in S$, an element $\gamma(v_n^\alpha) \in \mathrm{DOM}(\alpha)$.

We inductively define $\mathrm{val}(M, t, \gamma)$, where M is a σ structure, t is a σ term, and γ is a σ assignment. Note that $\mathrm{val}(M, t, \gamma)$ may or may not be defined.

i. Let v be a σ variable. $\mathrm{val}(M, v, \gamma) = \gamma(v)$.
ii. Let $c \in \mathrm{CS}(\sigma)$. $\mathrm{val}(M, c, \gamma) = c^*$.
iii. Let $F(s_1, \ldots, s_k)$ be a σ term. $\mathrm{val}(M, F(s_1, \ldots, s_k), \gamma) = F^*(\mathrm{val}(M, s_1, \gamma), \ldots, \mathrm{val}(M, s_k, \gamma))$ if defined; undefined otherwise.

Thus in order for $\mathrm{val}(M, F(s_1, \ldots, s_k, \gamma))$ to be defined, we require that $\mathrm{val}(M, s_1, \gamma), \ldots, \mathrm{val}(M, s_k, \gamma)$ be defined.

We inductively define $\mathrm{sat}(M, \varphi, \gamma)$, where M is a σ structure, φ is a σ formula, and γ is a σ assignment.

i. $\mathrm{sat}(M, s\uparrow, \gamma)$ if and only if $\mathrm{val}(M, s, \gamma)$ is undefined.
ii. $\mathrm{sat}(M, s\downarrow, \gamma)$ if and only if $\mathrm{val}(M, s, \gamma)$ is defined.
iii. $\mathrm{sat}(M, s = t, \gamma)$ if and only if $\mathrm{val}(M, s, \gamma) = \mathrm{val}(M, t, \gamma)$. Here we require that both sides be defined.
iv. $\mathrm{sat}(M, s \cong t, \gamma)$ if and only if $\mathrm{val}(M, s, \gamma) = \mathrm{val}(M, t, \gamma)$ or $\mathrm{val}(M, s, \gamma)$, $\mathrm{val}(M, t, \gamma)$ are both undefined.
v. $\mathrm{sat}(M, R(s_1, \ldots, s_k))$ if and only if $R^*(\mathrm{val}(M, s_1, \gamma), \ldots, \mathrm{val}(M, s_k, \gamma))$. Note that condition implies that each $\mathrm{val}(M, s_i, \gamma)$ is defined.
vi. $\mathrm{sat}(M, \neg\varphi, \gamma)$ if and only if not $\mathrm{sat}(M, \varphi, \gamma)$.
vii. $\mathrm{sat}(M, \varphi \wedge \psi, \gamma)$ if and only if $\mathrm{sat}(M, \varphi, \gamma)$ and $\mathrm{sat}(M, \psi, \gamma)$.
viii. $\mathrm{sat}(M, \varphi \vee \psi, \gamma)$ if and only if $\mathrm{sat}(M, \varphi, \gamma)$ or $\mathrm{sat}(M, \psi, \gamma)$.
ix. $\mathrm{sat}(M, \varphi \to \psi, \gamma)$ if and only if either not $\mathrm{sat}(M, \varphi, \gamma)$ or $\mathrm{sat}(M, \psi, \gamma)$.
x. $\mathrm{sat}(M, \varphi \leftrightarrow \psi, \gamma)$ if and only if either ($\mathrm{sat}(M, \varphi, \gamma)$ and $\mathrm{sat}(M, \psi, \gamma)$) or (not $\mathrm{sat}(M, \varphi, \gamma)$ and not $\mathrm{sat}(M, \psi, \gamma)$).
xi. $\mathrm{sat}(M, (\forall v_n^\alpha)(\varphi), \gamma)$ if and only if for all $x \in \mathrm{DOM}(\alpha)$, $\mathrm{sat}(M, \varphi, \gamma[v_n^\alpha | x])$. Here $\gamma[v_n^\alpha | x]$ is the σ assignment resulting from changing the value of γ at v_n^α to x.
xii. $\mathrm{sat}(M, (\exists v_n^\alpha)(\varphi), \gamma)$ if and only if there exists $x \in \mathrm{DOM}(\alpha)$ such that $\mathrm{sat}(M, \varphi, \gamma[v_n^\alpha | x])$.

We say that a σ structure M satisfies a formula φ of σ if and only if $\mathrm{sat}(M, \varphi, \gamma)$, for all σ assignments γ. We say that a σ structure M satisfies a set T of σ formulas if and only if M satisfies every element of T.

We now give a complete set of axioms and rules of inference for σ. It is required that v is a σ variable, c is a σ constant, $s, t, r, s_1, \ldots, s_k, t_1, \ldots, t_k$ are σ terms, φ, ψ, ρ are σ formulas, v is not free in φ, and t is substitutable for v

in ρ. It is also required that each line be a σ formula.

i. All tautologies.
ii. $v\downarrow, c\downarrow$.
iii. $t\uparrow \leftrightarrow \neg t\downarrow$, where t is a σ term.
iv. $t\downarrow \leftrightarrow t = t$.
v. $s \cong t \leftrightarrow (s = t \vee (s\uparrow \wedge t\uparrow))$.
vi. $F(s_1, \ldots, s_k)\downarrow \rightarrow (s_1\downarrow \wedge \cdots \wedge s_k\downarrow)$.
vii. $R(s_1, \ldots, s_k) \rightarrow (s_1\downarrow \wedge \cdots \wedge s_k\downarrow)$.
viii. $s = t \leftrightarrow t = s$.
ix. $(s = t \wedge t = r) \rightarrow s = r$.
x. $(s_1 = t_1 \wedge \cdots \wedge s_k = t_k) \rightarrow F(s_1, \ldots, s_k) \cong F(t_1, \ldots, t_k)$.
xi. $(s_1 = t_1 \wedge \cdots \wedge s_k = t_k) \rightarrow (R(s_1, \ldots, s_k) \rightarrow R(t_1, \ldots, t_k))$.
xii. $(t\downarrow \wedge (\forall v)(\rho)) \rightarrow \varphi[v/t]$.
xiii. $(t\downarrow \wedge \rho[v/t]) \rightarrow (\exists v)(\rho)$.
xiv. From $\varphi \rightarrow \psi$ derive $\varphi \rightarrow (\forall v)(\psi)$.
xv. From $\psi \rightarrow \varphi$ derive $(\exists v)(\psi) \rightarrow \varphi$.
xvi. From φ and $\varphi \rightarrow \psi$, derive ψ.

A theory is a pair T, σ, where σ is a signature, and T is a set of σ formulas.

Let T be a theory with signature σ. A proof from T is a nonempty finite sequence of σ formulas, where each entry lies in T, falls under i–xiii, or follows from previous entries by xiv, xv, or xvi.

A proof from T of φ is a proof from T whose last entry is φ.

We have the following completeness theorem.

THEOREM 1.1. *Let T be a theory in many sorted free logic with signature σ. Let φ be a σ formula. The following are equivalent.*

a. *Every σ structure satisfying T, also satisfies φ.*
b. *There is a proof from T of φ.*

§2. Interpretations, conservative extensions, synonymy. Let σ, τ be signatures in many sorted free logic, and S, T be theories with signatures σ, τ, respectively. We want to define what we mean by an interpretation of S in T.

We will first present a semantic formulation of this notion. We then discuss syntactic formulations.

It is convenient to first define what we mean by an interpretation π of σ in τ. This notion is used for both semantic and syntactic formulations.

We then define what we mean by an interpretation of S in T.

The notion of interpretation of σ in τ is quite weak; *e.g.*, there is no requirement that the interpretation of function symbols be partial functions.

π is an interpretation of σ in τ if and only if π consists of the following data.

i. For each sort $\alpha \in \mathrm{SRT}(\sigma)$, π assigns a τ defined set $\pi(\alpha)$ of tuples of objects of various nonzero lengths and various sorts in $\mathrm{SRT}(\tau)$. Only finitely many lengths are allowed, and separate formulas are needed for each length. We also need separate formulas for each sequence of sorts used. Also, π assigns a τ defined binary relation $= (\alpha)$ which is formally set up to hold only of pairs drawn from $\pi(\alpha)$. Again, separate formulas are needed for every pair of lengths. We allow prospective parameters, so that a finite list of distinguished free variables is given, for each α, which are for the prospective parameters.

ii. Since we are allowing parameters, there is no need to assign data for any $c \in \mathrm{CS}(\sigma)$ of sort $\alpha \in \mathrm{SRT}(\sigma)$. However, we will be interested in the notion of parameterless interpretation. So it is best to have π assign data to $c \in \mathrm{CS}(\sigma)$. π assigns a τ defined set $\pi(c)$, with distinguished variables for prospective parameters.

iii. For each $R \in \mathrm{RS}(\sigma)$ of sort $(\alpha_1, \ldots, \alpha_k)$, π assigns a τ defined set $\pi(R)$ of k tuples (of tuples of various lengths), with distinguished variables for prospective parameters.

iv. For each $F \in \mathrm{FS}(\sigma)$ of sort $(\alpha_1, \ldots, \alpha_{k+1})$, π assigns a τ defined set $\pi(F)$ of $k + 1$ tuples (of tuples of various lengths), with distinguished variables for prospective parameters.

Let S, T be theories in many sorted free logic, with signatures σ, τ. We now define the notion of interpretation. We say that π is an interpretation of S in T if and only if

i. π is an interpretation of σ in τ.
ii. Let $M \models T$. There exists a choice of parameters from the various domains of M such that π defines an actual model of S, with the proviso that equality in each sort be interpreted as the associated binary relation in π (often called a weak model of S when giving, say, the Henkin completeness proof for predicate calculus with equality).

We now give the natural equivalent syntactic notion of interpretation of S in T, in case S is a finite theory in a finite signature.

Let π be an interpretation of σ in τ. Since we are assuming that S is finite, there are only finitely many distinguished variables $v_1, \ldots, v_k$ used for prospective parameters, in π. Let φ be a σ sentence. Then $\pi\varphi$ is the τ formula with free variables $v_1, \ldots, v_k$ that asserts that

i. π defines a σ structure.
ii. φ holds in this σ structure.

The requirement is that if φ is the universal closure of an axiom of S, then $(\exists v_1, \ldots, v_k)(\pi\varphi)$ is provable in T.

It is obvious by the completeness theorem that this is equivalent to the original semantic definition, provided S is finite.

More generally, the semantic notion has a natural syntactic equivalent if

i. The relational type of S is finite; and
ii. We do not allow parameters.

NOTE. From now on we will only consider *finite* theories S, T, and only interpretations *without parameters*.

Let S, T be theories in many sorted free logic. We say that T is a conservative extension of S if and only if

i. The signature τ of T extends the signature σ of S.
ii. S, T prove the same σ formulas.

We say that T is a definitional extension of S if and only if

i. S, T have the same sorts.
ii. T is logically equivalent to an extension of S that is obtained only by adding axioms which explicitly define the new symbols in T by means of formulas in the signature of S

We say that π is a faithful interpretation of S in T if and only if π is an interpretation of S in T, where for all sentences φ in the signature of S, S proves φ if and only if T proves $\pi\varphi$.

We consider two important conditions on a pair of interpretations π of S in T, and π' of T in S.

The first condition, which we call weak synonymy, asserts that for all models M of S, $\pi\pi' M \approx M$, and for all models M of T, $\pi'\pi M \approx M$. Here $\approx$ is isomorphism.

THEOREM 2.1. *Let π, π' be a weak synonymy of S, T. Then*

i. *Let $M, M^* \models S$. Then $M \approx M^* \leftrightarrow \pi' M \approx \pi' M^*$.*
ii. *Let $M, M^* \models T$. Then $M \approx M^* \leftrightarrow \pi M \approx \pi M^*$.*
iii. *π is a faithful interpretation of S in T.*
iv. *π' is a faithful interpretation of T in S.*

PROOF. Let π, π' be a weak synonymy of S, T. For i, assume $M, M^* \models S$, $\pi' M \approx \pi' M^*$. Then $\pi\pi' M \approx \pi\pi' M^* \approx M \approx M^*$.

For iii, assume T proves $\pi\varphi$. Let $M \models S$. Then $\pi' M \models T$, and hence $\pi' M \models \pi\varphi$. Hence $\pi\pi' M \models \varphi$. Therefore $M \models \varphi$. Hence S proves φ.

Claims ii, iv are by symmetry. ⊣

The second condition is even stronger, and makes sense when S, T have the same sorts. We say that π, π' are a synonymy of S, T if and only if

i. π, π' are domain preserving interpretations of S in T and of T in S, respectively.
ii. For $M \models S$, $\pi\pi' M = M$.
iii. For $M \models T$, $\pi'\pi M = M$.

We now show that this notion is the same as another notion that is commonly used to mean synonymy. The first is also model theoretic.

We say that S, T are (weakly) synonymous if and only if there is a (weak) synonymy π, π' of S, T.

THEOREM 2.2. *Let S, T be two theories with the same sorts, but where the symbols have been renamed, if necessary, so that S, T have no symbols in common. There is a synonymy of S, T if and only if S, T have a common definitional extension.*

PROOF. Let π, π' be a synonymy of S, T, where S, T have the same sorts, and the symbols have been disjointified. Let R consist of

i. the axioms of S.
ii. the axioms of T.
iii. axioms defining the symbols of T by formulas in the signature of S, via π'.
iv. axioms defining the symbols of S by formulas in the signature of T, via π.

We claim that R is a definitional extension of S. To see this, it suffices to show that i, iii logically imply ii, iv. We now argue in i, iii. To obtain ii, we need only obtain the interpretations of the axioms of T by π'. But these interpretations are provable in S.

We now have to obtain iv. A typical instance of iv would take the form

1) $(\forall x_1, \dots, x_k)(P(x_1, \dots, x_k) \leftrightarrow \pi P(x_1, \dots, x_k))$.

From iii, we obtain

2) $\pi P(x_1, \dots, x_k) \leftrightarrow \pi' \pi P(x_1, \dots, x_k)$.

Since π, π' are a synonymy,

3) S proves $\pi' \pi P(x_1, \dots, x_k) \leftrightarrow P(x_1, \dots, x_k)$

by the completeness theorem.

Hence by i, iii, we obtain 1).

By the symmetric argument, we also see that R is a definitional extension of T.

Conversely, let R be a definitional extension of both S, T, where the symbols of S, T have been disjointified. We have two axiomatizations of R. The first corresponds to R as a definitional extension of T, and the second corresponds to R as a definitional extension of S. Let π be a definition of the symbols of S by formulas in the signature of T, viewed as a potential interpretation of S in T. Let π' be a definition of the symbols of T by formulas in the signature of S, viewed as a potential interpretation of T in S.

Let $M \models T$. Then $(\pi M, M)$ satisfies the first axiomatization of R. In particular, $\pi M \models S$. Also $(\pi M, M)$ satisfies the second axiomatization of R. Therefore $\pi' \pi M = M$.

Let $M \models S$. Then $(M, \pi' M)$ satisfies the second axiomatization of R. In particular, $\pi' M \models T$. Also $(M, \pi' M)$ satisfies the second axiomatization of R. Therefore $\pi \pi' M = M$. ⊣

THEOREM 2.3. *Let S, T be two theories with the same sorts. There is a synonymy of S, T if and only if there are interpretations π from S in T and*

π' from T in S, such that the following holds. For all formulas φ in the signature of S and ψ in the signature of T, S proves $\varphi \leftrightarrow \pi'\pi\varphi$, and T proves $\psi \leftrightarrow \pi\pi'\psi$.

PROOF. Let π, π' be a synonymy of S, T. Let $M \models S$, and $M \models \varphi[\alpha]$. Since π' is domain preserving, $\pi'M \models T$, and $\pi'M \models \pi\varphi[\alpha]$. Since π is domain preserving, $\pi\pi'M \models S$, and $\pi\pi'M \models \pi'\pi\varphi[\alpha]$. Hence $M \models \pi'\pi[\alpha]$. Therefore $M \models \varphi \leftrightarrow (\pi'\pi)[\alpha]$. Since α is an arbitrary assignment for the signature of S, and M is an arbitrary model of S, we have that $\varphi \leftrightarrow \pi'\pi\varphi$ is provable in S.

Conversely, assume that S, T, π, π' are as given. By applying the conditions to atomic formulas φ, ψ, we obtain that for all $M \models S$, $\pi\pi'M = M$, and for all $M \models T$, $\pi'\pi M = M$. ⊣

If S, T obey the equivalent conditions in Theorems 2.2 and 2.3, then we say that S, T are synonymous. If there is a weak synonymy of S, T, then we say that S, T are weakly synonymous.

We will also use the following well known result. A theory is said to be decidable if its set of consequences (in its own signature) is recursive.

THEOREM 2.4. *Suppose S is interpretable in T, where T is consistent and decidable. Then S has a consistent decidable extension with the same signature as S.*

PROOF. Let π be an interpretation of S in T, where T is decidable. Let S' consist of the sentences φ in the signature of S such that T proves $\pi\varphi$. Clearly S' extends S, and S' is a recursive set. Deductive closure and consistency are obvious. ⊣

§3. PFA(N), EFA(N, exp), logical strength.

We now present two very basic and well studied systems of arithmetic. The most comprehensive current reference to fragments of arithmetic is [HP98].

PFA(N), EFA(N, exp) are based on the set N of all nonnegative integers. In the later sections, with the exception of Section 4, we focus on systems based on the set Z of all integers.

PFA abbreviates "polynomial function arithmetic", and EFA abbreviates "exponential function arithmetic".

PFA(N), EFA(N, exp) build on an earlier system due to R.M. Robinson, called Q (see [Ro52]). We use the notation $Q(N)$, to emphasize that Q is based on N and not on Z.

The signature of $Q(N)$ is $L(N)$. $L(N)$ is one sorted, with 0, S, $+$, $\cdot$, $<$, $=$. The standard model for $L(N)$ is the usual N, 0, S, $+$, $\cdot$, $<$, $=$.

The signature of PFA(N) is also $L(N)$. The signature of EFA(N, exp) is $L(N, \exp)$, which is one sorted, with 0, S, $+$, $\cdot$, exp, $<$, $=$, where exp is a binary function symbol. The standard model for $L(N, \exp)$ is the usual N, 0, S, $+$, $\cdot$, $<$, $=$, where we take $\exp(0, 0) = 1$.

The nonlogical axioms of $Q(N)$ are as follows.

Q1. $\neg Sx = 0$.
Q2. $Sx = Sy \rightarrow x = y$.
Q3. $(\neg x = 0) \rightarrow (\exists y)(x = Sy)$.
Q4. $x + 0 = x$.
Q5. $x + Sy = S(x + y)$.
Q6. $x \cdot 0 = 0$.
Q7. $x \cdot Sy = (x \cdot y) + x$.
Q8. $x < y \leftrightarrow (\exists z)(z + Sx = y)$.

Note that in free logic, these axioms logically imply that $0\downarrow$, $Sx\downarrow$, $x + y\downarrow$, $x \cdot y\downarrow$.

The $\Sigma_0(N)$ ($\Sigma_0(N, \exp)$) formulas are the formulas of $L(N)$ ($L(N, \exp)$) defined as follows.

i) every atomic formula of $L(N)$ ($L(N, \exp)$) is in $\Sigma_0(N)$ ($\Sigma_0(N, \exp)$);
ii) if φ, ψ are in $\Sigma_0(N)$ ($\Sigma_0(N, \exp)$) then so are $\neg\varphi$, $\varphi \wedge \psi$, $\psi \vee \psi$, $\varphi \rightarrow \psi$, $\varphi \leftrightarrow \psi$;
iii) if φ is $\Sigma_0(N)$ ($\Sigma_0(N, \exp)$) and x is a variable not in the term t of $L(N)$ ($L(N, \exp)$), then $(\exists x)(x < t \wedge \varphi)$ and $(\forall x)(x < t \rightarrow \varphi)$ are in $\Sigma_0(N)$ ($\Sigma_0(N, \exp)$).

In [HP98], the terms t in bounded quantification are required to be variables. This is a minor difference.

PFA(N) is essentially the same as $I\Sigma_0$ in [HP98, page 29]. The nonlogical axioms of PFA(N) are as follows.

1. The axioms of $Q(N)$.
2. $(\varphi[x/0] \wedge (\forall x)(\varphi \rightarrow \varphi[x/Sx])) \rightarrow \varphi$, where φ is $\Sigma_0(N)$.

EFA$(N, \exp)$ is essentially the same as $I\Sigma_0(\exp)$ in [HP98, page 37], (although there only base 2 exponentiation is used). The nonlogical axioms of EFA$(N, \exp)$ are as follows.

1. The axioms of Q.
2. $\exp(x, 0) = S0$, $\exp(x, Sy) = \exp(x, y) \cdot x$.
3. $(\varphi[x/0] \wedge (\forall x)(\varphi \rightarrow \varphi[x/Sx])) \rightarrow \varphi$, where φ is in $\Sigma_0(N, \exp)$.

We introduced the one sorted system EFA $=$ EFA$(N, \exp)$ in [Fr80]. It was also used in the exposition of our work on Translatability and Relative Consistency, in [Sm82]. See [HP98, page 405], second paragraph, regarding some historical points.

EFA$(N, \exp)$ represents the minimum level of formal arithmetic where standard coding mechanisms in arithmetic can be done naturally without worry. For example, we do not have to worry about how to code sets of binary relations on $[0, n]$.

In fact, EFA$(N, \exp)$ appears to be quite strong from the mathematical viewpoint. We conjecture that EFA$(N, \exp)$ is sufficient to prove any normal theorem of number theory that is adequately formalizable in its language. We can be liberal about "formalizable" here, using the various natural codings available in $\mathrm{EFA}(N, \exp)$.

For example, we conjecture that Fermat's Last Theorem is provable in $\mathrm{EFA}(N, \exp)$. This has never been established. This conjecture captured the imagination of Jeremy Avigad who wrote extensively about it, and related issues, in [Av03].

Accordingly, we now make the following definition.

T has logical strength if and only if
$\mathrm{EFA}(N, \exp)$ is interpretable in T.

The main point of this paper is the presentation of strictly mathematical theories with logical strength. See Section 7, and Corollary 11.12.

§4. Five related systems of arithmetic with N. We now introduce six systems of arithmetic on N that are closely related to $\mathrm{PFA}(N)$ and $\mathrm{EFA}(N, \exp)$.

LEMMA 4.1. *There is a $\Sigma_0(N)$ formula* $\mathrm{Exp}(x, y, z)$ *with only the distinct free variables shown such that the following is provable in* $\mathrm{PFA}(N)$.

i) $\mathrm{Exp}(x, 0, z) \leftrightarrow z = S0$;
ii) $\mathrm{Exp}(x, Sy, z) \leftrightarrow (\exists v)(\mathrm{Exp}(x, y, v) \wedge z = v \cdot x)$;
iii) $(\mathrm{Exp}(x, y, z) \wedge \mathrm{Exp}(x, y, w)) \rightarrow z = w$.

PROOF. See [HP98, page 299]. ⊣

LEMMA 4.2. *Suppose* $\mathrm{Exp}(x, y, z)$ *and* $\mathrm{Exp}'(x, y, z)$ *satisfy the condition in Lemma* 4.1. *Then* $\mathrm{PFA}(N)$ *proves their equivalence.*

PROOF. Let $\mathrm{Exp}(x, y, z)$, $\mathrm{Exp}'(x, y, z)$ obey the conditions in Lemma 4.1. Let n, m, r be such that $\mathrm{Exp}(n, m, r) \wedge \neg \mathrm{Exp}'(n, m, r)$. Fix n, r, and let m be least such that $(\exists s \leq r)(\mathrm{Exp}(n, m, s) \wedge \neg \mathrm{Exp}'(n, m, s))$. Let $\mathrm{Exp}(n, m, s)$, $\neg \mathrm{Exp}'(n, m, s)$, $s \leq r$. Clearly $m > 0$. Let $\mathrm{Exp}(n, m - 1, t)$, $s = t \cdot n$. Then $\neg \mathrm{Exp}'(n, m - 1, t)$. Also $n = 0 \vee t \leq s \leq r$. The latter is impossible by the choice of m. Hence $n = 0$, $s = 0$. Since $m > 0$, $\mathrm{Exp}'(n, m, s)$. This is a contradiction. ⊣

The sentence $\mathrm{EXP}(N)$ is taken to be $(\forall x, y)(\exists z)(\mathrm{Exp}(x, y, z))$, where Exp is any formula satisfying the conditions of Lemma 4.1. By Lemma 4.2, this defines $\mathrm{EXP}(N)$ up to provable equivalence in $\mathrm{PFA}(N)$.

Let $\mathrm{CM}(N)$ = "common multiples" be the following sentence in $L(N)$.

$\mathrm{CM}(N)$: For all $n > 0$, the integers $1, 2, \ldots, n$ have a positive common multiple.

The five relevant fragments of arithmetic considered here are as follows.

$$Q(N),\ \mathrm{PFA}(N),\ \mathrm{PFA}(N) + \mathrm{EXP}(N),\ \mathrm{PFA}(N) + \mathrm{CM}(N),\ \mathrm{EFA}(N, \exp).$$

Note that the signature of all of these systems is $L(N)$, except for the last, which has signature $L(N, \exp)$.

The most basic relationships between these theories are well known, and summarized in the following two theorems.

THEOREM 4.3. $Q(N) \subseteq \text{PFA}(N) \subseteq \text{PFA}(N) + \text{CM}(N) = \text{PFA}(N) + \text{EXP}(N) \subseteq \text{EFA}(N, \exp)$. *These* $\subseteq$ *are all proper.*

PROOF. Assume $\text{PFA}(N) + \text{EXP}(N)$. Write 2^x, $x \geq 0$, according to $\text{EXP}(N)$. Fix $n \geq 1$. We prove by induction on $1 \leq m \leq n$ that $1, \ldots, n$ have a positive common multiple $\leq 2^{m^2}$. This is obvious for $m = 1$. Let $1 \leq m < n$, and x be a positive common multiple of $1, \ldots, m$, $x \leq 2^{m^2}$. Then $x(m + 1) \leq (2^{m^2})(2^m) \leq 2^{(m+1)^2}$. This establishes $\text{CM}(N)$.

Now assume $\text{PFA}(N) + \text{CM}(N)$. Fix $n, m \geq 2$. Let x be a positive common multiple of $1, \ldots, nm$. We can assume that x is the least positive common multiple of $1, \ldots, nm$. Show that every prime factor of x is $\leq nm$. Show that $x + 1, 2x + 1, \ldots, (nm)x + 1$ are pairwise relatively prime. Let y be a positive common multiple of $1, \ldots, nx + 1$. Code n-tuples as Gödel did, $\leq y$, in order to develop the geometric progression $1, n, n^2, \ldots, n^m$. This establishes $\text{EXP}(N)$.

To see that $\text{PFA}(N) + \text{EXP}(N) \subseteq \text{EFA}(N, \exp)$, write $\text{Exp}(n, m, r)$ for the internal exponentiation relation (in $L(N)$). Argue that $\text{Exp}(n, m, r) \leftrightarrow \exp(n, m) = r$ using $\Sigma_0(\exp)$ induction, exactly as in the proof of Lemma 4.2.

To see that $Q(N)$ does not prove $\text{PFA}(N)$, consider all of the polynomials in one variable x with integer coefficients, which have a positive leading coefficient, or is 0. These form a model of $Q(N)$ under the usual 0, S, $+$, $\cdot$, $=$, with $\leq$ defined according to axiom Q8 of $Q(N)$. This model of $Q(N)$ does not satisfy, for example, $(\exists y)(x = 2y \vee x = 2y + 1)$.

To see that $\text{PFA}(N)$ does not prove $\text{EXP}(N)$, let M be a nonstandard model of $\text{PFA}(N)$. Let x be a positive nonstandard integer in M, and let M' be the restriction of M to the integers of M whose magnitude is at most x^n, for some standard $n \geq 1$. Them M' is a model of $\text{PFA}(N)$. It is easily verified that 2^x does not exist in M. ⊣

THEOREM 4.4. $\text{EFA}(N, \exp)$ *is a definitional extension of* $\text{PFA}(N) + \text{EXP}(N)$. $\text{PFA}(N)$ *is interpretable in* $Q(N)$. $\text{PFA}(N) + \text{EXP}(N)$ *is not interpretable in* $\text{PFA}(N)$.

PROOF. For the first claim, note that by the proof of Theorem 4.3 (that $\text{PFA}(N) + \text{EXP}(N) \subseteq \text{EFA}(N, \exp)$), $\text{EFA}(N, \exp)$ proves that $\exp(n, m) = r \leftrightarrow \text{Exp}(n, m, r)$.

It remains to show that every axiom of $\text{EFA}(N, \exp)$ is provable in $\text{PFA}(N) + \text{EXP}(N) + (\forall n, m, r)(\exp(n, m) = r \leftrightarrow \text{Exp}(n, m, r))$. This is clear by inspection.

The second claim is credited to Wilkie, in the sharp form on p. 367 of [HP98].

For the third claim, see [Wi86], and [HP98, page 391]. ⊣

§5. Five related systems of arithmetic with Z. We now introduce six systems of arithmetic on Z that are closely related to the six systems of Section 4. These are parallel to those systems introduced in Section 4, and move us closer to the strictly mathematical theories of Section 7.

We first introduce LOID(Z). LOID abbreviates "linearly ordered integral domain". According to Theorem 5.3 below, LOID is an extremely robust strictly mathematical theory.

The signature of LOID(Z) is $L(Z)$. $L(Z)$ is one sorted, with 0, 1, $+$, $-$, $\cdot$, $<$, $=$, where $-$ is unary. The standard model for $L(Z)$ is the usual Z, 0, 1, $+$, $-$, $\cdot$, $<$, $=$.

The nonlogical axioms of LOID(Z) are

a. $x + 0 = x$.
b. $x + y = y + x$.
c. $x + (y + z) = (x + y) + z$.
d. $x + (-x) = 0$.
e. $x \cdot 1 = x$.
f. $x \cdot y = y \cdot x$.
g. $x \cdot (y \cdot z) = (x \cdot y) \cdot z$.
h. $x \cdot (y + z) = (x \cdot y) + (x \cdot z)$.
i. $x \cdot y = 0 \rightarrow (x = 0 \vee y = 0)$.
j. $0 < 1$.
k. $\neg x < x$.
l. $(x < y \wedge y < z) \rightarrow x < z$.
m. $x < y \vee x = y \vee y < x$.
n. $(0 < x \wedge 0 < y) \rightarrow (0 < x + y \wedge 0 < x \cdot y)$.
o. $x < y \leftrightarrow -y < -x$.

Note that in free logic, these axioms imply $0\downarrow$, $1\downarrow$, $x + y\downarrow$, $-x\downarrow$, $x \cdot y\downarrow$.

How do we know that we have included all appropriate axioms in LOID(Z)? We first present some basic development of LOID(Z). Define $x > y \leftrightarrow x < y$, $x \neq 0 \leftrightarrow \neg x = 0$.

LEMMA 5.1. *The following are provable in* LOID(Z).

i. $- - x = x$.
ii. $-(x + y) = -x + -y$.
iii. $x + x = 0 \leftrightarrow x = 0$.
iv. $-x = (-1) \cdot x$.
v. $(-1) \cdot (-1) = 1$.
vi. $x, y < 0 \rightarrow x \cdot y > 0$.
vii. $x < 0 \wedge y > 0 \rightarrow x \cdot y < 0$.

PROOF.

i. By d, $-x + - - x = 0$. Hence $x = x + (-x + - - x) = (x + -x) + - - x = 0 + - - x = - - x$.

ii. By d, $x + y + -(x + y) = 0$. Hence $-x + -y + x + y + -(x + y) = -x + -y = -(x + y)$.
iii. Use m. If $x = 0$ then we are done. Assume $0 < x$. Then by n, $0 < x + x$, and so $x + x \neq 0$, $x \neq 0$. Assume $x < 0$. By o, $0 < -x$, $0 < -x + -x = -(x + x)$. Hence $-(x + x) \neq 0$, $x \neq 0$.
iv. $(-1) \cdot x + 1 \cdot x = 0 = (-1) \cdot x + x$. Hence $0 + -x = (((-1) \cdot x) + x) + -x = (-1) \cdot x$.
v. $(-1) \cdot (-1) = - - 1 = 1$.
vi. Assume $x, y < 0$. By axioms n, o, $0 < -x, -y$, $(-x) \cdot (-y) > 0$. Now $(-x) \cdot (-y) = (-1) \cdot x \cdot (-1) \cdot y = (-1) \cdot (-1) \cdot x \cdot y = 1 \cdot x \cdot y = x \cdot y$.
vii. Assume $x < 0$, $y > 0$. By axioms n, o, $0 < -x, y$, $0 < (-x) \cdot y$. Since $(-x) \cdot y = (-1) \cdot x \cdot y = -(x \cdot y)$, we have $0 < -(x \cdot y)$, $x \cdot y < 0$. ⊣

The official definition of an ordered field is given in, say, [Ja85, page 307]:

An ordered field (F, P) is a field F together with a subset P (the set of positive elements) of F such that

i. $0 \notin P$.
ii. $a \in F \rightarrow (a \in P \vee a = 0 \vee -a \in P)$.
iii. $a, b \in P \rightarrow (a + b \in P \wedge a \cdot b \in P)$.

Lemma 5.2. *Let* $M = (D, 0, 1, +, -, \cdot, <)$ *be a model of* LOID(Z). *There is an ordered field* (F, P) *and an isomorphism* $j : (D, 0, 1, +, -, \cdot) \rightarrow F$ *such that for all* $x \in D$, $jx \in P \leftrightarrow x > 0$.

Proof. Let M be as given. By axioms a-i, M is an integral domain. Hence the fraction field construction results in a field F and a canonical isomorphism $j : (D, 0, 1, +, -, \cdot) \rightarrow F$.

Recall that F consists of the equivalence classes of ordered pairs (x, y), where $x, y \in D$, $y \neq 0$, under the equivalence relation $(x, y) \approx (z, w) \leftrightarrow xw = yz$. Define $jx = [(x, 1)]$. Obviously j is an isomorphism from M into F.

Define $P = \{[(x, y)] : x, y > 0 \vee x, y < 0\}$. *I.e.*, $P = \{[(x, y)] : x, y$ have the same nonzero sign$\}$. It is obvious that $0^F = [(0, 1)] \notin P$.

We claim independence of representatives, in the sense that for all x, y, z, w, if $[x, y] \approx [z, w]$, then

*) x, y have the same nonzero sign $\leftrightarrow$ z, w have the same nonzero sign.

To see this, assume $[x, y] \approx [z, w]$. Then

$$x \cdot w = y \cdot z, \qquad y, w \neq 0.$$

Case 1. $x \neq 0$. Then $x \cdot w \neq 0$, $y \cdot z \neq 0$, $z \neq 0$. By inspection, using Lemma 5.1, vi), vii), we see that *) holds.

Case 2. $x = 0$. Then $y \cdot z = 0$, $z = 0$. By inspection using Lemma 5.1, vi), vii), we see that *) holds.

Now let $[(x, y)] \in F$. If x, y have the same nonzero sign then $[(x, y)] \in P$. If x, y have opposite nonzero signs, then $-x, y$ have the same nonzero sign, and hence $-[(x, y)\} = [(-x, y)] \in P$. Finally, if $x = 0$ then $[(x, y)] = 0^F$.

Now let $[(x, y)], [(z, w)] \in P$. By the independence of representatives (claim above), we can assume that $x, y, z, w > 0$. Note that

$$[(x, y)] + [(z, w)] = [(x \cdot w + y \cdot z, y \cdot w)] \in P.$$
$$[(x, y)] \cdot [(z, w)] = [(x \cdot z, y \cdot w)] \in P.$$

This establishes that (F, P) is an ordered field.

Now let $x \in D$. If $jx = [x, 1] \in P$ then obviously $x > 0$. If $jx = [x, 1] \notin P$ then $\neg x > 0$, using the independence of representatives. ⊣

THEOREM 5.3. *A purely universal sentence in $L(Z)$ is true in the ordered field of real numbers if and only if it is provable in* LOID(Z). LOID(Z) *can be axiomatized as the set of all quantifier free formulas in $L(Z)$ which are universally true in the ordered field of real numbers.*

PROOF. It suffices to show that every purely existential sentence in $L(Z)$ that is true in some model of LOID is true in the ordered field of real numbers.

Let M be a model of LOID satisfying the purely existential sentence φ in $L[Z]$. By Lemma 5.2, let (F, P) be an ordered field extending M, with an isomorphism $j : (D, 0, 1, +, -, \cdot) \to F$, and for all $x \in D$, $jx \in P \leftrightarrow x > 0$.

We define $<$ on F by

$$x < y \leftrightarrow y - x \in P.$$

We claim that

$$j : (D, 0, 1, +, -, \cdot, <) \to (F, <)$$

is an isomorphism. To see this, let $x < y$ in M. Then $y - x > 0$, and so $j(y - x) \in P$. Hence $j(y) - j(x) \in P$, and so $x < y$ in $(F, <)$. Hence φ is preserved, and so φ holds in $(F, <)$.

Now every ordered field $(F, <)$, where $<$ is defined as above from P, extends to an ordered real closed field, whose $<$ agrees with the $<$ of $(F, <)$. Hence φ holds in some ordered real closed field. Since φ is a first order sentence, φ holds in all ordered real closed fields. Hence φ holds in the ordered field of real numbers.

The final claim follows from the previous claim using the observation that the axioms of LOID(Z) are quantifier free formulas in $L(Z)$ which are universally true in the ordered field of real numbers. ⊣

The $\Sigma_0(Z)$ ($\Sigma_0(Z, \exp)$) formulas are the formulas of $L(Z)$ ($L(Z, \exp)$) defined as follows.

i) every atomic formula of $L(Z)$ ($L(Z, \exp)$) is in $\Sigma_0(Z)$ ($\Sigma_0(Z, \exp)$);
ii) if φ, ψ are in $\Sigma_0(Z)$ ($\Sigma_0(Z, \exp)$) then so are $\neg\varphi$, $\varphi \wedge \psi$, $\psi \vee \psi$, $\varphi \to \psi$, $\varphi \leftrightarrow \psi$;

iii) if φ is $\Sigma_0(Z)$ ($\Sigma_0(Z, \exp)$) and x is a variable not in the term t of $L(Z)$ ($L(Z, \exp)$), then $(\exists x)(-t < x < t \wedge \varphi)$ and $(\forall x)(-t < x < t \rightarrow \varphi)$ are in $\Sigma_0(Z)$ ($\Sigma_0(Z, \exp)$).

Henceforth, we will use $x \leq y$ as an abbreviation for $x < y \vee x = y$, and $x \geq y$ as an abbreviation for $y < x \vee y = x$.

The signature of PFA(Z) is $L(Z)$. The nonlogical axioms of PFA(Z) are

1. LOID(Z).
2. $(\varphi[x/0] \wedge (\forall x)(\varphi \rightarrow \varphi[x/Sx])) \rightarrow (x \geq 0 \rightarrow \varphi)$, where φ is in $\Sigma_0(Z)$.

Note that PFA(Z) proves the axiom of discreteness: there is nothing in $(0, 1)$. To see this, let $\varphi = x \geq 0 \wedge (x < 1 \rightarrow x \leq 0)$. Use 2 for φ, to obtain $(x \geq 0 \wedge x < 1) \rightarrow x \leq 0$. Now suppose $x < 1$. If $x > 0$ then $x \leq 0$. Hence $x \leq 0$.

$L(Z, \exp)$ is one sorted, with 0, 1, $+$, $-$, $\cdot$, exp, $<$, $=$, where $-$ is unary and exp is binary. The standard model for $L(Z, \exp)$ is the usual Z, 0, 1, $+$, $-$, $\cdot$, exp, $<$, $=$, where $\exp(x, y)$ is the usual x^y, which is defined if and only if $y \geq 0$, and where $x^0 = 1$.

The signature of EFA(Z, exp) is $L(Z, \exp)$. The nonlogical axioms of EFA(Z, exp) are

1. LOID(Z).
2. $\exp(x, 0) = 1$.
3. $y \geq 0 \rightarrow (\exp(x, y + 1) = \exp(x, y) \cdot x \wedge \exp(x, -y - 1)\uparrow)$.
4. $(\varphi[x/0] \wedge (\forall x)(\varphi \rightarrow \varphi[x/x + 1])) \rightarrow (x \geq 0 \rightarrow \varphi)$, where φ is in $\Sigma_0(Z, \exp)$.

Lemma 5.4. *There is a* $\Sigma_0(Z)$ *formula* $\mathrm{Exp}(x, y, z)$ *with only the distinct free variables shown such that the following is provable in* PFA(Z).

i) $\mathrm{Exp}(x, 0, z) \leftrightarrow z = S0$;
ii) $y \geq 0 \rightarrow (\mathrm{Exp}(x, y + 1, z) \leftrightarrow (\exists v)(\mathrm{Exp}(x, y, v) \wedge z = v \cdot x))$;
iii) $(\mathrm{Exp}(x, y, z) \wedge \mathrm{Exp}(x, y, w)) \rightarrow z = w$;
iv) $\mathrm{Exp}(x, y, z) \rightarrow y \geq 0$.

Proof. This is a straightforward adaptation of Lemma 4.1 to PFA(Z). ⊣

Lemma 5.5. *Suppose* $\mathrm{Exp}(x, y, z)$ *and* $\mathrm{Exp}'(x, y, z)$ *satisfies the condition in Lemma* 5.4. *Then* PFA(Z) *proves their equivalence.*

Proof. This is a straightforward adaptation of Lemma 4.2 to PFA(Z). ⊣

The sentence EXP(Z) is taken to be $(\forall x, y)(\exists z)(\mathrm{Exp}(x, y, z))$, where Exp is any formula satisfying the conditions of Lemma 5.4. By Lemma 5.5, this defines EXP(Z) up to provable equivalence in PFA(Z).

Let CM(Z) = "common multiples" be the following sentence in $L(Z)$.

CM(Z): For all $n > 0$, the integers $1, 2, \ldots, n$ have a positive common multiple.

Note that $\mathrm{CM}(Z)$ is formally the same as $\mathrm{CM}(N)$, but it is still convenient to use the notation $\mathrm{CM}(Z)$, $\mathrm{CM}(N)$.

The five relevant fragments of arithmetic considered here are as follows.

$$\mathrm{LOID}(Z),\ \mathrm{PFA}(Z),\ \mathrm{PFA}(Z)+\mathrm{EXP}(Z),\ \mathrm{PFA}(Z)+\mathrm{CM}(Z),\ \mathrm{EFA}(Z,\exp).$$

The most basic relationships between these theories are summarized in the following three theorems.

THEOREM 5.6. $\mathrm{LOID}(Z) \subseteq \mathrm{PFA}(Z) \subseteq \mathrm{PFA}(Z)+\mathrm{CM}(Z) = \mathrm{PFA}(Z)+\mathrm{EXP}(Z) \subseteq \mathrm{EFA}(Z,\exp)$. *These* $\subseteq$ *are all proper.*

THEOREM 5.7. $\mathrm{EFA}(Z,\exp)$ *is a definitional extension of* $\mathrm{PFA}(Z)+\mathrm{EXP}(Z)$. $\mathrm{PFA}(Z)+\mathrm{EXP}(Z)$ *is not interpretable in* $\mathrm{PFA}(Z)$. $\mathrm{PFA}(Z)$ *is not interpretable in* $\mathrm{LOID}(Z)$.

PROOF. Theorem 5.6, 5.7, with the exception of the final claim of Theorem 5.7, can be proved by an adaptation of the corresponding proofs of Theorems 4.3, 4.4.

For the final claim of Theorem 5.7, the essence of the matter is that $Q(N)$ is not interpretable in the theory of ordered real closed fields, ORCF. If this were not the case, then, since the theory of ordered real closed fields is decidable, by Theorem 2.4 we have $Q(N)$ has a consistent decidable extension with signature $L(N)$. This contradicts the well known essential undecidability of $Q(N)$; see [Ro52]. Obviously $\mathrm{PFA}(Z)$ is not interpretable in $\mathrm{LOID}(Z)$, since $Q(N)$ is trivially interpretable in $\mathrm{PFA}(Z)$. ⊣

§6. Arithmetic on N and arithmetic on Z. We establish some relationships between the five systems of Section 4,

$$Q(N),\ \mathrm{PFA}(N),\ \mathrm{PFA}(N)+\mathrm{EXP}(N),\ \mathrm{PFA}(N)+\mathrm{CM}(N),\ \mathrm{EFA}(N,\exp).$$

and the five systems of Section 5,

$$\mathrm{LOID}(Z),\ \mathrm{PFA}(Z),\ \mathrm{PFA}(Z)+\mathrm{EXP}(Z),\ \mathrm{PFA}(Z)+\mathrm{CM}(Z),\ \mathrm{EFA}(Z,\exp).$$

In Sections 4, 5, we have discussed the relationships between the theories in each of the two groups separately.

We can interpret $\mathrm{PFA}(N)$ in $\mathrm{PFA}(Z)$, by taking the domain to be the nonnegative elements, and defining $0, S, +, \cdot, <, =$ in the obvious way. We call this interpretation $\pi(N,Z)$.

We can interpret $\mathrm{PFA}(Z)$ in $\mathrm{PFA}(N)$ by taking the domain to be the pairs $(n,0)$, $n>0$, and $(n,1)$. Here $(n,0)$ represents the negative integer $-n$, and $(n,1)$ represents the nonnegative integer n. We define $0, 1, +, -, \cdot, <, =$ in the obvious way. We call this interpretation $\pi(Z,N)$.

THEOREM 6.1. $\pi(N,Z)$, $\pi(Z,N)$ *is a weak synonymy of* $\mathrm{PFA}(N)$, $\mathrm{PFA}(Z)$, *and is also a weak synonymy of* $\mathrm{PFA}(N)+\mathrm{EXP}(N)$, $\mathrm{PFA}(Z)+\mathrm{EXP}(Z)$.

PROOF. It is obvious that $\pi(N, Z)$ is an interpretation of PFA(N) in PFA(Z), and $\pi(Z, N)$ is an interpretation of PFA(Z) in PFA(N). It is also obvious that $\pi(N, Z)$ is an interpretation of PFA(N)+EXP(N) in PFA(Z)+EXP(Z), and $\pi(Z, N)$ is an interpretation of PFA(Z) + EXP(Z) in PFA(N) + EXP(N).

For weak synonymy, let $M \models$ PFA(N), and within M, form the $(n, 0)$, $(n, 1)$ interpretation of PFA(Z), according to π', obtaining $\pi' M \models$ PFA(Z). Within $\pi' M$, form the nonnegative element interpretation of PFA(N), according to π, obtaining $\pi\pi' M$. The nonnegative element interpretation just uses the $(n, 1)$. Clearly we have an isomorphism from $\pi\pi' M$ onto M by sending each $(n, 1)$ to n.

Let $M \models$ PFA(Z), and within M, form the nonnegative element interpretation of PFA(N), according to π, obtaining $\pi M \models$ PFA(N). Within πM, form the $(n, 0)$, $(n, 1)$ interpretation of PFA(Z), according to π', obtaining $\pi' \pi M$. Clearly we have an isomorphism from $\pi' \pi M$ onto M by sending each negative n to $(n, 0)$, and each nonnegative n to $(n, 1)$. ⊣

We extend $\pi(N, Z)$ and $\pi(Z, N)$ in the obvious way to $\pi(N, Z; \exp)$, $\pi(Z, N; \exp)$.

THEOREM 6.2. *$\pi(N, Z, \exp)$, $\pi(Z, N, \exp)$ is a weak synonymy of* EFA(N, exp), EFA(Z, exp).

PROOF. Argue as for Theorem 6.1. ⊣

Note that $\pi(N, Z)$ and $\pi(Z, N)$ are not domain preserving, and so we cannot use them to establish synonymy. We give new interpretations for this purpose.

We can interpret PFA(N) in PFA(Z), by taking the N to be

$$0, 1, -1, 2, -2, \dots$$

with the obvious corresponding definition of $0, S, +, \cdot, <, =$. Specifically, we first define, in PFA(Z), the function $f : Z \to Z$ by $f(x) =$ the position in the above sequence $= 0$ if $x = 0$; $2x - 1$ if $x > 0$; $-2x$ if $x < 0$. Then we define $0', S', +', \cdot', <', =$, uniquely, in such a way that f is an isomorphism from Z, $0', S', +', \cdot', <', =$ onto $\{x : x \geq 0\}, 0, +1, +, \cdot, <, =$. Call this $\pi'(N, Z)$.

We can interpret PFA(Z) in PFA(N), by taking the Z to be

$$\dots 6, 4, 2, 0, 1, 3, 5, \dots$$

with the obvious corresponding definition of $0, 1, +, -, \cdot, <, =$. Specifically, we first define, in PFA(N), the function $g : N \to N \times \{0, 1\}$ by $g(2n + 1) = (1, n + 1)$, $g(2n + 2) = (0, n + 1)$, $g(0) = 0$. Then we define $0', 1', +', -', \cdot'$, $<', =$, uniquely, in such a way that g is an isomorphism from $N, 0', 1', +', -'$, $\cdot', <, =$ onto $\{(x, 0) : x > 0\} \cup \{(x, 1) : x \geq 0\}$ with its usual $0^*, 1^*, +^*, -^*$, $\cdot^*, <^*, =$, that makes it look like the arithmetic of Z. Call this $\pi'(Z, N)$.

LEMMA 6.3. *$\pi'(N, Z)$, $\pi'(Z, N)$ is a synonymy of* PFA(N), PFA(Z), *and also of* PFA(N) + EXP(N), PFA(Z) + EXP(Z).

PROOF. It is obvious that $\pi'(N, Z)$ is an interpretation of PFA(N) in PFA(Z), and $\pi'(Z, N)$ is an interpretation of PFA(Z) in PFA(N). This is also obvious with EXP(N) and EXP(Z).

For synonymy, let $M \models$ PFA(N), and within M, form the $\dots 6, 4, 2, 0, 1, 3, 5, \dots$ interpretation of PFA(Z), according to π', obtaining $\pi' M \models$ PFA(Z). Within $\pi' M$, form the $0, 1, -1, 2, -2, \dots$ interpretation of PFA(N), according to π, obtaining $\pi\pi' M$. Note that in $\pi' M$, $0, 1, -1, 2, -2, \dots$ is the $0, 1, 2, 3, \dots$ of M. Hence $\pi\pi' M = M$.

Let $M \models$ PFA(Z), and within M, form the $0, 1, -1, 2, -2, \dots$ interpretation of PFA(N), according to π, obtaining $\pi M \models$ PFA(N). Within πM, form the $\dots 6, 4, 2, 0, 1, 3, 5, \dots$ interpretation of PFA(Z), according to π', obtaining $\pi'\pi M$. Note that in πM, $\dots 6, 4, 2, 0, 1, 3, 5, \dots$ is the $\dots, -3, -2, -1, 0, 1, 2, 3, \dots$ of M. Hence $\pi'\pi M = M$. ⊣

We extend $\pi(N, Z)$ and $\pi(Z, N)$ in the obvious way to $\pi(N, Z; \exp)$, $\pi(Z, N; \exp)$.

LEMMA 6.4. $\pi'(N, Z, \exp)$, $\pi'(Z, N, \exp)$ *is a synonymy of* EFA(N, exp), EFA(Z, exp).

PROOF. Argue as for Lemma 6.3. ⊣

We now construct a certain model M of $Q(N)$. The domain will consist of certain polynomials in variables x_α, $\alpha < \omega_1$, with integer coefficients. We will not be using the ordering of variables.

Let P be such a polynomial. The maximal monomials of P are the monomials of P that are maximal with respect to the divides relation. Note that if P is not the trivial polynomial, 0, then P has at least one maximal monomial.

We take dom(M) to be these polynomials which are either 0, or whose maximal monomials all have positive coefficients.

For M, we use the ordinary $0, S, +, \cdot$. We define $<$ as in axiom Q8 of $Q(N)$.

LEMMA 6.5. *M is a model of $Q(N)$.*

PROOF. We first need to verify that dom(M) is closed under $+, \cdot$. Let $P, Q \in$ dom(M). We can assume that P, Q are not the 0 polynomial. Let α be a maximal monomial of $P + Q$. If α occurs in P and not in Q, or in Q but not in P, then it retains its coefficient, which must be positive. If α occurs in P and Q, then its coefficient in $P + Q$ is positive. Hence $P + Q \in$ dom(M).

It is trickier to establish that $PQ \in$ dom(M). Let $\alpha_1, \dots, \alpha_n$ be the monomials of P, and $\beta_1, \dots, \beta_m$ be the monomials of Q. Since we are assuming that neither P nor Q is 0, we have $n, m \geq 1$. Let S be the set of all $\alpha_i\beta_j$ that are maximal among all of the $\alpha_i\beta_j$ (even if the coefficient of $\alpha_i\beta_j$ in PQ is ≤ 0). Suppose $\alpha_i\beta_j \in S$. Then obviously α_i is maximal in P and β_j is maximal in Q. Now any $\alpha_p\beta_q = \alpha_i\beta_j$, where α_p is a monomial in P and β_q is a monomial in Q, must have that α_p is maximal in P and β_q is maximal in Q. Hence the coefficient of $\alpha_i\beta_j$ in PQ must be positive (since it is the sum of the coefficients contributed by each of these $\alpha_p\beta_q = \alpha_i\beta_j$). Therefore $\alpha_i\beta_j$ is a monomial of PQ with positive coefficient.

We now claim that every maximal monomial $\alpha_i\beta_j$ of PQ lies in S. To see this, let $\alpha_i\beta_j$ be a maximal monomial of PQ, where $\alpha_i\beta_j \notin S$. Let $\alpha_p\beta_q \in S$ be a proper multiple of $\alpha_i\beta_j$. By the previous paragraph, $\alpha_p\beta_q$ is a monomial of PQ (in fact, with positive coefficient), contradicting that $\alpha_i\beta_j$ is a maximal monomial of PQ.

We have now shown that every maximal monomial $\alpha_i\beta_j$ of PQ has a positive coefficient. Thus $PQ \in \text{dom}(M)$.

The verification of Q2, Q4–Q7 is by the ring laws for polynomials. Q8 is by definition. Q1 follows from the fact that $-1 \notin \text{dom}(M)$. For Q3, let $P \in \text{dom}(M)$, P not 0. If P is nonconstant then $P - 1 \in \text{dom}(M)$. If P is constant then $P \geq 1$, and so $P - 1 \in \text{dom}(M)$. ⊣

Lemma 6.6. *$Q(N)$ and* PFA(N) *are not weakly synonymous.*

Proof. We use the model M of Lemma 6.5. Let M' be a model of PFA(N) defined in M without parameters. We show that M' is countable.

Recall that in interpretations, we allow the domain to consist of tuples of varying lengths. We also allow the equality relation to be interpreted by an equivalence relation. This equivalence relation must be definable in M without parameters.

We call two polynomials isomorphic if and only if they are identical up to a permutation of variables. We call two finite sequences of polynomials isomorphic if and only if they are coordinatewise isomorphic via a single permutation. We call the equivalence classes under this equivalence relation on the finite sequences of polynomials, shapes.

Note that by the symmetry of M, for any two tuples of polynomials that are isomorphic, one lies in dom(M') if and only if the other lies in dom(M').

Case 1. Any two tuples of polynomials in dom(M') that are isomorphic, and lie in dom(M'), are interpreted to be equal in M'. Since the number of shapes is countable, we see that dom(M') is countable.

Case 2. Let $(P_1, \dots, P_n)$, $(Q_1, \dots, Q_n)$ be isomorphic elements of dom(M'), which are not satisfied to be equal in M'. Let α be an automorphism of the variables $\{x_\alpha : \alpha < \omega_1\}$, that interchanges $(P_1, \dots, P_n)$ and $(Q_1, \dots, Q_n)$. Then α extends uniquely to an automorphism α^* of M of finite order, which in turn induces an automorphism β of M' of finite order. Since β interchanges the distinct elements $[(P_1, \dots, P_n)]$ and $[(Q_1, \dots, Q_n)]$ of dom(M'), we see that β has finite order. But no model of PFA(N) can have an automorphism of finite order because of the definable linear ordering $<$, with parameters. So this case is impossible.

Since M' is countable, and M is uncountable, it is clear that we cannot define, in M', an isomorphic copy of M. Hence $Q(N)$, PFA(N) are not weakly synonymous. ⊣

We summarize the synonymy and mutual interpretability results.

THEOREM 6.7. PFA(N), PFA(Z) *are synonymous.* PFA(N) + EXP(N), EFA(N, exp), PFA(Z) + EXP(Z), EFA(Z, exp) *are synonymous. There are no other synonymy, or even weak synonymy, relations between the* 10 *systems.* $Q(N)$, PFA(N), PFA(Z) *are mutually interpretable.* PFA(N) + EXP(N), EFA(N, exp), PFA(Z) + EXP(Z), EFA(Z, exp) *are mutually interpretable. There are no other mutual interpretability relations between the* 10 *systems.* LOID(Z) *is interpretable in* $Q(N)$, *but not vice versa.*

PROOF. The first claim is by Lemma 6.3. For the second claim, PFA(N) + EXP(N), PFA(Z) + EXP(Z) are synonymous by Lemma 6.3. PFA(N) + EXP(N), EFA(N, exp) are synonymous by Theorems 4.4 and 2.2. PFA(Z) + EXP(Z), EFA(Z, exp) are synonymous by Theorems 5.7 and 2.2. EFA(N, exp), EFA(Z, exp) are synonymous by Lemma 6.4.

For the third claim, PFA(N) + EXP(N) is not interpretable in PFA(N) by Theorem 4.4. That $Q(N)$, PFA(N), PFA(Z) are not interpretable in LOID(Z) comes from (the proof of) Theorem 5.7. That $Q(N)$, PFA(N) are not weakly synonymous comes from Lemma 6.6.

For the fourth claim, use the first claim together with the interpretability of PFA(N) in $Q(N)$, from Theorem 4.4.

The fifth claim follows from the second claim.

The sixth claim follows from PFA(N) + EXP(N) not interpretable in PFA(N), and $Q(N)$ not interpretable in LOID(Z). The former is by Theorem 4.4, and the latter is by the proof of Theorem 5.7.

The seventh claim is by the proof of Theorem 5.7, and the interpretability of ORCF in $Q(N)$ in [FF02]. ⊣

§7. Seven strictly mathematical theories. Among the twelve theories considered in Section 6, only $Q(N)$ and LOID(Z) are strictly mathematical. The rest rely on induction stated for all bounded formulas. However, $Q(N)$ and LOID(Z) do not have logical strength, in the sense used in this paper (see the end of Section 3).

We now present six strictly mathematical theories. We will extend the one sorted signatures from Sections 3–5,

$$L(N),\ L(Z),\ L(N, \mathrm{exp}),\ L(Z, \mathrm{exp}),$$

with the new many sorted signatures

$$L(Z, \mathrm{fst}),\ L(Z, \mathrm{fsq}),\ L(Z, \mathrm{fst}, \mathrm{fsq}),\ L(Z, \mathrm{exp}, \mathrm{fst}),$$
$$L(Z, \mathrm{bexp}, \mathrm{fst}),\ L(Z, \mathrm{exp}, \mathrm{fsq}).$$

Here fst abbreviates "finite sets of integers", and fsq abbreviates "finite sequences of integers". Also bexp abbreviates "binary exponentiation".

$L(Z, \exp, \text{fst})$ is two sorted. We use Z for sort 1, and fst for sort 2. Here fst abbreviates "finite sets of integers". We use 0, 1, $+$, $-$, $\cdot$, exp, $<$, $=$ on the Z sort. We use $\in$ between sort Z and sort fst.

$L(Z, \exp, \text{fsq})$ is two sorted. We use Z for sort 1 and fsq for sort 2. Here fsq abbreviates "finite sequences of integers ". We use 0, 1, $+$, $-$, $\cdot$, exp, $<$, $=$ on the Z sort. We use the unary function symbol lth from sort fsq into sort Z. We use the binary function symbol val from sort fsq cross sort Z, into sort Z.

The standard model for $L(Z, \exp, \text{fst})$ has first sort Z, with 0, 1, $+$, $-$, $\cdot$, $<$, $=$ as usual, and $\exp(n, m) = r$ if and only if $n^m = r \wedge m \geq 0$, where $n^0 = 1$. Thus $\exp(n, m)$ is defined if and only if $m \geq 0$. The second sort, fst, consists of the finite subsets of Z, where $\in$ is as usual.

The standard model for $L(Z, \exp, \text{fsq})$ has domain Z, with 0, 1, $+$, $-$, $\cdot$, $<$, $=$ as usual, and $\exp(n, m) = r$ if and only if $n^m = r \wedge m \geq 0$, where $n^0 = 1$. Thus $\exp(n, m)$ is defined if and only if $m \geq 0$. The second sort, fsq, consists of the finite sequences from Z, where lth is the length function, which takes values in the nonnegative elements of Z. Also $\text{val}(x, n)$ is the n-th term of x, counting from 1, and so is defined if and only if $1 \leq n \leq \text{lth}(x)$.

We also work with the elimination of exp in $L(Z, \exp, \text{fst})$, $L(Z, \exp, \text{fsq})$.

The signature of FSTZ is $L(Z, \text{fst})$. The nonlogical axioms of FSTZ are stated informally as follows.

1. Linearly ordered integral domain axioms.
2. Finite interval. $[x, y]$ exists.
3. Boolean difference. $A \backslash B = \{x \in A : x \notin B\}$ exists.
4. Set addition. $A + B = \{x + y : x \in A \wedge x \in B\}$ exists.
5. Set multiplication. $A \cdot B = \{x \cdot y : x \in A \wedge x \in B\}$ exists.
6. Least element. Every nonempty set has a least element.

The signature of FSQZ is $L(Z, \text{fsq})$. The nonlogical axioms of FSQZ are stated informally as follows.

1. Linearly ordered integral domain axioms.
2. $\text{lth}(\alpha) \geq 0$.
3. $\text{val}(\alpha, n)\downarrow \leftrightarrow 1 \leq n \leq \text{lth}(\alpha)$.
4. The finite sequence $(0, \ldots, n)$ exists.
5. $\text{lth}(\alpha) = \text{lth}(\beta) \rightarrow -\alpha, \alpha + \beta, \alpha \cdot \beta$ exist.
6. The concatenation of α, β exists.
7. For all $n \geq 1$, the concatenation of α, n times, exists.
8. There is a finite sequence enumerating the terms of α that are not terms of β.
9. Every nonempty finite sequence has a least term.

Before giving formal versions of these axioms, we make some remarks about the nonlogical axioms of FSQZ.

a. $\downarrow$ indicates "is defined". See Section 1.
b. Axioms 4–8 are presented in terms of the length and values of the finite sequence that is asserted to exist. In the case of axiom 8, this involves the ring operations.
c. Axiom 7 uses n as a variable (not a standard integer).

We now give formal presentations of FSTZ and FSQZ.

The nonlogical axioms of FSTZ are given formally as follows.

1. Linearly ordered commutative ring axioms.
2. Finite interval.

$$(\exists A)(\forall x)(x \in A \leftrightarrow (y \leq x \wedge x \leq z)).$$

3. Boolean difference.

$$(\exists C)(\forall x)(x \in C \leftrightarrow (x \in A \wedge \neg(x \in B))).$$

4. Set addition.

$$(\exists C)(\forall x)(x \in C \leftrightarrow (\exists y)(\exists z)(y \in A \wedge z \in B \wedge x = y + z)).$$

5. Set multiplication.

$$(\exists C)(\forall x)(x \in C \leftrightarrow (\exists y)(\exists z)(y \in A \wedge z \in B \wedge x = y \cdot z)).$$

6. Least element.

$$(\exists x)(x \in A) \rightarrow (\exists x)(x \in A \wedge (\forall y)(y \in A \rightarrow y \leq x)).$$

The nonlogical axioms of FSQZ are given formally as follows.

1. The above linearly ordered commutative ring axioms for Z.
2. $0 \leq \mathrm{lth}(\alpha)$.
3. $\mathrm{val}(\alpha, n)\downarrow \leftrightarrow 1 \leq n \leq \mathrm{lth}(\alpha)$.
4. The finite sequence $(0, \ldots, n)$ exists.

$$(\exists \alpha)(\mathrm{lth}(\alpha) = n + 1 \wedge (\forall k)(1 \leq k \leq n + 1 \rightarrow \mathrm{val}(\alpha, k) = k + 1)).$$

5. $\mathrm{lth}(\alpha) = \mathrm{lth}(\beta) \rightarrow -\alpha, \alpha + \beta, \alpha \cdot \beta$ exist.

$$\begin{aligned} \mathrm{lth}(\alpha) = \mathrm{lth}(\beta) \rightarrow\ & (\exists \gamma)(\forall n)(\mathrm{val}(\gamma, n) \cong -\mathrm{val}(\alpha, n)) \\ & \wedge (\exists \gamma)(\forall n)(\mathrm{val}(\gamma, n) \cong \mathrm{val}(\alpha, n) + \mathrm{val}(\beta, n)) \\ & \wedge (\exists \gamma)(\forall n)(\mathrm{val}(\gamma, n) \cong \mathrm{val}(\alpha, n) \cdot \mathrm{val}(\beta, n)). \end{aligned}$$

6. The concatenation of α, β exists.

$$\begin{aligned} (\exists \gamma)(\forall k, n)((1 \leq k \leq \mathrm{lth}(\alpha) \rightarrow \mathrm{val}(\gamma, k) = \mathrm{val}(\alpha, k)) \\ \wedge (1 \leq n \leq \mathrm{lth}(\beta) \rightarrow \mathrm{val}(\gamma, \mathrm{lth}(\alpha) + n) = \mathrm{val}(\beta, n))). \end{aligned}$$

7. For all $n \geq 1$, the concatenation of α, n times, exists.

$$\text{lth}(\alpha) = n \rightarrow (\exists\beta)(\text{lth}(\beta) = n \cdot m \wedge (\forall q, r)(0 \leq q < m \wedge 1 \leq r \leq n \rightarrow \text{val}(\beta, n \cdot q + r) = \text{val}(\alpha, r))).$$

8. There is a finite sequence enumerating the terms of α that are not terms of β.

$$(\exists\gamma)(\forall k)((\exists n)(\text{val}(\gamma, n) = k) \leftrightarrow ((\exists n)(\text{val}(\alpha, n) = k) \wedge \neg(\exists n)(\text{val}(\beta, n) = k))).$$

9. Every nonempty finite sequence has a least term.

$$1 \leq \text{lth}(\alpha) \rightarrow (\exists k)(\forall i)(1 \leq i \leq \text{lth}(\alpha) \rightarrow \text{val}(\alpha, i) \leq \text{val}(\alpha, k)).$$

In axiom 5 above, we use the symbol $\cong$ from free logic, which means "either both undefined, or equal". See Section 1.

The signature of FSTZEXP is $L(Z, \text{exp}, \text{fst})$. FSTZEXP extends FSTZ by

i. $\text{exp}(n, 0) = 1$.
ii. $m \geq 0 \rightarrow (\text{exp}(n, m + 1) = \text{exp}(n, m) \cdot n \wedge \text{exp}(n, -m - 1)\uparrow)$.
iii. The finite set $\{\text{exp}(n, 0), \ldots, \text{exp}(n, m)\}$ exists.

We will find that FSTZEXP is quite weak. We let FSTZEXP′ extend FSTZ by

i. $\text{exp}(n, 0) = 1$.
ii. $m \geq 0 \rightarrow (\text{exp}(n, m + 1) = \text{exp}(n, m) \cdot n \wedge \text{exp}(n, -m - 1)\uparrow)$.
iii. $n \geq 2 \wedge 0 \leq m < r \rightarrow \text{exp}(n, m) < \text{exp}(n, r)$.
iv. The finite set $\{\text{exp}(n, 0) + 0, \text{exp}(n, 1) + 1, \ldots, \text{exp}(n, m) + m\}$ exists.

The signature of FSQZEXP is $L(Z, \text{exp}, \text{fsq})$. FSQZEXP extends FSQZ by

i. $\text{exp}(n, 0) = 1$.
ii. $m \geq 0 \rightarrow (\text{exp}(n, m + 1) = \text{exp}(n, m) \cdot n \wedge \text{exp}(n, -m - 1)\uparrow)$.
iii. The finite sequence $(\text{exp}(n, 0), \ldots, \text{exp}(n, m))$ exists.

Recall $\text{CM}(Z)$ from Section 3, stated in $L(Z)$.

Thus the seven strictly mathematical theories considered here are

$$\textit{FSTZ},\ \textit{FSQZ},\ \textit{FSTZ} + \text{CM}(Z),\ \textit{FSQZ} + \text{CM}(Z),\ \textit{FSTZEXP},\ \textit{FSTZEXP}',\ \textit{FSQZEXP}$$

in the respective signatures

$$L(Z, \text{fst}),\ L(Z, \text{fsq}),\ L(Z, \text{fst}),\ L(Z, \text{fsq}),\ L(Z, \text{exp}, \text{fst}),\ L(Z, \text{exp}, \text{fst}),\ L(Z, \text{exp}, \text{fsq}).$$

We offer the following remarks in comparing the strictly mathematical nature of FSQZ and FSTZ.

i. Finite sequences of integers are more commonplace in mathematics than finite sets of integers.

ii. The pointwise ring operations on finite sequences of integers, and the concatenation of finite sequences of integers (including indefinite concatenation), is more commonplace in mathematics than the Boolean ring operations on finite sets of integers, and the addition and multiplication of finite sets of integers.

§8. FSTZ. In this section we show that FSTZ is a conservative extension of PFA(Z). This follows from the particularly convenient axiomatization of FSTZ given in Theorem 8.28:

THEOREM 8.28. *FSTZ can be axiomatized as follows.*

1. LOID(Z).
2. $(\exists A)(\forall x)(x \in A \leftrightarrow (y < x < z \wedge \varphi))$, *where* $\varphi \in \Sigma_0(Z, \mathrm{fst})$ *and* A *is not free in* φ.
3. *Every nonempty set has a least element.*

Recall the axioms of FSTZ.

1. *Linearly ordered integral domain axioms* (LOID(Z)).
2. *Finite interval.*
3. *Boolean difference.*
4. *Set addition.*
5. *Set multiplication.*
6. *Least element.*

We will often use scalar addition and scalar multiplication. We write $A+x = x + A = A + \{x\}$, and $A \cdot x = x \cdot A = A \cdot \{x\}$.

In Lemmas 8.1–8.27, it is understood that we are asserting provability within FSTZ.

LEMMA 8.1.

i) $\neg(x < y \wedge y < x + 1)$;
ii) (a, b), $[a, b)$, $(a, b]$ *exist*;
iii) $\emptyset$, $\{x\}$ *exists*;
iv) $x \cdot A = \{x \cdot y : y \in A\}$ *exists*;
v) *every nonempty set has a greatest element*;
vi) *every set is included in some interval* $[a, b]$;
vii) *sets are closed under pairwise union and pairwise intersection*;
viii) *for standard* $n \geq 0$, $\{x_1, \dots, x_n\}$ *exists*;
ix) *the set of all positive (negative, nonnegative, nonpositive) elements of any set exists.*

PROOF. For i), assume $0 < x < 1$. By LOID(Z), $0 = 0 \cdot x < x \cdot x < 1 \cdot x = x$. Hence there is no least y such that $0 < y < 1$. By finite interval, $(0,1)$ exists. By least element, there is a least y such that $0 < y < 1$. This is a contradiction. So $\neg(0 < x < 1)$. Now suppose $x < y < x + 1$. Then $0 < y - x < 1$, which is a contradiction.

For ii), note that by i), $(a,b) = [a+1, b-1]$, $[a,b) = [a,b-1)$, $(a,b] = [a+1,b]$.

For iii), note that $\emptyset$ is the interval (x,x), and by i), $\{x\}$ is the interval $[x,x]$.

For iv), note that $x \cdot A = \{x\} \cdot A$, and apply set multiplication.

For v), Let A be nonempty. Then $-A = \{-1\} \cdot A$ has a least element x. Clearly $-x$ is the greatest element of A.

For vi), let A be given. Then $A \subseteq [\min(A), \max(A)]$.

For vii), note that $A \cap B = A \backslash (A \backslash B)$. Also note that $A \cup B = C \backslash (C \backslash A \cap C \backslash B)$, where $A, B \subseteq [\min(\min(A), \min(B)), \max(\max(A), \max(B))]$.

For viii), note that $\{x_1, \ldots, x_n\} = \{x_1\} \cup \cdots \cup \{x_n\}$.

For ix), let A be given. By vi), let $A \subseteq [a,b]$. Then the set of positive elements of A is $A \cap [1,b]$. The other cases are handled similarly. ⊣

We write $-A$ for $(-1) \cdot A$, and $A - B$ for $A + (-B)$.

Lemma 8.2. *Let $d \geq 1$ and x be an integer. There exists unique q, r such that $x = dq + r$ and $0 \leq r < d$.*

Proof. For uniqueness, let $x = dq + r = dq' + r'$, $0 \leq r, r' < d$. Then $d(q-q') + r - r' = 0$, $d(q-q') = r' - r$. Hence $d|q-q'| = |r'-r| < d$. So $|q - q'| < 1$, and hence $q = q'$. Therefore $0 = r' - r$, and so $r = r'$.

For existence, fix d, x as given, and first assume $x > 0$. Let $A = \{x - dq : q \in [0,x]\} = \{x\} - d \cdot [0,x]$. By Lemma 8.1 ix), let A' be the set of all nonnegative elements of A. Then A' is nonempty since $x - dq > 0$ for $q = -x$. Choose q such that $\min(A') = x - dq$. Obviously $0 \leq x - dq$ and $q \in [0,x]$.

If $q = x$ then $d = 1$ and $x - dq = 0$, in which case we are done. So we can assume that $q < x$.

Suppose $x - dq \geq d$. Then $x - d(q+1) \geq 0$ and $q+1 \in [0,x]$, contradicting the choice of q. Hence $0 \leq x - dq < d$. Set $r = x - dq$. Then $x = dq + r$ and $0 \leq r < d$.

We still have to handle the case $x \leq 0$. The case $x = 0$ is trivial, and so we assume $x < 0$. Write $-x = dq + r$, $0 \leq r < d$. Then $x = d(-q) - r$. If $r = 0$ then we are done, and so we assume $0 < r < d$. Then $x = d(-q-1) + d - r$, $0 \leq d - r < d$. ⊣

Lemma 8.3. *Let $k \geq 0$. The following is provable in FSTZ. For all $r \geq 2$, the elements of $[0, r^{k+1})$ have unique representations of the form $n_0 r^0 + \cdots + n_k r^k$, where each n_i lies in $[0,r)$. If $n_0 r^0 + \cdots + n_k r^k = m_0 r^0 + \cdots + m_k r^k$ and each n_i lies in $(-r/2, r/2)$, then each $n_i = m_i$.*

Proof. It is important to note that k is treated as a standard integer.

For uniqueness, suppose $n_0 r^0 + \cdots + n_k r^k = m_0 r^0 + \cdots + m_k r^k$, where each $n_i, m_i \in [0,r)$. Let i be greatest such that $n_i \neq m_i$. We can assume that $n_i < m_i$. Here we think of i as a standard integer defined by a large number of cases.

Now subtract the second representation from the first. Then we obtain an inequality of the form

$$p_0 r^0 + \cdots + p_{i-1} r^{i-1} \geq r^i,$$

where $p_0, \ldots, p_{i-1} \in (-r, r)$.

Note that $p_0 r^0 + \cdots + p_{i-1} r^{i-1} \leq (r-1)(r^0 + \cdots + r^{i-1}) = r^i - 1$. This is the desired contradiction.

The second claim can be established in the same way by subtraction, since any two elements of $(-r/2, r/2)$ must differ by $< r$, and hence at most $r - 1$.

For existence, we proceed by external induction on k. The case $k = 0$ is trivial. Suppose existence for all $r \geq 2$ and $x \in [0, r^{k+1})$, has been proved for a given k, where $k \geq 0$. Let $r \geq 2$ and $x \in [0, r^{k+2})$. Write $x = r^{k+1} n_{k+1} + y$, $0 \leq y < r^{k+1}$. Note that $0 \leq n_{k+1} < r$. By induction hypothesis, write $y = n_0 r^0 + \cdots + n_k r^k$, $n_0, \ldots, n_k \in [0, r)$. Then $x = n_0 r^0 + \cdots + n_k r^k + r^{k+1} n_{k+1}$, $n_0, \ldots, n_{k+1} \in [0, r)$. ⊣

Until the end of the proof of Lemma 8.12, we fix a standard integer $k > 0$.

Lemma 8.4. *For all $r > 1$, $S[r] = \{n_0 r^0 + n_1 r^2 + \cdots + n_i r^{2i} + \cdots + n_k r^{2k} : n_0, \ldots, n_k \in [0, r)\}$ exists. Every element of $S[r]$ is uniquely written in the displayed form.*

Proof. $S[r] = [0, r) \cdot r^0 + [0, r) \cdot r^2 + \cdots + [0, r) \cdot r^{2k}$. The second claim follows immediately from Lemma 8.3. ⊣

For $x \in S[r]$, we write $x[i]$ for n_i in this unique representation.

Lemma 8.5. *For all $r > 1$ and $i \in [0, k]$, $\{x \in S[r] : x[i] = 0\}$ and $\{x \in S[r] : x[i] = 1\}$ exist.*

Proof. The first set is

$$[0, r) \cdot r^0 + \cdots + [0, r) \cdot r^{2i-2} + [0, r) \cdot r^{2i+2} + \cdots + [0, r) \cdot r^{2k}.$$

The second set is

$$[0, r) \cdot r^0 + \cdots + [0, r) \cdot r^{2i-2} + r^{2i} + [0, r) \cdot r^{2i+2} + \cdots + [0, r) \cdot r^{2k}.$$

⊣

Lemma 8.6. *Let $r > 1$ and $i, j, p \in [0, k]$. Then $\{x \in S[r] : x[i] + x[j] = x[p]\}$ exists.*

Proof. Let r, i, j, p be as given. If $i = p$ then $\{x \in S[r] : x[i] + x[j] = x[p]\} = \{x \in S[r] : x[j] = 0\}$. If $j = p$ then $\{x \in S[r] : x[i] + x[j] = x[p]\} = \{x \in S[r] : x[i] = 0\}$. Both of these cases are covered by Lemma 8.5.

We now handle the case $i = j \neq p$. We wish to show that $A = \{x \in S[r] : 2x[i] = x[p]\}$ exists, where $i \neq p$.

Now $D = \{x : (\exists a \in [0, r))(x = ar^{2i} + 2ar^{2p})\}$ exists, since $D = [0, r) \cdot (r^{2i} + 2r^{2p})$.

Let T be the sum of the sets $[0, r) \cdot r^{2q}$, where $q \in [0, k] \backslash \{i\}$. We claim that $A = (D + T) \cap S[r]$.

To see this, obviously every element of A lies in $(D + T) \cap S[r]$. On the other hand, let $x \in (D + T) \cap S[r]$. Write

$$x = ar^{2i} + 2ar^{2p} + y$$

where $a, b \in [0, r)$ and $y \in T$. Since $2a < r^2$, this is the representation of $x \in S[r]$. Evidently, $x \in A$.

We now handle the case $i \neq j \neq p$. We wish to show that $B = \{x \in S[r] : x[i] + x[j] = x[p]\}$ exists.

Now $E = \{x : (\exists a, b \in [0, r))(x = ar^{2i} + br^{2j} + (a + b)r^{2p})\}$ exists, since $E = \{x : (\exists a, b \in [0, r))(x = a(r^{2i} + r^{2p}) + b(r^{2j} + r^{2p}))\} = ([0, r) \cdot (r^{2i} + r^{2p}) + [0, r) \cdot (r^{2j} + r^{2p}))$.

Let V be the sum of the sets $[0, r) \cdot r^{2q}$, where $q \in [0, k] \backslash \{i, j, p\}$. We claim that $B = (E + V) \cap S[r]$.

To see this, obviously every element of B lies in $(E + V) \cap S[r]$. On the other hand, let $x \in (E + V) \cap S[r]$. Write

$$x = ar^{2i} + br^{2j} + (a + b)r^{2p} + y$$

where $a, b \in [0, r)$ and $y \in V$. Since $a + b < r^2$, this is the representation of $x \in S[r]$. Evidently, $x \in B$. ⊣

We define $a|b \leftrightarrow (\exists c)(b = a \cdot c)$.

Lemma 8.7. *For all $r > 1$ and $i, j \in [0, k]$, $\{x \in S[r] : x[i]|x[j]\}$ exists.*

Proof. If $i = j$ then $\{x \in S[r] : x[i]|x[j]\} = S[r]$, which is handled by Lemma 8.4. Assume $i \neq j$. We want to prove that $A = \{x \in S[r] : x[i]|x[j]\}$ exists.

Now $E = \{x : (\exists a, b \in [0, r))(x = ar^{2i} + abr^{2j})\}$ exists, since $E = \{x : (\exists a, b \in [0, r))(x = a(r^{2i} + br^{2j}))\} = [0, r) \cdot (r^{2i} + [0, r) \cdot r^{2j})$.

Let D be the sum of the sets $[0, r) \cdot r^{2q}$, where $q \in [0, k] \backslash \{i, j\}$. We claim that $A = (E + D) \cap S[r]$.

To see this, obviously every element of A lies in $(E + D) \cap S[r]$. On the other hand, let $x \in (E + D) \cap S[r]$. Write

$$x = ar^{2i} + abr^{2j} + y$$

where $a, b \in [0, r)$ and $y \in D$. Since $ab < r^2$, this is the representation of $x \in S[r]$. Evidently $x \in A$. ⊣

Lemma 8.8. *For all $r > 1$, $i \in [0, k]$, and $A \subseteq [0, r)$, $\{x \in S[r] : x[i] \in A\}$ exists.*

Proof. Note that $\{x \in S[r] : x[i] \in A\}$ is $[0, r) \cdot r^0 + \cdots + [0, r) \cdot r^{2i-2} + A \cdot r^{2i} + [0, r) \cdot r^{2i+2} + \cdots + [0, r] \cdot r^{2k}$. ⊣

Lemma 8.9. *Let φ be a propositional combination of formulas $x_i = 0$, $x_i = 1$, $x_i + x_j = x_p$, $x_i|x_j$, $x_i \in A_j$, where $i, j, p \in [0, k]$. The following is provable in FSTZ. For all $r > 1$ and $A_0, \dots, A_k \subseteq [0, r)$, $\{x_0 r^0 + \cdots + x_k r^{2k} : \varphi \wedge x_0, \dots, x_k \in [0, r)\}$ exists.*

PROOF. For atomic φ, this follows from Lemmas 8.4–8.8. The propositional combinations are handled by the fact that the subsets of $S[r]$ form a Boolean algebra. ⊣

LEMMA 8.10. *For all* $r > 1$, $i \in [0,k]$, *and* $E \subseteq S[r]$, $\{x \in S[r] : (\exists y \in E)(\forall j \in [0,k]\backslash\{i\})(x[j] = y[j])\}$ *exists.*

PROOF. We first claim that $A = \{x \in S[r] : (\exists y \in E)(\forall j \in [0,k]\backslash\{i\})(x[j] = y[j])\} \subseteq E + (-r,r) \cdot r^{2i}$. To see this, suppose $x \in S[r]$, $y \in E$, and $\forall j \in [0,k]\backslash\{i\}, x[j] = y[j]$. Since the coefficients of r^{2i} in x and y both lie in $[0,r)$, we see that $x - y \in (-r,r) \cdot r^{2i}$. Hence $x \in (-r,r) \cdot r^{2i} + y \subseteq E + (-r,r) \cdot r^{2i}$.

We claim that $A = (E + (-r,r) \cdot r^{2i}) \cap S[r]$. To see this, let $x \in (E + (-r,r) \cdot r^{2i}) \cap S[r]$. Write

$$x = y + a \cdot r^{2i}$$

where $y \in E$ and $a \in (-r,r)$. In this equation, the coefficient of r^{2i} is the coefficient of r^{2i} in y plus a, which must lie in $(-r,2r)$. Hence this must be the representation of $x \in S[r]$. Evidently, x agrees with an element of E (namely y) at all positions other than at r^{2i}. ⊣

LEMMA 8.11. *Let φ be a propositional combination of formulas* $x_i = 0$, $x_i = 1$, $x_i + x_j = x_p$, $x_i | x_j$, $x_i \in A_j$, *where* $i, j, p \in [0,k]$. *Let* $m \in [1,k]$. *Let* $\psi = (Q_m x_m \in [0,r)) \dots (Q_k x_k \in [0,r))(\varphi)$. *The following is provable in FSTZ. For all* $A_0, \dots, A_k \subseteq [0,r)$, $\{x_0 r^0 + \cdots + x_{m-1} r^{2m-2} : \psi \wedge x_0, \dots, x_{m-1} \in [0,r)\}$ *exists.*

PROOF. Here Q_i is $\forall$ or $\exists$. Lemma 8.9 handles φ. Lemma 8.10 handles existential quantifiers. Universal quantifiers are taken care of by relative complementation. ⊣

LEMMA 8.12. *Let* $r > 1$, $E \subseteq S[r]$, $i_1 < \cdots < i_p \in [0,k]$, *and* $x_1, \dots, x_p \in [0,r)$. *Then* $\{y \in S[r] : y[i_1] = x_1 \wedge \cdots \wedge y[i_p] = x_p\}$ *exists.*

PROOF. Note that this set is $A \cap B_1 \cap \cdots \cap B_p$, where for all $j \in [1,p]$, $B_j = \{y \in S[r] : y[i_j] = x_j\} = [0,r) \cdot r^0 + \cdots + [0,r) \cdot r^{2k}$ where the term with exponent $2j$ is replaced by $x_j r^{2j}$. ⊣

We now release the fixed standard integer k. For formulas φ without bound set variables, and integer variables z not in φ, let φ^z be the result of relativizing all quantifiers in φ to $[-z,z]$.

LEMMA 8.13. *Let φ be a formula without bound set variables whose atomic subformulas are of the form* $x_i = 0$, $x_i = 1$, $x_i + x_j = x_p$, $x_i | x_j$, $x_i \in A_j$. *Let* y, z *be distinct integer variables, where* z *does not appear in* φ. *Then FSTZ proves that* $\{y \in [0,z] : \varphi^z\}$ *exists. Also FSTZ proves that* $\{y \in [-z,z] : \varphi^z\}$ *exists.*

PROOF. Note that the conclusion should be viewed as a separation principle with parameters (represented by the free variables of φ other than y).

By changing variables, we can assume that y is x_0, the free variables of φ are among $x_0, \dots, x_{m-1}$, and the quantified variables are among $x_m, \dots, x_k$. Also replace the relativizations to $[-z, z]$ with relativizations to $[0, z]$, by appropriately modifying the formula.

Now apply Lemma 8.11 with $r = z+1$. We obtain $\{x_0 r^0 + \dots + x_{m-1} r^{2m-2} : \varphi' \wedge x_0, \dots, x_{m-1} \in [0, z]\}$. Now apply Lemma 8.12 with $p = m-1$, $i_1, \dots, i_p = 1, \dots, m-1$, and $r = z+1$. We obtain $\{x_0 \in [0, z] : \varphi'\} = \{y \in [0, z] : \varphi^z\}$.

The second claim follows from the first. ⊣

LEMMA 8.14. *Let φ be a formula without bound set variables whose atomic subformulas are of the form $s = t$, $s < t$, $s|t$, or $t \in A_j$, where s, t are terms without $\cdot$. Let y, z be distinct integer variables, where z does not appear in φ. Then FSTZ proves that $\{y \in [-z, z] : \varphi^z\}$ exists.*

PROOF. By inductively introducing existential quantifiers needed to unravel the terms. A bound can be placed on the existential quantifiers introduced which depends only on φ and the value of the bound z. Since the terms do not use $\cdot$, the expansion stays within the form in Lemma 8.13. ⊣

Formulas of the form in Lemma 8.14 are called special formulas.

Note that we do not allow $\cdot$ in special formulas. We first need to use Lemma 8.14 to obtain some basic number theory before we can handle $\cdot$ appropriately.

LEMMA 8.15. $x, y \neq 0 \rightarrow \gcd(x, y), \operatorname{lcm}(x, y)$ *exist.* $x > 1 \rightarrow x$ *is divisible by a prime.*

PROOF. For the first claim, let $x, y \neq 0$. By Lemma 8.14, $\{a \in [1, |xy|] : a|x \wedge a|y\}$ exists. Then $\gcd(x, y)$ is its greatest element. By Lemma 8.14, $\{a \in [1, |xy|] : x|a \wedge y|a\}$ exists. Then $\operatorname{lcm}(x, y)$ is its least element.

For the second claim, let $x > 1$. By Lemma 8.14, $\{p \in [2, x] : p|x\}$ exists. Let p be the least element. Then p is a prime divisor of x. ⊣

LEMMA 8.16. *Suppose $x, y > 1$ and $ax + by = 1$. Then there exists $cx + dy = 1$, where $c \in (0, y)$, $d \in (-x, 0)$. Suppose $x, y > 0$ and $ax + by = 1$. Then there exists $cx + dy = 1$, where $c \in [0, y]$, $d \in [-x, 0]$.*

PROOF. Let x, y, a, b be as given. By symmetry we can assume that $a \geq 0$.

Let $A = \{s \in [0, ax] : (\exists t \in [1 - ax, 0])(x|s \wedge y|t \wedge s + t = 1)\} = \{s \in [0, ax] : (\exists t)(x|s \wedge y|t \wedge s + t = 1)\} = \{s \in [0, ax] :$ there is a multiple of x and a multiple of y, which add up to $1\}$. Note that A exists by Lemma 8.14, and A is nonempty since it includes $s = ax$, with $t = by$. Let cx be the least element of A.

Write $cx + dy = 1$. Note that $(c - y)x + (d + x)y = 1$. By the choice of c, $\neg(0 \leq c - y < c)$, and so $c - y < 0$ or $c - y \geq c$. Hence $c \in [0, y)$.

Note that $1 = cx + dy \leq xy + dy = (x + d)y$. Hence $x + d > 0$, and so $d > -x$. Hence $d \in (-x, 0]$.

Note that $c \neq 0$ and $d \neq 0$ because of $cx + dy = 1$, $x, y > 1$.

For the second claim, we need only consider the case $(x = 1 \vee y = 1)$. By symmetry, assume $x = 1$. Then take $c = 1$ and $d = 0$. ⊣

We say that x, y are relatively prime if and only if $x, y \neq 0$ and the only common divisors of x, y are 1 and -1.

LEMMA 8.17. *Let x, y be relatively prime. Then there exists a solution to $ax + by = 1$.*

PROOF. We fix a positive integer t. We wish to show by induction (equivalently, least element principle) that the following holds for every $0 < s \leq t$. For all $0 < x, y \leq s$, if x, y are relatively prime then $ax + by = 1$ has a solution.

We need to express this condition by a special formula.

$(\forall x, y \in [-t, t])((0 < x, y \leq s \wedge x, y$ relatively prime$) \rightarrow ax + by = 1$ has a solution$)$.

$(\forall x, y \in [-t, t])((0 < x, y \leq s \wedge$ nothing in $[1, x]$ divides both $x, y) \rightarrow ax + by = 1$ has a solution in $[-s, s])$.

Here we have used Lemma 8.16, which provides a bound on solutions to $ax + by = 1$.

The basis case $s = 1$ is trivial. Suppose true for a fixed $s \geq 1$. Let $x, y \leq s + 1$ be relatively prime. We can assume $1 < y < x = s + 1$. Write $x = qy + r$, $0 \leq r < y$. Since x, y are relatively prime, we have $0 < r < y$.

Note that y, r are relatively prime and positive. Hence by induction hypothesis write $cy + dr = 1$. Now $dx + (c - dq)y = 1$.

We still have to consider the case where x or y is negative. But then we can merely change the sign or signs of one or more of a, b. ⊣

LEMMA 8.18. *Let p be a prime and suppose $p|xy$. Then $p|x$ or $p|y$.*

PROOF. Let p, x, y be as given. Suppose the contrary. Then $x, y \neq 0$, and p, x are relatively prime, and p, y are relatively prime. By Lemma 8.17, write $ap + bx = 1$, $cp + dy = 1$. Then $apcp + apdy + bxcp + bxdy = 1$. Note that p divides every summand, and so p divides 1, which is a contradiction. ⊣

LEMMA 8.19. *Let x, y be relatively prime and let x, z be relatively prime. Suppose $x|yz$. Then $x = 1$ or -1.*

PROOF. Let x, y, z be as given. Write $ax + by = 1$ and $cx + dz = 1$. Then $axcx + axdz + bycx + bydz = 1$. Since x divides every summand, x divides 1. Hence $x = 1$ or -1. ⊣

LEMMA 8.20. *Let x, y be relatively prime and $x|yz$. Then $x|z$.*

PROOF. Let x, y, z be as given. We can assume that $z \neq 0$. It suffices to prove this for $x, y, z > 0$.

Now $x/\gcd(x, z)$ divides $y(z/\gcd(x, z))$ via the integer factor yz/x. Also note that $x/\gcd(x, z)$ and y are relatively prime.

We claim that $x/\gcd(x, z)$ and $z/\gcd(x, z)$ are relatively prime. To see this, suppose they have a common factor $u > 1$. Then $\gcd(x, z)u$ is a factor of x and also a factor of z, contradicting that $\gcd(x, z)$ is the greatest common factor of x, z.

By Lemma 8.19, $x/\gcd(x, z) = 1$. *I.e.*, $\gcd(x, z) = x$. So $x|z$. ⊣

LEMMA 8.21. *Let a, b be relatively prime. Then the least positive common multiple of a, b is ab.*

PROOF. Let a, b be as given, and let x be a positive common multiple of a, b. Write $x = ay$.

Since $b|ay$, by see by Lemma 8.20 that $b|y$. Hence $b \leq y$. Therefore $x = ay \geq ab$ as required. ⊣

Lemmas 8.22, 8.23 finally tell us how to handle $\cdot$ appropriately.

LEMMA 8.22. *There is a special formula φ with free variables among x, y such that the following is provable in FSTZ. For all z there exists $z' > z$ such that $(\forall x, y \in [-z, z])(x = y^2 \leftrightarrow \varphi^{z'})$.*

PROOF. Let φ express $x + y = \operatorname{lcm}(y, y + 1)$. Let z be given. If $y \notin [-1, 0]$ then $\gcd(y, y + 1) = 1$, and hence by Lemma 8.10, $\operatorname{lcm}(y, y + 1) = y(y + 1)$. Therefore $(\forall x, y \in [-z, z]\backslash[-1, 0])(\varphi \leftrightarrow x + y = y(y + 1))$. Hence $(\forall x, y \in [-z, z]\backslash[-1, 0])(\varphi \leftrightarrow x = y^2)$. The quantifiers in φ can be bounded to an integer z' that depends only on z.

We still have to modify φ in order to handle $[-1, 0]$. Take φ' to be $(\varphi \wedge x, y \notin [-1, 0]) \vee x = y = 0 \vee (x = 1 \wedge y = -1)$. ⊣

LEMMA 8.23. *There is a special formula ψ with free variables among u, v, w, such that the following is provable in FSTZ. For all z there exists $z' > z$ such that $(\forall u, v, w \in [-z, z])(u \cdot v = w \leftrightarrow \psi^{z'})$.*

PROOF. Let $\psi = (\exists x, y, a, b)(x = y^2 \wedge y = u + v \wedge a = u^2 \wedge b = v^2 \wedge 2w = x - a - b)$. Let z be given. Then $(\forall u, v, w \in [-z, z])(u \cdot v = w \leftrightarrow \psi)$. Use the φ from Lemma 8.22 to remove the first, third, and fourth displayed equations, to make ψ special. The quantifiers can be bounded to $z' > z$, where z' depends only on z. ⊣

We now extend Lemma 8.14.

LEMMA 8.24. *Let φ be a formula without bound set variables whose atomic subformulas are of the form $x_i = 0$, $x_i = 1$, $x_i + x_j = z$, $x_i \cdot x_j = x_p$, $x_i \in A_j$. Let y, z be distinct integer variables, where z does not appear in φ. Then FSTZ proves that $\{y \in [-z, z] : \varphi^z\}$ exists.*

PROOF. Let φ be as given. Replace each atomic subformula of the form $x \cdot y = z$ by the ψ of Lemma 8.23, with an appropriate change of variables. Call this expansion ρ. Let z be given. Then there exists $z' > z$ depending only on z such that for all $y \in [-z, z]$, $\varphi^z \leftrightarrow \rho^{z'}$. By Lemma 8.14, $\{y \in [-z', z'] : \rho^{z'}\}$ exists. Hence $\{y \in [-z', z'] : \varphi^z\}$ exists. Hence $\{y \in [-z, z] : \varphi^z\}$ exists. ⊣

LEMMA 8.25. *Let φ be a formula without bound set variables. Let y, z be distinct integer variables, where z does not appear in φ. Then FSTZ proves that $\{y \in [-z, z] : \varphi^z\}$ exists.*

PROOF. Let φ be as given, and let z be given. Expand the terms appearing in φ using existential quantifiers. Apply Lemma 8.24 with appropriately chosen z', where z' depends only on z and the terms that appear. ⊣

We now define the class of formulas of FSTZ, $\Sigma_0(Z, \text{fst})$.

i) every atomic formula of FSTZ is in $\Sigma_0(Z, \text{fst})$;
ii) if φ, ψ are in $\Sigma_0(Z, \text{fst})$, then so are $\neg\varphi$, $\varphi \wedge \psi$, $\psi \vee \psi$, $\varphi \rightarrow \psi$, $\varphi \leftrightarrow \psi$;
iii) if φ is in $\Sigma_0(Z, \text{fst})$, x is an integer variable, s, t are integer terms, x not in s, t, then $(\exists x \in [s, t])(\varphi)$ and $(\forall x \in [s, t])(\varphi)$ are in $\Sigma_0(Z, \text{fst})$.

LEMMA 8.26. *Let φ be in $\Sigma_0(Z, \text{fst})$. Let $x_1, \ldots, x_k$ be an enumeration without repetition of at least the free variables of φ. The following is provable in FSTZ. Let $r > 1$. Then $\{x_1 r^1 + \cdots + x_k r^k : x_1, \ldots, x_k \in [0, r) \wedge \varphi\}$ exists.*

PROOF. By induction on φ. Let φ be atomic. Then this follows from Lemma 8.25. Suppose this is true for φ, ψ in $\Sigma_0(Z, \text{fst})$. Let ρ be among $\neg\varphi$, $\varphi \vee \psi$, $\varphi \wedge \psi$, $\varphi \rightarrow \psi$, $\varphi \leftrightarrow \psi$. Then obviously this holds for ρ.

Now suppose this holds for φ in $\Sigma_0(Z, \text{fst})$. Let $\psi = (\exists x \in [s, t])(\varphi)$. Let $x_1, \ldots, x_k$ be an enumeration without repetition of at least the free variables of ψ. Then $x_1, \ldots, x_k$, x is an enumeration without repetition of at least the free variables of φ.

We want to show that

$$A = \{x_1 r^1 + \cdots + x_k r^k : x_1, \ldots, x_k \in [0, r) \wedge (\exists x \in [s, t])(\varphi)\}$$

provably exists for all $r > 1$. We know that

$$B = \{x_1 r^1 + \cdots + x_k r^k + x r^{k+1} : x_1, \ldots, x_k, x \in [0, r) \wedge \varphi\}$$

provably exists for all $r > 1$. We can define A from B appropriately so that we can simply apply Lemma 8.25. ⊣

LEMMA 8.27. *Let φ lie in $\Sigma_0(Z, \text{fst})$. Let z be an integer variable that does not appear in φ. Then FSTZ proves that $\{y \in [-z, z] : \varphi\}$ exists.*

PROOF. From Lemmas 8.25 and 8.26. ⊣

THEOREM 8.28. *FSTZ can be axiomatized as follows.*

1. LOID(Z).
2. $(\exists A)(\forall x)(x \in A \leftrightarrow (y < x < z \wedge \varphi))$, *where $\varphi \in \Sigma_0(Z, \text{fst})$ and A is not free in φ.*
3. *Every nonempty set has a least element.*

PROOF. Axiom scheme 2 is derivable from FSTZ by Lemma 8.27. For the other direction, first note that we can derive $(\forall A)$ $(-A$ exists). Hence every set has a greatest element. Then it is easy to see that finite interval,

Boolean difference, set addition, and set multiplication are special cases of axiom scheme 2 above. ⊣

THEOREM 8.29. *FSTZ is a conservative extension of* PFA(Z).

PROOF. By Theorem 8.28, FSTZ proves PFA(Z). It now suffices to show that any model M of PFA(Z) can be expanded by attaching sets to form a model of FSTZ. Take the sets of integers to be those sets of the form

$$\{x \in [-n, n] : \varphi\}$$

where φ is a formula in $\Sigma_0(Z)$ with parameters allowed, interpreted in the model M. The verification of FSTZ in the expansion is straightforward. ⊣

§9. **FSTZ = FSTZD = FSTZS.** We show that FSTZ is equivalent to two interesting weakenings of FSTZ. These results are of independent interest, and not central to the paper.

The signature of FSTZD is the same as that of FSTZ, which is $L(Z, \text{fst})$. FSTZD = "finite sets of integers with duplication". The nonlogical axioms of FSTZD are as follows.

1. Linearly ordered commutative ring axioms.
2. Finite interval.
3. Boolean difference.
4. Duplicate set addition.

$$(\exists B)(\forall x)(x \in B \leftrightarrow (\exists y)(\exists z)(y \in A \wedge z \in A \wedge x = y + z)).$$

5. Duplicate set multiplication.

$$(\exists B)(\forall x)(x \in B \leftrightarrow (\exists y)(\exists z)(y \in A \wedge z \in A \wedge x = y \cdot z)).$$

6. Every set has a least and greatest element.

Axiom 4 asserts the existence of $A + A$, and axiom 5 asserts the existence of $A \cdot A$. "Duplicate" refers to the use of $A + A$, $A \cdot A$, rather than $A + B$, $A \cdot B$.

Lemmas 9.1–9.8 refer to provability in FSTZD.

LEMMA 9.1. i)–iii), v)–ix) *of Lemma* 8.1.

PROOF. Straightforward. ⊣

LEMMA 9.2. *Let* $A \subseteq [-x, x]$, $x \geq 0$, *and* $|y| > 3x$. *Then* $A + y$ *exists.*

PROOF. Let A, x be as given. By Lemma 9.1, let $B = A \cup \{y\}$. Then $B + B$ is composed of three parts: $A + A$, $A + y$, $\{2y\}$. We don't know yet that the second part is a set. Note that $A + A \subseteq [-2x, 2x]$ and $A + y \subseteq [-x + y, x + y]$.

First assume $y > 0$. Note that $2x < -x + y$ and $x + y < 2y$. Hence these three parts are pairwise disjoint. Since $B + B$ and the first and third parts exist, clearly the second part exists.

Now assume $y < 0$. Note that $-2x > x + y$ and $-x + y > 2y$. Hence these three parts are pairwise disjoint. So the second part exists. ⊣

LEMMA 9.3. *Let* $A \subseteq [7z/8, 9z/8]$, $z > 0$, $w < -z/2$. *Then* $A + w$ *exists.*

PROOF. Let A, z, w be as given. Let $B = A \cup \{w\}$. Then $B + B$ is composed of three parts: $A + A$, $A + w$, $\{2w\}$. Note that $A + A \subseteq [7z/4, 9z/4]$ and $A + w \subseteq [7z/8 + w, 9z/8 + w]$.

Note that $9z/8 + w < 7z/4$. Hence the first two parts are disjoint. Therefore $A + w$ is among $B + B \backslash A + A$, $B + B \backslash A + A \cup \{2w\}$, $B + B \backslash A + A \backslash \{2w\}$. Hence $A + w$ exists. ⊣

LEMMA 9.4. *$A + y$ exists.*

PROOF. Let $A \subseteq [-x, x]$, $x > 0$, and y be given. We can assume that y is nonzero. Write $y = z + w$, where $z > 3x$, $A + z \subseteq [7z/8, 9z/8]$, $w < -z/2$. By Lemma 9.2, $A + z$ exists. By Lemma 9.3, $A + z + w$ exists. But $A + z + w = A + y$.

It remains to show how z, w can be chosen. Set $z = 9\max(x, |y|)$. Set $w = y - z = y - 9\max(x, |y|)$. Note that $A + z \subseteq [-x + z, x + z] \subseteq [7z/8, 9z/8]$.

We have only to verify that $w < -z/2$. *I.e.*, $y - 9\max(x, |y|) < -9\max(x, |y|)/2$, which is $y < 9\max(x, |y|)/x$. This follows from $x > 0$ and $y \neq 0$. ⊣

LEMMA 9.5. *$A + B$ exists.*

PROOF. Let A, B be given. Let $A, B \subseteq [-x, x]$, $x \geq 0$. By Lemma 9.2, let $C = B + 4x$. Consider $A \cup C + A \cup C$. This is composed of three parts: $A + A$, $A + C$, $C + C$. We don't know yet that the second part is a set.

Note that $A + A \subseteq [-2x, 2x]$, $A + C \subseteq [3x, 5x]$, $C + C \subseteq [6x, 10x]$. Hence these three parts are pairwise disjoint. Since $A \cup C + A \cup C$ and the first and third parts exist, clearly the second part exists. *I.e.*, $A + C$ exists.

Observe that $A + C = A + B + 4x$, and so $A + B = A + C + -4x$, which exists by Lemma 9.4. ⊣

LEMMA 9.6. *$-A$ exists.*

PROOF. Let A be given. First assume that $A \subseteq [1, x]$, $x > 1$. Let $B = A \cup \{-1\}$. Note that $B \cdot B = A \cdot A \cup \{1\} \cup -A$, where we don't know yet that $-A$ exists. However, $-A$ is disjoint from $A \cdot A \cup \{1\}$. Hence $-A$ exists.

Now assume that $A \subseteq [-x, -1]$, $x > 1$. Using Lemmas 9.1 and 9.4, let $B = A + -x^3$. Consider $B \cup \{-1\} \cdot B \cup \{-1\}$. This is composed of three parts: $B \cdot B$, $-B$, $\{1\}$, where we don't know yet that $-B$ exists. Note that $B \cdot B \subseteq [(x^3 + 1)^2, (x^3 + x)^2]$, $-B \subseteq [1 + x^3, x + x^3]$. Hence the three parts are pairwise disjoint. Therefore $-B$ exists.

Finally, let A be arbitrary. Write $A = A^+ \cup A^- \cup A^0$, where A^+ is the positive part of A, A^- is the negative part of A, and A^0 is the 0 part of A, which is $\{0\}$ if $0 \in A$ and $\emptyset$ if $0 \notin A$.

Note that $-A = -(A^+) \cup -(A^-) \cup A^0$, and so $-A$ exists. ⊣

LEMMA 9.7. *$A \cdot x$ exists.*

PROOF. First assume $A \subseteq [y^2, y^3)$, $y > x > 1$. Consider $A \cup \{x\} \cdot A \cup \{x\}$. This is composed of three parts: $A \cdot A$, $A \cdot x$, $\{x^2\}$, where we don't know yet that $A \cdot x$ exists. Note that $A \cdot A \subseteq [y^4, y^6]$, $A \cdot x \subseteq [xy^2, xy^3]$. Hence the three parts are pairwise disjoint. Therefore $A \cdot x$ exists.

Now assume A is arbitrary and $x > 1$. We can choose $y > x$ such that $B \subseteq [y^2, y^3]$, where B is a translation of A. Then $B \cdot x$ exists.

Let $A = B + c$. Then $A \cdot x = (B + c) \cdot x = B \cdot x + \{cx\}$. Therefore $A \cdot x$ exists.

The case $x = 0$ is trivial. Finally suppose A is arbitrary and $x < -1$. Then $A \cdot x = -(A \cdot -x)$, and $-x > 1$. Therefore $A \cdot x$ exists. ⊣

LEMMA 9.8. *$A \cdot B$ exists.*

PROOF. Let A, B be given. We first assume that $A, B \subseteq [1, x]$, $x > 1$. Let $C = A \cup -B$. Then $C \cdot C$ exists. Its negative part is obviously $A \cdot -B$, which therefore exists. Note that $A \cdot B = -(A \cdot -B)$, and therefore $A \cdot B$ exists.

For the general case, write $A = A^+ \cup A^- \cup A^0$, $B = B^+ \cup B^- \cup B^0$. Then $A \cdot B$ is the union of the nine obvious cross products. There are only three of them that we have to check exist, the other six obviously existing. These are $A^+ \cdot B^-$, $A^- \cdot B^+$, $A^- \cdot B^-$. However, it is easy to see that these are, respectively, $-(A \cdot B)$, $-(A \cdot B)$, $A \cdot B$, and therefore exist. ⊣

THEOREM 9.9. *FSTZ and FSTZD are equivalent.*

PROOF. By Lemmas 9.5 and 9.8. ⊣

We now present another variant of FSTZ which we call FSTZS = "finite sets of integers with scalars and squares". Here we replace $A \cdot B$ in favor of scalar multiplication and squares.

The signature of FSTZS is $L(Z, \text{fst})$. The nonlogical axioms of FSTZS are as follows.

1. Linearly ordered ring axioms.
2. Finite interval.
3. Boolean difference.
4. Duplicate set addition.
5. Scalar multiplication.

$$(\exists B)(\forall x)(x \in B \leftrightarrow (\exists y)(y \in A \wedge x = y \cdot z)).$$

6. Squares.

$$(\exists A)(\forall x)(x \in A \leftrightarrow (\exists y)(0 < y \wedge y < z \wedge x = y^2)).$$

7. Least element.

Axiom 5 asserts that each $c \cdot A$ exists. Axiom 6 asserts that each $\{1^2, 2^2, \ldots, n^2\}$, $n \geq 0$, exists.

Lemmas 9.10–9.27 refer to provability in FSTZS.

LEMMA 9.10. i)–ix) *of Lemma* 8.1. *$A + B$ exists.*

PROOF. For the first claim, we need only observe that by scalar multiplication, $-A$ exists. From this we obtain that every nonempty set has a greatest element. For the second claim, we can repeat the proof of Lemma 9.5. ⊣

To show that FSTZS is equivalent to FSTZ, it suffices to prove that $A \cdot B$ exists in FSTZS. We do not know a clean way of doing this. Instead, we recast the proof of Lemma 8.23 for FSTZS in order to derive that $A \cdot B$ exists. Much of the proof will be the same. The key point is to avoid use of | in the auxiliary languages, and instead use a monadic predicate for "being a square".

LEMMA 9.11. *Let $d \geq 1$ and x be an integer. There exists unique q, r such that $x = dq + r$ and $0 \leq r < d$.*

PROOF. See Lemma 9.2. ⊣

LEMMA 9.12. *Let $k \geq 0$. The following is provable in T_2. For all $r \geq 2$, the elements of $[0, r^{k+1})$ have unique representations of the form $n_0 r^0 + \cdots + n_k r^k$, where each n_i lies in $[0, r)$. If $n_0 r^0 + \cdots + n_k r^k = m_0 r^0 + \cdots + m_k r^k$ and each n_i lies in $(-r/2, r/2)$, then each $n_i = m_i$.*

PROOF. See Lemma 9.3. ⊣

Until the end of the proof of Lemma 8.21, we fix a standard integer $k > 0$.

LEMMA 9.13. *For all $r > 1$, $S[r] = \{n_0 r^0 + n_1 r^2 + \cdots + n_i r^{2i} + \cdots + n_k r^{2k} : n_0, \ldots, n_k \in [0, r)\}$ exists. Every element of $S[r]$ is uniquely written in the displayed form.*

PROOF. See Lemma 9.4. ⊣

LEMMA 9.14. *For all $r > 1$ and $i \in [0, k]$, $\{x \in S[r] : x[i] = 0\}$ and $\{x \in S[r] : x[i] = 1\}$ exist.*

PROOF. See Lemma 9.5. ⊣

LEMMA 9.15. *For all $r > 1$ and $i, j, p \in [0, k]$, $\{x \in S[r] : x[i] + x[j] = x[p]\}$ exists.*

PROOF. See Lemma 9.6. ⊣

Note that we cannot use Lemma 9.7 here since it involves multiplication of sets, as opposed to just scalar multiplication of sets.

LEMMA 9.16. *For all $r > 1$, $i \in [0, k]$, and $A \subseteq [0, r)$, $\{x \in S[r] : x[i] \in A\}$ exists.*

PROOF. See Lemma 9.8. ⊣

LEMMA 9.17. *For all $r > 1$ and $i \in [0, k]$, $\{x \in S[r] : x[i] \text{ is a square}\}$ exists.*

PROOF. Use Lemma 9.16 with $A = \{1^2, \ldots, r^2\}$. ⊣

LEMMA 9.18. *Let φ be a propositional combination of formulas $x_i = 0$, $x_i = 1$, $x_i + x_j = x_p$, $\mathrm{Sq}(x_i)$, $x_i \in A_j$, where $i, j, p \in [0, k]$. The following is provable in T_4. For all $A_0, \ldots, A_k \subseteq [0, r)$, $\{x_0 r^0 + \cdots + x_k r^{2k} : \varphi \wedge x_0, \ldots, x_k \in [0, r)\}$ exists.*

PROOF. See Theorem 9.9. $\mathrm{Sq}(x_i)$ means "x_i is a square". ⊣

LEMMA 9.19. *For all $r > 1$ and $i \in [0, k]$ and $E \subseteq S[r]$, $\{x \in S[r] : (\exists y \in E)(\forall j \in [0, k]\backslash\{i\})(x[j] = y[j])\}$ exists.*

PROOF. See Lemma 9.10. ⊣

LEMMA 9.20. *Let φ be a propositional combination of formulas $x_i = 0$, $x_i = 1$, $x_i + x_j = x_p$, $\mathrm{Sq}(x_i)$, $x_i \in A_j$, where $i, j, p \in [0, k]$. Let $m \in [1, k]$. Let $\psi = (Q_m x_m \in [0, r)) \dots (Q_k x_k \in [0, r))(\varphi)$. The following is provable in T_4. For all $A_0, \dots, A_k \subseteq [0, r)$, $\{x_0 r^0 + \cdots + x_{m-1} r^{2m-2} : \psi \wedge x_0, \dots, x_{m-1} \in [0, r)\}$ exists.*

PROOF. See Lemma 9.11. ⊣

LEMMA 9.21. *Let $r > 1$, $E \subseteq S[r]$, $i_1 < \cdots < i_p \in [0, k]$, and $x_1, \dots, x_p \in [0, r)$. Then $\{y \in S[r] : y[i_1] = x_1 \wedge \cdots \wedge y[i_p] = x_p\}$ exists.*

PROOF. See Lemma 9.12. ⊣

We now release the fixed standard integer k.

LEMMA 9.22. *Let φ be a formula without bound set variables whose atomic subformulas are of the form $x_i = 0$, $x_i = 1$, $x_i + x_j = x_p$, $\mathrm{Sq}(x_i)$, $x_i \in A_j$. Let y, z be distinct integer variables, where z does not appear in φ. Then FSTZS proves that $\{y \in [0, z] : \varphi^z\}$ exists. Also T_4 proves that $\{y \in [-z, z] : \varphi^z\}$ exists.*

PROOF. See Lemma 9.13. ⊣

LEMMA 9.23. *Let φ be a formula without bound set variables whose atomic subformulas are of the form $s = t$, $s < t$, $\mathrm{Sq}(t)$, or $t \in A_j$, where s, t are terms without $\cdot$. Let y, z be distinct integer variables, where z does not appear in φ. Then FSTZS proves that $\{y \in [-z, z] : \varphi^z\}$ exists.*

PROOF. See Lemma 9.14. ⊣

We call the formulas given in Lemma 9.23, good formulas.

LEMMA 9.24. *Let $x = y^2$, $y \geq 0$. Then the next square after x is $(y + 1)^2$, and this is at most $3x + 1$.*

PROOF. Suppose $y^2 < z^2 < (y + 1)^2$. We can assume $z \geq 0$. Clearly $y < z < y + 1$ since squaring is strictly increasing on the nonnegative integers. For the second claim, first note that $y \leq x$. Then observe that $(y + 1)^2 = y^2 + 2y + 1 = x + 2y + 1 \leq 3x + 1$. ⊣

LEMMA 9.25. *$x = y^2$ if and only if x is a square and the next square after x is $x + 2y + 1$. The next square after x is at most $2x + 1$.*

PROOF. The forward direction is by Lemma 9.24. For the reverse direction, let x be a square and the next square after x is $x + 2y + 1$. Let $x = z^2$. Then the next square after x is $(z + 1)^2$. So $(z + 1)^2 = z^2 + 2y + 1 = z^2 + 2z + 1$. Hence $y = z$. ⊣

LEMMA 9.26. *There is a good formula* φ *with at most the free variables among* x, y, *such that the following is provable in FSTZS. For all* z *there exists* $z' > z$ *such that* $(\forall x, y \in [-z, z])(x = y^2 \leftrightarrow \varphi^{z'})$.

PROOF. Let z be given. We can assume that $z \geq 0$. Let $\varphi(x, y)$ be $(y \geq 0 \wedge \mathrm{Sq}(x) \wedge \mathrm{Sq}(x + 2y + 1) \wedge (\forall w)(\mathrm{Sq}(w) \rightarrow \neg(x < w < x + 2y + 1)))$. Note that φ expresses that x is a square, $y \geq 0$, and $x + 2y + 1$ is the next square after x. Note also that when bounded to $[-3z+1, 3z+1]$, the meaning remains unchanged. This works for $x, y \in [0, z]$, and can be easily modified to work for $x, y \in [-z, z]$. ⊣

LEMMA 9.27. *There is a good formula* ψ *with at most the free variables* u, v, w, *such that the following is provable in FSTZS. For all* z *there exists* $z' > z$ *such that* $(\forall x, y, z \in [-z, z])(u \cdot v = w \leftrightarrow \psi^{z'})$.

PROOF. Let z be given. As in the proof of Lemma 9.23, use $\psi = (\exists x, y, a, b)(x = y^2 \wedge y = u + v \wedge a = u^2 \wedge b = v^2 \wedge 2w = x - a - b)$ and Lemma 9.26. We can easily bound the quantifiers to an appropriately chosen $[-z', z']$. ⊣

THEOREM 9.28. *FSTZ, FSTZD, FSTZS are equivalent.*

PROOF. It suffices to show that $A \cdot B$ exists within FSTZS. Let A, B be given, where $A, B \subseteq [-z, z]$. Then $A \cdot B \subseteq [-z^2, z^2]$, but we don't know yet that $A \cdot B$ exists.

Let z' be according to Lemma 9.27 for z^2. Then $A \cdot B = \{y \in [-z^2, z^2] : (\exists u, v, w)(u \in A \wedge v \in B \wedge \psi^{z'})\} = \{y \in [-z^2, z^2] : (\exists u, v, w)(u \in A \wedge v \in B \wedge \psi)^{z'}\}$ which exists by Lemma 9.23.

The second claim follows immediately from the first. ⊣

§10. FSQZ.

We now give a very simple interpretation of FSTZ in FSQZ, which is the identity on the Z sort. It follows immediately that FSQZ proves PFA(Z). We then show that FSQZ is a conservative extension of PFA(Z).

Recall the axioms of FSQZ.

1. Linearly ordered integral domain axioms.
2. $\mathrm{lth}(\alpha) \geq 0$.
3. $\mathrm{val}(\alpha, n)\downarrow \leftrightarrow 1 \leq n \leq \mathrm{lth}(\alpha)$.
4. The finite sequence $(0, \ldots, n)$ exists.
5. $\mathrm{lth}(\alpha) = \mathrm{lth}(\beta) \rightarrow -\alpha, \alpha + \beta, \alpha \cdot \beta$ exist.
6. The concatenation of α, β exists.
7. For all $n \geq 1$, the concatenation of α, n times, exists.
8. There is a finite sequence enumerating the terms of α that are not terms of β.
9. Every nonempty finite sequence has a least term.

The interpretation of the integer part is the identity. The interpretation of the sets of integers in T_0 are the sequences of integers in FSZ. The $\in$ relation is interpreted as

$$n \in x \text{ if and only if } n \text{ is a term of } \alpha.$$

We write $(n \text{ upthru } m)$ for the finite sequence α, if it exists, such that

i. $\text{lth}(\alpha) = \max(0, m - n + 1)$.
ii. For all $1 \leq i \leq \text{lth}(\alpha)$, $\text{val}(\alpha, i) = n + i - 1$.

LEMMA 10.1. *The empty sequence exists. For all n, (n) exists.*

PROOF. The empty sequence is from axiom 9 of FSQZ.

Clearly (0) exists by axiom 5. Let $n > 0$. Form $(0, \ldots, n-1)$, $(0, \ldots, n)$ by axiom 5, and delete the latter from the former by axiom 9, to obtain (n). If $n < 0$ then form $(-n)$, and then form (n) by axiom 6. ⊣

LEMMA 10.2. *For all n there is a sequence consisting of all n's of any length ≥ 0.*

PROOF. Let n be given. Form (n) by Lemma 10.1. Let $k \geq 1$. The sequence consisting of all n's of length k is obtained by axiom 8. ⊣

LEMMA 10.3. *For all n, m, $(n \text{ upthru } m)$ exists. The interpretation of Finite Interval holds.*

PROOF. Let n, m be given. We can assume that $n \leq m$. By axiom 5, form $(0, \ldots, m - n)$. By Lemma 10.2, form $(n, \ldots, n)$ of length $m - n + 1$. By axiom 6, form $(0, \ldots, m - n) + (n, \ldots, n) = (n \text{ upthru } m)$. ⊣

LEMMA 10.4. *The interpretation of Boolean Difference holds.*

PROOF. Let x, y be given. By axiom 9, we obtain the required sequence. ⊣

LEMMA 10.5. *The interpretation of Least Element holds.*

PROOF. Let α be nonempty. Apply axiom 10. ⊣

We now come to the most substantial part of the verification - Set Addition and Set Multiplication.

We first need to derive QRT = quotient remainder theorem. This asserts that

$$\text{for all } d \geq 1 \text{ and } n, \text{ there exists unique } q, r$$
$$\text{such that } n = dq + r \wedge 0 \leq r < d.$$

LEMMA 10.6. *For all $d \geq 1$ and n, there is at most one q, r such that $n = dq + r$.*

PROOF. Let $d \geq 1$, $dq + r = dq' + r'$, and $0 \leq r, r' < d$. Then $d(q - q') = r' - r$. Suppose $q \neq q'$. By discreteness, $|q - q'| \geq 1$, and so $|d(q - q')| \geq d$. However, $|r' - r| < d$. This is a contradiction. Hence $q = q'$. Therefore $r' - r = 0$, $r = r'$. ⊣

LEMMA 10.7. *Let $d \geq 1$ and $n \geq d$. There is a greatest multiple of d that is at most n.*

PROOF. By Lemma 10.3, form $(d, d, \ldots, d)$ of length n. By Lemma 10.3 and axiom 6, form $(1, 2, \ldots, n) \cdot (d, \ldots, d) = (d, 2d, \ldots, dn)$ of length n. By Lemma 10.3 and axiom 5, form $(n, \ldots, n)$ and $(-d, -2d, \ldots, -dn)$. By axiom 5 form $(n - d, n - 2d, \ldots, n - dn)$.

We now wish to delete the negative terms from $(n - d, n - 2d, \ldots, n - dn)$. If $n - dn \geq 0$ then there is nothing to delete. Assume $n - dn < 0$. Obviously, the negative terms of $(n - d, n - 2d, \ldots, n - dn)$ are in $[n - dn, -1]$. Form $-(1, \ldots, dn - n)$ by Lemma 10.3 and axiom 6. By axiom 9, delete $-(0, \ldots, dn - n)$ from $(n - d, n - 2d, \ldots, n - dn)$. The result lists the nonnegative $n - id$, $1 \leq i \leq n$. By axiom 10, let $n - id$ be least among the nonnegative $n - id$ with $1 \leq i \leq n$. Then id is the greatest multiple of d that is at most n. ⊣

LEMMA 10.8. *The Quotient Remainder Theorem holds.*

PROOF. Let $d \geq 1$ and n be given.

CASE 1. $n \geq d$. By Lemma 10.7, let dq be the greatest multiple of d that is at most n. Then $n - dq \geq 0$. If $n - dq \geq d$, then $d(q + 1) = dq + d$ is a greater multiple of d that is at most n. This is a contradiction. Hence $0 \leq n - dq < d$, and set $r = n - dq$.

CASE 2. $0 \leq n < d$. Set $q = 1, r = n$.

CASE 3. $-d \leq n < 0$. Set $q = -1, r = n + d$.

CASE 4. $n < -d$. Then $-n > d$. By case 1, write $-n = dq + r$, where $0 \leq r < d$. Then $n = d(-q) - r = d(-q - 1) + d - r$, and $0 < d - r \leq d$. If $r = 0$ write $n = d(-q) + 0$. Otherwise write $n = d(-q - 1) + d - r$, $0 < d - r < d$. ⊣

LEMMA 10.9. *Let* $\mathrm{lth}(\alpha) = n$ *and* m *be given. The sequence* $\alpha^{(m)}$ *given by axiom* 8 *is unique and has the same terms as* α.

PROOF. Recall from axiom 8 that $\alpha^{(m)}$ has length nm, and for all q, r with $0 \leq q < m \wedge 1 \leq r \leq n$, $\mathrm{val}(\alpha^{(n)}, n \cdot q + r) = \mathrm{val}(x, r)$.

According to the QRT, this defines all values at all positions of $\alpha^{(m)}$. Thus $\alpha^{(m)}$ is unique and obviously has the same terms as x. ⊣

LEMMA 10.10. *Let* $\mathrm{lth}(\alpha) = n + 1$ *and* $\mathrm{lth}(\beta) = n$, *where* $n \geq 1$. *Let* $\alpha^{(n)}$ *and* $\beta^{(n+1)}$ *be given by axiom* 8. *For all* $1 \leq i \leq j \leq n$, $\mathrm{val}(\alpha^{(n)}, jn - in + j) = \mathrm{val}(\alpha, i)$ *and* $\mathrm{val}(\beta^{(n+1)}, jn - in + j) = \mathrm{val}(\beta, j)$.

PROOF. Let α, n, i, j be as given. Note that $jn - in + j = (j - i)(n + 1) + i = (j - i)(n) + j$. Since $0 \leq j - i < n$,

$$\mathrm{val}(\alpha^{(n)}, jn - in + j) = \mathrm{val}(\alpha^{(n)}, (j - i)(n + 1) + i) = \mathrm{val}(\alpha, i).$$

$$\mathrm{val}(\beta^{(n+1)}, jn - in + j) = \mathrm{val}(\beta^{(n+1)}, (j - i)(n) + j) = \mathrm{val}(\beta, j).$$ ⊣

LEMMA 10.11. *Let α, β be nonempty. There exists γ, δ such that*

i. *α, γ have the same terms.*
ii. *β, δ have the same terms.*
iii. $\text{lth}(\gamma) = \text{lth}(\delta) + 1$.
iv. *Let a be a term of α and b be a term of β. Then there exists* $1 \leq i \leq j < \text{lth}(\delta)$ *such that* $\text{val}(\gamma, i) = a$ *and* $\text{val}(\delta, j) = b$.

PROOF. Let α, β be nonempty, $\text{lth}(\alpha) = n$, $\text{lth}(\beta) = m$. Let u be the last term of α and v be the last term of β. Let α' be αu^m, and $\beta' = \beta v^n$, where here multiplication is concatenation. Then $\text{lth}(\alpha') = \text{lth}(\beta')$, α' has the same terms as α, and β' has the same terms as β. Finally, let $\gamma = \alpha'\alpha' u$, and $\delta = \beta'\beta'$. Obviously $\text{lth}(\gamma) = \text{lth}(\delta + 1)$. Clearly every term in α is a term of the first α' in $\alpha'\alpha' u$, and every term in β is a term in the second β' in $\beta'\beta'$. We never have to use the last term of δ since the last two terms of β' are the same. ⊣

LEMMA 10.12. *The interpretations of Set Addition and Set Multiplication hold.*

PROOF. Let α, β be given. We can assume that α, β are nonempty. Let γ, δ be as given by Lemma 10.11, say with lengths $n + 1$, n. Let a be a term of α and b be a term of β. By Lemma 10.11, there exists $1 \leq i \leq j \leq n$ such that $\text{val}(\gamma, i) = a$ and $\text{val}(\delta, j) = b$. By Lemma 10.10, there exists k such that $\text{val}(\gamma^{(n)}, k) = a$ and $\text{val}(\delta^{(n+1)}, k) = b$. Hence

$$a + b \text{ is a term of } \gamma^{(n)} + \delta^{(n+1)}.$$

$$a \cdot b \text{ is a term of } \gamma^{(n)} \cdot \delta^{(n+1)}.$$

On the other hand, by Lemma 10.9, $\gamma^{(n)}$ has the same terms as α and $\delta^{(n+1)}$ has the same terms as β. Hence

i. the terms of $\gamma^{(n)} + \delta^{(n+1)}$ are exactly the result of summing a term of α and a term of β.
ii. the terms of $\gamma^{(n)} \cdot \delta^{(n+1)}$ are exactly the result of multiplying a term of α and a term of β.

Thus

iii. $\gamma^{(n)} + \delta^{(n+1)}$ witnesses Set Addition for α, β.
iv. $\gamma^{(n)} \cdot \delta^{(n+1)}$ witnesses Set Multiplication for α, β.

⊣

THEOREM 10.13. *The interpretation of every axiom of FSTZ is a theorem of FSQZ.*

PROOF. By Lemmas 10.3, 10.4, 10.7, and 10.12. ⊣

THEOREM 10.14. *FSQZ proves* PFA(Z).

PROOF. Since the interpretation of FSTZ in FSQZ used here is the identity on the Z sort, the result follows immediately from Theorem 8.29 and Theorem 10.13. ⊣

THEOREM 10.15. *FSQZ is a conservative extension of* PFA(Z).

PROOF. By Theorem 10.14, it suffices to expand any model M of PFA(Z) to a model of FSQZ. Use the bounded $\Sigma_0(Z)$ binary relations of M which are univalent, with domain some $\{1, \ldots, n\}$, as the finite sequences. ⊣

§11. Conservative extensions, interpretability, synonymy, and logical strength. The ten systems of arithmetic considered here are

$$Q(N), \text{PFA}(N), \text{PFA}(N) + \text{EXP}(N), \text{PFA}(N) + \text{CM}(N), \text{EFA}(N, \exp).$$

$$\text{LOID}(Z), \text{PFA}(Z), \text{PFA}(Z) + \text{EXP}(Z), \text{PFA}(Z) + \text{CM}(Z), \text{EFA}(Z, \exp).$$

These were presented in Sections 4, 5, and relationships between these twelve systems were established in Sections 4, 5, 6 - especially see Theorem 6.7.

The seven strictly mathematical theories considered here were presented in Section 7:

$$FSTZ, FSQZ, FSTZ + \text{CM}(Z), FSQZ + \text{CM}(Z),$$
$$FSTZEXP, FSTZBEXP, FSQZEXP.$$

Recall that FSTEXP extends FSTZ by

i. $\exp(n, 0) = 1$.
ii. $m \geq 0 \rightarrow (\exp(n, m + 1) = \exp(n, m) \cdot n \wedge \exp(n, -m - 1)\uparrow)$.
iii. The finite set $\{\exp(n, 0), \ldots, \exp(n, m)\}$ exists.

FSTZEXP′ extends FSTZ by

i. $\exp(n, 0) = 1$.
ii. $m \geq 0 \rightarrow (\exp(n, m + 1) = \exp(n, m) \cdot n \wedge \exp(n, -m - 1)\uparrow)$.
iii. $n \geq 2 \wedge 0 \leq m < r \rightarrow \exp(n, m) < \exp(n, r)$.
iv. The finite set $\{\exp(n, 0) + 0, \exp(n, 1) + 1, \ldots, \exp(n, m) + m\}$ exists.

FSQZEXP extends FSQZ by

i. $\exp(n, 0) = 1$.
ii. $m \geq 0 \rightarrow (\exp(n, m + 1) = \exp(n, m) \cdot n \wedge \exp(n, -m - 1)\uparrow)$.
iii. The finite sequence $(\exp(n, 0), \ldots, \exp(n, m))$ exists.

LEMMA 11.1. *FSTZEXP′ proves* PFA(Z)$+ (\forall n)(\{\exp(n, 0), \ldots, \exp(n, m)\}$ *exists*).

PROOF. Recall from Theorem 8.28 that FSTZ proves bounded $\Sigma_0(Z, \text{fst})$ separation:

*) $(\exists A)(\forall x)(x \in A \leftrightarrow (y < x \wedge x < z \wedge \varphi))$,

where $\varphi \in \Sigma_0(Z, \text{fst})$ and A is not free in φ.

We argue in FSTZEXP′. Fix $n \geq 0$. Let $A = \{\exp(n, 0) + 0, \ldots, \exp(n, m+1)+m+1\}$. Note that for all $0 \leq r \leq m$, the next element of A after

$\exp(n,r)+r$ is $\exp(n,r+1)+r+1$. Let B be the set of successive differences of elements of A. Then $B = \{\exp(n,r+1)+r+1-(\exp(n,r)+r) : 0 \le r \le m\} = \{\exp(n,r)+1 : 0 \le r \le m\}$. Hence $B - \{1\} = \{\exp(n,r) : 0 \le r \le m\}$. ⊣

We use n^m for the partial exponential function, according to PFA(Z). By [HP98, page 299], the relation $r = n^m$ is given by a bounded formula in PFA(Z). Note that the relation "r is of the form n^m" is also given by a bounded formula in PFA(Z).

LEMMA 11.2. *FSTZEXP′ proves that for $n, m \ge 0$, every $\exp(n,m)$ is of the form n^r.*

PROOF. Fix $n, m \ge 0$. By Lemma 11.1, let $A = \{\exp(n,0), \dots, \exp(n,m)\}$. Let B be the set of all elements of A that are not of the form n^r. Let $\exp(n,t)$ be the least element of B. Then $t > 0$ and $\exp(n,t-1)$ is of the form n^r. Hence $\exp(n,t)$ is of the form n^r. This is a contradiction. ⊣

LEMMA 11.3. *FSTZEXP′ proves $n, m \ge 0 \rightarrow \exp(n,m) = n^m$.*

PROOF. We can assume that $n \ge 2$ and $m \ge 0$. Let $A = \{\exp(n,0)+0, \dots, \exp(n,m)+m\}$. We first show the following. Let $\exp(n,r-2)+r-2$, $\exp(n,r-1)+r-1$ both be of the form n^s+s, where $r \ge 2$. Then $\exp(n,r) = n^r$.

By Lemma 11.2, write

$$\exp(n,r-2) = n^p.$$
$$\exp(n,r-1) = n^{p+1}.$$
$$n^p + r - 2 = n^s + s.$$
$$n^{p+1} + r - 1 = n^t + t.$$
$$s < t.$$

Hence

$$n^{p+1} + r - 1 - (n^p + r - 2) = n^{p+1} - n^p + 1 = n^t + t - (n^s + s)$$
$$= n^t - n^s + t - s.$$

Also by $n^p + r - 2 = n^s + s$, we have $p \le s$.

CASE 1. $p+1 < t$. Then $n^{p+1} \le n^t - n^s < n^t - n^s + t - s = n^{p+1} - n^p + 1 \le n^{p+1}$, which is a contradiction.

CASE 2. $t \le p$. Then $n^t - n^s + t - s \le n^p - n^p + p - p = 0 < n^{p+1} - n^p + 1$, which is a contradiction.

CASE 3. $t = p + 1$. The only possible case.

So $t = p+1$, $r-1 = t$, $p = r-2$, $\exp(n,r-2) = n^{r-2}$. Hence $\exp(n,r) = n^r$.

Next we claim that every element of A is of the form $n^s + s$. Suppose $\exp(n,r)+r$ is the least element of A that is not of the form n^s+s. Clearly $r \ge 2$ and $\exp(n,r-2)+r-2$, $\exp(n,r-2)+r-1$ are both of the form

$n^s + s$. By the claim, we have $\exp(n, r) = n^r$, and so $\exp(n, r) + r = n^r + r$. This is a contradiction.

In particular, $\exp(n, m - 2)$ and $\exp(n, m - 1)$ are of the form $n^s + s$, and so by the claim, $\exp(n, m) = n^m$. ⊣

LEMMA 11.4. *FSTZEXP′ is a definitional extension of FSTZ* $+ \mathrm{CM}(Z)$.

PROOF. By Lemma 11.3, FSTZEXP′ proves FSTZ + "exponentiation is total". Hence FSTZEXP′ proves *FSTZ* $+ \mathrm{CM}(Z)$. Also, by Lemma 11.3, FSTZEXP′ proves $\exp(n, m) = n^m$, defining exp. ⊣

LEMMA 11.5. *FSQZEXP proves* $n \geq 0 \rightarrow \exp(n, m) = n^m$.

PROOF. By Theorem 10.14, FSQZEXP proves $\mathrm{PFA}(Z)$. We now argue in FSQZEXP. Fix $n \geq 0$. Let α be the sequence $(\exp(n, 0), \dots, \exp(n, m))$. By using the ring operation axioms of FSQZ, we obtain the sequence

$$\beta = (\langle 0, \exp(n, 0)\rangle, \langle 1, \exp(n, 1)\rangle, \dots, \langle n, \exp(n, m)\rangle)$$

where $\langle x, y\rangle = (x + y)^2 + x$.

By the separation in Theorem 8.28, and Theorem 10.14, we obtain a sequence γ whose terms comprise the terms of β which are not of the form $\langle t, n^t\rangle$. (Only bounded quantifiers are involved in this construction). Let $\langle k, \exp(n, k)\rangle$ be the least term of γ. Then $k > 0$, and $\langle k - 1, \exp(n, k - 1)\rangle$ is of the form $\langle t, nt\rangle$. *I.e.*, $\exp(n, k - 1) = n^{k-1}$. Therefore $\exp(n, k) = n^k$, and $\langle k, \exp(n, k)\rangle$ is a term of β of the form $\langle k, n^k\rangle$. This is a contradiction. Hence γ is empty. Therefore every term of β is of the form $\langle t, n^t\rangle$. In particular, $\langle m, \exp(n, m)\rangle$ is of the form $\langle t, n^t\rangle$. Therefore $\exp(n, m) = n^m$. ⊣

LEMMA 11.6. *FSQZEXP is a definitional extension of FSQZ* $+ \mathrm{CM}(Z)$. *FSQZEXP is a conservative extension of* $\mathrm{EFA}(Z, \exp)$.

PROOF. The first claim is immediate from Lemma 11.5 and Theorem 10.14. For the second claim, first note that FSQZEXP proves $\mathrm{EFA}(Z, \exp)$. This is because given any formula in $\Sigma_0(Z, \exp)$, we can replace all occurrences of exp in favor of internal exponentiation, using Lemma 11.5, thereby obtaining a $\Sigma_0(Z)$ formula, to which we can apply induction in $\mathrm{PFA}(Z) \subseteq$ *FSQZEXP*.

Now let M be a model of $\mathrm{EFA}(Z, \exp)$. We can expand M to M' by adding the finite sequences, and associated apparatus, that is internal to M. Then $M' \models$ *FSQZEXP*. ⊣

LEMMA 11.7. *FSTZEXP is a conservative extension of FSTZ.*

PROOF. Let M be a model of FSTZ. For $n \geq 0$, define $\exp(0, n) = 1$ if $n = 0$; 0 otherwise, and $\exp(1, n) = 1$. Now let $n \geq 2$. Clearly $\{n^m : m \geq 0\}$ is unbounded. If n^m exists, define $\exp(n, m) = n^m$. If nm does not exist, $m \geq 0$, then define $\exp(n, m) = 0$. Note that the sets $\{\exp(n, 0), \dots, \exp(n, m)\}$ are already present in M. Hence $(M, \exp)$ is a model of FSTZEXP. ⊣

LEMMA 11.8. $Q(N)$, *FSTZ, FSQZ, FSTZEXP are mutually interpretable.*

PROOF. Since $Q(N)$ and PFA(Z) are mutually interpretable, it suffices to show that PFA(Z), FSTZ, FSQZ, FSTZEXP are mutually interpretable. Since PFA(Z) is provable in FSTZ, FSQZ, it suffices to show that FSTZ, FSQZ are interpretable in PFA(Z). It therefore suffices to show that FSTZ, FSQZ are interpretable in PFA(N).

Let M be a model of PFA(N). In M, look at the cut I of all n such that the internal 2^n exists. If I is a proper cut, then by cut shortening, we can assume that I forms a model of PFA(N). We can then expand I with all of the internal subsets of I bounded by an element of I, and all of the internal sequences from I whose length is an element of I, and whose terms are bounded by an element of I, to form the required models of FSTZ, FSQZ, FSTZEXP (using the proof of Lemma 11.7) ⊣

LEMMA 11.9. EFA$(N, \exp)$, *FSTZ*+CM(Z), *FSQZ*+CM(Z), *FSTZEXP*$'$, *FSQZEXP are mutually interpretable.*

PROOF. EFA$(N, \exp)$ is interpretable in PFA(Z)+CM(Z) by Theorems 5.6 and 6.7, which is provable in *FSTZ* + CM(Z) and *FSQZ* + CM(Z) by Theorems 8.28 and 10.14.

FSTZ + CM(Z) is provable in FSTZBEXP by Lemma 11.4. *FSQZ* + CM(Z) is provable in FSQZEXP by Lemma 11.6.

So it suffices to interpret FSTZBEXP, FSQZEXP in EFA$(N, \exp)$. Interpret the finite sets and finite sequences by finite coding in EFA$(N, \exp)$. ⊣

LEMMA 11.10. EFA$(N, \exp)$, *FSTZ* + CM(Z) *are synonymous.*

PROOF. By Theorem 6.7, EFA$(N, \exp)$ and EFA$(Z, \exp)$ are synonymous. It now suffices to show that EFA$(Z, \exp)$, *FSTZ* + CM(Z) are synonymous.

We interpret EFA$(Z, \exp)$ in *FSTZ* + CM(Z) by preserving the Z sort, and interpreting exp as internal exponentiation in PFA(Z)+CM(Z). We interpret *FSTZ* + CM(Z) in EFA$(Z, \exp)$ by preserving the Z sort and interpreting the finite sets by codes in EFA$(Z, \exp)$. Let M be a model of EFA$(Z, \exp)$. M will be sent to a model M' of *FSTZ* + CM(Z) with the same ordered ring, but where exp is gone. Since CM(Z) must still hold, we have an internal exponentiation in M', and so when going back, we recover the old exp. For the other compositional identity, let M be a model of *FSTZ* + CM(Z). Then the Z part of M is a model of PFA(Z)+CM(Z), and therefore has an internal exponentiation. M is sent to a model M' of EFA$(Z, \exp)$, where exp agrees with the internal exponentiation in M. When we go back, we must have the same ordered ring structure, and the sets are those given internally from the ordered ring structure of M.

Thus it suffices to verify that in M, the sets are exactly those sets coded internally in the ordered ring structure of M. By Theorem 8.28, FSTZ proves separation for formulas in $\Sigma_0(Z, \text{fst})$. We can use this to prove in *FSTZ* + CM(Z) = *FSTZ* + EXP(Z) that every set is coded internally in the ordered ring structure, as in the proof of Lemma 11.2. Recall that internal exponentiation is used in that argument to make sure that the induction or separation needed has only bounded quantifiers. ⊣

We now summarize these results.

THEOREM 11.11. *FSTZEXP′, FSQZEXP are definitional extensions of FSTZ* + CM(Z), *FSQZ* + CM(Z), *respectively. FSTZEXP is a definitional extension of FSTZ. FSQZEXP is a conservative extension of* EFA(Z, exp). EFA(N, exp), *FSTZ* + CM(Z), *FSQZ* + CM(Z), *FSTZEXP′, FSQZEXP are mutually interpretable.* EFA(N, exp), *FSTZ* + CM(Z) *are synonymous. FSTZ, FSQZ, FSTZEXP are conservative extensions of* PFA(Z). $Q(N)$, *FSQZ, FSTZ, FSTZEXP are mutually interpretable.*

COROLLARY 11.12. *FSTZ* + CM(Z), *FSQZ* + CM(Z), *FSQZEXP are strictly mathematical mutually interpretable theories with logical strength. I.e., they interpret* EFA(N, exp). *Moreover, they imply and are mutually interpretable with* EFA(Z, exp).

§**12. RM and SRM.** The Reverse Mathematics program originated with [Fr75, Fr76], and the widely distributed manuscripts [Fr75.76], which refer to some of our earlier insights from 1969 and 1974. Also see [FS00].

RM is the main focus of the highly recommended [Si99]. This book has unfortunately been out of print soon after it appeared, but there are ongoing efforts to have it reprinted.

In RM, the standard base theory, RCA_0, introduced in [Fr76], is not strictly mathematical. However, in RM, we add strictly mathematical statements to RM and classify the resulting theories according to implications, equivalences, and logical strengths.

Often, the formulations of the mathematical statements investigated in RM involve coding. Usually these codings are rather robust, but nevertheless constitute another place where elements that are not of a strictly mathematical nature appear.

Fortunately, there are substantial areas of mathematics and a substantial variety of mathematical statements whose formulations are sufficiently robust to support the vigorously active development of RM. RM has continued to grow very substantially since its inception in the 1970s. We fully expect an accelerating development of RM for the foreseeable future.

However, there is a much greater body of mathematical activity which is currently not in any kind of sufficiently robust logical form to support an RM treatment. The bulk of the relevant mathematical statements are probably too weak, in terms of logical strength, for an RM development, since RM starts at the logical strength level of PRA (primitive recursive arithmetic). PRA is far stronger, logically, than the nonzero level of logical strength on which this paper is based - that of EFA(N, exp), or equivalently, $I\Sigma_0(\exp)$, and lower.

We view this paper as an introduction to what we call Strict Reverse Mathematics, or SRM.

In SRM, the focus is on theories where all statements are strictly mathematical - including all axioms in any base theory. In a sense, SRM is RM with no base theories at all!

Here we have shown that one can achieve logical strength using only strictly mathematical statements. Without this fundamental fact, there cannot be any SRM.

The major goal of SRM is to rework and extend RM using only strictly mathematical statements. SRM should strive to create sensible logical structure out of a vastly increased range of mathematics, going far beyond what can be analyzed with conventional RM.

An integral part of SRM is to take the standard natural formal systems developed in the foundations of mathematics - whose axioms are very far from being strictly mathematical - and reaxiomatize them with strictly mathematical statements. Such axiomatizations may take the form of conservative extensions or mutually interpretable or synonymous systems, as we have done here for PFA(N) (*i.e.*, $I\Sigma_0$), and for EFA$(N, \exp)$ (*i.e.*, $I\Sigma_0(\exp)$).

In this vein, we mention some SRM challenges.

1. Find a strictly mathematical axiomatization of PFA(Z), in its signature $L(Z)$.
2. Find a strictly mathematical axiomatization of PFA(Z) + EXP(Z), in its signature $L(Z)$.

Some work in the direction of 1, 2 is contained in [Fr00].

3. Find a strictly mathematical axiomatization of EFA$(Z, \exp)$, in its signature $L(Z, \exp)$.
4. FSQZ appears to be too weak to be naturally synonymous, or even synonymous, with FSTZ. The same can be said of $FSQZ$ + CM(Z), $FSTZ$ + CM(Z), and also FSQZEXP.

However, we can extend FSQZ to FSQZ# and obtain synonymy. FSQZ# is axiomatized by

1. LOID(Z).
2. Discreteness.
3. $\text{lth}(\alpha) \geq 0$.
4. $\text{val}(\alpha, n)\downarrow \leftrightarrow 1 \leq n \leq \text{lth}(\alpha)$.
5. $\Sigma_0(Z, \text{fsq})$ comprehension for finite sequences.

$$(n \geq 0 \wedge (\forall i)(\exists! j)(\varphi)) \rightarrow (\exists \alpha)(\text{lth}(\alpha) = n \wedge (\forall i)(1 \leq i \leq n \rightarrow \varphi[j/\text{val}(\alpha, i)])),$$

   where φ is a $\Sigma_0(Z, \text{fsq})$ formula in which α is not free.
6. Every sequence of length ≥ 1 has a least term.

The challenge is to give a strictly mathematical axiomatizations of $FSQZ\#$, $FSQZ\# + \mathrm{EXP}(Z)$, $FSQZ\#EXP$, in their respective signatures $L(Z, \mathrm{fsq})$, $L(Z, \mathrm{fsq})$, $L(Z, \exp, \mathrm{fsq})$.

The notion "strictly mathematical" is sufficiently clear to support the SRM enterprise. However, there are still fine distinctions that can be profitably drawn among the strictly mathematical. We have drawn such distinctions in our discussion of the relative merits of FSTZ and FSQZ at the end of Section 7.

It is clear from the founding papers of RM, [Fr75.76, Fr75, Fr76], that we envisioned a development like SRM. We spoke of raw text, and our original axiomatizations of main base theory RCA_0 and our other principal systems WKL_0, ACA_0, ATR_0, and Π^1_1-CA_0, of RM, were considerably more mathematical than the formally convenient ones that are mostly used today. However, any major development of SRM before that of RM would have been highly premature.

[Fr01] and [Fr05a] are technical precursors of this paper, dealing with FSTZ and FSQZ, respectively. [Fr05] is a preliminary report on SRM, attempting to develop SRM at higher levels of strength, and in many ways goes beyond what we have done very carefully here. However, this earlier work will undergo substantial revisions and upgrading in light of this initial publication.

REFERENCES

[Av03] J. Avigad, *Number theory and elementary arithmetic*, ***Philosophia Mathematica. Philosophy of Mathematics, its Learning, and Its Application. Series III***, vol. 11 (2003), no. 3, pp. 257–284, http://www.andrew.cmu.edu/user/avigad/.

[Be85] M. J. Beeson, ***Foundations of Constructive Mathematics***, Ergebnisse der Mathematik und ihrer Grenzgebiete (3) [Results in Mathematics and Related Areas (3)], vol. 6, Springer-Verlag, Berlin, 1985, Metamathematical Studies.

[Fe75] S. Feferman, *A language and axioms for explicit mathematics*, ***Algebra and Logic (Fourteenth Summer Res. Inst., Austral. Math. Soc., Monash Univ., Clayton, 1974)***, Lecture Notes in Mathematics, vol. 450, Springer, Berlin, 1975, pp. 87–139.

[Fe79] ———, *Constructive theories of functions and classes*, ***Logic Colloquium '78 (Mons, 1978)***, Studies in Logic and the Foundations of Mathematics, vol. 97, North-Holland, Amsterdam, 1979, pp. 159–224.

[Fe95] ———, *Definedness*, ***Erkenntnis. An International Journal of Analytic Philosophy***, vol. 43 (1995), no. 3, pp. 295–320, Varia with a Workshop on the Foundations of Partial Functions and Programming (Irvine, CA, 1995), http://math.stanford.edu/~feferman/papers/definedness.pdf.

[FF02] A. M. Fernandes and F. Ferreira, *Groundwork for weak analysis*, ***The Journal of Symbolic Logic***, vol. 67 (2002), no. 2, pp. 557–578.

[Fr75.76] H. Friedman, *The Analysis of Mathematical Texts, and Their Calibration in Terms of Intrinsic Strength I, April 3, 1975, 7 pages. The Analysis of Mathematical Texts, and Their Calibration in Terms of Intrinsic Strength II, April 8, 1975, 5 pages. The Analysis of Mathematical Texts, and Their Calibration in Terms of Intrinsic Strength III, May 19, 1975, 26 pages. The Analysis of Mathematical Texts, and Their Calibration in Terms of Intrinsic Strength IV, August 15, 1975, 32 pages. The Logical Strength of Mathematical Statements, October 15,*

1975, 1 page. The Logical Strength of Mathematical Statements I, August, 1976, 20 pages, http://www.math.ohio-state.edu/~friedman/manuscripts.html.

[Fr75] ———, *Some systems of second order arithmetic and their use*, ***Proceedings of the International Congress of Mathematicians (Vancouver, B.C., 1974)***, vol. 1, Canadian Mathematical Congress, Montreal, Que, 1975, pp. 235–242.

[Fr76] ———, *Subsystems of second order arithmetic with restricted induction I, II, abstracts*, ***The Journal of Symbolic Logic***, vol. 1 (1976), no. 2, pp. 557–559.

[Fr80] ———, *A strong conservative extension of Peano arithmetic*, ***The Kleene Symposium (Proc. Sympos., Univ. Wisconsin, Madison, Wis., 1978)***, Studies in Logic and the Foundations of Mathematics, vol. 101, North-Holland, Amsterdam, 1980, pp. 113–122.

[Fr00] ———, *Quadratic Axioms*, draft, January 3 2000, 9 pages, http://www.math.ohio-state.edu/~friedman/.

[Fr01] ———, *Finite reverse mathematics*, draft, October 19 2001, 28 pages, http://www.math.ohio-state.edu/~friedman/.

[Fr05a] ———, *The inevitability of logical strength*, draft, May 31 2005, 13 pages, http://www.math.ohio-state.edu/~friedman/.

[Fr05] ———, *Strict reverse mathematics*, draft, January 31 2005, 24 pages, http://www.math.ohio-state.edu/~friedman/.

[FS00] H. Friedman and S. G. Simpson, *Issues and problems in reverse mathematics*, ***Computability Theory and Its Applications (Boulder, CO, 1999)***, Contemporary Mathematics, vol. 257, AMS, Providence, RI, 2000, pp. 127–144.

[HP98] P. Hájek and P. Pudlák, ***Metamathematics of First-Order Arithmetic***, Perspectives in Mathematical Logic, Springer-Verlag, Berlin, 1998.

[Ja85] N. Jacobson, ***Basic Algebra. I***, second ed., W. H. Freeman, New York, 1985.

[Pl68] R. A. Pljuskevicus, *A sequential variant of constructive logic calculi for normal formulas not containing structural rules*, ***Proceedings of the Steklov Institute of Mathematics***, vol. 98 (1968), pp. 175–229, In The Calculi of Symbolic Logic, I, AMS Translations 1971.

[Ro52] R. M. Robinson, *An essentially undecidable axiom system*, ***Proceedings of the 1950 International Congress of mathematicians, cambridge MA***, 1952, pp. 729–730.

[Si99] S. G. Simpson, ***Subsystems of Second Order Arithmetic***, Perspectives in Mathematical Logic, Springer-Verlag, Berlin, 1999.

[Sm82] C. Smoryński, *Nonstandard models and related developments*, ***Harvey Friedman's Research on the Foundations of Mathematics***, Studies in Logic and the Foundations of Mathematics, vol. 117, North-Holland, Amsterdam, 1985, pp. 179–229.

[Wi86] A. J. Wilkie, *On sentences interpretable in systems of arithmetic*, ***Logic Colloquium '84 (Manchester, 1984)***, Studies in Logic and the Foundations of Mathematics, vol. 120, North-Holland, Amsterdam, 1986, pp. 329–342.

THE OHIO STATE UNIVERSITY
DEPARTMENT OF MATHEMATICS
231 W. 18TH AVENUE
COLUMBUS, OHIO 43210
E-mail: friedman@math.ohio-state.edu
URL: http://www.math.ohio-state.edu/~friedman/

APPLICATIONS OF LOGIC IN ALGEBRA: EXAMPLES FROM CLONE THEORY

MARTIN GOLDSTERN

In this article, an "algebra" (sometimes also called a "universal algebra") is a nonempty set X together with a set of "basic operations"; each operation is a finitary total function on X, i.e., a function $f : X^n \to X$ for some $n \in \{1, 2, \dots\}$.[1] While our operations always have finite arity, we allow infinitely many basic operations in our algebras.

Clone theory investigates the structure of all algebras on a given set. It turns out that several questions originating in algebra can be solved (or at least attacked) with set-theoretic methods. This article collects a few examples of this approach, based on recent results of Pinsker, Shelah and the author.

§1. **Background.** A large part of universal algebra investigates the relations between the basic operations of an algebra; here we take a different point of view — all term operations are considered to live an an equal level; thus we will, for example, not distinguish between a Boolean algebra and the corresponding Boolean ring, because each ring operation can be written as composition of operations of the Boolean algebra, and conversely.

This point of view is motivated by the observation that many properties of algebras depend only on the clone of its term functions or polynomials, without any need to distinguish basic operations; for example, an equivalence relation is a congruence relation iff it is closed under all unary polynomials.

Notation. For any nonempty set X, we write $\mathcal{O}^{(n)}$ or $\mathcal{O}_X^{(n)}$ for the set X^{X^n} of n-ary functions, and $\mathcal{O}_X = \mathcal{O} = \bigcup_{n=1}^{\infty} \mathcal{O}^{(n)}$.

Whenever $f_1, \dots, f_k : I \to X$ are functions, and $h : X^k \to X$, we write $h(f_1, \dots, f_k)$ for the function $i \mapsto h(f_1(i), \dots, f_k(i))$.

A function $f : X^n \to X$ is called *essentially unary* iff there is a function $F : X \to X$ and a coordinate k such that $f(x_1, \dots, x_n) = F(x_k)$ for all

Supported by the Austrian Science foundation: FWF grant P17627-N12. I am grateful to the organizers of LC'06 for the opportunity to present this talk.

[1] It is notationally more convenient to interpret zeroary functions as unary constant functions.

Logic Colloquium '06
Edited by Barry Cooper, Herman Geuvers, Anand Pillay, and Jouko Väänänen
Lecture Notes in Logic, 32

$x_1, \dots, x_n \in X$. We will occasionally identify the essentially unary functions in $\mathcal{O}^{(n)}$ with the corresponding functions in $\mathcal{O}^{(1)}$.

Term functions. For any algebra $\mathfrak{X} = (X, f, g, \dots)$ and any natural number $n \geq 1$, the set of n-ary term functions is the smallest set of n-ary operations that contains all projections and is closed under the operations of X (which are understood to be taken pointwise). Each n-ary term function can be specified by a term, that is, a well-formed expression involving variables among $x_1, \dots, x_n$ and symbols for the operations $f, g, \dots$ of $\mathfrak{X}$.

Note that the set of term functions on any algebra is always closed under substitution; that is, if f is a k-ary term function and $h_1, \dots, h_k$ are n-ary term functions, then also the function

$$f(h_1, \dots, h_k) : \vec{x} \in X^n \mapsto f(h_1(\vec{x}), \dots, h_k(\vec{x})) \in X$$

is an n-ary term function; conversely, if $C \subseteq \mathcal{O}$ is a set closed under substitution and contains all projections, then C is the set of term functions of some algebra (for example, the algebra in which all functions of C are basic operations).

Definition 1. Let X be a nonempty set. A *clone* on X is a set $C \subseteq \mathcal{O}$ which contains all projections and is closed under substitution.

For any set $F \subseteq \mathcal{O}$ we write $\langle F \rangle$ for the smallest clone containing F.

It is often instructive to restrict attention to binary functions only. A set $C \subseteq \mathcal{O}^{(2)} = X^{X^2}$ is called a *binary clone* if C contains the two projection π_1^2, π_2^2 and has the following closure property: for all $f, g, h \in C : f(g, h) \in C$. Equivalently, C is a binary clone iff C is the binary fragment of a clone D: $C = D \cap \mathcal{O}^{(2)}$.

(This is a natural analog of the notion of a submonoid of X^X; such a submonoid might be called "unary clone" in our context.)

Remark 2. In general we do not require that a clone contain all constant functions; however, most of the clones that we consider will contain them.

Clone theory analyzes the relationships between different clones on the same set. For example, how "rich" in term functions is a ring $(X, +, \cdot)$ when compared with the underlying group $(X, +)$? This is really a comparison between the clones $\langle + \rangle$ and $\langle +, \cdot \rangle$. The distance between these two clones can be measured in several ways: From one point of view, it is very small, as the second clone is finitely generated over the first; another point of view would try to count the number of clones that lie strictly between those two clones, or use other cardinal characteristics of this partially ordered set (such as the size of the largest chain/antichain, etc.).

For a fixed set X, the set $\boldsymbol{CLONE}(X)$ of clones on X is naturally ordered by inclusion; with this relation, $\boldsymbol{CLONE}(X)$ is easily seen to be a complete lattice (as the intersection of any number of clones is again a clone).

Clones over finite sets are well investigated, but there are still many open problems. The book [10] contains many basic as well as advanced results about clones; [11] and the very recent [7] concentrate on clones on finite sets. More results about clones on infinite sets, as well as an extensive bibliography, can be found in [2].

It is clear that for finite sets the set $\mathcal{O}_X$ is countable; but already the case $|X| = 3$ holds a surprise:

FACT 3.
- If $|X| = 1$, then $\mathcal{O}_X$ is trivial.
- If $|X| = 2$, then $\boldsymbol{CLONE}(X)$ is countable, and completely understood. ("Post's lattice")
- If $3 \leq |X| < \aleph_0$, then $|\boldsymbol{CLONE}(X)| = 2^{\aleph_0}$, and not well understood.

If X is infinite, then

- $|\mathcal{O}_X| = 2^{|X|}$,
- $|\boldsymbol{CLONE}(X)| = 2^{2^{|X|}}$,
- and only little is known about the structure of $\boldsymbol{CLONE}(X)$.

Since $\boldsymbol{CLONE}(X)$ is so large for infinite X, investigations of the structure of $\boldsymbol{CLONE}(X)$ often concentrate on interesting subsets, such as the set of atoms or coatoms of this lattice, or intervals.

Algebraic lattices. A lattice $(L, \vee, \wedge)$ is called "complete" if every subset $X \subseteq L$ has a least upper bound $\sup X$.

We call an element $x \in L$ *compact* if x has the following property:

> Whenever $Y \subseteq L$ has a least upper bound, and $x \leq \sup Y$, then there is a finite $Y_0 \subseteq Y$ with $x \leq \sup Y_0$.

For example, in the lattice of substructures of a given algebraic structure A, the compact elements are exactly the finitely generated substructures. Hence, in this lattice every element is the supremum of its compact lower bounds.

A lattice L is called *algebraic* if every $x \in L$ is the supremum of all compact $y \leq x$.

The name "algebraic" is motivated by the fact that every algebraic lattice is isomorphic to the lattice of substructures of some algebra.

It is easy to prove the following:

FACT 4. For any set X, the lattice $\boldsymbol{CLONE}(X)$ is a complete algebraic lattice.

§2. Descriptive set theory.

A simple example: Intervals. The following example is rather trivial from the set-theoretic point of view, but it already shows that set-theoretic methods can be used even when investigating the clone lattice over a finite set.

THEOREM 5. *Let X be a finite set, and let $C \subseteq D$ be clones on X. Then the interval $[C, D]$ in $\mathbf{CLONE}(X)$ is*

- *either finite,*
- *or countable,*
- *or of size $2^{\aleph_0}$.*

PROOF. Recall that $\mathcal{O}_X$ is countable, so every $C \subseteq \mathcal{O}_X$ can be viewed as a real number.

The set $[C, D]$ is a Borel set (even closed) in the natural Polish topology.

It is well-known that all uncountable Borel sets contain a perfect set, and are hence equinumerous with the continuum. ⊣

REMARK 6. All these possibilities are realized: $1, 2, \ldots, \aleph_0, 2^{\aleph_0}$. (In fact, already Post's lattice $\mathbf{CLONE}(\{0, 1\})$ contains intervals which are chains of any finite length.)

More intervals. The above theorem cannot be generalized to infinite sets. Michael Pinsker [9] proved the following:

THEOREM 7. *Let X be an infinite set of cardinality λ. Then for every $\kappa \leq 2^\lambda$ there is an interval $[C, D]$ of cardinality κ, and an interval $[E, F]$ of cardinality 2^κ in $\mathbf{CLONE}(X)$. (Moreover, this intervals can be chosen in such a way that all clones in the interval have the same unary fragment.)*

One might ask if not all cardinals $\leq 2^{2^\kappa}$ can appear as cardinalites of intervals; but this is consistently false, by the following theorem of Kunen:

THEOREM 8. *The following is consistent:*

(A) $2^{\aleph_0} = \aleph_1$.
(B) $2^{\aleph_1} > \aleph_2$ *(in fact, $2^{\aleph_1}$ can be arbitrarily large).*
(C) *Whenever T is a tree with $\aleph_1$ nodes of height $\aleph_1$, then the number of ω_1-branches of T is either $\leq \aleph_1$ or $\geq 2^{\aleph_1}$.*

Uri Abraham has pointed out that (A)+(B)+(C) imply

(D) Whenever $\mathcal{F} \subseteq \mathcal{P}(\omega_1)$ is a family which is closed unter arbitrary intersections and arbitrary increasing unions, then $\mathcal{F}$ has either $\leq \aleph_1$ elements, or $\geq 2^{\aleph_1}$ elements.

which in turn implies that intervals in the clone lattice on ω always have cardinality $\leq \aleph_1$ or $\geq 2^{\aleph_1}$.

However, this is really a remark about cardinalities of algebraic lattices, so the right question to ask is the following:

QUESTION 9. Is every (infinite) complete algebraic lattice isomorphic to an interval in a clone lattice?

The following variant is easier (assuming that the answer is "yes"):

QUESTION 10. Is every (infinite) complete algebraic lattice equipotent to an interval in a clone lattice?

Completeness. We start with a well-known example from logic.

Example 11. The functions $\wedge, \vee, \mathtt{true}, \mathtt{false}$ do not generate all operations on $\{\mathtt{true}, \mathtt{false}\}$.

Proof. All these functions are monotone, and $\neg$ is not. ⊣

We can generalize this example to any partially ordered set X:

Example 12. Assume that $\leq$ is a nontrivial partial order on X, and that all functions in $C \subseteq \mathcal{O}$ are monotone with respect to $\leq$.
Then $\langle C \rangle \neq \mathcal{O}$.

However, the properties of $\leq$ play no role here, so we arrive at the following concept of "Polymorphisms":

Definition and Fact 13. Let X be any set, I any index set (often finite), and $\rho \subseteq X^I$. We call a function $f : X^n \to X$ a "ρ-polymorphism", or we say that "f respects ρ" if:

Whenever $\vec{x}_1, \vec{x}_2, \dots \vec{x}_n$ are I-tuples in ρ,
then also $f(\vec{x}_1, \vec{x}_2, \dots \vec{x}_n) := (f(x_1(i), \dots, x_n(i)) : i \in I)$ is in ρ.

(If $\rho \subseteq X^2$ is a partial order, then "respecting ρ" just means "monotone".)

We write $\mathrm{Pol}(\rho)$ for the set of all polymorphisms of ρ. Clearly, $\mathrm{Pol}(\rho)$ is a clone.

The above observation can be formulated as follows: Let F be a set of operations, ρ a relation. If $F \subseteq \mathrm{Pol}(\rho)$, and $\mathrm{Pol}(\rho) \neq \mathcal{O}$, then $\langle F \rangle \neq \mathcal{O}$ (or in other words: F is not complete). It is also easy to prove a converse:

Fact 14. Let X be nonempty, and $F \subseteq \mathcal{O}_X$. Then $\langle F \rangle \neq \mathcal{O}$ iff there exists an index set I and a relation $\rho \subseteq X^I$ with $F \subseteq \mathrm{Pol}(\rho) \subsetneq \mathcal{O}_X$. (Recall that $F \subseteq \mathrm{Pol}(\rho)$ just means that all functions in F preserve the relation ρ.)

In other words, the clones of the form $\mathrm{Pol}(\rho)$ are cofinal in the set $\boldsymbol{CLONE}(X) \setminus \{\mathcal{O}\}$.

The proof is disappointingly shallow: We can choose $\rho := \langle C \rangle \cap \mathcal{O}^{(2)} \subseteq X^{X^2}$, and use the fact that $\langle \mathcal{O}^{(2)} \rangle = \mathcal{O}$.

Definition 15. We call a clone $C \neq \mathcal{O}$ *precomplete* iff there is no clone C' with $C \subsetneq C' \subsetneq \mathcal{O}$ (i.e., if C is a coatom or dual atom of the lattice $\boldsymbol{CLONE}(X)$).

By the above remark, every precomplete clone must be of the form $\mathrm{Pol}(\rho)$ for some relation ρ.

Using the fact that $\mathcal{O} = \langle \mathcal{O}^{(2)} \rangle$, it is easy to see that on a finite set X, every clone $C \neq \mathcal{O}$ is included in a precomplete clone. On arbitrary sets X one can use Zorn's lemma to prove the following:

FACT 16. Let X be any set, and let $C_1 \subseteq C_2$ be clones on X such that C_2 is finitely generated over C_1 (i.e., there is a finite set F with $C_2 = \langle C_1 \cup F \rangle$).

Then every clone $C \neq C_2$ in the interval $[C_1, C_2]$ is contained in some coatom of this interval.

However, an easy counting argument shows that $\mathcal{O}_X$ is not finitely generated if X is infinite, so this fact does not imply that every clone $\neq \mathcal{O}$ is contained in a precomplete clone.

For every finite $n \in \omega$ there are only finitely many precomplete clones in $\boldsymbol{CLONE}(\{1, \ldots, n\})$. Rosenberg's theorem gives an explicit list $\rho_{n,1}, \ldots, \rho_{n,r_n}$ of them. Hence, fact 14 can be reformulated as follows:

THEOREM 17. *If* $|X| = n$, *then for any set* $F \subseteq \mathcal{O}_X$ *we have*

$$\langle F \rangle \neq \mathcal{O} \text{ iff there is } i \in \{1, \ldots, r_n\} \text{ with } F \subseteq \mathrm{Pol}(\rho_{n,i})$$

This shows that example 12 is really typical for the situation on finite sets.

§3. **Infinite Combinatorics.** A possible approach to analyzing all clones C on a set X is to first partition them according to their unary fragment $C \cap \mathcal{O}^{(1)}$, and then to analyze each equivalence class separately. The following is easy to see:

FACT 18. Let $M \subseteq X^X = \mathcal{O}^{(1)}$ be a monoid (with composition of functions). Then the set

$$\{C \in \boldsymbol{CLONE}(X) : C \cap \mathcal{O}^{(1)} = M\}$$

(called the "monoidal interval" of M) is equal to the interval $[\langle M \rangle, \mathrm{Pol}(M)]$.

Here, $\langle M \rangle = \{f \circ \pi^n_k : f \in M,\ 1 \leq k \leq n \in \omega\}$ is the set of all functions which are essentially in M, and $\mathrm{Pol}(C) = \bigcup_{n \geq 1} \{g : \forall f_1, \ldots, f_n \in M\ (g(f_1, \ldots, f_n) \in M)\}$.

We will consider the monoidal interval of the full monoid X^X.

Partition and anti-partition. Let $\mathcal{U}$ be the clone of all essentially unary functions. We are interested in the interval $[\mathcal{U}, \mathcal{O}]$. So in some sense we are investigating the complexity of n-ary functions, when the unary functions are treated as trivial or just "given".

- If the base set X is finite with k elements, then $[\mathcal{U}, \mathcal{O})$ is a chain of k elements; the last one (the *unique* maximal element in this interval) is the set of all functions that are essentially unary or not surjective. (See [11].)
- If the base set X is countable or weakly compact, then $[\mathcal{U}, \mathcal{O})$ has *two* maximal elements. (Gavrilov's theorem, see [4]. They will be described below.)
- If the cardinality of the base set X satisfies a certain strong negative partition relation, then $[\mathcal{U}, \mathcal{O})$ has *very many* maximal elements [4]; specifically, the number of such elements is $2^{2^{|X|}}$.

The clone T_1.

DEFINITION 19. Let κ be a regular cardinal. We say that a function $f : \kappa^k \to \kappa$ is almost unary iff there is a unary function $F : \kappa \to \kappa$ and an index $i \in \{1, \dots, k\}$ such for all $x_1, \dots, x_k \in \kappa$ we have

$$f(x_1, \dots, x_k) \leq F(x_i)$$

We write T_1 or $T_1(\kappa)$ for the set of all binary functions which are almost unary.

It is easy to see that the set

$$\mathrm{Pol}(T_1) := \{g : \forall f_1, \dots, f_n \in T_1 \, (g(f_1, \dots, f_n) \in T_1)\}$$

is the largest clone whose binary fragment is T_1, and that the functions in $\mathrm{Pol}(T_1)$ are exactly the almost unary functions.

Moreover, $\mathrm{Pol}(T_1)$ is precomplete.

The clone T_2.

DEFINITION 20. For any cardinal κ, we write $\Delta := \Delta_\kappa = \{(x, y) \in \kappa \times \kappa : x > y\}$, and $\nabla = \{(x, y) \in \kappa \times \kappa : x < y\}$.

DEFINITION 21. We say that a function $f : \kappa \times \kappa \to \kappa$ is "somewhere 1-1" if there are unary functions u and v such that at least one of the following holds:

- $f(u(x), v(y)) \restriction \nabla$ is 1-1,
- $f(u(x), v(y)) \restriction \Delta$ is 1-1.

We write T_2 for the set of binary functions which are nowhere 1-1 (i.e., not "somewhere 1-1").

We say that a function $f : \kappa^n \to \kappa$ (for $n \geq 2$) is somewhere 1-1 if there are essentially unary functions $u_1, \dots, u_n : \kappa \times \kappa \to \kappa$ such that $f(u_1, \dots, u_n) \restriction \Delta$ is 1-1.

In general the set of nowhere 1-1 functions is not a clone. However, if κ is countable or weakly compact, then we can use the following "canonisation theorem":

FACT 22. Assume that the partition relation $\kappa \to (\kappa)^2_2$ holds. Then for every function $f : \kappa \times \kappa \to \kappa$ we can find sets A, B of cardinality κ such that $f \restriction (A \times B) \cap \nabla$

- is injective,
- or depends injectively only on x: $f(x, y) = h_1(x)$, h_1 1-1,
- or depends injectively only on y: $f(x, y) = h_2(y)$, h_2 1-1,
- or is constant.

The binary clone T_2 is now just the set of all functions for which the first case never happens.

As a corollary, we get

FACT 23. Assume $\kappa \to (\kappa)^2_2$ holds. Then:

- The set T_2 is a binary clone,
- The set of functions which are "nowhere 1-1" is a clone; in fact it is equal to $\mathrm{Pol}(T_2)$, the largest clone whose binary fragment is T_2.

The reason we are interested in T_2 and $\mathrm{Pol}(T_2)$ is the following theorem, which was (for $\kappa = \omega$) first proved by Gavrilov [1]:

THEOREM 24. *Consider the interval $[\mathcal{U}, \mathcal{O}]$ in the clone lattice* $\boldsymbol{CLONE}(X)$ *over a countable set X (or over a set X whose cardinality is weakly compact).*

Then this interval has exactly 2 *coatoms*: $\mathrm{Pol}(T_1)$ *and* $\mathrm{Pol}(T_2)$.

Similarly, in the lattice of binary *clones the sets T_1 and T_2 are the only coatoms above the binary clone $\mathcal{U} \cap \mathcal{O}^{(2)}$ of essentially unary functions.*

The definition of T_2 may seem rather complicated. In a certain sense, this is unavoidable:

FACT 25. Let $\kappa = \omega$. [So both $\mathcal{O}^{(2)} = \omega^{\omega^2}$ and $\mathcal{O}$ are homeomorphic to Baire space; clones (and binary clones) can hence be viewed as sets of reals.]

Then the set T_1 is a Borel set, while T_2 is a Π^1_1-complete set. [6]

This fact can be used to show part 2 of the following proposition (which can again be seen as an application of a concept from set theory to a question originating in universal algebra):

FACT 26. We consider the two binary clones T_1 and T_2 over a countable set.

1. T_1 is finitely generated over $\mathcal{U}$.
   In fact, let $p : \omega \times \omega \to \omega$ be constant on Δ and 1-1 on ∇, then T_1 is (as a binary clone) generated by $\mathcal{U} \cup \{p\}$.
2. T_2 is not finitely generated over $\mathcal{U}$.
   In fact, for every analytic set $B \subseteq \mathcal{O}$ the clone $\langle B \rangle$ and the binary clone $\langle B \rangle \cap \mathcal{O}^{(2)}$ are analytic, hence different from T_2 and from $\mathrm{Pol}(T_2)$.

An open question concerning T_2. We consider clones on a fixed countable set X.

T_1 and T_2 are the only precomplete binary clones containing all unary functions; $\mathrm{Pol}(T_1)$ and $\mathrm{Pol}(T_2)$ are the only precomplete clones containing all unary functions.

The relation between T_1 and $\mathrm{Pol}(T_1)$ is known from [8]:

THEOREM 27. *The interval $[\langle T_1 \rangle, \mathrm{Pol}(T_1)]$ of all clones whose binary fragment is T_1 is a countable chain, anti-isomorphic to $\omega + 1$. Every clone in this interval is finitely generated over T_1.*

This theorem makes the following question all the more vexing:

OPEN QUESTION 28. Does T_2 generate $\mathrm{Pol}(T_2)$? (See Definition 21 and Fact 23 for definitions.)

If not, is $\mathrm{Pol}(T_2)$ finitely generated over T_2?

What is the cardinality (and structure) of the interval $[\langle T_2\rangle, \mathrm{Pol}(T_2)]$?

§4. **Structure or nonstructure?** We continue our investigation of intervals in the clone lattice.

Fix a set X. Let $M = \{\mathrm{id}\}$ be the trivial monoid. The largest clone whose unary fragment is M is the "idempotent" clone

$$C_{\mathrm{ip}} = \{f : \forall x\, f(x, \dots, x) = x\}$$

It turns out that the structure of the interval $[C_{\mathrm{ip}}, \mathcal{O}]$ is highly nontrivial, but in some sense also well understood:

For every filter D (including the trivial filter $\mathcal{P}(X)$) on X let C_D be clone of D-idempotent functions.

$$C_D = \{f : f(x, \dots, x) = x \ D\text{-almost everywhere}\}$$

THEOREM 29. *Every clone in the interval* $[C_{\mathrm{ip}}, \mathcal{O}]$ *is of the form* C_D *for some* D.

Hence, the interval $[C_{\mathrm{ip}}, \mathcal{O}]$ *is (as a lattice) isomorphic to the family of open subsets of* βX. [6]

While this does not solve the problem of analyzing the structure of this interval, it does translate it from algebra to topology.

§5. **Forcing.** We continue the discussion of precomplete clones from Section 2.

As remarked above, the knowledge of all precomplete clones on a finite set can be translated into a completeness criterion. But this translation depends on the fact that every clone $\neq \mathcal{O}$ is below a precomplete clone.

It was unknown for a long time whether a similar fact is also true on infinite sets. Recently, [3] and [5] showed the following:

THEOREM 30. *Let* κ *be a regular cardinal with* $2^\kappa = \kappa^+$.

Then the lattice $\mathbf{CLONE}(\kappa)$ *is not* dually atomic; *that is, there is a clone* $C \neq \mathcal{O}$ *such there is no precomplete clone above* C.

In fact, the interval $[C, \mathcal{O})$ *will contain a cofinal linearly ordered set of order type* κ^+.

While the proofs of Theorem 30 for the cases $\kappa = \omega$ and $\kappa > \omega$ were found at different times, and also have some differences, the structure of the proofs is similar.

The first step is a reduction to unary clones, based on the following easy fact: Let C_0 be the clone on κ consisting of all functions $f : \kappa^n \to \kappa$ which are bounded by the max function. Then the map $C \mapsto C \cap \mathcal{O}^{(1)}$ is 1-1 on the interval $[C_0, \mathcal{O}]$.

The second step is a reduction from the clone structure to properties of filters:

DEFINITION 31. For any infinite $A \subseteq \kappa$ we define $h_A(i) = \min\{j \in A : j > i\}$.

Let D be a filter on κ. C_D is the clone of all functions which are bounded by some iterate of h_A, $A \in D$:

$$C_D := \{f \in \mathcal{O} : \exists k\, \exists A \in D : f(\vec{x}) \leq h_A^{(k)}(\max(\vec{x}))\}$$

For unary functions f, g we define $f \leq_D g$ iff $f \in \langle C_D \cup \{g\}\rangle$. It turns out to be sufficient to restrict attention to monotone functions; for those functions, the following purely combinatorial characterization of $\leq_D$ is more practical:

$f \leq_D g$ iff f is bounded by a finite composition of functions from $\{g, h_A\}$ for some $A \in D$.

Assuming $2^\kappa = \kappa^+$, it is now possible to construct an ultrafilter D such that $\leq_D$ is on the one hand, highly nontrivial (in particular: without last element), but on the other hand, well understood (e.g., a linear order, or at least an order with a linearly ordered cofinal subset).

Clones above C_D will then be (more or less) initial segments in the Dedekind completion of $\leq_D$.

The ultrafilter D can be constructed with the help of a sufficiently generic filter on a cleverly chosen $\leq\kappa$-complete forcing notion.

OPEN QUESTION 32. Is it a theorem of ZFC that ***CLONE***(κ) is not dually atomic for some (all? all regular?) κ? Can ***CLONE***(κ) fail to be dually atomic for singular κ?

REFERENCES

[1] G. P. GAVRILOV, *On functional completeness in countably-valued logic*, ***Probl. Kibernetiki***, vol. 15 (1965), pp. 5–64.

[2] MARTIN GOLDSTERN and MICHAEL PINSKER, *A survey of clones on infinite sets*, ***Algebra Universalis***, vol. 59 (2008), pp. 365–403, `http://www.arxiv.org/math.RA/0701030`.

[3] MARTIN GOLDSTERN and SAHARON SHELAH, *Clones from creatures*, ***Transactions of the American Mathematical Society***, vol. 357 (2005), no. 9, pp. 3525–3551, `math.RA/0212379`.

[4] ———, *Clones on regular cardinals*, ***Fundamenta Mathematicae***, vol. 173 (2002), no. 1, pp. 1–20, `math.RA/0005273`.

[5] ———, *All creatures large and small*, preprint.

[6] ———, *Large intervals in the clone lattice*, ***Algebra Universalis***, `math.RA/0208066`.

[7] DIETLINDE LAU, ***Function Algebras on Finite Sets***, Springer Monographs in Mathematics, Springer-Verlag, Berlin, 2006, A basic course on many-valued logic and clone theory.

[8] M. PINSKER, *Clones containing all almost unary functions*, ***Algebra Universalis***, vol. 51 (2004), no. 2-3, pp. 235–255, `math.RA/0401102`.

[9] ———, *Monoidal intervals of clones on infinite sets*, ***Discrete Mathematics***, vol. 308 (2008), no. 1, pp. 59–70, `math.RA/0509206`.

[10] R. PÖSCHEL and L. A. KALUŽNIN, ***Funktionen- und Relationenalgebren***, Mathematische Monographien [Mathematical Monographs], vol. 15, VEB Deutscher Verlag der Wissenschaften, Berlin, 1979, Ein Kapitel der diskreten Mathematik. [A chapter in discrete mathematics].

[11] ÁGNES SZENDREI, ***Clones in Universal Algebra***, Séminaire de Mathématiques Supérieures [Seminar on Higher Mathematics], vol. 99, Presses de l'Université de Montréal, Montreal, QC, 1986.

DISCRETE MATHEMATICS AND GEOMETRY
VIENNA UNIVERSITY OF TECHNOLOGY
WIEDNER HAUPTSTRASSE 8-10 / 104
1040 WIEN, AUSTRIA
E-mail: Martin.Goldstern@tuwien.ac.at

ON FINITE IMAGINARIES

EHUD HRUSHOVSKI

§1. Introduction. In their beautiful paper [4], Cluckers and Denef study actions of linear algebraic groups on varieties over local fields of high residue characteristic. The orbits of these actions, known to be finite in number, can be viewed as imaginary elements of the theory HF_0 of Henselian fields of residue characteristic zero. Cluckers and Denef relate them to finite imaginaries of a certain extension, $T_\infty^{(d)}$ (cf. 2.2). They formulate a "tameness" conjecture 2.16 on the nature of such imaginaries, and show that it is true for imaginaries arising from algebraic group actions over number fields [4, Theorem 1.1]. From this they obtain consequences for orbital integrals (Theorem 1.2).

Sections 3 of this note contain a proof of Conjecture 2.16 in general (Theorem 2.3). This in turn is a special case of a more general description of imaginaries in this theory, and indeed in a wider class of theories of Henselian fields. However as Denef suggested finite imaginaries have a certain autonomy that permits their direct classification. The proof of the finite case follows the lines of the general (unpublished) proof, and has the merit of containing most the main ideas while avoiding technicalities. §2 contains some general observations on finite imaginaries, while §3 describes them for the theory $T_\infty^{(d)}$.

The rest of the paper contains two further comments on [4], from different points of view. Consider theories T of Henselian valued fields of residue characteristic 0. The data of [4] consists of an algebraic group G and a homogeneous space V for G, in the sense of algebraic geometry. This means that an action of G on V is given, and over an algebraically closed field the action is transitive. But for a given field $L \models T$ the action need not be transitive, and the question concerns the space of orbits of $G(L)$ on $V(L)$. The goal is to show that $V(L)/G(L)$ reduce to imaginaries of the residue field and the value group; and indeed to special, "tame" imaginaries (see below.)

The data G, V is geometric (i.e. quantifier-free) and group-theoretic. Both of these qualities are lost in the reductions of $G(F)/V(F)$ to the residual sorts given by one of the above methods. In §4 we describe another method that does not lose track of the group theory, and is geometric in the sense of being independent of a particular completion of the theory of Henselian fields.

Supported by ISF grant 1048-07.

Logic Colloquium '06
Edited by Barry Cooper, Herman Geuvers, Anand Pillay, and Jouko Väänänen
Lecture Notes in Logic, 32

We use the theory ACVF; but we cannot simply interpret the implied quantifiers of G-conjugacy on V in terms of the quantifier elimination of ACVF, since the resulting quotient would be trivial. Reformulating the question in terms of groupoids does allow us to work in ACVF without trivializing the problem. We illustrate this in the special case $V = G/T$, where T is a torus.

In [4] the residue field is pseudo-finite, the value group essentially $\mathbb{Z}$-like, and tameness is defined in concrete terms; an imaginary is tame if in the Denef-Pas language it is definable over the sorts $\boldsymbol{k}$ (the residue field) and $\Gamma/n\Gamma$ (where Γ is the value group); but not Γ itself. In the final section we attempt to understand the role of tameness from a more abstract viewpoint. In [3], Cherlin, Van den Dries and Macintyre consider imaginaries coding the Galois groups of finite extensions of a field F.[1] In the context of HF_0, we note that Galois imaginaries satisfy a strong (and more symmetric) form of tameness. We also show that $V(L)/G(L)$ is analyzable over the Galois sorts, and hence over the tame imaginaries.[2] We leave open the question of whether analyzability can be replaced by internality. In an appendix, we discuss Galois sorts for theories more general than fields.

§1–3, §4, and §5 can each be read independently of the others, and of [4]; however §5 (like [4]) relies on [11]. *In this paper, by "definable" we mean 0-definable, i.e. invisible parameters are not allowed.*

§2. Finite sorts. We will consider first order theories T in many-sorted languages L. T need not be complete but everything we do can easily be reduced to the complete case. Terms like "surjective" applied to definable relations and functions mean: provably in T. Thus a definable surjection $D \to D'$ means a definable $R \subset D \times D'$, such that in any model $M \models T$, $R(M)$ is the graph of a surjection $D(M) \to D'(M)$.

A sort S of L is called *finite* if $S(M)$ is finite for all $M \models T$.

A family $\mathcal{F}$ of sorts is said to be *closed under products* if the product of two sorts in $\mathcal{F}$ is definably isomorphic to a third.

Definition 2.1. If S is a sort of L and $\{S_j\}_{j\in J}$ is a family sorts, we say S is *dominated* by $\{S_j\}_{j\in J}$ if there exists a definable $D \subseteq S_{j_1} \times \cdots \times S_{j_k}$ and a definable surjective map $f : D \to S$, with $j_1, \ldots, j_k \in J$.

Equivalently, in any $M \models T$, $S(M) \subseteq \mathrm{dcl}(\cup_{j\in J} S_j(M))$.

In this language, Conjecture 2.16 can be stated as follows.

Definition 2.2. [4] Let $T_\infty^{(d)}$ be the theory of Henselian fields with pseudo-finite residue field of characterstic 0, and (dense) value group elementarily equivalent to $\{\frac{a}{b} : a \in \mathbb{Z}, b \in \mathbb{N}, (b,d) = 1\} \leq (\mathbb{Q}, <, +)$.

[1] cf. §1 of [2] for an account and further references. In these references, the field F is PAC, but this condition is irrelevant here.

[2] It follows that in the Denef-Pas language they are definable over the sorts $\Gamma/n\Gamma$ and $\boldsymbol{k}$. Asides from this remark, we use the intrinsic valued field structure in this paper, and do not split RV.

THEOREM 2.3. *Every finite imaginary sort of* $T_\infty^{(d)}$ *is dominated by the set of sorts consisting of the residue field* $\mathbf{k}$ *and the finite value group quotient sorts* $\Gamma/n\Gamma$.

This will be proved at the end of §3.

DEFINITION 2.4. Let S, F be sorts of T with F finite. Define the imaginary sort $S^{\subseteq F}$ to be the sort of partial functions $F \to S$.

Explanation: by definition, for some $N \in \mathbb{N}$, $T \models |F| \leq N$. Then $S^{\subseteq F} = \cup_{n \leq N} S^{\subseteq F}{}_n$ where $S^{\subseteq F}{}_n$ is the sort of partial functions $F \to S$ with n element domain. By identifying a function with its graph, an n-element set of pairs, and then identifying a set of pairs with a tuple of pairs up to $Sym(n)$, we see that $S^{\subseteq F}{}_n$ embeds naturally into the imaginary sort $(F \times S)^n/Sym(n)$. Observe in particular that $S^{\subseteq F}$ is dominated by S, F.

We view an element of $S^{\subseteq F}$ as a tuple of elements of S, indexed by a finite subset of F in place of a finite subset of $\mathbb{N}$. We thus refer to the sorts $S^{\subseteq F}$ as $(Aut(F)$-$)$ twisted Cartesian powers of S.

LEMMA 2.5. *Let T be a theory. Assume given a family $\mathcal{S}$ of sorts, and a family $\mathcal{F}$ of finite sorts, closed under products. Let T' be obtained from T by naming the elements of each $S \in \mathcal{F}$. I.e. $T' = Th((M, c_j)_{j \in J})$ for some $M \models T$ and enumeration $(c_j : j \in J)$ of $\mathcal{F} = \cup_{S \in \mathcal{F}} S(M)$. If T' eliminates imaginaries to the sorts $\mathcal{S}$, then T eliminates imaginaries to the sorts $\{S^{\subseteq F} : S \in \mathcal{S}, F \in \mathcal{F}\}$*

PROOF. Observe that $(S_1 \times S_2)^{\subseteq F}$ embeds naturally into to $S_1{}^{\subseteq F} \times S_2{}^{\subseteq F}$. Thus we may assume $\mathcal{S}$ closed under Cartesian products.

Let $M \models T$ and let e be an imaginary element of M. We have to find $h \in S^{\subseteq F}$ for some S, F with $\mathrm{dcl}(e) = \mathrm{dcl}(h)$.

Let M' be an expansion of M to a model of T'. By assumption, in M', $\mathrm{dcl}_{M'}(e) = \mathrm{dcl}_{M'}(g)$ for some tuple g from the sorts $\mathcal{S}$; so we may assume $g \in S(M)$, $S \in \mathcal{S}$. It follows that there exists $F \in \mathcal{F}$ and $d \in F(M)$ such that $\mathrm{dcl}(e, d) = \mathrm{dcl}(g, d)$ in the sense of M. So $g = H(e, d)$ and $e = G(g, d)$ for some definable functions G, H. We can restrict the domains of H, G to any given definable set containing (e, d) (respectively (g, d).) So we may assume that $G(H(x, y), y) = x$. Let $H_e(y) = H(e, y)$. So H_e is a function with nonempty domain contained in F, and range in S. So H_e is coded by some e- definable element h of $S^{\subseteq F}$. On the other hand e is determined by H_e, in fact $e = G(H_e(y), y)$ for any $y \in \mathrm{dom}(H_e)$. So $\mathrm{dcl}(e) = \mathrm{dcl}(h)$. ⊣

LEMMA 2.6. *Suppose $\mathcal{F}$ is a collection of finite cyclic groups; and the relation: "there exists a definable surjective homomorphism $A \to B$" is a directed partial ordering on $\mathcal{F}$, i.e. any two groups in $\mathcal{F}$ are definable homomorphic images of a third. Then the condition of being closed under products in Lemma 2.5 can be dispensed with.*

PROOF. Let $B_1, \ldots, B_n \in \mathcal{F}$. There exists $B \in \mathcal{F}$ and definable surjective maps $h_i : B \to B_i$. For any $S \in \mathcal{S}$, this gives rise to a surjective

$h : B^n \to \Pi_{i=1}^n B_i$ and hence a definable injection $S^{B_1 \times \cdots \times B_n} \to S^{B^n}$, $y \mapsto y \circ h$. Thus a twisted powers of a set S by a product of sorts of $\mathcal{F}$ embeds into a twisted power X^{B^k} by a power of single such B.

Now we use the fact that B is cyclic, say of order d. Then the map $B \times [1, \dots, d]^k \to B^k$, $(b, (a_1, \dots, a_k)) \mapsto (b^{a_1}, \dots, b^{a_k})$, is a surjective map. Let $N = d^k$ and fix a bijection $\{1, \dots, N\} \to [1, \dots, d]^k$. Then as above S^{B^k} embeds definably into $S^{B \times [1,\dots,d]^k} = (S^B)^N$, and similarly for partial functions. So the sorts S^B, $B \in \mathcal{F}$ dominate the sorts $S^{B_1 \times \cdots \times B_n}$, $B_1, \dots, B_n \in \mathcal{F}$. ⊣

LEMMA 2.7. *Let T_1, T_2 be two theories, and let $T = T_1 \times T_2$ be their Feferman-Vaught product, so that a model of T is a product of a model of T_1 with one of T_2, and a relation is a Boolean combination of products of relations. Assume T_1, T_2 eliminate imaginaries and that every sort of T_2 is linearly ordered. Then T eliminates imaginaries.*

PROOF. Let $M \models T$ and let R be an M- definable relation on M. Say e is a canonical code for R, possibly imaginary. Then $M = M_1 \times M_2$, $M_i \models T_i$, and R is a finite disjoint union of products of a definable set of M_1 with a definable set of M_2. Let B_2 be the Boolean algebra of definable subsets of M_2 generated by the sections $R(a) = \{y : (a, y) \in R\}$. This algebra is finite; let $\{R_2^i : i = 1, \dots, k\}$ be an enumeration of the atoms (without repetitions.) Let e_2^i be a canonical parameter for R_2^i. Then the set $\{e_2^i\}$ is e-definable. Since we assumed each sort of T_2 is linearly ordered, each e_2^i is actually e-definable. Let $R_1^i = \{x : \{x\} \times R_2^i \subseteq R\}$. Then R_1^i is clearly e-definable, and $R = \cup_i R_1^i \times R_2^i$. Let e_1^i be a canonical code for R_1^i. Then $\mathrm{dcl}(e) = \mathrm{dcl}((e_1^i, e_2^i)_i)$. So e is coded by a real element. ⊣

2.8. Elimination of finite imaginaries. Let T be a theory in a many-sorted language L, with sorts S_i $(i \in I)$. Let $I^- \subset I$, and let L^- be the language consisting of the sorts S_i $(i \in I^-)$ and the relations among them; let $T^- = T|L^-$. If $M \models T$, M^- will denote the restriction of M to the sorts $S_i (i \in I^-)$.

If $C \subseteq C' \subseteq M \models T$, we say C' is stationary over C if $\mathrm{dcl}^{eq}(C') \cap \mathrm{acl}^{eq}(C) = \mathrm{dcl}^{eq}(C)$, i.e. every imaginary element that is definable over C' and algebraic over C is definable over C. It is clear that if C'' is stationary over C', and C' over C, then so is C'' over C. A type $p = tp(c/C)$ is called *stationary* if $C \cup \{c\}$ is stationary over C.

LEMMA 2.9. (1) *A type p over C is stationary if and only if for any C-definable equivalence relation E on a C-definable set D with finitely many classes, if $p(x) \models D(x)$ then p chooses one of the classes of E, i.e. $p(x) \cup p(y) \models xEy$.*

(2) *If there exists a C-definable type $\bar{p}$ over M extending p, then p is stationary.*

PROOF. (1) If $p = tp(c/C)$ is not stationary, then for some C-definable function f (possibly with imaginary values), $f(c)$ is algebraic over C but not definable. Define $xEy \iff f(x) = f(y)$. Then E divides the solutions of p to finitely many classes, but more than one. Conversely if a C-definable equivalence relation E divides p into finitely many classes (but more than one), then c/E is an imaginary element in $\mathrm{acl}^{eq}(C) \setminus \mathrm{dcl}^{eq}(C)$.

(2) Let E be a C-definable equivalence relation with finitely many classes. Since $\bar{p}$ is a complete type over M there is a unique class D of E such that $(xEa) \in \bar{p}$ iff $a \in D$. By definability of the type, there exists a formula $\theta(y)$ over C such that $(xEa) \in \bar{p}$ iff $\theta(a)$. Thus D is C-definable, by θ; and $p \models D(x)$. ⊣

LEMMA 2.10. *Assume*

(1) *T admits elimination of S_i-quantifiers for each $i \in I \setminus I^-$. T^- is stably embedded in T.*
(2) *Let $M \models T$ be a countable model. Then there exists C containing M^- and stationary over M^-, such that* $\mathrm{acl}(C) \prec M$.
(3) *For $A \leq M \models T$, let $T_A = Th(M, a)_{a \in A}$. If F is a finite T_A-definable set then there exists a finite T_A-definable set F' of M^- and a definable bijection $g : F \to F'$.*

Then every finite imaginary sort of T is definably isomorphic to one of T^-.

PROOF. Quantifier-elimination will be used in the background, to avoid disagreement about the notion of a definable set between T and T^-.

Let $S = D/E$ be a finite imaginary sort, where D is a definable set in T; let $\pi : D \to S$ be the canonical map. We may assume all elements of S realize the same type. Let $M \models T$. View M^- as a subset of M, and let C be as in (2), and $N = \mathrm{acl}(C)$. Then by (2), $N \prec M$. Since S is finite, $S(N) = S(M)$, so each class of E has a representative in N. Thus there exists $e \in C$ and a finite e-definable set $H_e \subseteq D$, meeting every E-class. Using (3), let W be a finite e-definable subset of M^- and $h_e : W \to H_e$ a definable bijection.

By stable embeddedness (1), W is actually defined over some $e' \in M^-$. Write $W = H'_{e'}$. We have an induced e-definable surjection $\psi_{e'} : H'_{e'} \to S$. But there are only finitely many maps $H'_{e'} \to S$, hence all are algebraic over M^-; by the stationarity assumption (2), since $\psi_{e'}$ is C-definable it is also M^--definable. By enlarging e' we may assume it is e'-definable.

Let H' be a definable set of T^-, and $\psi : H' \to D$ a definable map, such that $H'_{e'} = \{x : (x, e') \in H', \psi_{e'}(x) = \psi(x, e')\}$. Then the composition $\pi \circ \psi : H' \to S$ is surjective. Let E' be the kernel, i.e. define $E'(x, y) \iff \pi \circ \psi(x) = \pi \circ \psi(y)$. Then we have a definable bijection $H'/E' \to S$. By stable embeddedness again, E' is T^--definable and H'/E' is an imaginary sort of T^-. ⊣

§3. Finite sorts of $T_\infty^{(d)}$. Let PF be the theory of pseudo-finite fields, $PF_0 = PF +$ "characteristic 0". Let F_n^{Gal} be a finite imaginary sort whose elements code the elements of the Galois group of the unique field extension of order n.

Let $\mathcal{F}^{Gal} = \{F_2, F_3, F_4, \dots\}$. Let PF' be the theory obtained from PF by naming the elements of $\mathcal{F}_n^{Gal}$ for each n. It was shown in [1] that PF' eliminates imaginaries. Since F_n^{Gal} surjects canonically onto F_m^{Gal} when m divides n, Lemma 2.6 applies. Hence, by Lemma 2.5 we have:

EXAMPLE 3.1. PF eliminates imaginaries to the level of $\mathcal{F}_n^{Gal}$-twisted powers of the field sort.

The finite group $\mathbb{Z}/n\mathbb{Z}$ admits elimination of imaginaries to the level of subsets of $\mathbb{Z}/n\mathbb{Z}$. To see this, it suffices to note that any subgroup H of the automorphism group G of $\mathbb{Z}/n\mathbb{Z}$ has the form $H = \{g \in G : gY = Y\}$ for some $Y \subseteq \mathbb{Z}/n\mathbb{Z}$. Indeed, we have $G = (\mathbb{Z}/n\mathbb{Z})^*$ so that $H \subseteq G \subset \mathbb{Z}/n\mathbb{Z}$, and we can let $Y = H$.

EXAMPLE 3.2. Let $T(d)$ be the theory of ordered Abelian groups, divisible by all primes q with $(q,d) = 1$, and such that for $p|d$ we have $\Gamma/p^m\Gamma \simeq \mathbb{Z}/p^m$, and moreover an isomorphism $\Gamma/p^m\Gamma \simeq \mathbb{Z}/p^m$ is given as part of the language (say by a predicate for the pullback to Γ of $1 \in \mathbb{Z}/p^m$; then each element of $\Gamma/p^m\Gamma$ becomes definable, as a multiple of the distinguished generator.) Then $T(d)$ admits elimination of imaginaries.

Let $\mathbb{Z}^{(d)} = \{a/b \in \mathbb{Q} : a, b \in \mathbb{Z}, b \neq 0, (b,d) = 1\}$. Then $Th(\mathbb{Z}^{(d)})$ in the ordered group language admits EI to the sorts Γ together with the Γ/n and the sort of subsets of Γ/n (where n can be taken to be a power of d.)

PROOF. The theory $T(d)$ admits elimination of quantifiers, and it is easy to classify the definable subsets of Γ and the definable functions $\Gamma \to \Gamma$ (in one variable) and see explicitly that they are coded. This suffices in general, cf. [5], and shows that $T(d)$ eliminates imaginaries.

It follows from Lemma 2.5 that $Th(\mathbb{Z}^{(d)})$ eliminates imaginaries to the level of twisted products $\Gamma^{\Gamma/n_1 \times \cdots \times \Gamma/n_2}$. This reduces to $\Gamma^{(\Gamma/n)^k}$ and again, by Lemma 2.6, to the sorts $\Gamma^{\Gamma/n}$. Now a function $\Gamma/n \to \Gamma$ carries the same information as a subset of Γ of size $\leq n$ (the image), together with a partial ordering of a subset Γ/n. As remarked above this reduces to subsets of Γ/n. ⊣

COROLLARY 3.3. *Let T be the model-theoretic disjoint sum of PF_0 and $Th(\mathbb{Z}^{(d)})$: T has two sorts $\mathbf{k}, \Gamma$, with relations $+, \cdot, 0, 1$ on $\mathbf{k}$, $+, <, 0$ on Γ; such that $\mathbf{k} \models PF_0$ and $\Gamma \models Th(\mathbb{Z}^{(d)})$. Then T eliminates imaginaries to the sorts Γ together with the sorts $S(n_1, n_2)$ for $n_1, n_2 \in \mathbb{N}$, where $S(n_1, n_2) = \mathbf{k}^{F_1 \times F_2}$, with $F_1 = F_{n_1}^{Gal}, F_2 = \Gamma/n_2$.*

PROOF. Let $\mathcal{S}$ be the set of Cartesian products of these sorts.

CLAIM. $\mathcal{S}$ is closed under twisted powers by Γ/n.

PROOF. Using the linear ordering, a function $\Gamma/n \to \Gamma$ can be coded by an n-tuple of elements of Γ together with a partial ordering on Γ/n, namely the pullback of the linear ordering on Γ. In turn this partial ordering can be coded by a function from Γ/n to a subset of the prime field of $\boldsymbol{k}$. This shows that functions $\Gamma/n \to \Gamma$ are coded in $\mathcal{S}$. On the other hand a function $\Gamma/n \to Y^{\Gamma/m}$ (with $Y = \boldsymbol{k}^{F_1}$) can be viewed as a function $\Gamma/n \times \Gamma/m \to Y$, and handled using Lemma 2.6. ⊣

Thus by Lemma 2.5 it suffices to prove that the theory T' obtained by naming the elements of each Γ/n eliminates imaginaries to the sorts Γ together with $\boldsymbol{k}^{F_n^{Gal}}$. But this follows from Lemma 2.7 together with Example 3.1 and the first part of Example 3.2. ⊣

3.4. Finite imaginaries in $T_\infty^{(d)}$. .

We take the theory $T_\infty^{(d)}$ with the sorts VF, $\boldsymbol{k}, \Gamma$, as well as the sorts $\Gamma/2, \Gamma/3, \ldots$ and all twisted product sorts $\boldsymbol{k}^{F_{n_1}^{Gal} \times \Gamma/n_2}$. The sort of these last two kinds will be called I_t. The language includes the field structure on VF, a valuation map $\mathrm{VF} \setminus \{0\} \to \Gamma$, predicates for the valuation ring $\mathcal{O}$ and its maximal ideal $\mathcal{M}$, the residue homomorphism $\mathcal{O}/\mathcal{M} \to k$, and finally a group isomorphism $\mathrm{VF}^*/(1+\mathcal{M}) \to (\boldsymbol{k}^* \times \Gamma)$ splitting the canonical maps $k^* \to \mathrm{VF}^*/(1+\mathcal{M}) \to \Gamma$. The projection to Γ is thus canonical, while the projection to $\boldsymbol{k}^*$ is the Denef-Pas "angular component" map.

Let $I^- = \{\Gamma\} \cup I_t$, $I = I^- \cup \{\mathrm{VF}\}$ where VF is the valued field sort. We will now show that the hypotheses (1–3) of Lemma 2.10 are valid for $T_\infty^{(d)}$.

Lemma 2.10(1) follows from Denef-Pas elimination of quantifiers, [9]. Stable embeddedness is clear from the form of the quantifier-elimination; see [5] for an identical proof in the case of ACVF.

Lemma 3.5. *Let $C \subseteq M \models T_\infty^{(d)}$. Assume $M^- \subseteq C$ and the maps* val, ac *restricted to C are onto the value group and residue field of M. Then* $\mathrm{acl}(C) \prec M$.

Proof. Let $N = \mathrm{acl}(C) \cap \mathrm{VF}(M)$. Then N is Henselian, so by the Ax-Kochen, Ershov principle for the Denef-Pas language applied to $T_\infty^{(d)}$, we see that $N, \mathrm{VF}(M)$ are elementarily equivalent, and by model completeness, $N \prec \mathrm{VF}_M$. It follows immediately from the surjectiveness that $\mathrm{acl}(C) \prec M$. ⊣

Now to prove Lemma 2.10(2), let $M \models T = T_\infty^{(d)}$. Let $\mathrm{RV} = \mathrm{VF}/(1+\mathcal{M})$, $\mathrm{rv} : \mathrm{VF} \to \mathrm{RV}$ the quotient map. Let $A^- = \{\mathrm{rv}(m) : m \in M\}$. From the point of view of M, $\mathrm{rv}(m)$ is equi-definable with the pair $(\mathrm{val}(m), \mathrm{ac}(m))$, and so A^- can be identified with $\mathrm{val}(M) \times \boldsymbol{k}^*(M) \subseteq \mathrm{dcl}(M^-)$. But A^- can also be considered as a substructure of a model of ACVF, in the language considered in [8].

Let C be a maximal subset of $\mathrm{VF}(M)$ such that $C \cup M^-$ is stationary over M^-. Let A be the definable closure of $A^- \cup C$ in the sense of ACVF.

We have to show that $\mathrm{acl}(C \cup M^-) \prec M$, and by Lemma 3.5 it suffices to show that $(\mathrm{val}, \mathrm{ac})|C$ is surjective. In other words given $m \in M$, to find $c \in C$ with $\mathrm{val}(c) = \mathrm{val}(m), \mathrm{ac}(c) = \mathrm{ac}(m)$. Let $\beta = \mathrm{rv}(m)$, $B_\beta = \mathrm{rv}^{-1}(\beta)$. We have to find $c \in B_\beta(C) := B_\beta \cap C$.

Claim 1. B_β is not transitive in ACVF_A. In other words some proper sub-ball of B_β is ACVF_A-definable.

Proof. Suppose for contradiction that B_β is not transitive in ACVF_A. In this case, for any polynomial F, $\mathrm{rv} \circ F$ is constant on B_β. [8, Lemma 3.47]. So $\mathrm{ac}(F)$ and $\mathrm{val}(F)$ are constant. By Denef-Pas quantifier elimination, we see that B_β is also transitive in $(T_\infty^{(d)})_{\{M^- \cup C\}}$. (Compare [8, Lemma 12.1]) Let p be the unique type over $M^- \cup C$ of elements of B_β.

Moreover there exists a β-definable type $p_\beta(x)$ over M concentrating on elements of B_β. Namely, first pick a $\mathrm{val}(m)$-definable type $r(t)$ of elements of Γ, concentrating on intervals $(\mathrm{val}(m), \mathrm{val}(m) + \varepsilon) \subseteq \Gamma$. Over M, the type r can be

$$\begin{aligned} r(t) = \{t \in \Gamma, t > \mathrm{val}(m)\} &\cup \{t < s : s \in \Gamma(M), s > \mathrm{val}(m)\} \\ &\cup \{(\exists t')(nt' = t) : n = 1, 2, 3, \dots\} \end{aligned}$$

Then let

$$\begin{aligned} p_\beta(x) = \{x \in B_\beta\} &\cup \{\mathrm{val}(x - u) \models r : u \in B_\beta(M)\} \\ &\cup \{\mathrm{ac}(x - u) = 1 : u \in B_\beta(M)\} \end{aligned}$$

It is clear that this is a complete, consistent type over M and is $(\mathrm{ac}(m), \mathrm{val}(m))$-definable.

By Lemma 2.9(2), $p = p_\beta | M^- \cup \{C\}$ is stationary. Choose any $c \in B_\beta(M)$. Then $tp(c/M^- \cup \{C\}) = p$. So $C \cup \{c\} \cup M^-$ is stationary over M^-, contradicting the maximality of C. ⊣

Hence B_β contains a proper ACVF_A-definable closed ball. In this case by [8, Lemma 3.39] B_β contains an ACVF_A-definable point d. So $d \in \mathrm{dcl}(M^- \cup \{C\})$ and hence $d \in C$.

This finishes the proof of (2).

(3) We may take $F \subset \mathrm{VF}^n$. $T_\infty^{(d)}$ is algebraically bounded in the sense of [12], so F is contained in a finite $ACVF_A$-quantifier-free definable set F'. This reduces us to the same lemma for ACVF; for a proof, see for example [8, Lemma 3.9].

Corollary 3.6. *Every finite definable imaginary set of $T_\infty^{(d)}$ can be definably embedded into some power of the twisted product sorts $\boldsymbol{k}^{F_{n_1}^{Gal}} \times \Gamma / n_2$*

Proof. Lemma 2.10 reduces this to imaginary sorts of $(T_\infty^{(d)})^-$; so by Corollary 3.3 it suffices to show this for a finite definable $D \subset \Gamma^m \times P$, where P is a power of twisted product sorts. We can use induction on the cardinality, so we may assume D is not the union of two proper definable subsets. Since Γ is linearly ordered, it follows that the projection $D \to \Gamma^m$ has a one-point image. Thus D projects injectively to a product of twisted product sorts. ⊣

By the remark below Definition 2.4, the twisted product sorts $\boldsymbol{k}^{F_{n_1}^{Gal}} \times \Gamma / n_2$ are dominated by the sorts $\boldsymbol{k}, \Gamma/2, \Gamma/3, \dots$, and F_n^{Gal}. These Galois sorts are dominated by $\boldsymbol{k}$; hence every finite definable imaginary set of $T_\infty^{(d)}$ is dominated by the tame sorts $\boldsymbol{k}$ and $\Gamma/2, \Gamma/3, \ldots$.

This proves Theorem 2.3.

§4. Groupoids. A *groupoid* is a category $\Gamma = (Ob_\Gamma, Mor_\Gamma)$ in which every morphism is invertible. We will consider definable groupoids with a single isomorphism type. See [7], though the use of definable groupoids there is different. A *morphism* between groupoids is a quantifier-free definable functor.

Given a groupoid Γ defined without quantifiers in a theory T, one obtains an equivalence relation E_Γ, defined uniformly over $T = Th(L)$ for any definably closed subset L of a model of T. Namely the equivalence relation of Γ-isomorphism on the objects of Γ: for $a, b \in Ob_\Gamma(L)$, $aE_\Gamma^L b$ iff $Mor(a,b)(L) \neq \emptyset$. Let $Iso(\Gamma; L)$ be the quotient $Ob_\Gamma(L)/E_\Gamma^L$, i.e. the set of isomorphism classes of $\Gamma(L)$.[3]

Let HF_0 be the theory of Henselian fields with residue field of characteristic 0. This can be viewed as the theory of definably closed substructures of models of $ACVF_{\mathbb{Q}}$ (with trivially valued $\mathbb{Q}$.) Fix $L \models HF_0$. We will use only quantifier-free formulas, and notions such as dcl will refer to $ACVF_L$.

We wish to reduce a given quantifier-free definable groupoid Γ over ACVF to a groupoid Γ' over RV, in a way that yields a reduction of the imaginaries $Iso(\Gamma; L)$ to imaginaries $Iso(\Gamma'; L)$, uniformly over Henselian valued fields L with various theories.

Note that a morphism $f : \Gamma \to \Gamma'$ yields, for any $L \models HF_0$, a map $f_* : Iso(\Gamma, L) \to Iso(\Gamma', L)$. We say that f is an *elementary reduction* of Γ to Γ' (respectively, of Γ' to Γ) if f_* is injective (resp., surjective.) A *reduction* is a finite sequence of elementary reductions.

Let G be a definable group acting on a definable set V. Define a groupoid $\Gamma = \Gamma(G, V)$ whose objects are the points of V. The morphisms $v \to v'$ are defined to be: $Mor_\Gamma(v, v') = \{g \in G : gv = v'\}$. Composition is multiplication in G. Then $Iso(\Gamma, L)$ is precisely the orbit space $V(L)/G(L)$.

Let Γ be a groupoid, with one isomorphism class. Then all isomorphism groups $G_a := Mor(a, a)$ are isomorphic to each other, non-canonnically: given a, b Ob$_\Gamma$, choose $f \in Mor(a, b)$; then $g \mapsto f \circ g \circ f^{-1}$ is an isomorphism $G_a \to G_b$. The isomorphism $G_a \to G_b$ is defined up to conjugation; if the G_a are Abelian, this isomorphism is canonical, so all G_a are canonically isomorphic to a fixed group H. (This is not essential to the discussion that follows, but simplifies it.) Let N be a normal subgroup of H. We define a quotient groupoid Γ/N. It has the same objects as Γ, but the morphism set is $Mor_{\Gamma/N}(a, b) = Mor_\Gamma(a, b)/N$. There is a natural morphism $\Gamma \to \Gamma/N$.

---

[3]This connection between groupoids and imaginaries is different from the one considered in [7]. The approach in this section is apparently in the spirit of stacks.

Our reductions will use a sequence of canonical normal subgroups of a torus T. First let $T = G_m^r$, a split torus. The valuation map induces a homomorphism $T \to \Gamma^n$, with kernel $(\mathcal{O}^*)^r$ (where $\mathcal{O}$ is the valuation ring.) Next, we have a reduction map $(\mathcal{O}^*)^r \to (\boldsymbol{k}^*)^r$, with kernel $(1+\mathcal{M})^r$. Now if T is any torus defined over a valued field F, by definition there exists a finite Galois extension L of F, and an isomorphism $f : T \to G_m^r$ defined over L. It is easy to see that $N := f^{-1}(\mathcal{O}^*)^r$ and $N^{-1} := f^{-1}(1+\mathcal{M})^r$ do not depend on f; so these are quantifier-free definable subgroups of T. The quotient T/N is internal to Γ, while N/N^- is internal to the residue field (and N is generically metastable.) Note that N^{-1} is a uniquely divisible Abelian group.

We will assume L is not trivially valued, so that $L^{alg} \models ACVF$. In particular all definable torsors have L^{alg}-definable points. The proof in the trivially valued case is left to the reader.[4]

THEOREM 4.1. *Let G be a linear algebraic group, $T \leq G$ a torus. Then $\Gamma(G, G/T)$ reduces to a groupoid defined over* RV.

PROOF. Let $V = G/T$, $\Gamma = \Gamma(G, V)$. Consider first Γ/N. Each automorphism group $Mor(a, a)$ is a uniquely divisible Abelian group (isomorphic, over additional parameters, to Γ^r.) Hence given a finite subset of $Mor(a, b)$, it is possible to take the average, obtaining a unique point. In this way we can find for each $a \in V$ a definable point $c(a) \in Mor(1, a)$ (where 1 is the image in G/T of $1 \in G$.) Given $a, b \in V$ let $c(a, b) = c(b)c(a)^{-1} \in Mor(a, b)$. Then we have a subgroupoid of Γ/N with the same objects, and whose only morphisms are the $c(a, b)$.

Let Γ_1 have the same objects as Γ, and $Mor_{\Gamma_1}(a, b) = \{f \in Mor_\Gamma(a, b) : fN = c(a, b)\}$. The inclusion morphism $\Gamma_1 \to \Gamma$ induces surjective maps $Iso(\Gamma_1, L) \to Iso(\Gamma_2, L)$ for any L, being surjective on objects. Thus Γ reduces to Γ_1.

Let $\Gamma_2 = \Gamma_1/N^-$. The automorphism groups of Γ_2 are isomorphic to $\mathrm{t} := N/N^-$, a torus over RV (i.e. a group isomorphic, with parameters, to $(\boldsymbol{k}^*)^r$.)

CLAIM 1. Let $L \models HF_0$. Let Y be principal homogeneous space for N (defined in $ACVF_L$). If Y/N^- has a point in $\mathrm{dcl}(L)$, then so does Y.

PROOF. Let Y^- be an L-definable point of Y/N^-, i.e. an L-definable N^--subtorsor of Y. As above, since N^- is uniquely divisible, Y^- has an L-definable point. ⊣

Applying this to $Mor_{\Gamma_1}(a, b)$, we see that if a, b are Γ_2-isomorphic then they are Γ_1-isomorphic; so the natural morphism $\Gamma_1 \to \Gamma_2$ is injective on isomorphism classes over any $L \models HF_0$. Thus Γ_1 reduces to Γ_2.

[4]If F is trivially valued, then any element of Γ defined over F equals zero; the only definable subgroups of the additive group G_a are thus $0, \mathcal{O}$ and G_a; the only definable $\mathcal{O}$-subtorsor of G_a is $\mathcal{O}$ (consider valuation of elements).

Finally we reduce the objects. We have $Mor_{\Gamma_2}(a,b) \subseteq \mathrm{RV}$. By stable embeddedness of RV there exists a definable map $j : \mathrm{Ob}_{\Gamma_2} \to Y \subseteq \mathrm{RV}$ such that $Mor(1,a)$ is $j(a)$-definable. It follows that $Mor(a,b) = Mor(1,a) \times_t Mor(1,b)$ is $j(a), j(b)$-definable. Let Γ_3 be the groupoid with objects $j(Ob_{\Gamma_2})$, and the same morphism sets as Γ_2. The natural morphism $\Gamma_2 \to \Gamma_3$ (j on objects, identity on morphisms) is bijective on *Iso*, since $Mor_{\Gamma_3}(j(a), j(b)) = Mor_{\Gamma_2}(a,b)$. Thus Γ_2 reduces to Γ_3. But Γ_3 is over RV. $\dashv$

Since we used only the quantifier elimination of ACVF, rather than HF, this method of investigation is not blocked in positive characteristic. Theorem 4.1 should go through for tori that split in a tamely ramified extension, replacing the unique divisibility argument for the additive groups by Hilbert 90 for the residue field, and an appropriate extension to RV. The right statement in the general case may give a lead with respect to motivic integration in positive characteristic.

§5. **Galois sorts.** Let F be any field, and consider the 2-sorted structure $(F, F^{alg}, +, \cdot)$. Working in the structure (F, F^{alg}) is convenient but harmless, since no new structure is induced on F[5].

Let T_0 be a theory of fields (possibly with additional structure.) Let T be the theory whose models have the form $(F, K, +, \cdots)$, with K an algebraically closed field and F a distinguished subfield (possibly with additional structure) such that $F \models T$. We can restrict attention to $K = F^{alg}$ since $(F, F^{alg}) \prec (F, K)$. In this section, T is fixed, and "definable" means: definable in T, imaginary sorts included. Let $F_0 = \mathrm{dcl}(\emptyset)_T$.

5.1. Definition of Galois sorts. Let $\mathcal{E}_n$ be the set of Galois extensions of F of degree n, within F^{alg}; this is clearly a definable set of imaginaries of (F, F^{alg}). For $e \in \mathcal{E}_n$ coding an extension F_e of F, let G_e be the Galois group $Aut(F_e/F)$. Let $\mathcal{G}_n$ be the disjoint union of the G_e; it comes with a map $\mathcal{G}_n \to \mathcal{E}_n$. Let $\mathcal{G} = (\mathcal{G}_n : n \in \mathbb{N})$. See the Appendix for a definition at a greater level of generality, including some definitions for Galois cohomology.

5.2. A finiteness statement for H^1. Let A an algebraic group defined over F_0, not necessarily commutative. We are interested in the first Galois cohomology set $H^1(F, A) = H^1(Aut(F^{alg}/F, A(F^{alg})))$, where $F \models T$.

To say that an object such as $H^1(F, A)$ is definable means that there exists a definable set H of T^{eq} and for any $F' \models T$, a canonical bijection $H(F') \to H^1(F', A)$. By standard methods of saturated models, such a definable set H, if it exists, is unique up to a definable bijection. Given a property P of definable sets (invariant under definable bijections), we say that $H^1(F, A)$ has P if H has P.

THEOREM 5.3. *Let F be a perfect field. If A is a linear group, then $H^1(F, A)$ is definable, and $\mathcal{G}$-analyzable.*

[5]as one easily sees by an automorphism argument.

This will be proved as Proposition 5.10 below.

In case the Galois group of L *property F* in the sense of [11], (or *bounded* in the sense of [6]), Theorem 5.3 says simply that $H^1(F, A)$ (resp., the kernel $H^1(F, A) \to H^1(F, G)$) is finite. This is Theorem 5 of [11], Chapter 3, §4.

QUESTION 5.4. Is $V(K)/G(K)$ in fact internal to $\mathcal{G}$?

Presumably it is not the case, in general, that $V(K)/G(K) \subseteq \mathrm{dcl}(\mathcal{G})$, even over $\mathrm{acl}(0)$; It would be good to have an example.

If L is a Galois extension of F, let $Z(L; A)$ be the set of maps $a : Aut(L/F) \to A(L)$ satisfying $a(st) = a(s)s(a(t))$. Two elements a, a' are cohomologous if $a'(s) = b^{-1}a(s)s(b)$ for some $b \in A(L)$. (Note $a'(1) = 1 = b^{-1}s(b)$ so $b \in Fix(s)$ whenever $a(s) = a'(s) = 1$.) The quotient of $Z(L, A)$ by this equivalence relation is denoted $H^1(L/F, A)$. If $L \leq L'$ we have a natural map $Z(L, A) \to Z(L', A)$, obtained by composing with the canonical quotient map $Aut(L'/F) \to Aut(L/F)$. This induces a map $H^1(L/F, A) \to H^1(L'/F, A)$, which is injective.

Let $Z(n; A)$ be the disjoint union of the sets $Z(L; A)$ over all Galois extensions L of F with $[L : F]|n$. Define an equivalence relation E on $Z(n; A)$: if $a_i \in Z(L_i; A)$ for $i = 1, 2$, write $a_1 \sim a_2$ if a_1, a_2 are cohomologous in L_1L_2; equivalently, for some Galois extension L of F containing L_1, L_2, there exists $b \in A(L)$ such that for $s \in Aut(L/F)$, $a_1(s|L_1) = b^{-1}a_2(s|L_2)s(b)$. The second formulation shows that $\sim$ is an equivalence relation; the first shows that $\sim$ is definable. Definability of $Z(n; A)$ is clear. Denote the quotient $Z(n; a)/\sim = H(n; A)$.

If $n|n'$ we obtain an injective map $H(n; A) \to H(n'; A)$. It is clearly definable. For any $F \models T$, $H(n; A)(F)$ is the set of elements of $H^1(F, A)$ in the image of $H^1(L/F, A)$ for some Galois extension L of degree n.

More generally, if $\{A_y\}$ is a definable family of algebraic groups, for b from F the set $H(n; A_b)(F)$ of elements of $H^1(F, A_b)$ in the image of $H^1(L/F, A)$ for some Galois extension L of degree n is definable uniformly in the parameter b.

PROPOSITION 5.5. *Let A be a linear algebraic group. Then $H^1(F, A)$ is definable; for some n, $H^1(F, A) = H(n; A)(F)$.*

PROOF. We have $A \leq GL_n$. By [10, Chapter X, Proposition 3], $H^1(F, GL_n) = 1$. The kernel of $H^1(F, A) \to H^1(F, GL_n)$ is canonically isomorphic to $GL_n(F)/A(F)$. Since GL_n/A is a definable set, by a standard compactness argument, $\lim_n H(n; A) = \cup_n H((n+1)!; A) \setminus H(n!; A)$ must also be a definable set, i.e. for large enough n the set $H((n+1)!; A) \setminus H(n!; A)$ must be empty. ⊣

COROLLARY 5.6. *Let A be a finite (linear algebraic) group. Then $H^1(F, A)$ is contained in* $\mathrm{dcl}(A, \mathcal{G})$.

PROOF. A function from a finite set into the finite set A is definable over the elements of A and the elements of the domain. Thus $Z(n; A) \subseteq \mathrm{dcl}(\mathcal{G}_n, A)$. ⊣

We give a second proof, similar to the proof of (a) implies (b) in [11, Chapter 4.1, Proposition 8]. This second proof goes through in a more general context, see the Appendix.

LEMMA 5.7. *Let A be a finite definable group, of order n. Then $H^1(F, A) = H(n!^{n!}; A)(F)$.*

PROOF. Let L_0 be the Galois extension of F generated by the n points of A. Since by a trivial estimate $|Aut(A)| \leq (n-1)!$, we have $[L_0 : F] \leq (n-1)!$. Let L be any Galois extension of F, containing L_0, $G = Aut(L/F)$, and let $a \in Z(L; A)$. We have to show that the class of a in $H^1(L/F, A)$ is in the image of some $H^1(L'/F, A)$ with $[L' : F] \leq (n!)^{n!}$. In fact we will prove this even at the level of cocycles. The restriction of a to $G_0 = Aut(L/L_0)$ is a homomorphism $a|G_0 : G_0 \to A$. Let G_1 be the kernel of $a|G_0$. Then $[G_0 : G_1] \leq n$, so $[G : G_1] \leq n!$. The number of conjugates of G_1 in G is thus $\leq n!$; their intersection G_2 has index $\leq (n!)^{n!}$. Let L' be the fixed field of G_2. Then a factors through a function a' on G/G_2, so $a' \in Z(L'/F, A)$, as required. ⊣

COROLLARY 5.8. *Let A be a torus, i.e. A becomes isomorphic to G_m^n after base change to some Galois extension F'. Say $[F' : F] = m$, and let $B = \{a \in A : a^m = 1\}$. Then $H^1(F, A) \subseteq \mathrm{dcl}(B, \mathcal{G})$. If F is perfect, then this holds for any connected solvable A (for an appropriate finite group B.)*

PROOF. The proof in [11], Chapter III, Theorem 4, goes through, showing that $H^1(F, A) \cong H^1(F, B)$. ⊣

REMARK 5.9. We also have $H^1(F, A) \subseteq \mathrm{dcl}(F)$, since the Galois group $Aut(F^{alg}/F)$ acts trivially on $H^1(F, A)$.

PROPOSITION 5.10. *Assume F is perfect. Let A be a linear algebraic group. Then $H^1(F, A)$ is $\mathcal{G}$-analyzable.*

PROOF. The proof of [11, Chapter III, Theorem 4] goes through. ⊣

REMARK 5.11. The proof of Proposition 5.5 shows that for any embedding of algebraic groups $A \to B$, the kernel of $H^1(F, A) \to H^1(F, B)$ is definable. It can be shown as in [11], chapter III, Theorem 5 that over a perfect field, this kernel is $\mathcal{G}$-analyzable.

COROLLARY 5.12. *Let L be a perfect field. Let G be a linear algebraic group over L, and let V be a homogeneous space for V defined over L, i.e. G acts on the variety V and $G(L^{alg})$ acts transitively on $V(L^{alg})$. Then in the structure $(L, L^{alg}, +, \cdot)$, $V(L)/G(L)$ is $\mathcal{G}_L$-analyzable.*

PROOF. Immediate from Proposition 5.10, since after picking any point $c \in V(L)$ and letting $H = \{g \in G : gc = c\}$, we have a c-definable injective map $V(L)/G(L) \to H^1(F, H)$. ⊣

5.13. Henselian fields. We now move to valued fields of residue characteristic zero.

LEMMA 5.14. *Let K be a Henselian field with residue field of characteristic* 0. *Let $\boldsymbol{k}$ denote the residue field, $\mu = \cup_n \mu_n$ the roots of unity in $\boldsymbol{k}^{alg}$, Γ he value group. Then in (K, K^{alg}) we have $\mathcal{G}_K \subseteq \mathrm{dcl}(\mathcal{G}_{\boldsymbol{k}} \cup \mu \cup \cup_n \Gamma/n\Gamma)$*

PROOF. Let L be a finite Galois extension of K. We have to show that $Aut(L/K) \subseteq \mathrm{dcl}(\mathcal{G}_{\boldsymbol{k}}, \mu, \Gamma/n\Gamma)$ for some n.

We will use some standard valuation theory. Call K' a *ramified root extension* of K if it is a finite purely ramified extension obtained by adding roots to some elements of K.

CLAIM 1. There exists a finite unramifield extension L' of L, an unramified Galois extension $K_u \leq L'$ of K, and a ramified root extension $K_{rr} \leq L'$ of K, such that L'/K_{rr} is unramified, and $Aut(L'/K) = Aut(L'/K_u)Aut(L'/K_{rr})$.

PROOF. The finite group $\Gamma(L)/\Gamma(K)$ is a direct sum of finite cyclic groups, $\oplus_{i=1}^{k} \mathbb{Z}/n_i\mathbb{Z}$; let $c_1, \ldots, c_k \in L$ be such that $\mathrm{val}(c_1)+\Gamma(K), \ldots, \mathrm{val}(c_k)+\Gamma(K)$ are generators for these cyclic groups. So $\mathrm{val}(c_i^{n_i}) = \mathrm{val}(d_i)$ for some $d_i \in K$. Let $e_i = c_i^{n_i}/d_i$, so that $\mathrm{val}(e_i) = 0$. In a Henselian field of residue characteristic prime to n, if an element f has $\mathrm{val}(f) = 0$, and $\mathrm{res}(f)$ has an n'th root, then by Hensel's lemma so does f. Hence in some unramifield Galois extension L' of L, each e_i has an n_i'th root. Clearly d_i has an n_i'th root $r_i \in L'$. Let $K_{rr} = K(r_1, \ldots, r_k)$. Let K_u be the maximal unramified subextension of L'. Then L' has the same residue field over K_u; L'/K_u is purely ramified. Since $\mathrm{val}(L') = \mathrm{val}(L) = \mathrm{val}(K_{rr})$, L'/K_{rr} is unramified. In particular $K_u \cap K_{rr} = K$, so $Aut(L'/K) = Aut(L'/K_u)Aut(L'/K_{rr})$ by Galois theory. ⊣

CLAIM 2. $Aut(L'/K_{rr}) \subseteq \mathrm{dcl}(\mathcal{G}_{\boldsymbol{k}})$.

PROOF. The canonical homomorphism $Aut(L'/K_{rr}) \to Aut(\boldsymbol{k}_{L'}/\boldsymbol{k})$ is an isomorphism, since L'/K_{rr} is unramified. This homomorphism is definable, hence embeds the elements of $Aut(L'/K_{rr})$ into $\mathcal{G}_{\boldsymbol{k}}$. ⊣

CLAIM 3. $Aut(L'/K_u) \cong Hom(\Gamma(L)/\Gamma(K), \mu_n)$ (canonically and definably.)

PROOF. Let $E = \{c \in L^* : c^n \in K^*\}$. Define a map $b : Aut(L'/K_u) \times E \to \mu_n$ by $b(\sigma, e) = \sigma(e)/e$. Clearly b is multplicative in the second variable. If $e \in E$ and $\mathrm{val}(e) = 0$, then $\mathrm{res}(e^n)$ has an n'th root in $\mathrm{res}(L')$ and hence in $\mathrm{res}(K_u)$, since L'/K_u is purely ramified. So e^n has an n'th root in K_u. Since all roots of unity in L' lie in K_u, we have $e \in K_u$; hence $b(\sigma, e) = 1$ for all σ. More generally if $e \in E$ and $\mathrm{val}(e) \in \Gamma(K)$, then $\mathrm{val}(e/c) = 0$ for some $c \in K^*$, so $b(\sigma, e/c) = 0$ and hence $b(\sigma, e) = 0$ for all σ. Thus b factors through $b' : Aut(L'/K_u) \times \Gamma(L')/\Gamma(K) \to \mu_n$. We obtain a homomorphism $Aut(L'/K_u) \to Hom(\Gamma(L')/\Gamma(K), \mu_n)$. It is injective since if $b(\sigma, r_i) = 1$

for each i, then $\sigma(r_i) = r_i$ so $\sigma = 1$. Surjectivity comes from cardinality considerations (but will not be needed.) ⊣

CLAIM 4. $Aut(L'/K_u) \subseteq \mathrm{dcl}(\mu_n, \Gamma/n\Gamma)$ where $n = [L' : K_u]$.

PROOF. By Claim 3, $Aut(L'/K_u)$ is definably isomorphic to $Hom(\Gamma(L)/\Gamma(K), \mu_n)$. Now $\Gamma(L)/\Gamma(K)$ is a finite subgroup of $(1/n)\Gamma(K)/\Gamma(K)$, hence isomorphic to a finite subgroup $S \leq \Gamma(K)/n\Gamma(K)$. Each element of S lies in $\Gamma/n\Gamma$, and so a homomorphism $S \to \mu_n$ is definable over teh elements of S and the elements of μ_n, and the claim is proved. ⊣

The lemma follows from Claims 1, 2 and 4. ⊣

THEOREM 5.15. *Let K be a Henselian field with residue field of characteristic 0. Let G be an algebraic group over K, acting on a variety V; assume $G(K^{alg})$ acts transitively on $V(K^{alg})$. Let $V(K)/G(K)$ be the orbit space. Then for some n, $V(K)/G(K)$ is analyzable over sorts $\Gamma/n\Gamma$ and the n-th Galois sort $\mathcal{G}_n(\boldsymbol{k})$ of the residue field.*

PROOF OF THEOREM 5.15. Given $c \in V(K)$, $H = \{g : gc = c\}$, there is a canonical c-definable bijection between $V(K)/G(K)$ and the kernel of $H^1(K, H) \to H^1(K, G)$. Hence the theorem follows from Theorem 5.3 and Lemma 5.14. ⊣

REMARK 5.16. If we add to the structure the Denef-Pas splitting, (at least when e.g. Γ is a $\mathbb{Z}$-group, but probably in general), it follows that $V(K)/G(K)$ is tame, i.e. $V(K)/G(K) \subseteq \mathrm{dcl}(\boldsymbol{k}, \Gamma/n\Gamma)$. *Indeed for every imaginary definable set D_0 of Γ, either $D_0 \subseteq \mathrm{dcl}(\Gamma/n\Gamma)$ for some n, or else $D \subseteq \mathrm{dcl}(D_0)$ for some infinite definable $D \subseteq \Gamma$. But it is easy to see that D is not internal over $\boldsymbol{k} \cup \Gamma/m\Gamma$. Hence the definable sets internal over $\boldsymbol{k} \cup \Gamma/m\Gamma$ are contained in $\mathrm{dcl}(\boldsymbol{k} \cup \Gamma/n\Gamma)$ for some n, and by induction the same goes for analyzability.*

§6. Appendix.

It may seem at first sight that Galois sorts are peculiar to fields; but in fact they can be defined for any theory eliminating imaginaries on its finite subsets, see below.

In particular, the Galois sorts of RV will be defined. This will permit the remark that, for a theory T of Henselian fields of residue characteristic 0, if T_{RV} is the induced theory on RV, we have $\mathcal{G}_T \cong \mathcal{G}_{T_{\mathrm{RV}}}$; which, along with Theorem 5.3, clarifies the reductions of [4] and the present paper.

Fix a language L. Let $\mathfrak{T}$ be a theory admitting elimination of imaginaries with respect to finite sets of tuples. Thus for any finite product S of sorts of L, and any $m \in \mathbb{N}$, we have given a sort $S[\leq m]$ and a function $c_{S,m} : S^m \to S[\leq m]$, such that

$$\mathfrak{T} \models \left(c_{S,m}(x) = c_{S,m}(y) \iff \bigwedge_{\sigma \in Sym(m)} x^\sigma = y \right).$$

Let $S[m]$ be the image of the distinct m-tuples.

For simplicity we assume also that $\mathfrak{T}$ eliminates quantifiers, and that any quantifier-free definable function of $\mathfrak{T}$ is given by a term (piecewise), so that substructures are definably closed. Let T_0 be the theory of substructures of models of $\mathfrak{T}$. If $M \models T_0$, let M^{alg} denote the algebraic closure of M within some model of $\mathfrak{T}$. The Galois imaginaries, strictly speaking, belong to $T_0^a = Th(\{(M, M^{alg}) : M \models T_0\})$. We will also note some related imaginary sorts of T_0 itself. In practice it seems more convenient to use T_0^a, then note by considerations of stable embeddedness that definable sets on which the Galois group acts trivially belong to T_0. For instance this is the case with the cohomology sets $H^1(Aut(M^{alg}/M), A)$ considered below.

Below we abuse notation, identifying elements of $S[m]$ with m-element sets. Note however that if $M \models T$, then $S[m](M)$ corresponds to m-element subsets of M^{alg}, not of M.

We define the *Galois sorts* $\mathcal{G}$ of T_0. These are certain definable sets of imaginaries (not quantifier-free.)

We have a relation $(\subseteq) \subset S[\leq m]^2$, corresponding to inclusion of finite sets. Let $S_{irr}[m]$ be the set of *irreducible* elements of $S[m]$:

$$x \in S_{irr}[m] \iff \left(x \in S[m] \wedge \bigwedge_{1 \leq k < m} \neg(\exists y \in S[k])(y \subseteq x) \right).$$

Thus if $M \models T_0$, then $S_{irr}[m](M)$ corresponds to the set of orbits of $Aut(M^{alg}/M)$ on M^{alg} of size m. If $k \leq m$, for $x \in S_{irr}[k], y \in S_{irr}[m]$ we let $Mor(x, y)$ be the set of codes of functions $y \to x$ (viewed as subsets of $y \times x$.) This makes $\cup_{S,m} S_{irr}[m]$ into a category $\mathcal{C}$.

If $a \in S(M^{alg})$ has m conjugates under $Aut(M^{alg}/M)$, let $s(a)$ be the (code for) the set of conjugates. Then $s(a) \in S_{irr}[m](M)$, and every element of $S_{irr}[m](M)$ arises in this way. A $\mathcal{C}$-morphism $s(b) \to s(a)$ exists iff $M(a) \leq M(b)$. In particular, $s(a), s(b)$ are $\mathcal{C}$-isomorphic iff $M(a) = M(b)$, i.e. a, b generate the same substructure of M^{alg} over M. Isomorphism classes of $\mathcal{C}[M]$ correspond to finitely generated extensions of M within M^{alg}.

If $M \models T_0$ and $a \in S_{irr}[k](M), b \in S_{irr}[m](M)$, then $Mor(a, b)(M)$ is a finite set, possibly empty. By irreducibility, any function coded in $Mor(a, b)[M]$ must be surjective. In particular if $k = m$ it must be bijective, so the subcategory with objects $S_{irr}[m](M)$ is a groupoid (every morphism is invertible.) For $s \in S_{irr}[m]$, let $H_s = Mor(s, s)$. Let $S_{gal}[m]$ be set of $s \in S_{irr}[m]$ such that H_s acts regularly on s. Let $\mathcal{E}_{S,m}$ be the set of $\mathcal{C}$- isomorphism classes of objects in $S_{gal}[m]$. If $s \in S_{gal}[m]$ codes a set D_s, let $Gal(s) = Aut(D_s; H_s)$ be the opposite group to H_s, i.e. the group of permutations of D_s commuting with each element of H_s.

Let $E_s = Gal(s)^m/ad$ be the set of $Gal(s)$-conjugacy classes on $Gal(s)^m$. (More canonically, we could look at $E_s^k = Gal(s)^k/ad$ for all $k \in \mathbb{N}$, but all E_s^k are definable from $E_s^m = E_s$.)

The definition of $Gal(s)$ requires knowledge of D_s, i.e. of a particular choice of an algebraic closure M^{alg} of M. But E_s is canonical and does not depend on this choice.

If s, s' are isomorphic, i.e. $Mor(s, s') \neq \emptyset$, the choice of $f \in Mor(s, s')$ yields an isomorphism $Gal(s) \to Gal(s')$ (namely $\tau \mapsto f \circ \tau \circ f^{-1}$), and this isomorphism does not depend on the choice of f. Indeed let $M(s)$ be the substructure of M^{alg} generated over M by any element of s; the substructures $M(s), M(s')$ are equal, and $Gal(s) = Aut(M(s)/M)$. The induced bijection $E_s \to E_{s'}$ depends therefore neither on this choice nor on a choice of M^{alg}.

Let $\underset{\mathcal{C}}{\sim}$ denote $\mathcal{C}$- isomorphism, and let $\mathcal{E}_{S,m} = S_{gal}[m]/\underset{\mathcal{C}}{\sim}$. then if $e = s/\underset{\mathcal{C}}{\sim} \in \mathcal{G}_{S,m}$ we may define $Gal(e) = Gal(s), M(e) = M(s), E_e = E_s$. Let $\mathcal{G}_{S,m}$ be the disjoint union over $e \in \mathcal{E}_{S,m}$ of the groups $Gal(e)$, and let $\mathcal{G}_{S,m} \to \mathcal{E}_{S,m}$ be the natural map. Similarly let $\check{\mathcal{G}}$ be the direct limit over all S, m of $\mathcal{E}_{S,m}$.

Let $\check{\mathcal{G}}$ denote the family of all sorts $\check{\mathcal{G}}_{S,m}$ and $\mathcal{E}_{S,m}$. These will be called the *Galois sorts of* T_0. Let T_0^{gal} consist of all sorts interpretable over these sorts. (I.e. close under quotients by definable equivalence relations.) Note that the groups $Gal(e)$ are not themselves part of $\check{\mathcal{G}}$, in general.

Remark 6.1. The Galois group $G = Aut(M^{alg}/M)$ is the projective limit over $\mathcal{E}_{S,m}$ of all groups $Gal(e)$. An element $e \in \mathcal{G}_{S,m}(M)$ corresponds to a normal open subgroups $N(e)$ of G, and we have $Gal(e) \cong G/N(e)$ canonically. Similarly for any k the space G^k/ad_G of conjugacy classes of G on G^k can be deduced from $\check{\mathcal{G}}_{S,m}$ by an appropriate projective limit.

Remark 6.2. Let $M \models T$. Let $G = Aut(M^{alg}/M)$, C the centralizer of G in $Aut(M^{alg})$. Then $Aut(M/\check{\mathcal{G}}(M))$ is the image of C in $Aut(M)$. Equivalently, $\sigma \in Aut(M/\check{\mathcal{G}}(M))$ iff every extension τ of σ to $Aut(M^{alg})$ normalizes $Aut(M^{alg}/M)$, and induces an inner automorphism of this group. It follows in particular that any such τ preserves each Galois extension of M.

Remark 6.3. Assume (PE): L has a sort S_{basic} for which the primitive element theorem is valid, i.e. any finite extension of $M \models T_0$ is generated by a single element of S_{basic}. Then only the Galois sorts $\mathcal{E}_m = \mathcal{E}_{S_{basic},m}$ and $\check{\mathcal{G}}_m = \mathcal{E}_{S_{basic},m}$ need be considered; any Galois sort is in definable bijection with a definable subset of these.

6.4. Galois cohomology. Let G be a profinite group, acting continuously on a discrete group A. We write ${}^g a$ for the action.

Recall the definition of $H^1(G, A)$ [11, Chapter I, Subsection 5.1]. A 1-cocyle is a continuous map $a : G \to A$ satisfying $a(st) = a(s)^s a(t)$. The set of 1-cocycles is denoted $Z^1(G, A)$. Two cocycles a, a' are cohomologous if $a'(s) = b^{-1}a(s)^s b$ for some $b \in A$. The quotient of $Z^1(G, A)$ by this equivalence relation is denoted $H^1(G, A)$. The action of G on $Z^1(G, A)$, induced from the actions on G and on A, satisfies: ${}^g a(t) = b^{-1} a(t)^t b$, where $b = a(g)$; hence ${}^g a$ is cohomologous to a, so G acts trivially on $H^1(G, A)$.

When A is a definable group of $\mathfrak{T}$, for any $M \models T_0$ we have a profinite group $G_M := Aut(M^{alg}/M)$, and a continuous action of G_M on $A(M^{alg})$. We write G for the functor $M \mapsto G_M$.

LEMMA 6.5. *Assume* (PE) (*cf.* 6.3). *Let* A *be a definable finite group of* $\mathfrak{T}$. *Let* G_M *denote the automorphism group* $Aut(M^{alg}/M)$. *Then* $H^1(G, A)$ *is interpretable in the Galois sorts of* T_0. *In other words there exists a definable set* S *of* T_0^{gal} *and for any* $M \models T_0$ *a canonical bijection* $S(M) \to H^1(G_M, A)$.

PROOF. The proof of Lemma 5.7 goes through. ⊣

REFERENCES

[1] Z. CHATZIDAKIS and E. HRUSHOVSKI, *Model theory of difference fields*, ***Transactions of the American Mathematical Society***, vol. 351 (1999), no. 8, pp. 2997–3071.

[2] ———, *Perfect pseudo-algebraically closed fields are algebraically bounded*, ***Journal of Algebra***, vol. 271 (2004), no. 2, pp. 627–637.

[3] GREGORY CHERLIN, LOU VAN DEN DRIES, and ANGUS MACINTYRE, *Decidability and undecidability theorems for PAC-fields*, ***American Mathematical Society. Bulletin. New Series***, vol. 4 (1981), no. 1, pp. 101–104.

[4] R. CLUCKERS and J. DENEF, *Orbital integrals for linear groups*, ***Journal of the Institute of Mathematics of Jussieu***, vol. 7 (2008), no. 2, pp. 269–289.

[5] D. HASKELL, E. HRUSHOVSKI, and H. D. MACPHERSON, *Definable sets in algebraically closed valued fields: elimination of imaginaries*, ***Journal für die Reine und Angewandte Mathematik***, vol. 597 (2006), pp. 175–236.

[6] E. HRUSHOVSKI, *Pseudo-finite fields and related structures*, ***Model Theory and Applications***, Quaderni di Matematica, vol. 11, Aracne, Rome, 2002, pp. 151–212.

[7] ———, *Groupoids, imaginaries and internal covers*, `math.LO/0603413`.

[8] E. HRUSHOVSKI and D. KAZHDAN, *Integration in valued fields*, ***Algebraic Geometry and Number Theory***, Progress in Mathematics, vol. 253, Birkhäuser Boston, Boston, MA, 2006, pp. 261–405.

[9] JOHAN PAS, *Uniform p-adic cell decomposition and local zeta functions*, ***Journal für die Reine und Angewandte Mathematik***, vol. 399 (1989), pp. 137–172.

[10] J.-P. SERRE, ***Local Fields***, Graduate Texts in Mathematics, vol. 67, Springer-Verlag, New York, 1979, Translated from the French by Marvin Jay Greenberg.

[11] ———, ***Cohomologie Galoisienne***, fifth ed., Lecture Notes in Mathematics, vol. 5, Springer-Verlag, Berlin, 1994.

[12] LOU VAN DEN DRIES, *Dimension of definable sets, algebraic boundedness and Henselian fields*, ***Annals of Pure and Applied Logic***, vol. 45 (1989), no. 2, pp. 189–209.

INSTITUTE OF MATHEMATICS
THE HEBREW UNIVERSITY OF JERUSALEM
GIVAT RAM, JERUSALEM, 91904, ISRAEL
E-mail: ehud@math.huji.ac.il

STRONG MINIMAL COVERS AND A QUESTION OF YATES: THE STORY SO FAR

ANDREW E. M. LEWIS

Abstract. An old question of Yates as to whether all minimal degrees have a strong minimal cover remains one of the longstanding problems of degree theory, apparently largely impervious to present techniques. We survey existing results in this area, focussing especially on some recent progress.

§1. Introduction. By the 60's and 70's degree theorists had become concerned with some particular and fundamental questions of a global nature concerning the structure of the Turing degrees. In order to address issues regarding homogeneity and the decidability and degree of the theory, the approach taken at this time was to proceed through a deep analysis of the initial segments of the structure. Along these lines a technique for piecemeal construction of initial segments, even if only locally successful, would have been very useful and it was in this context that interest was first aroused in a question of Yates:

DEFINITION 1.1. A degree $\boldsymbol{b}$ is a strong minimal cover for $\boldsymbol{a}$ if $\mathcal{D}[< \boldsymbol{b}] = \mathcal{D}[\leq \boldsymbol{a}]$. A degree $\boldsymbol{a}$ is minimal if it is a strong minimal cover for $\boldsymbol{0}$.

QUESTION 1.1 (Yates). Does every minimal degree have a strong minimal cover?

In fact, the question of characterizing those degrees with strong minimal cover had already been raised by Spector in his 1956 paper [CS]. Certainly in $\mathcal{D}_m$—the structure of the many-one degrees, induced by a strengthening of the Turing reducibility—Lachlan's proof of the fact that every m-degree has a strong minimal cover played a vital role in Ershov's [YE] and Paliutin's [EP] results characterizing the structure and in showing, for instance, that $\boldsymbol{0}_m$ is

---

2000 *Mathematics Subject Classification.* 03D28.

The author was supported by Marie-Curie Fellowship MEIF-CT-2005-023657 and was partially supported by the NSFC Grand International Joint Project no. 60310213, New Directions in the Theory and Applications of Models of Computation.

Logic Colloquium '06
Edited by Barry Cooper, Herman Geuvers, Anand Pillay, and Jouko Väänänen
Lecture Notes in Logic, 32

the only definable singleton. In the Turing degrees, however, progress with global concerns of this nature was eventually achieved through other means—initially through complicated ad hoc initial segment embeddings, and more recently using simpler coding techniques. Yates' question remained behind, apparently largely impervious to existing techniques.

At the opposite end of the spectrum to those degrees with a strong minimal cover are those which satisfy the cupping property, where we say that a degree $\boldsymbol{a}$ satisfies the cupping property if for every $\boldsymbol{b} > \boldsymbol{a}$ there exists $\boldsymbol{c} < \boldsymbol{b}$ with $\boldsymbol{a} \vee \boldsymbol{c} = \boldsymbol{b}$. We say that a degree $\boldsymbol{a}$ is generalized low$_n$ (GL$_n$) if $\boldsymbol{a}^{(n)} = (\boldsymbol{a} \vee \boldsymbol{0}')^{(n-1)}$. Since Jockusch and Posner have shown that all degrees which are not GL$_2$ satisfy the cupping property, possession of a strong minimal cover may be seen as a property satisfied by degrees which are in some sense *low down* in the structure. Until very recently, however, results on the positive side of Yates' question were restricted to exhibiting particular examples of individual degrees which possess a strong minimal cover. As well as initial segment results along these lines, Cooper [BC3] showed that there exists a non-zero c.e. degree with a strong minimal cover—the motivation at that time being an approach towards the definability of $\boldsymbol{0}'$. Much later Kumabe [MK2] was able to show that there exists a 1-generic degree with this property.

It was not until 1999 that Ishmukhametov [SI] was able to provide an interesting *class* of degrees—the c.e. traceable degrees—every member of which has a strong minimal cover. In Sections 4 and 5 we shall provide an alternative proof of Ishmukhametov's result, appearing for the first time in this paper, by showing that the c.e. traceable degrees are actually a proper subclass of another very natural class of degrees all of which have strong minimal cover. In Section 3 we shall sketch an alternative proof of what may be regarded the strongest result on the negative side of Yates' question, the fact that there exists a hyperimmune-free degree with no strong minimal cover. Finally in section 5 we shall go on to discuss some other results recently obtained by the author in this area. In order to do so, it is necessary that we should begin with some discussion of the splitting tree technique of minimal degree construction.

§2. The splitting tree technique and strong minimal covers. The splitting tree technique is very much the standard approach to minimal degree construction. In fact really it is more than that—almost without exception, when a computability theorist wishes to construct a set of minimal degree it is the splitting tree technique that they will use. For a clear introduction to the techniques of minimal degree construction we refer the reader to any one of [RS, BC2, ML, PO1]. This paper requires knowledge only of those techniques for constructing a minimal degree below $\boldsymbol{0}''$.

Given any $T \subseteq 2^{<\omega}$ and $\tau \in T$, we shall say that τ is a leaf of T if it has no proper extensions in T. When $\tau, \tau' \in T$ and $\tau \subset \tau'$ we call τ' a successor of τ in T if there doesn't exist $\tau'' \in T$ with $\tau \subset \tau'' \subset \tau'$. Given any $T \subseteq 2^{<\omega}$

and $A \subseteq \omega$ we denote $A \in [T]$ and we say that A lies on T, if there exist infinitely many initial segments of A in T. We shall say that $\tau \in T$ is of level n in T if τ has precisely n proper initial segments in T and we shall say that finite T is of level n if all leaves of T are of level n in T. The following definition is slightly non-standard, but seems convenient where the discussion of splitting trees with unbounded branching is concerned: we say that $T \subseteq 2^{<\omega}$ is a c.e. tree if it has a computable enumeration $\{T_s\}_{s \geq 0}$ such that $|T_0| = 1$ and such that for all $s \geq 0$ if $\tau \in T_{s+1} - T_s$ then τ extends a leaf of T_s (and such that a finite number of strings are enumerated at any given stage). For any Turing functional Ψ, we say that two finite strings τ and τ' are a Ψ-splitting if $\Psi(\tau)$ and $\Psi(\tau')$ are incompatible. We say that $T \subseteq 2^{<\omega}$ is Ψ-splitting if whenever $\tau, \tau' \in T$ are incompatible these strings are a Ψ-splitting. We shall say that τ is $A\oplus$-compatible if, for all n such that $\tau(2n)\downarrow$, we have $\tau(2n) = A(n)$. For any $\tau \in 2^{<\omega}$ if $|\tau| > 0$ we define τ^- to be the initial segment of τ of length $|\tau| - 1$, and otherwise we define $\tau^- = \tau$. Given any Turing functional Ψ we define $\hat{\Psi}$ as follows. For all τ and all n, $\hat{\Psi}(\tau; n)\downarrow = x$ iff the computation $\Psi(\tau; n)$ converges in $< |\tau|$ steps, $\Psi(\tau; n) = x$ and $\hat{\Psi}(\tau^-; n')\downarrow$ for all $n' < n$. We say that $T \subseteq 2^{<\omega}$ with a single string of level 0 is 2-branching if every member of T has precisely two successors. We say non-empty $T \subseteq 2^{<\omega}$ is perfect if each $\tau \in T$ has at least two successors.

The basic problem which presents itself in attempting to construct a strong minimal cover for any given degree $\boldsymbol{a}$ is that if we simply relativize the standard minimal degree construction to $\boldsymbol{a}$ then, since the splitting trees concerned will now only be c.e. in $\boldsymbol{a}$, what results will certainly be a minimal cover for $\boldsymbol{a}$ but will not necessarily be a strong minimal cover—and where we say that $\boldsymbol{b}$ is a minimal cover for $\boldsymbol{a}$ if $\boldsymbol{b} > \boldsymbol{a}$ and there does not exist $\boldsymbol{c}$ with $\boldsymbol{a} < \boldsymbol{c} < \boldsymbol{b}$. As was observed in [AL1], however, a simple analysis of this construction can be used in order to provide a class of degrees for which the construction of a strong minimal cover is possible.

LEMMA 2.1. *Suppose* $\Psi = \hat{\Psi}$. *If* T_0 *is* A*-computable, 2-branching and* Ψ*-splitting, then* $T_1 = \{\Psi(\tau) : \tau \in T_0\}$ *is* A*-computable and 2-branching. Let* $T_2 \subseteq T_1$ *be* A*-computable and 2-branching. Then* $T_3 = \{\tau \in T_0 : \Psi(\tau) \in T_2\}$ *is* A*-computable and 2-branching with* $T_3 \subseteq T_0$.

PROOF. The proof is not difficult and is left to the reader. ⊣

Now let $\{\Psi_i\}_{i \geq 0}$ be some fixed effective listing of the Turing functionals and let us suppose that we wish to construct a strong minimal cover for $\boldsymbol{a}$. In order to do so we suppose given $A \in \boldsymbol{a}$ and we must construct $B \geq_T A$ so as to satisfy all requirements;

$\mathcal{R}_i : \Psi_i(B)$ total $\rightarrow (\Psi_i(B) \leq_T A$ or $B \leq_T \Psi_i(B))$ $\mathcal{P}_i : B \neq \Psi_i(A)$ A standard technique can be used in order to ensure that $B \geq_T A$, we simply insist that B should be an $A\oplus$-compatible string. Thus we begin with the restriction that B should lie on T_0 which contains all strings of even length which are $A\oplus$-compatible. T_0, then, is A-computable and 2-branching.

In order to meet all other requirements we might try to proceed with an easy forcing argument. We define B_0 to be the empty string. Suppose that by the end of stage s we have defined $B_s \in 2^{<\omega}$ on T_s, which is A-computable and 2-branching, in such a way that if B extends B_s and lies on T_s then all requirements $\mathcal{R}_i, \mathcal{P}_i$ for $i < s$ will be satisfied. At stage $s+1$ we might proceed, initially, just as if we were only trying to construct a minimal cover for $\boldsymbol{a}$. We can assume that $\Psi_s = \hat{\Psi}_s$, otherwise replace Ψ_s with $\hat{\Psi}_s$ in what follows. We ask the question, "does there exist $\tau \supseteq B_s$ on T_s such that no two strings on T_s extending τ are a Ψ_s-splitting?".

If so: then let τ be such a string. We can define $T_{s+1} = T_s$ and (just to make the satisfaction of $\mathcal{P}_s$ explicit) define B_{s+1} to be some extension of τ on T_s sufficient to ensure $\mathcal{P}_s$ is satisfied.

If not: then we can define T'_s to be an A-computable 2-branching Ψ_s-splitting subset of T_s having B_s as least element—the idea being that we shall eventually define T_{s+1} to be some subset of T'_s. If B lies on T'_s then we shall have that $B \leq_T \Psi_s(B) \oplus A$. Of course this does not suffice, since for the satisfaction of $\mathcal{R}_s$ we require that $B \leq_T \Psi_s(B)$. Suppose, however, that we know $\boldsymbol{a}$ is a *tree basis*:

DEFINITION 2.1. We say $deg(A)$ is a tree basis if for every perfect T computable in A there exists some perfect $T' \subseteq T$ computable in A such that every $C \in [T']$ computes A.

Lemma 2.1 then suffices to ensure that we can define T_{s+1} to be a subset of T'_s which is A-computable and 2-branching, and which satisfies the property that if $B \in [T_{s+1}]$ then $A \leq_T \Psi_s(B)$ so that, since $B \leq_T \Psi_s(B) \oplus A$, $B \leq_T \Psi_s(B)$. Then we can define B_{s+1} to be an extension of B_s lying on T_{s+1} sufficient to ensure the satisfaction of $\mathcal{P}_s$.

We therefore have:

THEOREM 2.1 (Lewis [AL1]). *Every tree basis has a strong minimal cover.*

2.1. Delayed splitting trees. Given the monopoly that the splitting tree technique presently has on minimal degree construction it seems an obvious question to ask the generality of this technique. Since it seems to be in the very least difficult to construct a negative solution to Yates' question using the splitting tree technique, it is a natural question to ask whether we should be looking for other techniques of minimal degree construction. So perhaps it is interesting to observe that in order to obtain a perfectly general technique of minimal degree construction all one need do is to use splitting trees with splitting delayed by one level at a time:

DEFINITION 2.2. We say that $T \subseteq 2^{<\omega}$ is a delayed Ψ-splitting c.e. tree if T is a c.e. tree and whenever $\tau_0, \tau_1 \in T$ are incompatible any $\tau_2, \tau_3 \in T$ properly extending τ_0 and τ_1 respectively are a Ψ-splitting.

THEOREM 2.2 (Lewis [AL1]). *A is of minimal degree iff A is noncomputable and, whenever $\Psi(A)$ is total and noncomputable, A lies on a delayed Ψ-splitting c.e. tree.*

Delayed splitting trees, then, provide a perfectly general technique of minimal degree construction in the sense that any set of minimal degree lies on trees of this kind. To put it another way, a trivial modification of the proof of Theorem 2.2 suffices to show that any set of minimal degree is generic for the associated notion of forcing. One might ask, however, whether we really need to use delayed splitting trees in order to get this result, perhaps it already holds for standard splitting trees? In fact this is not the case. In order to see this we need to consider the fixed point free degrees:

DEFINITION 2.3. Let ϕ_n denote the nth partial computable function according to some fixed effective listing. $A \subseteq \omega$ is of fixed point free (FPF) degree if there exists $f \leq_T A$ with $\phi_n \neq \phi_{f(n)}$ for all n.

It was another reasonably longstanding question regarding minimal degrees as to whether there exists a minimal degree which is FPF. Kumabe was able to answer this question in the affirmative:

THEOREM 2.3 (Kumabe [MK1]). *There exists a FPF minimal degree.*

In order to achieve this result Kumabe uses delayed splitting trees—in fact this is the only example in the literature of such a construction—and one can show that it is necessary that he should do so:

THEOREM 2.4 (Lewis [AL1]). *If A lies on a Ψ-splitting c.e. tree whenever $\Psi(A)$ is total and noncomputable, then A is not of FPF degree.*

§3. A hyperimmune-free degree with no strong minimal cover. Recall that A is of hyperimmune-free degree if for every $f \leq_T A$ there exists computable g which majorizes f i.e. such that $g(n) > f(n)$ for all n, and that a degree is PA if it computes a complete extension of Peano Arithmetic. In some ways the sets of hyperimmune-free degree and the sets of minimal degree can be seen as being very closely related, at least in the sense that the standard constructions are very similar. In fact the most primitive form of minimal degree construction uses perfect splitting trees and it is not difficult to show that if A satisfies the property that whenever $\Psi(A)$ is total and noncomputable A lies on a perfect c.e. Ψ-splitting tree, then A will automatically be of hyperimmune-free degree. Given this close relationship, perhaps the strongest result on the negative side of Yates' question is the fact that there exists a hyperimmune-free degree with no strong minimal cover.

THEOREM 3.1 (Kucera [AK]). *Every PA degree satisfies the cupping property.*

COROLLARY 3.1. *There exists a hyperimmune-free degree with no strong minimal cover.*

PROOF. There exists a PA degree which is hyperimmune-free. ⊣

DEFINITION 3.1. We say that $\Lambda \subseteq 2^{<\omega}$ is downward closed if whenever $\tau \in \Lambda$ and $\tau' \subset \tau$ we have $\tau' \in \Lambda$. $\mathcal{P} \subseteq 2^{\omega}$ is a Π^0_1 class if it is $[\Lambda]$ for some downward closed computable Λ.

ALTERNATIVE PROOF OF THEOREM 3.1 (SKETCH). In [AL1] the author described an alternative proof of Theorem 3.1 which does not require reasoning within PA like the original and which has subsequently turned out to be useful in proving other results. It is this alternative proof which we shall sketch here. ⊣

In fact, the statement of the theorem is easily seen to be equivalent to the following proposition: there exists a non-empty Π^0_1 class every member of which is of degree which satisfies the cupping property. In one direction this follows immediately because there exists a non-empty Π^0_1 class which contains only sets which effectively code a complete extension of PA. For the other direction it suffices to recall that A is of PA degree iff A computes a member of every non-empty Π^0_1 class and then to observe that the degrees which satisfy the cupping property form an upward closed class. It is this equivalent form of the theorem which we shall prove. In order to do so we make use of the following very simple lemma:

LEMMA 3.1. *If A computes 2-branching T such that no set lying on T computes A, then the degree of A satisfies the cupping property.*

PROOF. Suppose that A and T satisfy the hypothesis of the lemma and that we are given $B >_T A$. We define $C = \bigcup_s \tau_s$ such that $C <_T B$ and $B \leq_T C \oplus A$. We define τ_0 to be the string of level 0 in T. Given τ_s we define τ_{s+1} to be the left successor of τ_s in T if $B(s) = 0$ and the right successor otherwise. Then C lies on T so that $C <_T B$, since no set lying on T computes A. If we are given an oracle for A and an oracle for C then clearly we can compute B. ⊣

What we aim to achieve, then, is to construct downward closed computable Λ such that there exist infinite paths through Λ and such that if A is an infinite path through Λ then A computes some non-empty 2-branching T^A (let's say) such that no set lying on T^A computes A. In order to ensure that no set lying on T^A computes A it is convenient to construct a Turing functional Ψ such that no set on T^A computes $\Psi(A)$. In order to define T^A for any A which is an infinite path through Λ we shall define values T^τ for τ in Λ and then T^A will be defined to be the union of all T^τ such that $\tau \subset A$. Thus there are three different kinds of object under construction, Λ, T^A for A which is an infinite path through Λ and Ψ, and we must define these values in such a way that there exist infinite paths through Λ and so that the following requirements are satisfied:

$$\mathcal{N}_i : (A \in [\Lambda] \wedge C \in [T^A]) \rightarrow (\Psi_i(C; i) \neq \Psi(A; i)).$$

In fact, what we shall do here is just to consider how to satisfy a single requirement $\mathcal{N}_0$. We shall therefore only be concerned with the values $\Psi_0(C;0)$ and with defining Ψ on argument 0.

The most primitive form of the intuition runs as follows: if we are given four strings and we colour those four strings using two colours then there exists some colour such that at least two strings are not that colour (okay so actually we only need three strings but it is convenient here to do everything in powers of two). Now we extend this idea. First we define a certain set of strings T. The role of T is that it is the set of all strings that could possibly be in T^τ for some $\tau \in \Lambda$. In the case that we are only looking to satisfy a single requirement T is rather trivial, we just define T to be the set of all strings of even length. The important thing about T is that it is 4-branching. We let $T(n)$ denote the set of strings in T of level $\leq n$ in T. Next, we consider a certain form of finite subset of T:

DEFINITION 3.2. We say that finite $T' \subset T$ is $(T, 2)$-compatible if the strings of level n in T' are of length $2n$ and every string in T' which is not a leaf of T' has precisely two successors.

The role of these $(T, 2)$-compatible T' is that when we define T^τ for $\tau \in \Lambda$ actually we shall define this value to be some $(T, 2)$-compatible T'. The following lemma is what we need in order to satisfy the first requirement:

LEMMA 3.2. *For any finite $T' \subseteq 2^{<\omega}$ a 2-colouring of T' is an assignment of some col(σ) $\in \{0, 1\}$ to each leaf σ of T'. For any n and any 2-colouring of $T(n)$ there exists T' which is $(T, 2)$ compatible of level n and $d \in \{0, 1\}$ such that no leaf σ of T' has col(σ) $= d$.*

PROOF. This is easily seen by induction. The case $n = 0$ is trivial and, in fact, we have already seen the case $n = 1$. If we are given four strings and we colour those four strings using two colours then there exists some colour such that at least two strings are not that colour. Those two strings then define some $(T, 2)$-compatible T' of level 1. In order to see the induction step suppose we are given a 2-colouring of $T(n + 1)$. First we use this 2-colouring in order to define a 2-colouring of $T(n)$ as follows. Consider each leaf σ of $T(n)$. Such σ has precisely four successors in $T(n+1)$. If more than two of those successors are coloured 0 then colour σ with 0. If more than two of those successors are coloured 1 then colour σ with 1, and otherwise colour σ with 0. What this means is that if σ is not coloured d then at least two successors of σ are not coloured d. By the induction hypothesis there exists some $(T, 2)$-compatible T' of level n and there exists d such that no leaf of T' is coloured d. In order to define T'' of level $n + 1$ sufficient to complete the induction step all we need do is to choose two successors of each leaf of T' which are not coloured d. $\dashv$

Now we see how to use this lemma in order to satisfy $\mathcal{N}_0$ while also satisfying the condition that $[\Lambda]$ should be non-empty. Before defining Λ we define a set of strings $\Lambda^\star$. These are strings which may or may not be in Λ. We do not insist that $\Lambda^\star$ should be downward closed, in order to form Λ we shall later add strings in so that Λ is downward closed. For every n we let $\Lambda^\star(n)$ denote the set of strings in $\Lambda^\star$ which are of level n in $\Lambda^\star$. Thus for every n we must define the set of strings which are in $\Lambda^\star(n)$, for each such τ we must define a value T^τ which will be some $(T, 2)$-compatible T' of level n and we must also ensure that if $n > 0$ then $\Psi(\tau)$ is defined on argument 0. In order to satisfy this latter condition we can just ensure that $\Psi(\tau; 0)$ is defined for all τ of level 1 in $\Lambda^\star$ and then this task is done once and for all. We shall not describe here precisely how to define these values, but hopefully it is clear that we can do so in such a way that the following lemma is satisfied:

LEMMA 3.3. *For any $n > 0$, any $(T, 2)$-compatible T' of level n and any $d \in \{0, 1\}$ there exists $\tau \in \Lambda^\star(n)$ such that $T^\tau = T'$ and $\Psi(\tau; 0) = d$.*

The fact that we can satisfy Lemma 3.3 is really completely obvious. We are not insisting that $\Lambda^\star$ should be downward closed, so in order to ensure satisfaction of the lemma all we need do is to put enough strings into each $\Lambda^\star(n)$ so that all possibilities can be realised.

What this means is that if we define Λ by taking the strings in $\Lambda^\star$, adding in strings in order to make it downward closed and then removing any string τ (together with all extensions) for which it is the case that there exists $\sigma \in T^\tau$ with $\hat{\Psi}_0(\sigma; 0) \downarrow = \Psi(\tau; 0)$ then for every n we must be left with strings in $\Lambda^\star(n)$ which are in Λ. In order to see this suppose given $n > 0$. Then we can consider the values $\hat{\Psi}_0(\sigma; 0)$ for those $\sigma \in T(n)$ to define a 2-colouring of $T(n)$—where values are not defined to be either 0 or 1 we need not be concerned with them. Then Lemma 3.2 tells us that for this 2-colouring of $T(n)$ there exists some $(T, 2)$-compatible T' of level n and there exists $d \in \{0, 1\}$ such that no leaf of T' is coloured d. Fixing such T' and such d we may then apply Lemma 3.3 which tells us that there exists $\tau \in \Lambda^\star(n)$ with $T^\tau = T'$ and $\Psi(\tau; 0) = d$. This string τ, then, is a string in $\Lambda^\star(n)$ which is in Λ. By weak König's lemma the fact that there exist an infinite number of strings in Λ suffices to ensure that $[\Lambda]$ is nonempty.

In order to satisfy all requirements we must become a little more sophisticated—we need more colours and bushier trees—but the basic idea remains the same.

Upon seeing this result it seems a natural question to ask what we can say (to the opposite extreme) about Π^0_1 classes every member of which is of degree with a strong minimal cover. Of course it is a trivial matter to construct a non-empty class of this kind because all we need do is to include a single computable member. On the other hand we cannot hope to construct such a class of positive measure because any such class contains a member of every degree above $\boldsymbol{0}'$ and so certainly contains members of degree with no strong minimal cover. In a sense, then, perhaps the following result is the strongest we could hope for:

THEOREM 3.2 (Lewis [AL1]). *There exists a non-empty Π_1^0 class with no computable members, every member of which is of degree with a strong minimal cover.*

§4. **The c.e. traceable degrees.** As was outlined in the introduction, it seems fair to say that for a long time very little progress was made in the attempt to understand the issues surrounding Yates' problem. To a certain extent this changed relatively recently with Ishmukhametov's characterization of the c.e. degrees which have a strong minimal cover. In order to obtain this characterization Ishmukhametov built on previous work by Downey, Jockusch and Stob, who considered the array noncomputable degrees.

DEFINITION 4.1 (Downey, Jockusch, Stob [DJS]). A degree $\boldsymbol{a}$ is array noncomputable (a.n.c.) if for each $f \leq_{wtt} K$ there is a function g computable in $\boldsymbol{a}$ which is not dominated by f i.e. such that $g(n) \geq f(n)$ for infinitely many n.

Here f being *wtt* reducible to K just means that f is computable in the halting problem and that there is a computable bound on the *use* of the computation i.e. the number of bits of the oracle tape scanned on any given argument. It is an interesting characteristic of the a.n.c. degrees that they satisfy many of the properties of the degrees which are not generalized low$_2$. The following result, in particular, is the one that is of relevance to us now:

THEOREM 4.1 (Downey,Jockusch,Stob [DJS]). *Given* $\boldsymbol{a}$ *which is a.n.c.*:

1. $\boldsymbol{a}$ *is not minimal,*
2. $\boldsymbol{a}$ *satisfies the cupping property.*

In fact, it is not difficult to provide an alternative proof of Theorem 4.1 by showing that the a.n.c. degrees are a proper subclass of those degrees whose sets satisfy the hypothesis of Lemma 3.1:

THEOREM 4.2. *The a.n.c. degrees are a proper subclass of those degrees* $\boldsymbol{a}$ *satisfying* (†) *if* $A \in \boldsymbol{a}$ *then* A *computes 2-branching* T *such that no set lying on* T *computes* A.

PROOF. The proof of Theorem 3.1 described in [AL1] and sketched in the last section suffices to show that there exists a hyperimmune-free degree which satisfies (†). Since no hyperimmune-free degree can be a.n.c. we are left to show that any a.n.c. degree satisfies (†).

So suppose given A of a.n.c. degree. For any τ we define $f(\tau)$ as follows. For each $i \leq |\tau|$ let s_i be the least such that there exist τ_0, τ_1 extending τ of length s_i and which are a $\hat{\Psi}_i$-splitting and if there exist no such τ_0, τ_1 then let $s_i = 0$. Define $f(\tau) = \max\{s_i : i \leq |\tau|\}$. For every s define $f^\star(s) = \max\{f(\tau) : |\tau| = s\}$. Then $f^\star \leq_{wtt} K$ and since the degree of A is a.n.c. there exists $g \leq_T A$ with $g(s) \geq f^\star(s)$ for infinitely many s. We can assume that g is an increasing function.

We define $T = \bigcup_s T_s$ as follows.

Stage 0. We define T_0 to be $\{\lambda\}$ (and where λ is the string of length 0).

Stage $s + 1$. For each leaf τ of T_s and each $i \leq |\tau|$ such that $\hat{\Psi}_i(\tau)$ is compatible with A, check to see whether there exists a $\hat{\Psi}_i$-splitting τ_0, τ_1 such that both these strings extend τ and are of length $\leq g(s)$.

If so (for some $i \leq |\tau|$): then let i be the least such, let $\tau' \supset \tau$ be as short as possible such that $\hat{\Psi}(\tau')$ is incompatible with A and if $|\tau'| \leq s$ then enumerate the two one element extensions of τ' into T_{s+1}.

If not: then enumerate the two one element extensions of τ into T_{s+1}.

The verification that T is 2-branching and that no set lying on T computes A is not difficult and is left to the reader. ⊣

In order to show that Theorem 4.1 follows from Theorem 4.2 it remains to show that no a.n.c. degree is minimal. So suppose towards a contradiction that A is of minimal a.n.c. degree $\boldsymbol{a}$. Then clearly $\boldsymbol{a}$ is hyperimmune, so we can take increasing $f \leq_T A$ which is not dominated by any computable function. By Theorem 4.2 we can suppose given $T <_T A$ which is 2-branching and such that no set lying on T computes A. We define $C \leq_T A$ with $C \in [T]$ which is noncomputable. Since A is of minimal degree this suffices to ensure that $A \leq_T C$, which gives the required contradiction.

Let ϕ_i be the ith partial computable function according to some fixed effective listing. The e-state of τ at stage s is the binary string σ of length e defined as follows: for all $i < e$, $\sigma(i) = 1$ iff $\phi_i[s]$ is compatible with τ—and where $\phi_i[s]$ is the approximation to ϕ_i at stage s. For all s, $h(s)$ is defined to be the length of the longest string of level $s+1$ in T. We define $C = \bigcup_s \tau_s$ as follows.

Stage 0. Define τ_0 to be the string of level 0 in T.

Stage $s + 1$. Define τ_{s+1} to be the successor of τ_s in T which has the lexicographically least $(s + 1)$-state at stage $f(h(s + 1))$ (or to be the leftmost if both successors have the same $(s + 1)$-state at stage $f(h(s + 1)))$.

To prove that C is noncomputable is not difficult and is left to the reader.

Ishmukhametov then combined Theorem 4.1 to great effect with work on c.e. traceability:

Definition 4.2. $A \subseteq \omega$ is c.e. traceable if there is a computable function p such that for every function $f \leq_T A$ there is a computable function h such that $|W_{h(n)}| \leq p(n)$ and $f(n) \in W_{h(n)}$ for all $n \in \omega$.

How should one understand this definition? The function p here can be thought of as a *bounding* function and the function h can be thought of as a *guessing* function. Thus A is c.e. traceable if there exists some computable bounding function p such that for every $f \leq_T A$ there exists some computable guessing function h which makes at most $p(n)$ guesses as regards each value $f(n)$ and one of these guesses is always correct. Having observed that the class of c.e. traceable degrees (those whose sets are c.e. traceable) is complementary to the class of a.n.c. degrees in the c.e. degrees, Ishmukhametov [SI] was then able to show:

THEOREM 4.3 (Ishmukhametov [SI]). *All c.e. traceable degrees have a strong minimal cover.*

COROLLARY 4.1. *A c.e. degree has a strong minimal cover iff it is c.e. traceable.*

In fact, given Theorem 2.1, Theorem 4.3 follows as an immediate corollary of Theorem 4.4 below. In the next section we shall be able to observe that the c.e. traceable degrees are actually a proper subclass of those degrees which are a tree basis.

THEOREM 4.4. *Every c.e. traceable degree is a tree basis.*

PROOF. Terwijn and Zambella [TZ] have shown that being c.e. traceable is actually equivalent to satisfaction of the following condition:

($\dagger_0$) For every function $f \leq_T A$ there is a computable function h such that $|W_{h(n)}| \leq 2^n$ and $f(n) \in W_{h(n)}$ for all $n \in \omega$.

It is this characterization of the c.e. traceable sets that we shall use in what follows. For any finite T' we shall say that T' is 2-branching to level n if it is of level n and every string in T' which is not a leaf of T' has precisely two successors. In what follows we shall subdue mention of effective codings between strings and natural numbers and between finite sets of strings and natural numbers for the sake of readability. So now assume that A is c.e. traceable, $T \leq_T A$ and that T is perfect. In fact we may assume without loss of generality that T is 2-branching and that we are given Ψ such that:

- for any τ, the value $\Psi(\tau)$ is 2-branching to level n for some n and can be computed in $|\tau|$ many steps,
- $\Psi(A) = T$,
- for any $\tau \subset \tau'$, if $\sigma \in \Psi(\tau)$ then a) $\sigma \in \Psi(\tau')$ and b) if $\sigma' \subset \sigma$ and $\sigma' \notin \Psi(\tau)$ then $\sigma' \notin \Psi(\tau')$,
- if $\Psi(\tau)$ is of level $n > 0$ then there exists $\tau' \subset \tau$ such that $\Psi(\tau')$ is of level $n - 1$.

First we define the function $f \leq_T A$ as follows; for every n, $f(n)$ is the shortest initial segment of A, τ say, such that $\Psi(\tau)$ is of level $\Sigma_{i=0}^{n}(i+2)$. Since A is c.e. traceable we can then take computable h such that $|W_{h(n)}| \leq 2^n$ and $f(n) \in W_{h(n)}$ for all $n \in \omega$. We can assume that if $n > 0$ then τ is not enumerated into $W_{h(n)}$ until some initial segment of τ has been enumerated into $W_{h(n-1)}$ and unless $\Psi(\tau)$ is of level $\Sigma_{i=0}^{n}(i+2)$ and this is not the case for any proper initial segment of τ.

Next we proceed to computably enumerate various values $T_0(\tau)$ and $T_1(\tau)$ and axioms for a Turing functional Φ such that $T_0(A)$ is 2-branching and for every set C lying on $T_0(A)$ we have $\Phi(C) = A$. Initially all strings are available for use. There can be only a single string enumerated into $W_{h(0)}$, τ say. We define $T_1(\tau)$ to be the strings of level 0 and 2 in $\Psi(\tau)$ and we define $T_0(\tau)$ to be the string, σ say, of level 0 in $\Psi(\tau)$. We declare σ to be unavailable for use and enumerate the axiom $\Phi(\sigma) = \tau$. Whenever some τ is enumerated into $W_{h(n)}$ for $n > 0$ we shall already have enumerated a (unique and proper) initial segment of this string, τ' say, into $W_{h(n-1)}$. For each leaf σ of $T_0(\tau')$ there will be precisely 2^{n+1} successors in $T_1(\tau')$. Let σ_0 and σ_1 be the first two of these which are still available for use, enumerate these strings into $T_0(\tau)$ and enumerate all leaves σ' of $\Psi(\tau)$ which extend these two strings into $T_1(\tau)$ before enumerating the axioms $\Phi(\sigma_0) = \tau$, $\Phi(\sigma_1) = \tau$. Declare σ_0 and σ_1 to be unavailable for use.

In order to see that the axioms enumerated for Φ are consistent observe that when we enumerate an axiom $\Phi(\sigma) = \tau$, σ is a leaf of $T_0(\tau)$. The consistency of the axioms enumerated for Φ then follows from the fact that it is easily seen by induction on the stage of the construction that if τ and τ' are incompatible and we have defined values $T_0(\tau)$, $T_0(\tau')$ then the leaves of these two sets of strings are pairwise incompatible. The fact that $T_0(A)$ is 2-branching and that for every $C \in T_0(A)$ we have $\Phi(C) = A$ then follows immediately from the description of the construction. ⊣

Unfortunately c.e. traceability does not relate in such a tidy way where the minimal degrees are concerned. Gabay [YG] has shown that there exist minimal degrees which are not c.e. traceable and which have a strong minimal cover and that there exist minimal degrees below $\mathbf{0}'$ which are not c.e. traceable. Theorem 4.3 does suffice to show, however, that many of the minimal degrees which we typically construct will automatically have a strong minimal cover. Let us say that a splitting tree has bounded branching if there exists n such that each string in the tree has at most n successors. It is not difficult to show that if A satisfies the property that whenever $\Psi(A)$ is total and noncomputable A lies on a c.e. Ψ-splitting tree with bounded branching, then A is c.e. traceable. The vast majority of minimal degree constructions in the literature use splitting trees with bounded branching.

§5. **Further results.** Although we know that there exists a hyperimmune-free degree with no strong minimal cover, the example of such a degree which we have so far is PA and so certainly cannot be minimal. Of course, what we are really interested in right now is the minimal degrees, and in fact it is possible to achieve a strong result in the opposite direction:

THEOREM 5.1 (Lewis [AL1]). *Every hyperimmune-free degree which is not FPF is a tree basis and so has a strong minimal cover.*

Corollary 5.1. *Every hyperimmune-free degree bounded by a* 1*-generic has a strong minimal cover.*

Proof. No 1-generic is FPF and the degrees which are not FPF are downward closed. ⊣

Through an analysis of Gabay's proof [YG] that there exists a minimal degree which is not c.e. traceable, and using also Theorem 2.4, it is possible to show that there exist minimal degrees which are hyperimmune-free, not c.e. traceacble and not FPF. Thus the c.e. traceable degrees are actually a proper subclass of those degrees which are a tree basis. Since there are a number of different classes of degrees concerned now perhaps a picture is helpful—the diagram below shows the Turing degrees divided into four basic sections according to whether or not they are FPF and/or hyperimmune-free. The roughly triangular area on the top left (and with no intersection with the area below the horizontal dotted line) is the a.n.c. degrees and the area below the horizontal dotted line depicts the degrees which are c.e. traceable.

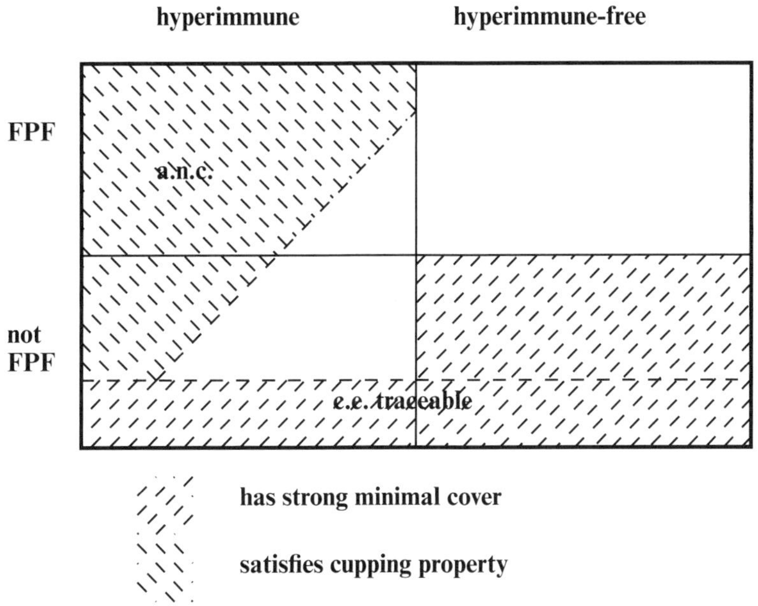

Figure 1. Degree classes with strong minimal cover.

No c.e. traceable degree is FPF and in Theorem 5.1 we have also required the condition of not being FPF in order to ensure the existence of a strong minimal cover. On the other hand, all PA degrees satisfy the cupping property and are all FPF. One begins, then, to question whether it might be the case that in fact all FPF degrees satisfy the cupping property, or at least fail to have a strong minimal cover. A positive answer to this question would have been very exciting. Theorem 2.3 would then imply a negative solution to Yates' question and we would also have a complete characterization of the hyperimmune-free degrees with strong minimal cover as those which are not FPF. Unfortunately, however, this question is answered in the negative:

THEOREM 5.2 (Lewis [AL2]). *There exists a Martin-Löf random degree with strong minimal cover.*

Since Kucera has shown that all Martin-Löf random degrees are FPF Theorem 5.2 proves the existence of a FPF degree with strong minimal cover. Since this random degree can also be made hyperimmune-free we therefore have:

COROLLARY 5.2 (Lewis [AL2]). *The hyperimmune-free degrees with strong minimal cover cannot be characterized as those which are not FPF.*

5.1. Other techniques of minimal degree construction. The fact that any set of minimal degree lies on delayed splitting trees is not to say that in order to construct a set of minimal degree we need necessarily think in terms of splitting trees. There *are* other ways of constructing a set of minimal degree (even if such methods are not used in practice). One such way involves the use of Cantor-Bendixson rank:

DEFINITION 5.1. For any $\Lambda \subseteq 2^{<\omega}$ we define $B([\Lambda])$, the Cantor-Bendixson derivative of $[\Lambda]$, to be the set of non-isolated points of $[\Lambda]$ according to the Cantor topology. The iterated Cantor-Bendixson derivative $B^\alpha([\Lambda])$ is defined for all ordinals α by the following transfinite recursion. $B^0([\Lambda]) = [\Lambda]$, $B^{\alpha+1}([\Lambda]) = B(B^\alpha([\Lambda]))$ and $B^\lambda([\Lambda]) = \bigcap_{\alpha<\lambda} B^\alpha([\Lambda])$ for any limit ordinal λ. A set A has Cantor-Bendixson rank α if α is the least ordinal such that for some Π^0_1 class $[\Lambda]$, $A \in B^\alpha([\Lambda]) - B^{\alpha+1}([\Lambda])$.

The following theorem was not claimed explicitly by Owen but follows immediately from his results:

THEOREM 5.3 (Owings [JO]). *If A is hyperimmune-free and of Cantor-Bendixson rank* 1 *then A is of minimal degree.*

This theorem then combines with a result of Downey's in order to provide us with a technique of minimal degree construction which does not explicitly make use of splitting trees:

THEOREM 5.4 (Downey [RD]). *There exists a set of hyperimmune-free degree and which is of rank one.*

Of course, the hope might be that using this alternative technique one might be able to avoid complications inherent in the use of the splitting tree technique and thereby produce a negative solution to Yates' question. Such hopes, however, are in vain. Any minimal degree constructed using this particular technique will have a strong minimal cover.

THEOREM 5.5 (Lewis [AL2]). *If A is of hyperimmune-free degree and is of rank 1 then the degree of A is not FPF and therefore has strong minimal cover.*

We close with a question:

QUESTION 5.1. Do the degrees with strong minimal cover form a downward closed class?

If one wishes to characterize the degrees with strong minimal cover then it would seem important to know whether or not these degrees form a downward closed class. Properties such as being c.e. traceable or (being hyperimmune-free and) not being FPF define downward closed classes of degrees. If the degrees with strong minimal cover are not downward closed then certainly no property along those lines is going to be sufficient.

REFERENCES

[BC3] S. B. COOPER, ***Degrees of Unsolvability***, Ph.D. thesis, University of Leicester, 1971.

[BC2] ———, ***Computability Theory***, Chapman & Hall/CRC, Boca Raton, FL, 2004.

[RD] R. DOWNEY, *On Π^0_1 classes and their ranked points*, ***Notre Dame Journal of Formal Logic***, vol. 32 (1991), no. 4, pp. 499–512.

[DJS] R. DOWNEY, C. G. JOCKUSCH, and M. STOB, *Array nonrecursive degrees and genericity*, ***Computability, Enumerability, Unsolvability***, London Mathematical Society Lecture Note Series, vol. 224, Cambridge University Press, Cambridge, 1996, pp. 93–104.

[YE] Y. L. ERSHOV, *The uppersemilattice of enumerations of a finite set*, ***Alg. Log.***, vol. 14 (1975), pp. 258–284.

[YG] Y. GABAY, ***Double Jump Inversion and Strong Minimal Covers***, Ph.D. thesis, Cornell University, 2004.

[SI] S. ISHMUKHAMETOV, *Weak recursive degrees and a problem of Spector*, ***Recursion Theory and Complexity*** (Arslanov and Lempp, editors), de Gruyter Series in Logic and its Applications, vol. 2, de Gruyter, Berlin, 1999, pp. 81–87.

[AK] A. KUČERA, *Measure, Π^0_1-classes and complete extensions of* PA, ***Recursion Theory Week (Oberwolfach, 1984)***, Lecture Notes in Mathematics, vol. 1141, Springer, Berlin, 1985, pp. 245–259.

[MK1] M. KUMABE, *A fixed point free minimal degree*, unpublished, 1993.

[MK2] ———, *A 1-generic degree with a strong minimal cover*, ***The Journal of Symbolic Logic***, vol. 65 (2000), no. 3, pp. 1395–1442.

[ML] M. LERMAN, ***Degrees of Unsolvability***, Perspectives in Mathematical Logic, Springer-Verlag, Berlin, 1983.

[AL1] A. E. M. LEWIS, *Π^0_1 classes, strong minimal covers and hyperimmune-free degrees*, ***Bulletin of the London Mathematical Society***, vol. 39 (2007), no. 6, pp. 892–910.

[AL2] ———, *A random degree with strong minimal cover*, ***Bulletin of the London Mathematical Society***, vol. 39 (2007), no. 5, pp. 848–856.

[PO1] P. G. ODIFREDDI, ***Classical Recursion Theory. Vol. I***, Studies in Logic and the Foundations of Mathematics, vol. 125, North-Holland, Amsterdam, 1989.

[JO] J. OWINGS, *Rank, join, and Cantor singletons*, ***Archive for Mathematical Logic***, vol. 36 (1997), no. 4-5, pp. 313–320.

[EP] E. PALIUTIN, *Addendum to the paper of Ershov*, ***Alg. Log.***, vol. 14 (1975), pp. 284–287.

[RS] R. I. SOARE, ***Recursively Enumerable Sets and Degrees***, Perspectives in Mathematical Logic, Springer-Verlag, Berlin, 1987.

[CS] C. Spector, *On degrees of recursive unsolvability*, ***Annals of Mathematics***, vol. 64 (1956), pp. 581–592.

[TZ] S. A. Terwijn and D. Zambella, *Computational randomness and lowness*, ***The Journal of Symbolic Logic***, vol. 66 (2001), no. 3, pp. 1199–1205.

DEPARTMENT OF PURE MATHEMATICS
UNIVERSITY OF LEEDS
LEEDS, ENGLAND, LS29JT, UK
E-mail: andy@aemlewis.co.uk

EMBEDDINGS INTO THE TURING DEGREES

ANTONIO MONTALBÁN

§1. Introduction. The structure of the Turing degrees was introduced by Kleene and Post in 1954 [KP54]. Since then, its study has been central in the area of Computability Theory. One approach for analyzing the shape of this structure has been looking at the structures that can be embedded into it. In this paper we do a survey of this type of results.

The Turing degree structure is a very natural object; it was defined with the intention of abstracting the properties of the relation "computable from", which is the most important notion in computability theory introduced by Turing in [Tur39]. It is defined as follows. Consider $\mathcal{P}(\omega)$, the set of sets of natural numbers. Given $A, B \in \mathcal{P}(\omega)$, we say that A is *computable from* B, if there is a computer program which, on input $n \in \omega$, decides whether $n \in A$ or not using B as an *oracle*. That means that the program is allowed to ask questions to the oracle of the form "does m belong to B?" We write $A \leqslant_T B$ if A is computable from B. The relation $\leqslant_T$ is a quasi-ordering on $\mathcal{P}(\omega)$. This quasi-ordering induces an equivalence relation on $\mathcal{P}(\omega)$, given by

$$A \equiv_T B \quad \Leftrightarrow \quad A \leqslant_T B \,\&\, B \leqslant_T A,$$

and a partial ordering on the equivalence classes. The equivalence classes are called *Turing degrees*. (The concept of Turing degree was introduced by Post [Pos44].) We use $\langle \boldsymbol{D}, \leqslant_T \rangle$ to denote this partial ordering. One of the main goals of Computability Theory is to understand the structure of $\langle \boldsymbol{D}, \leqslant_T \rangle$.

There are two basic but important remarks to make here. First, when we talk about a computer program, we are fixing a programming language, say for example the language of Turing machines, or Java. The notion of computability is independent of the programming language chosen. Second, we note that we chose to work with subsets of ω because every finite object can be encoded by a single number (using, for instance, the binary representation of the number). For example, strings, graphs, trees, simplicial complexes, group presentations, etc., if they are finite, they can be effectively coded by a natural number. There would be no essential difference if we had chosen to work with subsets of $\mathbb{Z}$, $2^{<\omega}$ or $V(\omega)$, instead of ω.

Before we go into the embeddability results, we will start by mentioning basic facts about the structure of the Turing Degrees. The embeddability

Logic Colloquium '06
Edited by Barry Cooper, Herman Geuvers, Anand Pillay, and Jouko Väänänen
Lecture Notes in Logic, 32

results are divided in four sections: embeddings of countable structures, initial segments embeddings, embeddings of larger structures, and embedding into the high/low hierarchy. Embeddability results are very closely related to decidability results, so we dedicate our last section to them.

No knowledge of Computability Theory is assumed. Basic references on the topic are [Ler83] and [Soa87]. Two nice surveys have been recently written. One is by Ambos-Spies and Fejer [ASF], where they describe the history of the Turing Degrees. The other one, by Shore [Sho06], describes the current situation of this research program, and also looks at its history and possible future directions. Our paper has something of both of those papers, but it concentrates just on embeddability results, and is mostly about the global structure. We will not mention results about other reducibilities, even though many have been considered and studied.

§2. Background.

2.1. First observations. Let us start by making the most basic observations about the structure of the Turing degrees.

There is a least Turing degree that we denote by $\boldsymbol{0}$. It is the degree whose members are the computable sets.

Every degree has at most countably many degrees below it. We call this property, the *countable predecessor property* or *c.p.p.* The reason is that there are only countably many programs one can write, so there are at most countably many sets that are computable from a fixed set. It also follows that each Turing degree contains at most countably many sets.

There are $2^{\aleph_0}$ many Turing degrees. Because there are $2^{\aleph_0}$ many subsets of ω and each equivalence class is countable.

The Turing degrees form an *upper semilattice*, or *usl*; that is, every pair of elements has a least upper bound. We denote the least upper bound of $\boldsymbol{a}$ and $\boldsymbol{b}$ by $\boldsymbol{a} \vee \boldsymbol{b}$, and we refer to it as the *join* of $\boldsymbol{a}$ and $\boldsymbol{b}$. Given $A, B \in \mathcal{P}(\omega)$, let

$$A \oplus B = \{2n : n \in A\} \cup \{2n + 1 : n \in B\}.$$

Is not hard to note that $A \leqslant_T A \oplus B$, $B \leqslant_T A \oplus B$, and that if both $A \leqslant_T C$ and $B \leqslant_T C$, then $A \oplus B \leqslant_T C$. We let $\boldsymbol{a} \vee \boldsymbol{b}$ be the degree of $A \oplus B$, where A and B are sets in $\boldsymbol{a}$ and $\boldsymbol{b}$ respectively.

2.2. Turing jump. There is another naturally defined operation on the Turing degrees called the *Turing jump* (or just *jump*). The jump of a degree $\boldsymbol{a}$, denoted $\boldsymbol{a}'$, is given by the degree of the *Halting Problem* relativized to some set in $\boldsymbol{a}$. Given $A \subseteq \mathcal{P}(\omega)$, we define A', the *Halting Problem relative to A*, as follows:

> A' is the set of codes for programs that, when run with oracle A, halt.

Note that a computer program is a finite sequence of characters and hence can be encoded by a natural number. It can be shown that the jump operation is strictly increasing and monotonic. That is, for every $\boldsymbol{a}, \boldsymbol{b} \in \boldsymbol{D}$,

(1) $\boldsymbol{a} <_T \boldsymbol{a}'$, and
(2) $\boldsymbol{a} \leqslant_T \boldsymbol{b} \Rightarrow \boldsymbol{a}' \leqslant_T \boldsymbol{b}'$.

The only non-trivial fact here is that $A' \not\leqslant_T A$, and it is proved the same way one proves that the Halting problem is not computable.

Definition 2.1. A *jump upper semilattice* is a structure

$$\mathcal{J} = \langle J, \leqslant_{\mathcal{J}}, \cup, \mathrm{j} \rangle$$

such that

- $\langle J, \leqslant_{\mathcal{J}} \rangle$ is a partial ordering,
- for all $x, y \in J$, $x \cup y$ it is the least upper bound of x and y, and
- $\mathrm{j}(\cdot)$ is a unary operation such that for all $x, y \in J$, $x <_{\mathcal{J}} \mathrm{j}(x)$; and if $x \leqslant_{\mathcal{J}} y$, then $\mathrm{j}(x) \leqslant_{\mathcal{J}} \mathrm{j}(y)$.

2.3. The picture. We have observed so far that $\mathcal{D} = \langle \boldsymbol{D}, \leqslant_T, \vee, ' \rangle$ is a jump upper semilattice of size $2^{\aleph_0}$, with a least element called $\boldsymbol{0}$, and with the countable predecessor property.

The next natural question is whether $\mathcal{D}$ is a lattice. The answer is no. Kleene and Post [KP54] proved that there exists degrees $\boldsymbol{a}$ and $\boldsymbol{b}$ with no greatest lower bound. There are aslo pairs of incomparable degrees which do have greatest lower bounds.

The only particular degree we have mentioned so far is $\boldsymbol{0}$. We have also mentioned the Halting problem, which has degree $\boldsymbol{0}'$. The structure of degrees below $\boldsymbol{0}'$, that we denote by $\mathcal{D}(\leqslant_T \boldsymbol{0}')$, is already very rich. For instance, all the computable enumerable sets are computable from $0'$. A set is *computable enumerable, or c.e.*, if there is a computer program that lists all its elements. The study of the structure of the c.e. degrees is also a topic where extensive research has been done.

$\boldsymbol{0}'$ is very low down inside the whole structure of the Turing degrees. We can start going up and construct a sequence of degrees $\boldsymbol{0} <_T \boldsymbol{0}' <_T \boldsymbol{0}'' <_T \dots$. This way we get all the way up the arithmetic hierarchy: It is not hard to show that the sets that are Turing below $0^{(n)}$ for some $n \in \omega$ are exactly the *arithmetic* ones, that is, the ones that can be defined by a formula of first order arithmetic. (We use $X^{(n)}$ to denote the nth iteration of the Turing jump.) Then, we can take the uniform join of all these sets and get $0^{(\omega)} = \{\langle n, m \rangle : m \in 0^{(n)}\}$, which is Turing equivalent to the set of sentences true in first order arithmetic. We can then continue taking jumps and define $0^{(\omega+1)} = 0^{(\omega)\prime}$, and even define $0^{(\alpha)}$ for any countable computable ordinal α by taking uniform joins at limit levels. The situation when α is a non-computable ordinal is a bit more delicate. A *computable ordinal* is one which can be presented as a computable ordering of a computable subset of the natural numbers. We use ω_1^{CK} to denote the first ordinal which does not have a computable presentation. A set which is computable in $0^{(\alpha)}$ for some $\alpha < \omega_1^{CK}$ is said to be *hyperarithmetic*. These are exactly the Δ_1^1 sets (Kleene and Suslin [Kle55]). Higher up comes Kleene's O, the set of computable indices (i.e. programs) of computable well-orderings. Kleene's O is Π_1^1-complete and computes all the hyperarithmetic sets. We could then take Kleene's O relative to Kleene's O, and so on. Much higher up is the set of true sentences of second order arithmetic, and there are still many more degrees higher up. Whenever we have a countable set of degrees, there exists a degree that bounds them all.

So far, our picture looks thin and tall. But actually, $\mathcal{D}$ is not taller than it is wide. Since $\mathcal{D}$ has the countable predecessor property, every chain in $\mathcal{D}$ can have size at most $\aleph_1$. However, it is known that there is an antichain that contains $2^{\aleph_0}$ minimal degrees (Lacombe [Lac54]). A degree $\boldsymbol{a} >_T \boldsymbol{0}$ is *minimal* if there is no degree $\boldsymbol{x}$, with $\boldsymbol{0} <_T \boldsymbol{x} <_T \boldsymbol{a}$. (The existence of minimal degrees is due to Spector [Spe56].)

§3. Embeddings of countable structures.

We now start analyzing the structures that embed into $\mathcal{D}$.

3.1. Upper semilattices. The first result in this direction was proved by Kleene and Post [KP54] in the same paper where they introduced the Turing degrees. They showed that there is an infinite independent set of degrees, that is, a set of degrees none of which can be computed from the other ones altogether. They prove it using the method of finite approximations. Today we would refer to such a construction as a forcing construction. The ideas in [KP54] can be easily extended to get the following result.

THEOREM 3.1 (Kleene and Post). *Every countable upper semilattice embeds into the Turing degrees.*

PROOF SKETCH. It is enough to show that the countable atomless Boolean algebra embeds into $\mathcal{D}$ since every countable upper semilattice embeds into it. Let $G \subseteq \omega$ be sufficiently generic. In other words, G meets the countably many dense open sets considered for the proof. Via a computable bijection between ω and the set of rational numbers $\mathbb{Q}$, think of G as a subset of $\mathbb{Q}$. It is well know that the countable atomless Boolean algebra is isomorphic to $\mathrm{Int}(\mathbb{Q})$, the interval algebra of $\mathbb{Q}$, that is, the algebra whose elements are the finite unions of closed-open intervals of $\mathbb{Q}$. Now, define $h\colon \mathrm{Int}(\mathbb{Q}) \to \mathcal{P}(\mathbb{Q})$ by $h(I) = G \cap I$. The proof that h preserves $\leqslant_T$, $\vee$ and 0 does not use the genericity of G and is quite simple. The genericity of G is used to show that h preserves $\nleqslant_T$. It can also be used to show that h preserves greatest lower bounds. ⊣

It also follows from the proof above that every countable distributive lattice can be embedded into $\mathcal{D}$, even preserving greatest lower bounds, since they can be embedded into the atomless Boolean algebra.

The fact that every countable lattice can be embedded into $\mathcal{D}$ preserving greatest lower bounds follows from a much stronger result of Lachlan and Lebeuf (see Theorem 4.1 below).

3.2. Local structures. A local structure is one of the form $\boldsymbol{D}(\leqslant_T \boldsymbol{a}) = \{\boldsymbol{x} \in \boldsymbol{D} : \boldsymbol{x} \leqslant_T \boldsymbol{a}\}$. There has been a lot of research done on local structures, and we will just quickly refer to some of it.

The first approach here is usually of the following sort. Theorem 3.1 says that every usl can be embedded into $\mathcal{D}$. Of course, the construction of this embedding cannot be computable, although, if we had an oracle smart enough, we could produce this embedding computably in the oracle. The question is how complex this oracle has to be. In the case of Theorem 3.1, a good answer is $0'$. A better answer is 1-generic. A 1-generic set is like a Cohen generic set, but it only needs to meet a small class of dense open sets: An infinite binary sequence $G \in 2^\omega$, is *1-generic* if for every Σ^0_1 set $S \subseteq 2^{<\omega}$ there exists a string $\sigma \subseteq G$ such that either $\sigma \in S$ or $\forall \tau \in 2^{<\omega}$ $(\tau \supseteq \sigma \Rightarrow \tau \notin S)$. (We are abusing notation and identifying 2^ω and $\mathcal{P}(\omega)$.) It is not hard to show that $\boldsymbol{0}'$ is able to compute a 1-generic set. (Moreover, any computable enumerable set or non-GL_2 set can compute a 1-generic.) So we get that every countable usl, and also every distributive lattice, embeds in $\boldsymbol{D}(\leqslant_T \boldsymbol{0}')$. Relativizing, one can get the whole embedding between $\boldsymbol{a}$ and $\boldsymbol{a}'$ for any $\boldsymbol{a} \in \boldsymbol{D}$.

For lattices in general it is not possible to get such a result. The reason is that there are $2^{\aleph_0}$ many lattices with finitely many generators [Sho82], but there are only countably many possibilities for those generators below $\boldsymbol{0}'$. However, Lerman [Ler83] proved that every computable presentable lattice embeds in $\boldsymbol{D}(\leqslant_T \boldsymbol{0}')$. Actually he proved this for $0''$-computable lattices and embedded them even as initial segments below $\boldsymbol{0}'$. Moreover, if $\boldsymbol{a}$ bounds a 1-generic degree, then every computable lattice embeds in $\boldsymbol{D}(\leqslant_T \boldsymbol{a})$ (Shore [Sho82]). This is not true if we also want to preserve top element. This follows from Kumabe's [Kum00] construction of a strong minimal cover of a 1-generic. However, it is true for $\boldsymbol{a} = \boldsymbol{0}'$, as it was proved by Fejer [Fej89]. Moreover, it is known that every computable lattice embeds in $\boldsymbol{D}(\leqslant_T \boldsymbol{a})$ preserving top element if $\boldsymbol{a}$ is non-GL_2 [Fej89], even array non-recursive (Downey, Jockusch and Stob [DJS96]) or if it is 1-generic (Greenberg and Montalbán [GM03]).

Since we are here, we should mention that the question of which lattices embed into the structure of the c.e. degrees is open, and a lot of effort has been put into it. (For a survey on this topic see Lempp, Lerman and Solomon [LLS06].)

3.3. Jump partial orderings. If we forget about joins but add jump to the language, we get the the following type of structure.

DEFINITION 3.2. A *jump partial ordering*, or *jpo*, is a structure

$$\mathcal{P} = \langle P, \leqslant_P, \mathrm{j} \rangle$$

such that

- $\langle P, \leqslant_P \rangle$ is a partial ordering, and
- j is a unary operation such that for all $x, y \in P$, $x <_P \mathrm{j}(x)$; and if $x \leqslant_P y$, then $\mathrm{j}(x) \leqslant_P \mathrm{j}(y)$.

A *jump partial ordering with* 0, or *jpo* $w/0$, is a structure $\mathcal{P} = \langle P, \leqslant_P, \mathrm{j}, 0 \rangle$, where $\mathcal{P} = \langle P, \leqslant_P, \mathrm{j} \rangle$ is a jump partial ordering and 0 is the least element.

As we mentioned in Section 2.3, if $\langle P, \leqslant_P \rangle$ is a well ordering and the jump function corresponds to the successor function on $\langle P, \leqslant_P \rangle$, then $\mathcal{P}$ can be embedded into $\mathcal{D}$. Such an embedding is called a *jump hierarchy*. Even if $\langle P, \leqslant_P \rangle \cong \omega_1^{CK}(1+\eta)$, we get the embeddability result, where ω_1^{CK} is the least non-computable ordinal and η is the order type of the rationals. Such an embedding is a *Harrison pseudo-hierarchy* [Har68].

If we let $\langle P, \leqslant_P \rangle \cong \mathbb{Z}$, the ordering of the integers, and we let $\mathrm{j}(n) = n+1$, the fact that $\mathcal{P}$ embeds into $\mathcal{D}$ follows from Harrison's pseudo-hierarchy theorem [Har68] and Friedberg's jump inversion theorem [Fri57]. Such an embedding has to be high up in $\mathcal{D}$; it can be proved that every degree in the image of such an embedding has to compute all the hyperarithmetic sets (Enderton and Putnam [EP70]). A curiosity, is that if we want to get an embedding $h\colon \mathbb{Z} \to \mathcal{P}(\omega)$ such that $h(n)' = h(n+1)$, (where equality here is as sets, not only as Turing degrees,) we cannot (Steel [Ste75]).

The most general theorem in this setting is the following one.

THEOREM 3.3 (Hinman and Slaman [HS91]). *Every countable jump partial ordering can be embedded into the Turing Degrees (of course, preserving order and jump).*

The proof is via a complicated forcing construction. Much more that 1-genericity is needed in this case. One needs to consider sets that are arithmetically generic over a Harrison pseudo-hierarchy.

3.4. Jump partial orderings with 0. If we add 0 to the language the problem becomes much more complicated, and very different techniques are required. The reason is that the constructions before used sets which are very generic and very far from arithmetically definable. But now, if for example we have that $x \leqslant_P \mathrm{j}^n(0)$, then we need to map x to a degree below $\boldsymbol{0}^{(n)}$, and hence to a set which is arithmetically definable, with no more than $n+1$ quantifiers.

Hinman and Slaman [HS91] started to look at the quantifier-free 1-types of jump partial orderings with 0 realizable in $\mathcal{D}$. Note that realizing a quantifier-free n-type is equivalent to embedding a jpo $w/0$ and with n many generators. They got some partial results, that were rounded off later by Hinman in [Hin99]. He showed that every quantifier-free 1-type $p(x)$ of jump partial orderings with 0, and with a formula of the form $x \leqslant_P \mathrm{j}^m(0)$, is realizable in $\mathcal{D}$. Then, Montalbán [Mon03], showed the same for 1-types $p(x)$ with no formula of the form $x \leqslant_p \mathrm{j}^m(0)$. Putting these results together we get the following one.

THEOREM 3.4 (Hinman [Hin99], Montalbán [Mon03]). *Every quantifier-free 1-type of jump partial orderings with* 0, *is realizable in* $\mathcal{D}$.

The following question remainds open.

QUESTION 3.5. Which quantifier-free n-type of jpo $w/0$ can be realized in $\mathcal{D}$?

Here is one of the difficulties to solve this question. The main tool used in Hinman's result about 1-types is the Shoenfield [Sho59] jump inversion theorem: If A is c.e.a. B', there there is a set C, c.e.a. B such that $C' \equiv_T A$. (By Y c.e.a. X we mean that Y computable enumerable in X and is Turing above X.) For n-types, we do not have such an inversion theorem. Worst than that, we have Shore's non-inversion theorem [Sho88]: There are sets B_0, B_1 and B_2 c.e. over $0'$ with $0' <_T B_0, B_1 \leqslant_T B_2$, for which there are no sets $A_0, A_1 \leqslant_T A_2 \leqslant_T 0'$ with $A_i' \equiv_T B_i$.

On the positive side, Montalbán [Mon03] showed that if every quantifier free n-type $p(x_1, \ldots, x_n)$ of jpo $w/0$, which contains a formula $x_1 \leqslant \mathrm{j}^m(0)$ & $\cdots$ & $x_n \leqslant \mathrm{j}^m(0)$ for some m, is realized in $\mathcal{D}$, then every quantifier free n-type of jpo $w/0$ is realized in $\mathcal{D}$.

A very nice result is the following one. In [LL96], Lempp and Lerman used their method of Iterated Trees of Strategies and showed that every quantifier free formula $\varphi(x_1, \ldots, x_n)$ consistent with the axioms jpo $w/0$ plus $x_1 \leqslant \mathrm{j}(0)$ & $\cdots$ & $x_n \leqslant \mathrm{j}(0)$ is realizable in $\mathcal{D}$, getting some interesting decidability results as corollaries[1].

3.5. Jump upper semilattices. The reader might be wondering by now what happens if we have both join and jump. We get the following extension of Hinman and Slaman's theorem.

THEOREM 3.6 (Montalbán [Mon03]). *Every countable jump upper semilattice can be embedded into the Turing Degrees (of course, preserving jump and join).*

The proof uses ideas from Hinman and Slaman [HS91], but it also needs a array of new ideas.

OUTLINE OF THE PROOF. The proof has two main steps. First, we introduce the notion of h-embeddable jusl. We say that a jusl $\mathcal{J}$ is *h-embeddable* ("h" for hierarchy) if there is a map $H\colon J \to \mathcal{P}(\omega)$ such that for all $x, y \in J$,

- if $x <_{\mathcal{J}} y$ then $H(x)' \leqslant_T H(y)$,
- $\mathcal{J} \leqslant_T H(y)$, and $\bigoplus_{x \leqslant_{\mathcal{J}} y} H(x) \leqslant_T H(y)$.

We call such a map H, a *jump hierarchy*.

[1] Lerman has recently produced a proof that every finite part of a jpo $w/0$ embeds in $\mathcal{D}$.

Via a forcing construction, we get that for every h-embeddable jusl $\mathcal{J}$, there is an embedding $f\colon \mathcal{J} \to \mathcal{D}$. Essentially, the forcing notion has to make sure that $x \leqslant_{\mathcal{J}} y \Rightarrow f(x) \leqslant_T f(y)$; $f(x \vee y) \equiv_T f(x) \vee f(y)$ and that $f(\mathrm{j}(x)) \leqslant_T (f(x))'$. Genericity is used to ensure that $x \nleqslant_{\mathcal{J}} y \Rightarrow f(x) \nleqslant_T f(y)$ and that $(f(x))' \leqslant_T f(\mathrm{j}(x))$. The jump hierarchy is used for this last reduction, $(f(x))' \leqslant_T f(\mathrm{j}(x))$. The point is to have that for every x, $f(x) \geqslant_T H(x)$, and use $H(\mathrm{j}(x))$ to decode $(f(x))'$ from $f(\mathrm{j}(x))$. There are many subtleties one has to worry about here.

Now we can embed a big family of jusl's. For instance, every well founded jusl is h-embeddable: If $\mathcal{J}$ and the rank function on it are computable in X, take $H(x) = X^{\mathrm{rk}(x)}$. However, there is no reason to believe that every jusl is h-embeddable.

The second step is to prove that every jusl embeds into an h-embeddable one. This part of the proof is more algebraic and uses Harrison linear orderings, Fraïssé limits and well-quasi-orderings. ⊣

As we mentioned right after Definition 3.2, even for simple jpo's such as $\mathbb{Z}$, these embeddings cannot be done inside the hyperarithmetic degrees. For the proof above, again one needs to consider a Harrison pseudo-hierarchy and a set arithmetically generic over it. These sets can be found below Kleene's $\mathcal{O}$, and even hyperarithmetically-low. So we get that every computable jusl embeds in $\mathcal{D}(\leqslant_T$ Kleene's $\mathcal{O})$.

3.6. Jump upper semilattices with 0. The situation when we add 0 to the language is again very different. In this case we get a negative answer right away.

THEOREM 3.7 (Montalbán [Mon03]). *Not every countable jusl $w/0$ can be embedded into $\mathcal{D}$. Indeed, there is a jusl $w/0$ and with only one generator* (*other than* 0) *which cannot be embedded in $\mathcal{D}$.*

IDEA OF THE PROOF. The reason is that there are $2^{\aleph_0}$ many jusl $w/0$ with one generator x satisfying $x \leqslant_{\mathcal{J}} \mathrm{j}^2(0)$. But there are only countably many degrees $\boldsymbol{x} \leqslant_T \boldsymbol{0}''$. ⊣

QUESTION 3.8. Is there a simple (say computable) jusl $w/0$ and with one generator that cannot be embedded into $\mathcal{D}$?

If we do not require the jump operation to be total, it makes sense to talk about finite jusl's. The problem of whether every finite jusl $w/0$ can be embedded into $\mathcal{D}$ is still open. It is believed that a positive answer could be achieved using Lempp and Lerman's method of Iterated Trees of Strategies (see for instance [LL96]). This method gives a general framework to do $0^{(n)}$-priority arguments and is very complicated.

§4. Initial segment embeddings. A completely different family of embeddability results are initial segment embeddings.

There is a long history of results in this area. We mention only some of them. Hugill [Hug69] showed that every countable linear ordering embeds into $\mathcal{D}$ as an initial segment. In [Lac68], Lachlan proved that every countable distributive lattice is isomorphic to an initial segment of $\mathcal{D}$. Then, in [Ler71], Lerman showed the same for every finite usl. A complete characterization of the countable initial segments of $\mathcal{D}$ was later given Lachlan and Lebeuf.

THEOREM 4.1 (Lachlan and Lebeuf [LL76]). *Every countable upper semilattice with least element is isomorphic to an initial segment of $\mathcal{D}$*

These embeddings can be done quite locally, as long as the usl is not too complex. Lerman [Ler83, XII] showed that every countable usl $w/0$ that is computable in $0''$ is isomorphic to an initial segment of $\mathcal{D}$ below $\boldsymbol{0}'$. This result was later extended by Kjos-Hanssen [KH03], who showed that a countable usl $w/0$ is isomorphic to an initial segment of $\mathcal{D}$ below $\boldsymbol{0}'$ if and only if it has a presentation c.e. in $0''$.

The methods used for this kind of results are forcing with computable perfect trees and lattice tables. Forcing with computable perfect trees, or Sacks forcing, was already used in the first construction of a minimal degree by Spector [Spe56], as noticed by Sacks [Sac71]. A more complex class of trees is necessary to get other initial segments results. Lerman's book [Ler83] contains all these embeddability results.

§5. Embeddings of larger structures.

Now we look at uncountable structures. Recall that $\mathcal{D}$ has the countable predecessor property (c.p.p.), and hence any subordering of it has to have it too.

5.1. Partial orderings. The first result of this sort is due to Sacks and the key step of his proof is the following extensions-of-embeddings lemma. The finite version of this lemma is due to Kleene and Post [KP54].

LEMMA 5.1 (Sacks [Sac61]). *Let $\mathcal{P} \subseteq \mathcal{Q}$ be two countable partial orderings such that $\mathcal{P}$ is downward closed in $\mathcal{Q}$ and for every $q \in Q$ we have that every two elements of P below q have an upper bound in P also below q. Then any embedding of $\mathcal{P}$ into $\mathcal{D}$ extends to an embedding of $\mathcal{Q}$ into $\mathcal{D}$.*

Sacks actually proved a slightly stronger lemma where $|\mathcal{P}| < 2^{\aleph_0}$, $|\mathcal{Q} \setminus \mathcal{P}| \leqslant \aleph_0$, and $\mathcal{Q}$ has the c.p.p.

THEOREM 5.2 (Sacks [Sac61]). *Every partial ordering of size $\aleph_1$ with the c.p.p. can be embedded into $\mathcal{D}$.*

PROOF. First extend the partial ordering to an usl, also with the c.p.p., and then decompose it as an increasing union of countable partial orderings so that we can apply the lemma above. ⊣

He also showed that there is a maximal independent set of degrees size $2^{\aleph_0}$. That is, a set $\{\boldsymbol{x}_\xi : \xi < 2^{\aleph_0}\}$ such that for every $\xi_0, \dots, \xi_k \in 2^{\aleph_0}$, $\boldsymbol{x}_{\xi_0} \not\leq_T (\boldsymbol{x}_{\xi_1} \vee \dots \vee \boldsymbol{x}_{\xi_k})$ (unless, of course, $\xi_0 = \xi_i$ for some $i = 1, \dots, k$). It followed that every partial ordering with the *finite predecessor property* and size $2_0^{\aleph}$ embeds into $\mathcal{D}$.

He made the following conjecture which is still unsolved.

CONJECTURE 5.3 (Sacks [Sac63]). Every partial ordering of size $2^{\aleph_0}$ with the c.p.p. embeds into $\mathcal{D}$.

Of course, the affirmative answer is consistent with ZFC, as it is implied by the theorem above if $2^{\aleph_0} = \aleph_1$. However, the lemma used to show Theorem 5.2 cannot be extended to higher cardinalities in ZFC. Groszek and Slaman [GS83] showed that it is consistent with ZFC that $2^{\aleph_0} \geqslant \aleph_2$ and there is an independent set of size $\aleph_2$ which cannot be extended to a larger independent set. In contrast, Simpson [Sim77] pointed out that if Martin's Axiom holds at κ, then there is no maximal independent set of size κ. (See, for instance, [Kun80] for information on Martin's Axiom.)

5.2. Upper semilattices. Sacks theorem was improved by Abraham and Shore the following way.

THEOREM 5.4 (Abraham and Shore [AS86]). *Every usl of size $\aleph_1$ with the c.p.p. and with* 0 *is isomorphic to an initial segment of* $\mathcal{D}$.

With respect to usl embeddings, this is as far we can go in ZFC. Slaman and Groszek [GS83] show that there is a model of ZFC where $2^{\aleph_0} \geqslant \aleph_2$, and there is an usl of size $\aleph_2$ with the c.p.p. which does not embed into $\mathcal{D}$ preserving joins.

5.3. Jump partial orderings. If we add jump to the language, we get the following negative result.

THEOREM 5.5 (Montalbán [Mon03]). *There is a jpo of size $2^{\aleph_0}$ and with the c.p.p. which cannot be embedded into $\mathcal{D}$.*

Montalbán [Mon03] also showed that if Martin's Axiom holds at κ, then every jusl with the c.p.p. and size $\leqslant \kappa$ can be embedded into $\mathcal{D}$. As a corollary we get that whether every jpo (or jusl) with the c.p.p. and size $\aleph_1$ is embeddable into $\mathcal{D}$ or not is independent of ZFC. It is false if $\aleph_1 = 2^{\aleph_0}$ and true if Martin's Axiom holds at $\aleph_1$.

§6. GH-embeddings. There are other very meaningful predicates on the Turing degrees that are defined in terms of $\leqslant_T$, $\vee$ and $'$. To understand these predicates better, we now look a embeddings of structures which preserve them.

6.1. The low/high hierarchy. The degrees below $\mathbf{0}'$ are classified depending on how close they are to being computable or how close they are to being complete (i.e. to compute $\mathbf{0}'$) via the low/high hierarchy. This classification has been extremely useful in the study of the degrees below $\mathbf{0}'$. We say that a degree $\mathbf{a} \leqslant_T \mathbf{0}'$ is *low*, if its jump is as low as it could be, that is, if $\mathbf{a}' \equiv_T \mathbf{0}'$. We say that $\boldsymbol{a} \leqslant_T \boldsymbol{0}'$ is *high*, if its jump is as high as it could be, that is, if $\boldsymbol{a}' \equiv_T \boldsymbol{0}''$. More generally:

DEFINITION 6.1 (Soare [Soa74], Cooper [Coo74]). A Turing degree $\boldsymbol{a} \leqslant_T \boldsymbol{0}'$ is

- low_n (L_n) if $\boldsymbol{a}^{(n)} = \boldsymbol{0}^{(n)}$.
- high_n (H_n) if $\boldsymbol{a}^{(n)} = \boldsymbol{0}^{(n+1)}$.
- intermediate (I) if $\forall n\ (\boldsymbol{0}^{(n)} <_T \boldsymbol{a}^{(n)} <_T \boldsymbol{0}^{(n+1)})$.

Note that for each n, $L_n \subseteq L_{n+1}$, $H_n \subseteq H_{n+1}$, and L_n, H_n and I are disjoint. These classes induce a partition, $\mathcal{C}^*$, of the degrees $\leqslant \boldsymbol{0}'$.

$$\mathcal{C}^* = \{L_1^*, L_2^*, \dots\} \cup \{I^*\} \cup \{H_1^*, H_2^*, \dots\},$$

where $L_1^* = L_1$, $H_1^* = H_1$, $I^* = I$ and for $n > 1$, $L_n^* = L_n \setminus L_{n-1}$, and $H_n^* = H_n \setminus H_{n-1}$. We define an ordering, $\prec$, on $\mathcal{C}^*$ as follows:

$$L_1^* \prec L_2^* \prec \cdots \prec I^* \prec \cdots \prec H_2^* \prec H_1^*.$$

It follows from the monotonicity of the jump that if $\boldsymbol{x} \leqslant_T \boldsymbol{y}$, $\boldsymbol{x} \in X \in \mathcal{C}^*$ and $\boldsymbol{y} \in Y \in \mathcal{C}^*$, then $X \preceq Y$. The following theorem of Lerman's helps us to understand how the degrees in the different classes of the hierarchy are located.

DEFINITION 6.2. An *H-poset* is a structure $\mathcal{P} = \langle P, \leqslant, 0, 1, f(\cdot)\rangle$, where $\langle P, \leqslant\rangle$ is a partial ordering, 0 and 1 are the least and greatest elements respectively, and f is a labeling function from P to $\mathcal{C}^*$ such that for every $x, y \in P$,

$$x \leqslant y \Rightarrow f(x) \preceq f(y),$$

$f(0) = L_1^*$ and $f(1) = H_1^*$.

THEOREM 6.3 (Lerman [Ler85]). *Every finite H-poset can be embedded into* $\mathcal{D}$ (*of course, preserving labels*).

6.2. The Generalized low/high hierarchy. As a generalization of this hierarchy to all the Turing degrees we get the *generalized high/low hierarchy*. In [JP78], Jockusch and Posner defined the *generalized high/low hierarchy* with the intention of classifying all the Turing degrees depending on how close a degree is to being computable, and on how close it is to computing the Halting Problem. This classification coincides with the High/Low hierarchy on the degrees below $\boldsymbol{0}'$.

DEFINITION 6.4. For $n \geqslant 1$ we say that a degree $\boldsymbol{x}$ is *generalized* $\textit{low}_n$, or GL_n, if $\boldsymbol{x}^{(n)} = (\boldsymbol{x} \vee 0')^{(n-1)}$. We say that a degree $\boldsymbol{x}$ is a *generalized* $\textit{high}_n$ *degree*, or GH_n, if $\boldsymbol{x}^{(n)} = (\boldsymbol{x} \vee 0')^{(n)}$, and it is *generalized intermediate*, or GI, if $\forall n\big((\boldsymbol{x} \vee 0')^{(n-1)} <_T \boldsymbol{x}^{(n)} <_T (\boldsymbol{x} \vee 0')^{(n)}\big)$.

This classification has also been very useful in the study of $\mathcal{D}$. Many order-theoretic properties of $\boldsymbol{0}'$ have been proven to hold for the members in the higher classes of this hierarchy. For instance, every non-GL_2 cups to every degree above it [JP78]; every GH_1 degree bounds a minimal degree [Joc77], but not every GH_2 does [Ler86]; and every GH_1 degree has the complementation property [GMS04]. Also, degrees that should not contain much information appear in the lower classes: every 1-generic set is GL_1 (see [Ler83, IV.2]); every 2-random real is GL_1 [Kau91]; every minimal degree is GL_2 [JP78]. So, one could argue that degrees in the upper classes of this hierarchy are more complex than the ones in the lower classes. One would think that generalized high degrees should be above generalized low degrees, or at least not below. However, there are generalized low degrees which compute generalized high degrees. (Take an H_1 degree $\boldsymbol{x} <_T \boldsymbol{0}'$. By the Posner and Robinson join theorem relative to $\boldsymbol{x}$ [PR81], there exists $\boldsymbol{y} >_T \boldsymbol{x}$ with $\boldsymbol{y} \vee \boldsymbol{0}' = \boldsymbol{y}' = \boldsymbol{x}' = \boldsymbol{0}''$. So we get that $\boldsymbol{y} >_T \boldsymbol{x}$, $\boldsymbol{x}$ is GH_1 and $\boldsymbol{y}$ is GL_1.) Moreover, we have the worst situation possible in this respect:

Let

$$\mathcal{G}^* = \{GL_1^*, GL_2^*, \dots\} \cup \{GI^*\} \cup \{GH_1^*, GH_2^*, \dots\},$$

where $GL_1^* = GL_1$, $GH_1^* = GH_1$, $GI^* = GI$ and for $n > 1$, $GL_n^* = GL_n \smallsetminus GL_{n-1}$, and $GH_n^* = GH_n \smallsetminus GH_{n-1}$.

A *GH-poset* is a structure $\mathcal{P} = \langle P, \leqslant, 0, f(\cdot)\rangle$, where $\langle P, \leqslant\rangle$ is a partial ordering, 0 is the least element and f is a function from P to $\mathcal{C}^*$ such that $f(0) = GL_1^*$. Note that no condition at all is imposed on the labels of a GH-poset except for $f(0) = GL_1^*$.

THEOREM 6.5 (Montalbán [Mon06]). *Every finite GH-poset can be embedded into $\mathcal{D}$.*

§7. **Decidability.** It is impossible to talk about embeddings, extensions of embeddings and initial segment results without mentioning decidability results. For instance, since every finite distributive lattice embeds into $\mathcal{D}$ as an initial segment [Lac68], we can reduce the theory of distributive lattices to the theory of $\langle \boldsymbol{D}, \leqslant_T\rangle$ (by quantifying over all the top elements of initial segments of $\mathcal{D}$ which are distributive). Since the theory of distributive lattices is undecidable (Ervsov and Tauĭclin [ET63]), we get the following theorem:

THEOREM 7.1 (Lachlan [Lac68]). *The theory of $\langle \boldsymbol{D}, \leqslant_T\rangle$ is undecidable.*

However, if we restrict ourself to certain classes of formulas, many decidability results have been proved. The next question is what fragments of the theory of $\langle \boldsymbol{D}, \leqslant_T\rangle$ are decidable.

7.1. Existential theories. The decidability of existential theories is closely related to embeddability results.

LEMMA 7.2. *Let $\mathcal{L}$ be a finite relational language, and let $\mathcal{F}$ be an $\mathcal{L}$-structure. Then the following are equivalent*

(1) *The $\exists$-theory of $\mathcal{F}$ in the language $\mathcal{L}$ is decidable;*
(2) *There is an algorithm that decides which finite $\mathcal{L}$-structures can be embedded into $\mathcal{F}$.*

PROOF. For the implication (1)$\Rightarrow$(2) note that for each finite $\mathcal{L}$-structure there is an existential formula in $\mathcal{L}$ which holds in $\mathcal{F}$ if and only if $\mathcal{P}$ embeds in $\mathcal{F}$.

For the other direction consider a existential sentence φ of $\mathcal{L}$. We can write φ as a disjunction of formulas of the form $\psi_j = \exists x_1, \dots, x_k\ (\varphi_{j,1} \ \&\ \cdots\ \&\ \varphi_{j,n})$, where each $\varphi_{j,i}$ is a literal (either an atomic formula or a negation of one). Note that $\mathcal{F} \models \psi_j$ if and only if there is an $\mathcal{L}$-structure of k elements which satisfies ψ_j and embeds into $\mathcal{F}$. All we have to do now is check all the $\mathcal{L}$-structures of size k. ⊣

We can now apply the embeddability results of Section 3 to get decidability results.

COROLLARY 7.3 (Kleene and Post [KP54]). *The $\exists$-theory of $\langle \boldsymbol{D}(\leqslant_T \boldsymbol{0}'), \leqslant_T, \vee\rangle$ is decidable.*

PROOF. Think of $\vee$ as a 3-ary relation and let $\mathcal{L}$ be the language with $\leqslant_T$ and $\vee$. From [KP54] it follows that a finite $\mathcal{L}$-structure embeds into $\mathcal{D}(\leqslant_T \boldsymbol{0}')$ if and only if it is a partial upper semilattice. ⊣

COROLLARY 7.4 (Montalbán [Mon03]). *The $\exists$-theory of $\mathcal{D} = \langle \boldsymbol{D}, \leqslant_T, \vee, ' \rangle$ is decidable.*

The results in Section 6 imply the following decidability results.

THEOREM 7.5. *(Lerman [Ler85]) The $\exists$-theory of $\langle \boldsymbol{D}, \leqslant_T, \boldsymbol{0}, \boldsymbol{0}', \mathcal{L}_1, \mathcal{L}_2, \dots, \mathcal{I}, \dots, \mathcal{H}_2, \mathcal{H}_1\rangle$ is decidable. (Montalbán [Mon06]) The $\exists$ theory of $\langle \boldsymbol{D}, \leqslant_T, \boldsymbol{0}, GL_1, GL_2, \dots, GI, \dots, GH_2, GH_1\rangle$ is decidable.*

7.2. Two quantifier theories and extensions of embeddings. When we look at $\forall\exists$-theories, more than embeddability results, we need extension of embedding results.

LEMMA 7.6. *Let $\mathcal{L}$ be a finite relational language, and let $\mathcal{F}$ be an $\mathcal{L}$-structure. Then, the following are equivalent*

(1) *The $\forall\exists$-theory of $\mathcal{F}$ in the language $\mathcal{L}$ is decidable;*
(2) *There is an algorithm such that given finite $\mathcal{L}$-structures $\mathcal{P}, \mathcal{Q}_1, \dots, \mathcal{Q}_l$, with $\mathcal{P} \subseteq \mathcal{Q}_i$ for each i, it decides whether every embedding of $\mathcal{P}$ into $\mathcal{F}$ extends to an embedding of $\mathcal{Q}_i$ for some $i \leqslant l$.*

We leave the proof to the reader.

To get such an algorithm to solve the finite-extensions-of-embeddings problem in $\langle \boldsymbol{D}, \leqslant_T\rangle$, the two main ingredients are: Kleene and Post [KP54] finite version of Lemma 5.1; and Lerman's theorem [Ler71] that every finite usl $\mathcal{P}$ embeds into $\mathcal{D}$ as an initial segment. This was used independently by Shore [Sho78] and Lerman [Ler83, VII.4] to get that the $\forall\exists$-theory of $\langle \boldsymbol{D}, \leqslant_T \rangle$ is decidable. But Kleene and Post's [KP54] finite-extensions-of-embeddings result is not sufficient to get the $\forall\exists$-theory of $\langle \boldsymbol{D}, \leqslant_T, \vee \rangle$. Jockusch and Slaman [JS83] used a different forcing technique to prove that if $\mathcal{P}$ and $\mathcal{Q}$ are countable usl's $w/0$ and $\mathcal{P}$ is downward closed in $\mathcal{Q}$, then every usl-embedding of $\mathcal{P}$ into $\mathcal{D}$ extends to an embedding of $\mathcal{Q}$ into $\mathcal{D}$. The finite version of this result, together with Lerman [Ler71] initial segment theorem, gives us the algorithm needed in (2).

THEOREM 7.7 (Jockusch and Slaman [JS83]). *The $\forall\exists$-theory of $\langle \boldsymbol{D}, \leqslant_T, \vee \rangle$ is decidable.*

The situation below $\boldsymbol{0}'$ is more complicated. Lerman and Shore [LS88] showed that the $\forall\exists$-theory of $\langle \boldsymbol{D}(\leqslant_T \boldsymbol{0}'), \leqslant_T \rangle$ is decidable. However, the following question is still open.

QUESTION 7.8. Is the $\forall\exists$-theory of $\langle \boldsymbol{D}(\leqslant_T \boldsymbol{0}'), \leqslant_T, \vee \rangle$ decidable?

Montalbán (2003, unpublished) made the following observation, which shows that, to solve the question above, it will be necessary to have more than an extensions-of-embeddings result like the one used for $\langle \boldsymbol{D}, \leqslant_T, \vee \rangle$ where only one $\mathcal{Q}$ is considered. For every $\boldsymbol{x}_1$ and $\boldsymbol{x}_2$ with $0 <_T \boldsymbol{x}_1 <_T \boldsymbol{x}_2 <_T \boldsymbol{0}'$, either there exists $\boldsymbol{y}$ such that $\boldsymbol{0} <_T \boldsymbol{y} <_T \boldsymbol{x}_1$, or there exists $\boldsymbol{y}$ such that $\boldsymbol{x}_1 <_T \boldsymbol{y} <_T \boldsymbol{0}'$ and $\boldsymbol{y} \vee \boldsymbol{x}_2 \equiv_T \boldsymbol{0}'$, but neither disjunct holds for every such $\boldsymbol{x}_1, \boldsymbol{x}_2$. (If $\boldsymbol{x}_1$ is a minimal degree, $\boldsymbol{0}'$ is high relative to it, and the existence of $\boldsymbol{y}$ follows from Posner and Robinson's join theorem [PR81]. To get $\boldsymbol{x}_1$ and $\boldsymbol{x}_2$ for which a $\boldsymbol{y}$ of the second type does not exists consider and c.e. operator which constructs a c.e. degree without the join property and then use Jockusch and Shore's pseudo-jump inversion theorem [JS83].)

As in a side, we should mention that it is also unknown whether the $\forall\exists$-theory of $\langle \mathcal{R} \leqslant_T, \vee \rangle$ is decidable, where $\mathcal{R}$ is the set of degrees of c.e. sets.

In larger languages, we do get to the boundary of decidability at the two quantifier level. For the theory of $\langle \boldsymbol{D}, \leqslant_T, \vee, ' \rangle$, Montalbán's result on its decidability is as far we can get.

THEOREM 7.9 (Slaman and Shore [SS06]). *The $\forall\exists$-theory of $\langle \boldsymbol{D}, \leqslant_T, \vee, ' \rangle$ is undecidable.*

We also get undecidability if we add greatest lower bounds instead of jump.

THEOREM 7.10 (Miller, Nies and Shore [MNS04]). *The $\forall\exists$-theory of $\langle \boldsymbol{D}, \leqslant_T, \vee, \wedge \rangle$ is undecidable, where $\wedge$ is any total extension of the infimum relation.*

The following seems to be a difficult open question.

QUESTION 7.11. Is the $\forall\exists$-theory of $\langle \boldsymbol{D}, \leqslant_T, ' \rangle$ decidable?

A positive answer would give a decidability procedure for $\exists$-theory of $\langle \boldsymbol{D}, \leqslant_T, ', \boldsymbol{0}\rangle$.

7.3. Other results. Three quantifiers is the end of the story in term of decidability results. Schmerl (see [Ler83, VII.4.6]; the proof there needs a small correction) extended Lachlan's Theorem 7.1, and showed that the $\forall\exists\forall$-theory of $\langle \boldsymbol{D}, \leqslant_T\rangle$ is undecidable.

A quite interesting result about the theory of $\langle \boldsymbol{D}, \leqslant_T\rangle$ is that it is Turing (actually one-to-one) equivalent to true second order arithmetic (Simpson [Sim77]). Shore [Sho81], then proved that the theory of $\langle \boldsymbol{D}(\leqslant_T \boldsymbol{0}'), \leqslant_T\rangle$ is Turing (actually one-to-one) equivalent to true first order arithmetic. Moreover, also in [Sho81], he proved this result for $\langle \boldsymbol{D}(\leqslant_T \boldsymbol{a}), \leqslant_T\rangle$ where $\boldsymbol{a}$ is arithmetic and above $\boldsymbol{0}'$, computable enumerable, or high. Greenberg and Montalbán [GM03] extended this result to $\boldsymbol{a}$ n-CEA, 1-generic and below $\boldsymbol{0}'$, 2-generic and arithmetic, or arithmetically generic.

Acknowledgments. The author's contribution to the field is contained in his Ph.D. thesis [Mon05], written under the supervision of Richard A. Shore, to whom he is very thankful. Part of this paper was written while visiting the National University of Singapore being partially supported by Grant NUS-WBS-146-000-054-123. I would like to thank them for their hospitality. This work is partially supported by NSF Grant DMS-0600824 and by the the Marsden Found of New Zealand, via postdoctoral fellowship. I would also like to thank Sheila Miller and Rod Downey for proofreading this paper.

REFERENCES

[AS86] U. Abraham and R. A. Shore, *Initial segments of the degrees of size $\aleph_1$*, ***Israel Journal of Mathematics***, vol. 53 (1986), no. 1, pp. 1–51.

[ASF] K. Ambos-Spies and P. A. Fejer, *Degrees of unsolvability*, to appear.

[Coo74] S. B. Cooper, *Minimal pairs and high recursively enumerable degrees*, ***The Journal of Symbolic Logic***, vol. 39 (1974), pp. 655–660.

[DJS96] R. Downey, C. G. Jockusch, and M. Stob, *Array nonrecursive degrees and genericity*, ***Computability, Enumerability, Unsolvability***, London Mathematical Society Lecture Note Series, vol. 224, Cambridge University Press, Cambridge, 1996, pp. 93–104.

[EP70] H. B. Enderton and H. Putnam, *A note on the hyperarithmetical hierarchy*, ***The Journal of Symbolic Logic***, vol. 35 (1970), pp. 429–430.

[ET63] Ju. L. Eršov and M. A. Taĭclin, *Undecidability of certain theories*, ***Akademiya Nauk SSSR. Sibirskoe Otdelenie. Institut Matematiki. Algebra i Logika***, vol. 2 (1963), no. 5, pp. 37–41.

[Fej89] P. A. Fejer, *Embedding lattices with top preserved below non-*GL_2 *degrees*, ***Zeitschrift für Mathematische Logik und Grundlagen der Mathematik***, vol. 35 (1989), no. 1, pp. 3–14.

[Fri57] R. Friedberg, *A criterion for completeness of degrees of unsolvability*, ***The Journal of Symbolic Logic***, vol. 22 (1957), pp. 159–160.

[GM03] N. Greenberg and A. Montalbán, *Embedding and coding below a 1-generic degree*, ***Notre Dame Journal of Formal Logic***, vol. 44 (2003), no. 4, pp. 200–216 (electronic) (2004).

[GMS04] N. Greenberg, A. Montalbán, and R. A. Shore, *Generalized high degrees have the complementation property*, ***The Journal of Symbolic Logic***, vol. 69 (2004), no. 4, pp. 1200–1220.

[GS83] M. J. GROSZEK and T. A. SLAMAN, *Independence results on the global structure of the Turing degrees*, ***Transactions of the American Mathematical Society***, vol. 277 (1983), no. 2, pp. 579–588.

[Har68] J. HARRISON, *Recursive pseudo-well-orderings*, ***Transactions of the American Mathematical Society***, vol. 131 (1968), pp. 526–543.

[Hin99] P. G. HINMAN, *Jump traces with large gaps*, ***Recursion Theory and Complexity (Kazan, 1997)*** (M. M. Arslanov and S. Lempp, editors), de Gruyter Series in Logic and Its Applications, vol. 2, de Gruyter, Berlin, 1999, pp. 71–80.

[HS91] P. G. HINMAN and T. A. SLAMAN, *Jump embeddings in the Turing degrees*, ***The Journal of Symbolic Logic***, vol. 56 (1991), no. 2, pp. 563–591.

[Hug69] D. F. HUGILL, *Initial segments of Turing degrees*, ***Proceedings of the London Mathematical Society. Third Series***, vol. 19 (1969), pp. 1–16.

[Joc77] C. G. JOCKUSCH, JR., *Simple proofs of some theorems on high degrees of unsolvability*, ***Canadian Journal of Mathematics. Journal Canadien de Mathématiques***, vol. 29 (1977), no. 5, pp. 1072–1080.

[JP78] C. G. JOCKUSCH, JR. and D. B. POSNER, *Double jumps of minimal degrees*, ***The Journal of Symbolic Logic***, vol. 43 (1978), no. 4, pp. 715–724.

[JS83] C. G. JOCKUSCH, JR. and R. A. SHORE, *Pseudojump operators. I. The r.e. case*, ***Transactions of the American Mathematical Society***, vol. 275 (1983), no. 2, pp. 599–609.

[Kau91] S. KAUTZ, ***Degrees of Random Sets***, Ph.D. thesis, Cornell University, Ithaca, New York, 1991.

[KH03] B. KJOS-HANSSEN, *Local initial segments of the Turing degrees*, ***The Bulletin of Symbolic Logic***, vol. 9 (2003), no. 1, pp. 26–36.

[Kle55] S. C. KLEENE, *Hierarchies of number-theoretic predicates*, ***Bulletin of the American Mathematical Society***, vol. 61 (1955), pp. 193–213.

[KP54] S. C. KLEENE and E. L. POST, *The upper semi-lattice of degrees of recursive unsolvability*, ***Annals of Mathematics. Second Series***, vol. 59 (1954), pp. 379–407.

[Kum00] M. KUMABE, *A 1-generic degree with a strong minimal cover*, ***The Journal of Symbolic Logic***, vol. 65 (2000), no. 3, pp. 1395–1442.

[Kun80] K. KUNEN, ***Set Theory: An Introduction to Independence Proofs***, Studies in Logic and the Foundations of Mathematics, vol. 102, North-Holland, Amsterdam, 1980.

[Lac68] A. H. LACHLAN, *Distributive initial segments of the degrees of unsolvability*, ***Zeitschrift für Mathematische Logik und Grundlagen der Mathematik***, vol. 14 (1968), pp. 457–472.

[LL76] A. H. LACHLAN and R. LEBEUF, *Countable initial segments of the degrees of unsolvability*, ***The Journal of Symbolic Logic***, vol. 41 (1976), no. 2, pp. 289–300.

[Lac54] D. LACOMBE, *Sur le semi-réseau constitué par les degrés d'indécidabilité récursive*, ***Les Comptes rendus de l'Académie des sciences Paris***, vol. 239 (1954), pp. 1108–1109.

[LL96] S. LEMPP and M. LERMAN, *The decidability of the existential theory of the poset of recursively enumerable degrees with jump relations*, ***Advances in Mathematics***, vol. 120 (1996), no. 1, pp. 1–142.

[LLS06] S. LEMPP, M. LERMAN, and R. SOLOMON, *Embedding finite lattices into the computably enumerable degrees — a status survey*, ***Logic Colloquium '02*** (Zoe Chatzidakis, Peter Koepke, and Wolfram Pohlers, editors), Lecture Notes in Logic, vol. 27, ASL, La Jolla, CA, 2006, pp. 206–229.

[Ler71] M. LERMAN, *Initial segments of the degrees of unsolvability*, ***Annals of Mathematics. Second Series***, vol. 93 (1971), pp. 365–389.

[Ler83] ———, ***Degrees of Unsolvability: Local and Global Theory***, Perspectives in Mathematical Logic, Springer-Verlag, Berlin, 1983.

[Ler85] ———, *On the ordering of classes in high/low hierarchies*, ***Recursion Theory Week (Oberwolfach, 1984)***, Lecture Notes in Mathematics, vol. 1141, Springer, Berlin, 1985, pp. 260–270.

[Ler86] ———, *Degrees which do not bound minimal degrees*, ***Annals of Pure and Applied Logic***, vol. 30 (1986), no. 3, pp. 249–276.

[LS88] M. LERMAN and R. A. SHORE, *Decidability and invariant classes for degree structures*, ***Transactions of the American Mathematical Society***, vol. 310 (1988), no. 2, pp. 669–692.

[MNS04] R. G. MILLER, A. O. NIES, and R. A. SHORE, *The* $\forall\exists$*-theory of* $\mathcal{R}(\leqslant, \vee, \wedge)$ *is undecidable*, ***Transactions of the American Mathematical Society***, vol. 356 (2004), no. 8, pp. 3025–3067.

[Mon05] A. MONTALBÁN, ***Beyond the Arithmetic***, Ph.D. thesis, Cornell University, Ithaca, New York, 1991.

[Mon03] ———, *Embedding jump upper semilattices into the Turing degrees*, ***The Journal of Symbolic Logic***, vol. 68 (2003), no. 3, pp. 989–1014.

[Mon06] ———, *There is no ordering on the classes in the generalized high/low hierarchies*, ***Archive for Mathematical Logic***, vol. 45 (2006), no. 2, pp. 215–231.

[PR81] D. B. POSNER and R. W. ROBINSON, *Degrees joining to* $\mathbf{0}'$, ***The Journal of Symbolic Logic***, vol. 46 (1981), no. 4, pp. 714–722.

[Pos44] E. L. POST, *Recursively enumerable sets of positive integers and their decision problems*, ***Bulletin of the American Mathematical Society***, vol. 50 (1944), pp. 284–316.

[Sac61] G. E. SACKS, *On suborderings of degrees of recursive unsolvability*, ***Zeitschrift für Mathematische Logik und Grundlagen der Mathematik***, vol. 7 (1961), pp. 46–56.

[Sac63] ———, ***Degrees of Unsolvability***, Princeton University Press, Princeton, NJ, 1963.

[Sac71] ———, *Forcing with perfect closed sets*, ***Axiomatic Set Theory (Proc. Sympos. Pure Math., Vol. XIII, Part I, Univ. California, Los Angeles, Calif., 1967)***, AMS, Providence, R.I., 1971, pp. 331–355.

[Sho59] J. R. SHOENFIELD, *On degrees of unsolvability*, ***Annals of Mathematics. Second Series***, vol. 69 (1959), pp. 644–653.

[Sho78] R. A. SHORE, *On the* $\forall\exists$*-sentences of* α*-recursion theory*, ***Generalized Recursion Theory, Vol. II (Proceedings of the Second Symposium, University of Oslo, Oslo, 1977)***, Studies in Logic and the Foundations of Mathematics, vol. 94, North-Holland, Amsterdam, 1978, pp. 331–353.

[Sho81] ———, *The theory of the degrees below* $\mathbf{0}'$, ***The Journal of the London Mathematical Society. Second Series***, vol. 24 (1981), no. 1, pp. 1–14.

[Sho82] ———, *Finitely generated codings and the degrees r.e. in a degree* $\mathbf{d}$, ***Proceedings of the American Mathematical Society***, vol. 84 (1982), no. 2, pp. 256–263.

[Sho88] ———, *A noninversion theorem for the jump operator*, ***Annals of Pure and Applied Logic***, vol. 40 (1988), no. 3, pp. 277–303.

[Sho06] ———, *Degree structures: local and global investigations*, ***The Bulletin of Symbolic Logic***, vol. 12 (2006), no. 3, pp. 369–389.

[SS06] R. A. SHORE and T. A. SLAMAN, *The* $\forall\exists$ *theory of* $\mathcal{D}(\leqslant, \vee, ')$ *is undecidable*, ***Logic Colloquium '03*** (V. Stoltenberg-Hansen and J. Väänänen, editors), Lecture Notes in Logic, vol. 24, ASL, La Jolla, CA, 2006, pp. 326–344.

[Sim77] S. G. SIMPSON, *First-order theory of the degrees of recursive unsolvability*, ***Annals of Mathematics. Second Series***, vol. 105 (1977), no. 1, pp. 121–139.

[Soa74] R. I. SOARE, *Automorphisms of the lattice of recursively enumerable sets*, ***Bulletin of the American Mathematical Society***, vol. 80 (1974), pp. 53–58.

[Soa87] ———, ***Recursively Enumerable Sets and Degrees***, Perspectives in Mathematical Logic, Springer-Verlag, Berlin, 1987, A study of computable functions and computably generated sets.

[Spe56] C. SPECTOR, *On degrees of recursive unsolvability*, ***Annals of Mathematics. Second Series***, vol. 64 (1956), pp. 581–592.

[Ste75] J. STEEL, *Descending sequences of degrees*, ***The Journal of Symbolic Logic***, vol. 40 (1975), no. 1, pp. 59–61.

[Tur39] A. M. TURING, *Systems of logic based on ordinals*, ***Proceedings of the London Mathematical Socicety. Second Series***, vol. 45 (1939), pp. 161–228.

DEPARTMENT OF MATHEMATICS
UNIVERSITY OF CHICAGO
5734 S. UNIVERSITY AVE
CHICAGO, IL 60637, USA
E-mail: antonio@math.uchicago.edu
URL: http://www.math.uchicago.edu/~antonio/index.html

RANDOMNESS — BEYOND LEBESGUE MEASURE

JAN REIMANN

Abstract. Much of the recent research on algorithmic randomness has focused on randomness for Lebesgue measure. While, from a computability theoretic point of view, the picture remains unchanged if one passes to arbitrary computable measures, interesting phenomena occur if one studies the the set of reals which are random for an arbitrary (continuous) probability measure or a generalized Hausdorff measure on Cantor space. This paper tries to give a survey of some of the research that has been done on randomness for non-Lebesgue measures.

§1. **Introduction.** Most studies on algorithmic randomness focus on reals random with respect to the uniform distribution, i.e. the $(1/2, 1/2)$-Bernoulli measure, which is measure theoretically isomorphic to Lebesgue measure on the unit interval. The theory of uniform randomness, with all its ramifications (e.g. computable or Schnorr randomness) has been well studied over the past decades and has led to an impressive theory.

Recently, a lot of attention focused on the interaction of algorithmic randomness with recursion theory: What are the computational properties of random reals? In other words, which computational properties hold effectively for almost every real? This has led to a number of interesting results, many of which will be covered in a forthcoming book by Downey and Hirschfeldt [14].

While the understanding of "holds effectively" varied in these results (depending on the underlying notion of randomness, such as computable, Schnorr, or weak randomness, or various arithmetic levels of Martin-Löf randomness, to name only a few), the meaning of "for almost every" was usually understood with respect to Lebesgue measure. One reason for this can surely be seen in the fundamental relation between uniform Martin-Löf tests and descriptive complexity in terms of (prefix-free) Kolmogorov complexity: A real is not covered by any Martin-Löf test (with respect to the uniform distribution) if and only if all of its initial segments are incompressible (up to a constant additive factor).

However, one may ask what happens if one changes the underlying measure. This question is virtually as old as the theory of randomness. Martin-Löf [45] defined randomness not only for Lebesgue measure but also for arbitrary

Logic Colloquium '06
Edited by Barry Cooper, Herman Geuvers, Anand Pillay, and Jouko Väänänen
Lecture Notes in Logic, 32

Bernoulli distributions. Levin's contributions in the 1970's [35, 36, 37, 78] extended this to arbitrary probability measures. His framework of *semimeasures* provided an elegant uniform approach, flanked by a number of remarkable results and principles such as the existence of uniform tests, conservation of randomness, and the existence of neutral measures. This essentially defined the "Russian" school of randomness in succession of Kolmogorov, to which Gacs, Muchnik, Shen, Uspensky, Vyugin, and many others have contributed.

Recently, partly driven by Lutz's introduction of effective fractal dimension concepts [42] and their fundamental connection with Kolmogorov complexity, interest in non-uniform randomness began to grow "outside" the Russian school, too. It very much seems that interesting mathematics arises out of combining non-uniform randomness and logical/computational complexity in the same way it did for Lebesgue measure.

The purpose of this article is to complement recent, far more complex endeavours of capturing research on randomness and computability (such as [14] or [52]) by focusing on randomness for non-Lebesgue measures. Of course, even this restricted plan is far too comprehensive for a rather short article. Therefore, the present paper tries to achieve mostly three things: First, give a uniform framework for randomness, in spirit of the Russian school, that however is not necessarily restricted to probability measures but also incorporates Hausdorff measures, and, besides, is presented in a style more approachable to readers unfamiliar with the Russian tradition. Second, apply this framework to present results that seem important to the recent lines of research on randomness. And third, give an overview over very recent results. For this purpose, a lot of work had to be left out, most notably Levin's and Gacs' research on uniform frameworks for randomness for probability measures. Also, we exclusively deal with randomness for reals, i.e. binary infinite sequences. The reader may refer to Gacs' recent paper [20] for randomness for probability measures on metric spaces.

It is the hope of the author that the restrictions pay off and the reader will find this presentation a stimulating description of fascinating topics.

The article is organized as follows. Section 2 deals with outer measures on Cantor space 2^{ω}. Our exposition adapts Rogers' presentation [63] based on *premeasures* to define a general framework for randomness. Section 3 presents the generalized theory of Martin-Löf randomness, extended to arbitrary outer measures. Section 4 then goes on to describe results in connection with *computable probability measures*, the "oldest" area of research on non-uniform randomness. This is followed in Section 5 by more recent results on *Hausdorff measures*. Finally, Section 6 surveys very recent research on randomness for arbitrary probability, results which indicate that questions on *algorithmic* randomness can extend far into other realms of mathematical logic.

I have tried to give short sketches of proofs wherever possible. A reader with some acquaintance with algorithmic randomness should be able to reconstruct most proofs. Throughout the text we assume some basic knowledge in mathematical logic, especially in computability theory.

§2. **Measures on cantor space.** In this section we introduce the basic notions of measure on the Cantor space 2^ω. We make use of the special topological structure of 2^ω to give a unified treatment of a large class of measures, not necessarily σ-finite. We follow Rogers' approach [63] based on premeasures, which combines well with the clopen set basis of 2^ω. This way, in the general framework of randomness, we do not have to distinguish between probability measures and Hausdorff measures, for instance.

2.1. The cantor space as a metric space. The *Cantor space* 2^ω is the set of all infinite binary sequences, also called *reals*. The mapping $x \mapsto \sum x(n)2^{-n}$ surjects 2^ω onto the unit interval [0, 1]. On the other hand, an element of 2^ω can be seen as the characteristic sequence of a subset of the natural numbers.

The usual metric on 2^ω is defined as follows: Given $x, y \in 2^\omega$, $x \neq y$, let $x \cap y$ be the longest common initial segment of x and y (possibly the empty string $\emptyset$). Define

$$d(x, y) = \begin{cases} 2^{-|x \cap y|} & \text{if } x \neq y, \\ 0 & \text{if } x = y. \end{cases}$$

Given a set $A \subseteq 2^\omega$, we define its *diameter* $d(A)$ as

$$d(A) = \sup \{d(x, y) : x, y \in A\}.$$

The metric d is compatible with the *product topology* on $\{0, 1\}^{\mathbb{N}}$, if $\{0, 1\}$ is endowed with the discrete topology.

2^ω is a compact Polish space. A countable basis is given by the *cylinder sets*

$$N_\sigma = \{x : x \lceil n = \sigma\},$$

where σ is a finite binary sequence. We will occasionally use the notation $N(\sigma)$ in place of N_σ to avoid multiple subscripts. $2^{<\omega}$ denotes the set of all finite binary sequences. If $\sigma, \tau \in 2^{<\omega}$, we use $\subseteq$ to denote the usual prefix partial ordering. This extends in a natural way to $2^{<\omega} \cup 2^\omega$. Thus, $x \in N_\sigma$ if and only if $\sigma \subset x$. Finally, given $U \subseteq 2^{<\omega}$, we write N_U to denote the open set induced by U, i.e. $N_U = \bigcup_{\sigma \in U} N_\sigma$.

2.2. Outer measures. A measure is a monotone, additive set function on a σ-algebra. Measures can be obtained from outer measures via restriction to a suitable family of sets. The Cantor space is a compact metric space, so we can follow the usual development of measure theory on locally compact spaces to introduce (outer) measures on 2^ω (see Halmos [21] or Rogers [63]). The following method to construct outer measures has been referred to as *Method I* [51, 63].

DEFINITION 2.1. Let $2^{<\omega}$ be the set of all finite binary sequences. A *premeasure* is a mapping $\rho : 2^{<\omega} \to \mathbb{R}^{\geq 0}$.

If ρ is a premeasure, define the set function $\mu_\rho^* : \mathcal{P}(2^\omega) \to \mathbb{R}^{\geq 0}$ by letting

$$\mu_\rho^*(A) = \inf \left\{ \sum_{\sigma \in U} \rho(\sigma) : A \subseteq N_U \right\}, \tag{2.1}$$

where we set $\mu_\rho^*(\emptyset) = 0$. It can be shown that μ_ρ^* is an outer measure. An *outer measure* is a function $\nu^* : \mathcal{P}(2^\omega) \to \mathbb{R}^{\geq 0} \cup \{\infty\}$ such that

(M1) $\nu^*(\emptyset) = 0$,
(M2) $\nu^*(A) \leq \nu^*(B)$ whenever $A \subseteq B$,
(M3) if (A_n) is a countable family of subsets of 2^ω, then

$$\nu^*\left(\bigcup_n A_n\right) \leq \sum_n \nu^*(A_n).$$

If we restrict an outer measure ν^* to sets E which satisfy

$$\nu^*(A) = \nu^*(A \cap E) + \nu^*(A \setminus E) \text{ for all } A \subseteq 2^\omega, \tag{2.2}$$

we obtain the ν^**-measurable sets*. The restriction of ν^* to the measurable sets is called a *measure*, and it will be denoted by ν. It can be shown that the measurable sets form a σ*-algebra*, i.e. they are closed under countable unions, complement, and the empty set is measurable.

In the course of this article, we will always assume that a measure ν is derived from an outer measure via (2.2), and that every outer measure in turn stems from a premeasure as in (2.1). (Rogers [63] studies in great detail the relations between measures, outer measures, and premeasures.)

Of course, the nature of the outer measure μ_ρ^* obtained via (2.1) and the μ_ρ-measurable sets will depend on the premeasure ρ. In the following, we will discuss the two most important kinds of outer measures studied in randomness theory: probability measures and Hausdorff measures.

2.3. Probability measures. A *probability measure* ν is any measure that is based on a premeasure ρ which satisfies $\rho(\emptyset) = 1$ and

$$\rho(\sigma) = \rho(\sigma^\frown 0) + \rho(\sigma^\frown 1) \tag{2.3}$$

for all finite sequences σ. The resulting measure μ_ρ preserves ρ in the sense that $\mu_\rho(N_\sigma) = \rho(N_\sigma)$ for all σ. This follows from the Caratheodory extension theorem. In the following, we will often identify probability measures with their underlying premeasure, i.e. we will write $\mu(\sigma)$ instead of $\mu(N_\sigma)$.

It is not hard to see that μ_ρ is a *Borel measure*, i.e. all Borel sets are measurable. It is also G_δ*-regular*, which means that for every measurable set A there exists a G_δ-set G such that $\mu_\rho(A) = \mu_\rho(G)$.

For $\rho(\sigma) = d(N_\sigma) = 2^{-|\sigma|}$ we obtain the *Lebesgue measure* λ on 2^ω, which is the unique translation invariant measure on 2^ω for which $\lambda(N_\sigma) = d(N_\sigma)$.

(Generalized) Bernoulli measures correspond to product measures on the space $\{0,1\}^{\mathbb{N}}$. Suppose $\bar{p} = (p_0, p_1, p_2, \dots)$ is a sequence of real numbers such that $0 \leq p_i \leq 1$ for all i. Let $\rho_i(1) = p_i$, $\rho_i(0) = 1 - p_i$, and set

$$\rho(\sigma) = \prod_{i=0}^{|\sigma|-1} \rho_i(\sigma(i)). \tag{2.4}$$

The associated measure μ_ρ will be denoted by $\mu_{\bar{p}}$. If $p_i = p$ for all i, we call the measure $\mu_{\bar{p}} = \mu_p$ simply a *Bernoulli measure*. Note that the Bernoulli measure with $p_i = 1/2$ for all i coincides with Lebesgue measure λ.

Dirac measures are probability measures concentrated on a single point. If $x \in 2^\omega$, we define

$$\rho(\sigma) = \begin{cases} 1 & \text{if } \sigma \subset x, \\ 0 & \text{otherwise.} \end{cases}$$

For the induced outer measure we obviously have $\mu_\rho(A) = 1$ if and only if $x \in A$, and $\mu_\rho(A) = 0$ if and only if $x \notin A$. The corresponding measure is usually denoted by δ_x.

2.4. Hausdorff measures. Hausdorff measures are of fundamental importance in geometric measure theory. They share the common feature that the premeasures they stem from only depend on the diameter of an open set. Therefore, the resulting measure will be translation invariant.

Assume h is a nonnegative, nondecreasing, continuous on the right function defined on all nonnegative reals. Assume, furthermore, that $h(t) > 0$ if and only if $t > 0$. Define the premeasure ρ_h as

$$\rho_h(N_\sigma) = h(d(N_\sigma)) = h(2^{-|\sigma|}).$$

The resulting measure μ_{ρ_h} will in general not be a Borel measure. Therefore, one refines the transition from a premeasure to an outer measure, also known as *Method II* [51, 63].

Given $\delta > 0$, define the set function

$$\mathcal{H}^h_\delta(A) = \inf\left\{ \sum_{\sigma \in U} \rho_h(N_\sigma) : A \subseteq N_U \text{ and } (\forall \sigma \in U)\, 2^{-|\sigma|} < \delta \right\}, \tag{2.5}$$

that is, we restrict the available coverings to cylinders of diameter less than δ. Now let

$$\mathcal{H}^h(A) = \lim_{\delta \to 0} \mathcal{H}^h_\delta(A).$$

Since, as δ decreases, there are fewer coverings available, $\mathcal{H}^h_\delta$ is nondecreasing, so the limit is defined, though may be infinite. It can be shown (see Rogers [63]) that the restriction of $\mathcal{H}^h$ to the measurable sets in sense of (2.2) is a Borel measure. It is called a *Hausdorff measure*. For $h(x) = x^t$, where $0 \leq t$, $\mathcal{H}^h$ is called the *t-dimensional Hausdorff measure* and is denoted by $\mathcal{H}^t$.

We will be mostly concerned with $\mathcal{H}^h$-nullsets. It is not hard to see that for any set A, $\mathcal{H}^h(A) = 0$ if and only if $\mu_{\rho_h}(A) = 0$, that is, the nullsets obtained from a premeasure via Method I and Method II coincide. Hence in the case of nullsets we can work with the less involved definition via Method I.

Due to the special nature of the metric d on 2^ω, only diameters of the form 2^{-n}, $n \in \mathbb{N}$, appear. So we can take any nondecreasing, unbounded function $h : \mathbb{N} \to \mathbb{R}^{\geq 0}$ and set $\rho_h(N_\sigma) = 2^{-h(|\sigma|)}$. The resulting Hausdorff measure will, in slight abuse of notation, also be denoted by $\mathcal{H}^h$.

Among the numerous Hausdorff measures, the family of t-dimensional Hausdorff measures $\mathcal{H}^t$ is probably most eminent. It is not hard to see that for any set A, $\mathcal{H}^s(A) < \infty$ implies $\mathcal{H}^t(A) = 0$ for all $t > s$. Likewise, $\mathcal{H}^r(A) = \infty$ for all $r < s$. Thus there is a critical value where $\mathcal{H}^s$ "jumps" from ∞ to 0. This value is called the *Hausdorff dimension* of A, written $\dim_{\mathrm{H}} A$. Formally,

$$\dim_{\mathrm{H}} A = \inf \{s : \mathcal{H}^s(A) = 0\}.$$

Hausdorff dimension is an important notion in fractal geometry, see [17].

2.5. Transformations, image measures, and semimeasures. One can obtain new measures from given measures by transforming them with respect to a sufficiently regular function. Let $f : 2^\omega \to 2^\omega$ be a function such that for every Borel set A, $f^{-1}(A)$ is Borel, too. Such functions are called *Borel (measurable)*. Every continuous function is Borel. If μ is a measure on 2^ω and f is Borel, then the *image measure* μ_f is defined by

$$\mu_f(A) = \mu(f^{-1}(A)).$$

It can be shown that every probability measure can be obtained from Lebesgue measure λ by means of a measurable transformation.

THEOREM 2.2 (folklore, see e.g. Billingsley [6]). *If μ is a Borel probability measure on 2^ω, then there exists a measurable $f : 2^\omega \to 2^\omega$ such that $\mu = \lambda_f$.*

PROOF. The proof uses a simple observation on distribution functions. For this purpose, we identify 2^ω with the unit interval. If g is the distribution function of μ, i.e.

$$g(x) = \mu([0, x]),$$

Let us define

$$f(x) = \inf \{y : x \leq g(y)\}.$$

g is nondecreasing and continuous on the right, so $\{y : x \leq g(y)\}$ is always an interval closed on the left. Therefore, $\{y : x \leq g(y)\} = [f(x), 1]$, so

$f(x) \leq y$ if and only if $x \leq g(y)$, so f can be seen as an inverse to g. Clearly, f is Borel measurable. We claim that $\lambda_f = \mu$. It suffices to show that for every y, $\lambda_f([0, y]) = \mu([0, y])$. We have

$$\begin{aligned}\lambda_f([0, y]) = \lambda(f^{-1}([0, y])) &= \lambda(\{x : f(x) \leq y\}) \\ &= \lambda(\{x : x \leq g(y)\}) = \mu([0, y]). \qquad \dashv\end{aligned}$$

We can use the representation of functions $2^\omega \to 2^\omega$ via mappings of finite strings to obtain a finer analysis of image measures.

Let S, T be trees on $2^{<\omega}$. A mapping $\phi : S \to T$ is called *monotone* if $\sigma \subseteq \tau$ implies $\phi(\sigma) \subseteq \phi(\tau)$. Typical examples of monotone mappings are *Turing operators*.

Monotone mappings of strings induce (partial) mappings of 2^ω. Given a monotone $\phi : S \to T$, let $D(\phi) = \{x \in [S] : \lim_n |\phi(x \lceil n)| = \infty\}$. Then define $\widehat{\phi} : D(\phi) \to T$ by $\widehat{\phi}(x) = \bigcup_n \phi(x \lceil n)$.

It is easy to see that $\widehat{\phi}$ is continuous. On the other hand, one can show that every continuous function on 2^ω has a representation via a monotone string function.

Proposition 2.3 (see [29]). *If $f : 2^\omega \to 2^\omega$ is continuous, then there exists a monotone mapping $\phi : 2^{<\omega} \to 2^{<\omega}$ such that $D(\phi) = 2^\omega$ and $f = \widehat{\phi}$.*

Using monotone functions, we can define a transformation of premeasures. Given a monotone ϕ and a premeasure ρ, let, for any $\tau \in 2^{<\omega}$, the set $\mathrm{Pre}(\tau)$ consist of all strings σ such that $\phi(\sigma) = \tau$, and no proper prefix of σ maps to τ. Define

$$\rho_\phi(\tau) = \sum_{\sigma \in \mathrm{Pre}(\tau)} \rho(\sigma),$$

where we let $\rho_\phi(\tau) = 0$ if $\mathrm{Pre}(\tau) = \emptyset$.

If μ is a probability measure and $\widehat{\phi}$ is total, then it is easy to see that μ_ϕ induces $\mu_{\widehat{\phi}}$. If $\widehat{\phi}$ is not total, we obtain a premeasure ρ_ϕ with the following properties.

$$\rho_\phi(\emptyset) \leq 1 \quad \text{and} \quad (\forall \sigma)\, \rho_\phi(\sigma) \geq \rho_\phi(\sigma^\frown 0) + \rho_\phi(\sigma^\frown 1).$$

Premeasures with this property have first been studied by Zvonkin and Levin [78] who introduced them as *(continuous) semimeasures*. Every semimeasure is the result of applying a monotone mapping to Lebesgue measure λ.

2.6. Effective transformations. In their seminal paper, Zvonkin and Levin [78] showed that an analysis of monotone functions (which they call *processes*) underlying continuous transformations yield a finer understanding of image measures. In particular, they showed that Theorem 2.2 holds for an "almost" continuous transformation. Furthermore, they showed that the transformation can be chosen to be at most as complex as the measures involved, in terms of their logical/computational complexity.

DEFINITION 2.4. Let ρ be a premeasure.

(1) ρ is *computable* if there exists a computable function $g : 2^{<\omega} \times \mathbb{N} \to \mathbb{Q}$ such that for all σ, n,
$$\left|\rho(\sigma) - g(\sigma, n)\right| \leq 2^{-n}.$$
A measure μ is *computable* if it is induced by a computable premeasure.

(2) ρ is *enumerable (from below)* or simply Σ^0_1 if its *left-cut* $\{(q, \sigma) \in \mathbb{Q} \times 2^{<\omega} : q < \rho(\sigma)\}$ is recursively enumerable.

To define effective monotone mappings, interpret them as relations $R \subseteq 2^{<\omega} \times 2^{<\omega}$ such that
$$\text{if } (\sigma, \tau) \in R, \text{ then } (\forall \sigma_0 \subseteq \sigma)(\exists \tau_0)\left[\tau_0 \subseteq \tau \wedge (\sigma_0, \tau_0) \in R\right].$$
This way we can speak of *(partial) recursive monotone mappings*, meaning that the underlying R is recursively enumerable. Accordingly, we say that a monotone function is *recursive in some* $x \in 2^\omega$ if R is r.e. in x. Obviously, enumerable monotone mappings are precisely the mappings induced by some *Turing operator* Φ. Therefore, we will henceforth denote recursive monotone mappings simply by the name *Turing operator*.

One can show that every enumerable semimeasure is the result of applying a Turing operator to Lebesgue measure λ. For probability measures, Levin was able to show that these can be effectively generated from λ by means of an "almost total" transformation.

THEOREM 2.5 (Levin, [78]). *Let* $\mu : 2^{<\omega} \to [0, 1]$ *be a probability measure recursive in* $x \in 2^\omega$

(1) *If* ϕ *is a computable monotone mapping such that* $\mu(D(\phi)) = 1$, *then* μ_ϕ *is a probability measure recursive in* x.

(2) *There exists a monotone* ϕ *recursive in* x *such that* $\mu = \lambda_\phi$. ϕ *can be chosen such that* $\lambda(D(\phi)) = 1$, *i.e.* $\widehat{\phi}$ *is total except on a set of* λ*-measure zero.*

PROOF. We will sketch Levin's proof of (2) for computable μ which easily relativizes. Essentially, we show that the mapping f in the proof of Theorem 2.2 can be obtained by some Turing operator ϕ. For this purpose, it is often convenient to identify 2^ω with the unit interval.

The measure μ is computable, so there exists a computable function $\gamma : 2^{<\omega} \times \mathbb{N} \to \mathbb{Q}$ such that, for all n,
$$\left|\mu(\sigma) - \gamma(\sigma, n)\right| \leq 2^{-n}.$$
We may assume that γ is approximating μ from above, so $\theta(\sigma, n) := \gamma(\sigma, n) - 2^{-n}$ is an approximation from below.

We construct ϕ as follows: Given a string $\sigma \in 2^{<\omega}$, $|\sigma| = n$, σ represents a binary interval $[a_\sigma, b_\sigma]$ of length 2^{-n} in $[0, 1]$. To compute f as in the proof of Theorem 2.2, we have to find an interval that is contained in $[c_\sigma, d_\sigma]$, where, if g is the distribution function of μ, $g(c_\sigma) = a_\sigma$, $g(d_\sigma) = b_\sigma$.

Define a set $Z = Z_\sigma$ of strings by selecting all those strings τ of length n for which

$$\sum_{\substack{|\xi|=n \\ \xi \leq \tau}} \gamma(\xi, 2n) \geq a_\sigma \quad \text{and} \sum_{\substack{|\zeta|=n \\ \zeta \leq \tau}} \theta(\zeta, 2n) \geq 1 - b_\sigma. \tag{2.6}$$

Here $\xi \leq \tau$ denotes the lexicographic ordering of strings. Let $\phi(\sigma)$ be the longest common initial segment of all strings in Z. We claim that if $x \in D(\phi)$, then $\widehat{\phi}(x) = f(x)$.

If $x \leq g(y)$, then every $x\lceil n$ is mapped to some string less or equal (lexicographically) $y\lceil n$, so $\widehat{\phi}(x) \leq y$ (as real numbers). On the other hand, since $\{y : x \leq g(y)\} = [f(x), 1]$, $z \leq f(x)$ implies $g(z) \leq x$, so an analogous argument yields $z \leq \widehat{\phi}(x)$.

It remains to show that $\lambda(D(\phi)) = 1$. There are three cases when $\widehat{\phi}$ might not be defined:

1. Suppose x is an atom of μ, i.e. $\mu(\{x\}) > 0$. This means that g has a discontinuity at x. In order to transform Lebesgue measure in to μ, $\widehat{\phi}$ must then map an interval to the single real x. Let $y < z$ be such that $f^{-1}(\{x\}) = (y, z]$. Suppose $y < x_0 < z$. From some n on the interval $[a_{x_0\lceil n} - 2^{-n}, b_{x_0\lceil n} + 2^{-n}]$ (as defined in the construction of ϕ) is contained in $[y, z]$. But then the only string to enter $Z_{x_0\lceil n}$ is $x\lceil n$, so for such x we have $\widehat{\phi}(x_0) = x$. $\widehat{\phi}$ might not be defined on y and z, but since μ is a finite measure, there can be at most countably many points x of positive μ-measure.
2. If there exists an interval $[y, z]$ such that $\mu([y, z]) = 0$, the distribution function g will remain constant on that interval, i.e. $g(x) = g(y)$ for all $y \leq x \leq z$. Thus $g(y)$ is mapped to an interval by $\widehat{\phi}$. However, as μ is finite, there can be at most countably many intervals of μ-measure zero.
3. Obviously, $\widehat{\phi}$ is also undefined if $f(x)$ is a dyadic rational number, for those numbers possess ambiguous dyadic representations, so the correspondent initial segments are always both included in Z. Thus, if $f(x) = m/2^k$, $\phi(x\lceil n)$ will map to a string of length less than k for all sufficiently large n. ⊣

By computing approximating g, the distribution function of μ through a monotone mapping, Levin was also able to show that every probability measure can be transformed into Lebesgue measure by means of a monotone function, taking into account possible complications given by cases (1)–(3).

THEOREM 2.6 (Levin, [78]). *Let μ be a probability measure recursive in $x \in 2^\omega$. Then there exists a monotone ψ recursive in x such that $\lambda = \mu_\psi$ and such that the complement of $D(\psi)$ contains only recursive reals or reals lying in intervals of μ-measure zero.*

2.7. Transformations of Hausdorff measures. For Hausdorff measures, transformations reflecting geometric properties are particularly interesting.

Generally, a *Hölder transformation* is a mapping h between metric spaces (X, d_X) and (Y, d_Y) such that for some constants $c, \alpha > 0$,

$$d_Y(h(x), h(y)) \leq c d_X(x, y)^\alpha.$$

In the Cantor space, this implies (recall the definition of metric d)

$$|h(x) \cap h(y)| \geq \alpha |x \cap y| + \log c,$$

The last formula suggests a generalization of Hölder mappings based on string functions.

DEFINITION 2.7. A monotone mapping $\phi : 2^{<\omega} \to 2^{<\omega}$ is *α-expansive*, $\alpha > 0$, if for all $y \in D(\phi)$,

$$\liminf_{n \to \infty} \frac{|\phi(y \lceil n)|}{n} \geq \alpha.$$

PROPOSITION 2.8. *Let $\phi : 2^{<\omega} \to 2^{<\omega}$ be α-expansive for some $\alpha > 0$. Then for all $B \subseteq D(\phi)$, and for all $s \geq 0$,*

$$\mathcal{H}^s(B) = 0 \Rightarrow \mathcal{H}^{s/\alpha}(\widehat{\phi}(B)) = 0.$$

The case where $\alpha = 1$ is especially important. Such functions are called *Lipschitz*, and if the inverse mapping is Lipschitz, too, an easy corollary of Proposition 2.8 yields that Hausdorff dimension is *invariant under bi-Lipschitz functions.*

It is also possible to consider more general Lipschitz-like conditions and prove a related version of Proposition 2.8, see [63, Theorem 29].

§3. Martin-Löf randomness. It was Martin-Löf's fundamental idea to define randomness by choosing a *countable family* of nullsets. For any non-trivial measure, the complement of the union of these sets will have positive measure, and any point in this set will be considered *random*. There are of course many possible ways to pick a countable family of nullsets. In this regard, it is very benefiting to use the framework of recursion theory and effective descriptive set theory.

3.1. Nullsets. Before we go on to define Martin-Löf randomness formally, we note that every nullset for a measure defined via Method I (and Method II, as is easily seen) is contained in a G_δ-nullset.

Proposition 3.1. *Suppose ρ is a premeasure. Then a set $A \subseteq 2^\omega$ is μ_ρ-null if and only if there exists a set $U \subseteq \mathbb{N} \times 2^{<\omega}$ such that for all n,*

$$A \subseteq N(U_n) \quad \text{and} \quad \sum_{\sigma \in U_n} \rho(N_\sigma) \leq 2^{-n}, \tag{3.1}$$

where $U_n = \{\sigma : (n, \sigma) \in U\}$.

Of course, the G_δ-cover of A is given by $\bigcap_n U_n$. There is an alternative way of describing nullsets which turns out to be useful both in the classical and algorithmic setting (see [63, Theorem 32] and [69]).

Proposition 3.2. *Suppose ρ is a premeasure. A set $A \subseteq 2^\omega$ is μ_ρ-null if and only if there exists a set $V \subseteq 2^{<\omega}$ such that*

$$\sum_{\sigma \in V} \rho(N_\sigma) < \infty, \tag{3.2}$$

and for all $x \in A$ there exist infinitely many $\sigma \in V$ such that $x \in N_\sigma$, or equivalently, $\sigma \subseteq x$.

3.2. Martin-Löf tests and randomness. Essentially, a Martin-Löf test is an effectively presented G_δ nullset (relative to some parameter z).

Definition 3.3. Suppose $z \in 2^\omega$ is a real. A *test relative to z*, or simply a *z-test*, is a set $W \subseteq \mathbb{N} \times 2^{<\omega}$ which is recursively enumerable in z. Given a natural number $n \geq 1$, an *n-test* is a test which r.e. in $\emptyset^{(n-1)}$, the $(n-1)$st Turing jump of the empty set. A real x *passes* a test W if $x \notin \bigcap_n N(W_n)$.

Passing a test W means not being contained in the G_δ set given by W. The condition "*r.e. in z*" implies that the open sets given by the sets W_n form a uniform sequence of $\Sigma^0_1(z)$ sets, and the set $\bigcap_n N(W_n)$ is a $\Pi^0_2(z)$ subset of 2^ω. To test for randomness, we want to ensure that W actually describes a nullset.

Definition 3.4. Suppose μ is a measure on 2^ω. A test W is *correct for μ* if

$$\sum_{\sigma \in W_n} \mu(N_\sigma) \leq 2^{-n}. \tag{3.3}$$

Any test which is correct for μ will be called a *test for μ*, or μ-test.

Now we can state Martin-Löf's definition of randomness for Lebesgue measure λ.

Definition 3.5 (Martin-Löf [45]). Suppose $z \in 2^\omega$. A real x is *Martin-Löf λ-random relative to z*, or simply *λ-z-random*, if x passes every z-test W for λ. Accordingly, x is *λ-n-random*, if it passes every n-test which for λ. Finally, x is *arithmetically λ-random* if it is λ-n-random for every $n \geq 1$.

Since there are only countably many Martin-Löf tests, it follows that the set of Martin-Löf random reals has Lebesgue measure 1. Martin-Löf showed that this set can be obtained as the complement of a *single* G_δ-nullset, a *universal test*.

Theorem 3.6 (Martin-Löf [45]). *For every $z \in 2^\omega$, there exists a Martin-Löf test U^z such that x is Martin-Löf random relative to z if and only if $x \notin \bigcap_n N(U_n^z)$. Furthermore, U^z can be obtained uniformly in z.*

The existence of a universal test is of great technical value. It facilitates a lot of proofs, since one has to consider only one test instead of a whole family of them.

3.3. Randomness for arbitrary measures. Martin-Löf defined randomness not only for Lebesgue measure, but also for arbitrary computable probability measures.

The problem of extending Definition 3.5 to other measures is that the measure itself may contain non-trivial information. If one defines randomness for an arbitrary measure μ simply by considering tests which are correct for μ, this works fine as long as μ is computable. However, if the measure itself contains additional algorithmic information, this leads to possibly unacceptable phenomena.

As an example, consider any real x. Define a premeasure on 2^ω by "perturbing" Lebesgue measure a little, so that the values $\rho(\sigma)$ remain rational and one can reconstruct x from them. If the perturbance is very small, the new measure μ_ρ will have the same nullsets as Lebesgue measure λ, and moreover it is possible to find for every λ-test a μ_ρ-test that covers the same reals, and vice versa. As a result, a real could be random with respect to a measure although it is computable from the (pre)measure.

Therefore, it seems worthwhile to incorporate the information given by the measure into the test notion. In the following, we will do this in a straightforward and most general way. This is followed by a discussion of advantages and drawbacks of this approach. Later on we will briefly address more refined concepts, which, however, due to topological reasons, have to be restricted to probability measures on 2^ω.

3.4. Representations of premeasures. To incorporate measures into the effective aspects of a randomness test we have to represent it in a form that makes it accessible for recursion theoretic methods. Essentially, this means to code a measure via an infinite binary sequence or a function $f : \mathbb{N} \to \mathbb{N}$.

The way we introduced it, an outer measure on 2^ω is completely determined by its underlying premeasure defined on the cylinder sets. It seems reasonable to represent these values via approximation by rational intervals.

Definition 3.7. Given a premeasure ρ, define its *rational representation* r_ρ by letting, for all $\sigma \in 2^{<\omega}$, $q_1, q_2 \in \mathbb{Q}$,

$$\langle \sigma, q_1, q_2 \rangle \in r_\rho \;\Leftrightarrow\; q_1 < \rho(\sigma) < q_2. \tag{3.4}$$

The real r_ρ encodes the complete information about the premeasure ρ in the sense that for each σ, the value $\rho(\sigma)$ is uniformly recursive in r_ρ. Therefore, every μ_ρ-nullset is $\Pi^0_2(r_\rho)$. This allows for a straightforward generalization of randomness tests relative to a given measure.

DEFINITION 3.8. Suppose ρ is a premeasure on 2^ω and $z \in 2^\omega$ is a real. A real is *Martin-Löf μ_ρ-random relative to z*, or simply *μ_ρ-z-random* if it passes all $r_\rho \oplus z$-tests which are correct for μ_ρ.

Hence, a real x is random with respect to an arbitrary measure μ_ρ if and only if it passes all tests which are enumerable in the representation r_ρ of the underlying premeasure ρ.

The representation r_ρ is a very straightforward approach to represent measures. As it turns out, for probability measures this representation corresponds to a canonical representation with respect to the weak topology.

Though this approach integrates the information presented in measures into tests, it resolves the "perturbance phenomenon" in a rather radical fashion—as the following example suggests.

Let $\bar{p} = (\frac{1}{2}+\beta_0, \frac{1}{2}+\beta_1, \frac{1}{2}+\beta_2, \dots)$ be a sequence of rational numbers such that the sequence of "biases" (β_i) is uniformly computable and $\sum \beta_i^2 < \infty$. We will see in Section 4.1 that a real is random for $\mu_{\bar{p}}$ if and only if it is random with respect to λ.

Now consider a second sequence $\bar{p}' = (\frac{1}{2}+\gamma_0, \frac{1}{2}+\gamma_1, \frac{1}{2}+\gamma_2, \dots)$ with $0 < \gamma_i^2 < \beta_i^2$, but this time the sequence of biases (γ_i) this time is not effective, for instance it codes some Martin-Löf λ-random real y. Then, according to Definition 3.8, y is not $\mu_{\bar{p}'}$-random, although in some sense the measure $\mu_{\bar{p}'}$ is "*closer*" to λ than $\mu_{\bar{p}}$.

For probability measures, Levin [38] has proposed alternative definitions of randomness using the topological properties of the space of probability measures on 2^ω which dwell further on this problem. However, to the authors knowledge the "naive" representation r_ρ is the only way to incorporate the information content of measures into Martin-Löf tests in a uniform way, valid for all premeasures alike.

Finally, it should be mentioned that Martin-Löf [45] already gave a definition of randomness for for arbitrary Bernoulli measures. His approach circumvents the difficulties presented by non-computable measures. He exploited the combinatorial properties one would expect from Bernoulli random reals. This way, he was able to give a *uniform test* for all Bernoulli measures.

THEOREM 3.9 (Martin-Löf). *There exists a test W_B such that W_B is correct for all Bernoulli measures μ_p.*

Martin-Löf showed that a real x which passes W_B is stochastic in the sense of Von-Mises-Wald-Church (see Ambos-Spies and Kučera [1] for details on stochasticity).

Levin [35] was able to strengthen Martin-Löf's result considerably using the topological structure of the space of probability measures on 2^ω (see Section 6).

THEOREM 3.10 (Levin [35]). *Let S be an effectively closed set of probability measures. Then there is a test W which is correct for all measures in S, and such that for every x that passes the test W, there is a measure $\mu \in S$ such that x is μ-random.*

3.5. Solovay tests. It is possible to base a definition of randomness on Proposition 3.2. This was suggested by Solovay [69]

DEFINITION 3.11. Suppose ρ is a premeasure on 2^ω and $z \in 2^\omega$. A *Solovay z-test* is a set V r.e. in z. A Solovay test is *correct for μ_ρ* if

$$\sum_{\sigma \in V} \rho(\sigma) < \infty.$$

A real x *passes a Solovay test V* if there exist only finitely many $\sigma \in V$ such that $x \in N_\sigma$. Finally, a real is *Solovay μ_ρ-random relative to z*, or simply *Solovay μ_ρ-z-random*, if it passes all $r_\rho \oplus z$-tests which are correct for μ_ρ.

Solovay [69] observed that for Lebesgue measure, a real is Solovay random if and only if it is Martin-Löf random. This result easily extends to all probability measures.

THEOREM 3.12 (Solovay). *If μ is a probability measure on 2^ω, and $z \in 2^\omega$, then x is Solovay μ-z-random if and only if it is Martin-Löf μ-z-random.*

PROOF. If U is a Martin-Löf test for μ_ρ, then $V = \bigcup U_n$ forms a Solovay test which is correct for μ_ρ. On the other hand, given a Solovay test V which is correct for μ, we can, by omitting finitely many elements, pass to a Solovay test V' for which $\sum_V \rho(\sigma) \leq 1$. Now define a set $U \subseteq \mathbb{N} \times 2^{<\omega}$ enumerable in V' as follows:

> Put (n, σ) into U if and only if at the stage when σ is enumerated into V',
> 1. no extension of σ has been enumerated into V' already, and
> 2. at least 2^{-n} predecessors of σ have been enumerated.

Then it is easy to see that $\sum_{U_n} \mu(N_\sigma) \leq 2^{-n}$, so U is a Martin-Löf test which is correct for μ. ⊣

If μ_ρ is not a probability measure, the above construction cannot be applied to yield that Martin-Löf tests and Solovay tests are equivalent. To see this, note that if μ_ρ is a probability measure, then, if $U \subseteq V$ are open sets represented by $C_U, C_V \subseteq 2^{<\omega}$, respectively, $\sum_{C_U} \rho(\sigma) \leq \sum_{C_V} \rho(\sigma)$. This allowed us to conclude in the preceding proof that $\mu(U_n) \leq 2^{-n}$. The same reasoning is, however, not possible if, for instance, $\rho(\sigma) < \rho(\sigma^\frown 0) + \rho(\sigma^\frown 1)$. In fact, Reimann and Stephan [62] were able to separate Martin-Löf tests and Solovay tests for a large family of premeasures.

DEFINITION 3.13. A *geometrical premeasure* is a premeasure ρ such that $\rho(\emptyset) = 1$ and there are (computable) real numbers p, q with

(1) $1/2 \leq p < 1$ and $1 \leq q < 2$;
(2) $\forall \sigma \in 2^{<\omega}\, \forall i \in \{0, 1\}\, [\rho(\sigma^\frown i) \leq p\rho(\sigma)]$;
(3) $\forall \sigma \in 2^{<\omega}\, [q\rho(\sigma) \leq \rho(\sigma^\frown 0) + \rho(\sigma^\frown 1)]$.

We will call such ρ a (p, q)-premeasure. ρ is called an *unbounded premeasure* if it is (p, q)-premeasure for some $q > 1$. A premeasure ρ is called *length-invariant* if

$$(\forall \sigma, \tau)\, \big[\, |\sigma| = |\tau| \;\Rightarrow\; \rho(\sigma) = \rho(\tau) \big].$$

Note that every premeasure $\rho(\sigma) = 2^{-|\sigma|s}$ (on which the Hausdorff measure $\mathcal{H}^s$ is based), $0 < s < 1$, is an unbounded, length invariant premeasure.

THEOREM 3.14 (Reimann and Stephan [62]). *For every computable, unbounded premeasure ρ there exists a real x which is Martin-Löf μ_ρ-random but not Solovay μ_ρ-random.*

This answered a question raised by Calude, Staiger, and Terwijn [8]. In particular, we see that for effective Hausdorff measures, Martin-Löf tests and Solovay tests do not yield the same notion of randomness.

§4. Computable probability measures. Most work on algorithmic randomness beyond Lebesgue measure has been done on computable probability measures. One reason for this can certainly be seen in the fact that Martin-Löf's approach carries over to arbitrary computable probability measures without facing the problem of representing measures described above.

Computable premeasures were defined in Section 2.6. A measure μ is *computable* if it is induced by a computable premeasure.

It is easy to see that a premeasure ρ is computable if and only if its rational representation r_ρ is computable. Therefore, a real x is μ_ρ-random by Definition 3.8 if and only if it passes every test which is correct for μ_ρ. (Recall that a test was defined to be just an r.e. subset of $\mathbb{N} \times 2^{<\omega}$.)

4.1. Equivalent measures and randomness. In our discussion of how to define randomness with respect to arbitrary measures we mentioned an invariance property of randomness due to Vovk [76]. If a generalized Bernoulli measure is close enough the uniform distribution, the corresponding sets of random reals coincide. The underlying dichotomy due to Kakutani [27] has been used by Shen to separate the notions of Martin-Löf randomness and Kolmogorov-Loveland stochasticity.

DEFINITION 4.1. Let μ, ν be two probability measures on 2^ω. μ is called *absolutely continuous* with respect to ν, written $\mu \ll \nu$, if every ν-nullset is also a μ-nullset. If two measures μ, ν are mutually absolutely continuous, we call them *equivalent* and write $\mu \sim \nu$. If on the other hand there exists a set A such that $\mu(A) = 0$ and $\nu(2^\omega \setminus A) = 0$, we call μ and ν *orthogonal*, written $\mu \perp \nu$.

The relation $\sim$ is an equivalence relation on the space of probability measures on 2^ω. For (generalized) Bernoulli measures, Kakutani [27] obtained a fundamental result concerning equivalence of measures.

THEOREM 4.2 (Kakutani [27]). *Let $\mu_{\bar{p}}$ and $\mu_{\bar{q}}$ be two generalized Bernoulli measures with associated sequences $\bar{p} = (p_i)$ and $\bar{q} = (q_i)$, respectively, such that for some $\varepsilon > 0$, $p_i, q_i \in [\varepsilon, 1 - \varepsilon]$ for all i.*

(1) *If $\sum_i (p_i - q_i)^2 < \infty$, then $\mu_{\bar{p}} \sim \mu_{\bar{q}}$.*
(2) *If $\sum_i (p_i - q_i)^2 = \infty$, then $\mu_{\bar{p}} \perp \mu_{\bar{q}}$.*

Vovk [76] showed that this dichotomy holds effectively.

THEOREM 4.3 (Vovk [76]). *Let $\mu_{\bar{p}}$ and $\mu_{\bar{q}}$ as in Theorem 4.2, and suppose that in addition $\mu_{\bar{p}}$ and $\mu_{\bar{q}}$ are computable.*

(1) *If $\sum_i (p_i - q_i)^2 < \infty$, then a real x is $\mu_{\bar{p}}$-random if and only if it is $\mu_{\bar{q}}$-random.*
(2) *If $\sum_i (p_i - q_i)^2 = \infty$, then no real is random with respect to both $\mu_{\bar{p}}$ and $\mu_{\bar{q}}$.*

Let $\mu_{\bar{p}}$ be a computable generalized Bernoulli measure induced by $\bar{p} = (\frac{1}{2} + \beta_0, \frac{1}{2} + \beta_1, \frac{1}{2} + \beta_2, \dots)$ with $\beta_i \in [\varepsilon, 1 - \varepsilon]$ for some $\varepsilon > 0$ and all i, and $\lim_i \beta_i = 0$. Shen [67] was able to show that if x is $\mu_{\bar{p}}$-random, then it is Kolmogorov-Loveland stochastic. (For a definition of Kolmogorov-Loveland stochasticity refer to Ambos-Spies and Kučera [1] or Muchnik, Semenov, and Uspensky [50].) However, by Theorem 4.3, if $\sum_i \beta_i^2 = \infty$, x cannot be λ-random. It follows that Martin-Löf randomness is a stricter notion than Kolmogorov-Loveland stochasticity.

One can ask whether Vovk's result holds in larger generality. Bienvenu [3] showed that if two computable probability measures have exactly the same set of random reals, then they must be equivalent. However, Bienvenu and Merkle [5] were able to show that the converse does not hold.

THEOREM 4.4 (Bienvenu and Merkle [5]). *There exists a computable probability measure μ and a real x such that $\mu \sim \lambda$ and x is λ-random but not μ-random. In fact, x can be chosen to be Chaitin's Ω.*

PROOF IDEA. For measures μ, ν, and $k \in \mathbb{N} \cup \{\infty\}$, define

$$\mathcal{L}^k_{\mu/\nu} = \left\{ x \in 2^\omega : \sup_n \frac{\mu(x \lceil n)}{\nu(x \lceil n)} \right\}$$

(define $0/0 = 1$, and $c/0 = \infty$ for $c > 0$). It holds that $\mu \sim \nu$ if and only if $\mu(\mathcal{L}^\infty_{\mu/\nu}) = \nu(\mathcal{L}^\infty_{\nu/\mu}) = 0$.

Let x be a λ-random real x in Δ^0_2 such as Chaitin's Ω. Use a computable approximation to x to define a computable measure μ such that $\mathcal{L}^\infty_{\mu/\lambda} = \emptyset$ and

$\mathcal{L}^{\infty}_{\lambda/\mu} = \{x\}$. Then $\mu \sim \lambda$, but the fact that μ along x converges much faster to 0 than λ, while on all other paths it behaves like λ, up to a multiplicative constant, can be used to define a μ-test that covers x. ⊣

4.2. Proper reals. One may ask whether a given real is random with respect to some computable probability measure. This question was first considered by Zvonkin and Levin [78]. The called reals that are random with respect to some computable probability measure *proper*. Muchnik et al. [50] used the name *natural*.

First note that a real x is trivially random with respect to a measure μ if the set $\{x\}$ does not have μ-measure 0, i.e. if x is an *atom* of μ. It is not hard to see that every atom of a computable probability measure is recursive.

PROPOSITION 4.5 (Levin, 1970). *If μ is a computable probability measure and if $\mu(\{x\}) > 0$ for $x \in 2^{\omega}$, then x is recursive.*

PROOF. Suppose $\mu(\{x\}) > c > 0$ for some computable μ and rational c. Let g be a computation function for μ, i.e. g is recursive and for all σ and n, $|g(\sigma, n) - \mu(\sigma)| \leq 2^{-n}$. Define a recursive tree T by letting $\sigma \in T$ if and only if $g(\sigma, |\sigma|) \geq c - 2^{-|\sigma|}$. By definition of T and the fact that μ is a probability measure, it holds that for sufficiently large m,

$$\left|\{\sigma : \sigma \in T \wedge |\sigma| = m\}\right| \leq \frac{1}{c - 2^{-m}}.$$

But this means that every infinite path through T is isolated, i.e. if x is an infinite path through T, there exists a string σ such that for all $\tau \supseteq \sigma$, $\tau \in T$ implies $\tau \subset x$. Furthermore, every isolated path through a recursive tree is recursive, and hence x is recursive. ⊣

On the other hand, if x is recursive and not a μ-atom, where μ is a computable probability measure, then one can easily use the recursiveness of x to devise a μ-test that covers x.

Concerning non-recursive reals, examples of non-proper reals can be obtained using *arithmetic Cohen forcing* (see for instance Odifreddi [54] for an introduction).

THEOREM 4.6 (Muchnik, 1998). *If x is Cohen* 1*-generic, it cannot be proper.*

PROOF IDEA. It suffices to show that for every $\sigma \in 2^{<\omega}$ and for every $n \in \mathbb{N}$, we can effectively find an extension $\tau \supseteq \sigma$ such that $\mu(N_\tau) \leq 2^{-n}$. This, however, follows easily by induction, since either $\mu(N_{\sigma^\frown 0}) \leq \mu(N_\sigma)/2$ or $\mu(N_{\sigma^\frown 1}) \leq \mu(N_\sigma)/2$, so we can use the computability of μ to search effectively for a suitable extension τ. ⊣

It is straightforward to prove the slightly more general result that any real that has a 1-generic real as a recursive subsequence cannot be proper.

The next example is probably more unexpected.

THEOREM 4.7 (Levin, 1970). *The halting problem $\emptyset'$ is not proper.*

PROOF SKETCH. Let μ a computable probability measure. Given a set $S \subseteq 2^\omega$ and $n \in \mathbb{N}$, let $S_{n,i} = \{y \in S : y(n) = i\}$.

Use the recursion theorem to construct an r.e. set W_e as follows. Set $F_0 = 2^\omega$. Given F_n, let $n \in W_e$ if and only if

$$\mu\big(F_n \cap S_{\langle e,n\rangle,1}\big) < \mu\big(F_n \cap S_{\langle e,n\rangle,0}\big).$$

Let $F_{n+1} = F_n \cap S_{\langle e,n\rangle, W_e(n)}$. Hence, W_e picks its values such that the restriction of F_n to paths x which satisfy $x(\langle e,n\rangle) = W_e(n)$ has minimal measure, at most half as large as the measure of F_n.

Since $\emptyset' = \{\langle e,n\rangle : n \in W_e\}$, W_e can be used to define a Martin-Löf μ-test for $\emptyset'$. ⊣

4.3. Computable probability measures and Turing reducibility. We saw in Section 2.6 that Turing reductions transform measures effectively. Every computable probability measure on 2^ω is the result of transforming Lebesgue measure by means of an almost everywhere defined Turing operator. On the other hand, every computable probability measure can be mapped effectively to Lebesgue measure.

Levin formulated the *principle of randomness conservation*: If a μ-random real is transformed by means of an effective continuous mapping f, then the result should be random with respect to the image measure μ_f.

The results of Section 2.6 easily yield that conservation of randomness holds for computable probability measures.

PROPOSITION 4.8. *Let μ be a computable probability measure and ϕ a Turing operator such that $\mu(D(\phi)) = 1$. If x is μ-random, then $\widehat{\phi}(x)$ is μ_ϕ-random.*

PROOF IDEA. If $U = \{U_n\}$ is a test for μ_ϕ, then $V = \{V_n\}$ with $V_n = \{\phi^{-1}(\sigma) : \sigma \in U_n\}$ is a μ-test. ⊣

A finer analysis of Theorems 2.5 and 2.6 yields a much stronger result. The operator $\widehat{\phi}$ is undefined only on a set of effective λ-measure 0. Hence it is defined on every λ-random real. Likewise, the operator $\widehat{\psi}$ is undefined only on recursive reals or reals lying in intervals of μ-measure 0. We obtain the following result, which says that with regard to Turing reducibility, proper reals have the same computational power as the standard Martin-Löf, i.e. λ-random reals. It has independently been proved by Kautz [28]. (His approach is also presented in Downey and Hirschfeldt [14].)

THEOREM 4.9 (Levin, [78]; Kautz [28]). *Let μ be a computable probability measure. If x is μ-random and non-recursive, then x is Turing equivalent to some λ-random real R.*

This result can be used to obtain a number of interesting consequences. Demuth [12] observed that every real which is tt-reducible to some λ-random real is in the same Turing degree with some λ-random real.

THEOREM 4.10 (Demuth [12]). *If $x \leq_{tt} R$ and R is λ-random, then there exists some $y \in 2^\omega$ such that y is λ-random and $y =_T x$.*

PROOF. If $x \leq_{tt} R$ via Φ, conservation of randomness implies that x is λ_Φ-random, where λ_Φ is a computable probability measure. Now apply Theorem 4.9. ⊣

It is known that below a hyperimmune-free Turing-degree, i.e. a degree $\boldsymbol{a}$ such that every function f recursive in $\boldsymbol{a}$ is majorized by some recursive function, Turing and truth-table reducibility coincide. Applying the *hyperimmune-free basis theorem* [25] to a Π^0_1 class containing only random reals, we obtain a degree below which every (non-zero) degree contains a λ-random real.

THEOREM 4.11 (Kautz [28]). *There exists a Turing degree $\boldsymbol{a} > \boldsymbol{0}$ such that any degree $\boldsymbol{b}$ with $\boldsymbol{0} < \boldsymbol{b} \leq \boldsymbol{a}$ contains a λ-random real.*

Another straightforward application concerns the halting problem $\emptyset'$. Bennett [2] investigated the notion of *logical* or *computational depth*. A main result of this investigation was that the halting problem $\emptyset'$ is not truth-table reducible to any λ-random real. (Note however that, by results of Kučera [33] and Gács [19], it is Turing reducible to some λ-random real.) This result can be derived easily from Theorem 4.7 and conservation of randomness.

THEOREM 4.12 (Bennett [2], see also Juedes, Lathrop, and Lutz [26]). *The halting problem $\emptyset'$ is not truth-table reducible to a λ-random real.*

§5. Hausdorff measures.

Recently, a lot of research on randomness for non-Lebesgue measures focused on *Hausdorff measures*. One reason for this can certainly be seen in Lutz's introduction of *effective fractal dimension concepts* [41, 42]. Close connections between Kolmogorov complexity and Hausdorff dimension had been known to exist for for quite some time, e.g. through works of Ryabko [64, 65], Staiger [72, 70], or Cai and Hartmanis [7]. But Lutz's concepts brought these together with the topics and techniques that had been developed in resource-bounded measure theory and the investigation of computational properties of random reals.

We can and will not cover these new developments in full breadth, for this purpose the reader may refer to survey articles [22, 40] or the author's PhD-thesis [57]. Instead, we will focus on a few recent results which suit well in the line of this article.

5.1. Effective Hausdorff measures and Kolmogorov complexity.

A lot of interesting recent research on effective dimension concepts is based on a fundamental correspondence between Hausdorff measures and Kolmogorov complexity. Although the general framework of randomness from Section 3.4 extends to arbitrary Hausdorff premeasures, investigations focused on effective Hausdorff measures.

Let $h : \mathbb{N} \to \mathbb{R}^{\geq 0}$ be a nondecreasing, unbounded function. In connection with randomness, such functions were studied by Schnorr [66], without explicit reference to Hausdorff measures. Schnorr called such functions *orders* or *order functions*. He used them to classify growth rates of martingales and give a martingale characterization of the randomness concept that is now known as *Schnorr randomness*.

If h is an order function, recall that $\mathcal{H}^h$ denotes the Hausdorff measure induced by the premeasure $2^{-h(|\sigma|)}$. Schnorr's characterization of λ-randomness via Kolmogorov complexity can be extended to $\mathcal{H}^h$-random reals. We assume the reader is familiar with the basic definitions of Kolmogorov complexity, as presented in the books by Li and Vitányi [39] and Downey and Hirschfeldt [14].

THEOREM 5.1 (Tadaki [74]; Reimann [57]). *If h is a computable order function, a real $x \in 2^\omega$ is $\mathcal{H}^h$-random if and only if there exists a constant c such that for all n,*

$$\mathrm{K}(x \lceil n) \geq h(n) - c,$$

where K *denotes prefix-free Kolmogorov complexity.*

PROOF SKETCH. If x is not $\mathcal{H}^h$-random, choose an r.e. test W that covers $\{x\}$ and is correct for $2^{-h(|\sigma|)}$. Define functions $m_n : 2^{<\omega} \to \mathbb{Q}$ by

$$m_n(\sigma) = \begin{cases} n2^{-h(|\sigma|)} & \text{if } \langle n, \sigma \rangle \in W, \\ 0 & \text{otherwise,} \end{cases}$$

and let

$$m(\sigma) = \sum_{n=1}^{\infty} m_n(\sigma).$$

Obviously, all m_n and thus m are enumerable from below. Furthermore, it is not hard to see that

$$\sum_{\sigma \in 2^{<\omega}} m(\sigma) < \infty,$$

hence m is an *enumerable discrete semimeasure*. Apply the *coding theorem* (see [39]) to obtain a constant c_m such that $-\log m(\sigma) \geq \mathrm{K}(\sigma) - c_m$ for all σ. By definition of m, for every n there exists some l_n such that $m(x \lceil l_n) \geq n\, 2^{-h(l_n)}$, which implies $\mathrm{K}(x \lceil l_n) - c_m \leq -\log m(x \lceil l_n) \leq h(l_n) - n$.

For the other direction, we use a result by Chaitin [11] which establishes that for any l,

$$|\{\sigma \in \{0,1\}^n : \mathrm{K}(\sigma) \leq n + \mathrm{K}(n) - l\}| \leq 2^{n+C-l}, \tag{5.1}$$

where C is a constant independent of n, l. (Here the natural numbers are identified with their binary representation.)

Assume that the complexity of x is not bounded from below by $h(n) - c$ for any constant c. Define

$$W_n = \{\sigma \in 2^{<\omega} : \mathrm{K}(\sigma) \leq h(|\sigma|) - n - C\}.$$

Then the test W covers x, since for every l there is some prefix σ of x such that $\mathrm{K}(\sigma) \leq h(|\sigma|) - l$. Furthermore, W is r.e., since K is enumerable from above. Finally, using (5.1), we have for each n,

$$\sum_{\sigma \in W_n} 2^{-h(|\sigma|)} = \sum_{k=0}^{\infty} \sum_{\substack{\sigma \in W_n \\ |\sigma|=k}} 2^{-h(|\sigma|)} = \sum_{k=0}^{\infty} 2^{-h(k)} |\{0,1\}^k \cap W_n|$$

$$\leq 2^{-n} \sum_{k=0}^{\infty} 2^{-\mathrm{K}(k)} \leq 2^{-n}. \qquad \dashv$$

The concept of an *effective* $\mathcal{H}^h$*-nullset* leads in straightforward way to *effective Hausdorff dimension*. Given $x \in 2^\omega$, let

$$\dim_{\mathrm{H}}^1 x = \inf\{s \geq 0 : x \text{ is not } \mathcal{H}^s\text{-random}\}.$$

This was first defined by Lutz [42] via a variant of martingales (*gales*) under the name *constructive dimension*. Effective Hausdorff dimension has an elegant characterization via Kolmogorov complexity, which follows easily from Theorem 5.1.

THEOREM 5.2. *For every* $x \in 2^\omega$,

$$\dim_{\mathrm{H}}^1 x = \liminf_{n \to \infty} \frac{\mathrm{K}(x\lceil n)}{n}.$$

The theorem was first explicitly proved by Mayordomo [47]. However, as Staiger [71] pointed out, much of it was present in earlier work by Ryabko [64, 65], Staiger [72, 70], or Cai and Hartmanis [7]. Essentially, the characterization is a consequence of the correspondence between semimeasures and Kolmogorov complexity established by Levin [78].

Another important feature of effective dimension is the *stability property*. Although we defined effective dimension only for single reals, it is easy to use effective $\mathcal{H}^s$-nullsets (i.e. correct $\mathcal{H}^s$-tests) to define the effective Hausdorff dimension of a set A of reals, denoted by $\dim_{\mathrm{H}}^1 A$.

THEOREM 5.3 (Lutz [42]). *For every* $A \subseteq 2^\omega$,

$$\dim_{\mathrm{H}}^1 A = \sup\{\dim_{\mathrm{H}}^1 x : x \in A\}.$$

This means that, with respect to dimension, every set of reals has to contain an element of accordant complexity, measured in terms of asymptotic algorithmic complexity, as given by Theorem 5.2, where the correspondence is exact for effective dimension. This can be seen as a generalization of the fact that any set of positive Lebesgue measure contains a λ-random real.

Geometric measure theory knows a multiplicity of dimension notions besides Hausdorff dimension (see e.g. Falconer [17]). Many of these can be effectivized, most notably *packing dimension*, and be related to Kolmogorov complexity. We will not address this here, but instead refer to the aforementioned sources.

5.2. Hausdorff measures and probability measures. So far, there are few types of examples of reals which are random for some Hausdorff measure. All of them are derived from randomness for probability measures.

(1) If $0 < r < 1$ is rational, let $Z_r = \{\lfloor n/r \rfloor : n \in \mathbb{N}\}$. Given a λ-random real x, define x_r by
$$x_r(m) = \begin{cases} x(n) & \text{if } m = \lfloor n/r \rfloor, \\ 0 & \text{otherwise.} \end{cases}$$
Using Theorem 5.2, it is easy to see that $\dim^1_{\mathrm{H}} x_r = r$. This technique can be refined to obtain sets of effective dimension s, where $0 \leq s \leq 1$ is any Δ^0_2-computable real number (see e.g. Lutz [43]), or reals which are $\mathcal{H}^h$-random, where h is a computable order function.

(2) Given a Bernoulli measure μ_p with bias $p \in \mathbb{Q} \cap (0,1)$, the effective dimension of any set that is Martin-Löf random with respect to μ_p equals the entropy of the measure $\mathrm{H}(\mu_p) = -[p \log p + (1-p)\log(1-p)]$ (Lutz [41]). This is an effectivized version of a classical theorem due to Eggleston [16].

(3) Let U be a universal, prefix-free machine. Given a computable real number $0 < s \leq 1$, the binary expansion of the real number
$$\Omega^{(s)} = \sum_{\sigma \in \mathrm{dom}(U)} 2^{-|\sigma|/s}$$
has effective Hausdorff dimension s. This was shown by Tadaki [74]. For $s = 1$, we obtain Chaitin's Ω, which is λ-random [10]. The effective dimension of $\Omega^{(s)}$ is linked to the behavior of nullsets under Hoelder transformations as described in Section 2.7.

The first two examples are random with respect to a computable probability measure. However, this is not the case for every real which is $\mathcal{H}^h$-random for some order function h.

THEOREM 5.4 (Reimann [57]). *For every order function h there exists a real x such that x is not proper but $\mathcal{H}^h$-random.*

PROOF IDEA. As in example (1), recursively join a real y of low complexity and a λ-random real with the appropriate density, given by h. We can choose y to be 1-generic real, which is not proper by Theorem 4.6. ⊣

Nevertheless, every $\mathcal{H}^h$-random real *is* random with respect to some probability measure. The following result can be seen as an effective variant of

Frostman's Lemma in geometric measure theory, which establishes a close connection between Hausdorff dimension and capacity (see e.g. [46]).

THEOREM 5.5 (Reimann [58]). *If $x \in 2^\omega$ is $\mathcal{H}^h$-random, where h is an order function, then there exists a probability measure μ such that x is μ-random and there exists a c such that for all σ,*

$$\mu(\sigma) \leq c2^{-h(|\sigma|)}.$$

5.3. The computational power of Hausdorff randomness. It is a question of apparently intriguing difficulty to determine the computational power of reals of non-trivial Hausdorff dimension. The examples (1)–(3) are all Turing equivalent to a λ-random real. For (1) this is obvious, for (2) this follows from Theorem 4.9. For (3), this follows from a different property of $\mathcal{H}^h$-random reals, which we will address further below.

This observation might suggest to conjecture that every real of positive effective Hausdorff dimension, or more generally, every real that is $\mathcal{H}^h$-random for some computable order function h, computes a λ-random real, or at least a real of dimension 1 (or arbitrarily close to 1).

It turns out that this is in general not the case for strong reductions, and not true with respect to Turing reducibility for every computable order function h. But it remains an open question whether such an "*extraction of randomness*" via Turing reductions is possible for higher levels of entropy, e.g. for reals of positive Hausdorff dimension (see Reimann [57] and Miller and Nies [48]).

We first address the results for strong reducibilities. Reimann and Terwijn [57] showed that a many-one reduction cannot increase the entropy of a real x random for a Bernoulli measure μ_p, p rational. It follows that every real m-reducible to x has effective dimension at most $\mathrm{H}(\mu_p)$.

However, this result does not extend to weaker reducibilities such as truth-table reducibility, since for Bernoulli-measures μ_p with $p \in (0, 1)$ the Levin-Kautz result (Theorem 4.9) holds for a total Turing reduction.

Using a different approach, Stephan [73] was able to construct an oracle relative to which there exists a wtt-lower cone of positive effective dimension at most $1/2$. A most general unrelativized result was obtained by Nies and Reimann [53].

THEOREM 5.6. *For each rational r, $0 \leq r \leq 1$, there is a real $x \leq_{\mathrm{wtt}} \emptyset'$ such that $\dim^1_{\mathrm{H}} x = r$ and for all $z \leq_{\mathrm{wtt}} x$, $\dim^1_{\mathrm{H}} z \leq r$.*

PROOF IDEA. We construct x satisfying the requirements

$$R_{\langle e,j\rangle} : z = \Psi_e(x) \Rightarrow \exists (k \geq j)\ \mathrm{K}(z \lceil k) \leq \big(r + 2^{-j}\big)k + \mathcal{O}(1)$$

where (Ψ_e) is a uniform listing of wtt reduction procedures. We can assume each Ψ_e also has a certain (non-trivial) lower bound on the use g_e, because otherwise the reduction would decrease complexity anyway.

To ensure that x has dimension r we construct it inside the Π^0_1 class

$$P = \{y : (\forall n \geq n_0)\ \mathrm{K}(y \lceil n) \geq \lfloor rn \rfloor\}$$

where n_0 is chosen so that $\lambda(P) \geq 1/2$. P is given as an effective approximation through clopen sets P_s.

We approximate longer and longer initial segments σ_j of x, where σ_j is a string of length m_j, both σ_j, m_j controlled by R_j.

Define a length k_j where we intend to compress z, and let $m_j = g_e(k_j)$. Define σ_j of length m_j in a way that, if $\tau = \Psi_e^{\sigma_j}$ is defined then we compress it down to $(\alpha + 2^{-b_j})k_j$, by constructing an appropriate Martin-Löf test L.

The "opponent's" answer could be to remove σ_j from P. (σ_j is not of high dimension.) In this case, the capital he spent for this removal exceeds what we spent for our request, so we can account our capital against his. Of course, usually σ_j is much longer than x. So we will only compress x when the measure of oracle strings computing it is large. The advantage we have in measure is reflected by the following lemma.

LEMMA 5.7. *Let $C \subseteq 2^\omega$ be clopen such that $C \subseteq P_s$ and $C \cap P_t = \emptyset$ for stages $s < t$. Then*

$$\Omega_t - \Omega_s \geq (\lambda C)^r.$$

Here Ω_s is the (rational valued) approximation to Chaitin's Ω at stage s.

In the course of the construction, some R_j might have to pick a new σ_j. In this case we have to initialize all R_n of lower priority ($n > j$).

We have to make sure that this does not make us enumerate too much measure into L. Therefore, we have to assign a new length k_n to the strategies R_n. ⊣

In the course of the construction, it is essential that we know the use of the reduction related to R_j, so that we can assign proper new lengths. This is the reason why the construction does not extend to the Turing case.

However, there exists a non-extractability result for Turing reducibility.

THEOREM 5.8. *There exists a computable order function h and an $\mathcal{H}^h$-random real x such that no real $y \leq_{\mathrm{T}} x$ is λ-random.*

The result was independently proved by Kjos-Hanssen, Merkle, and Stephan [30] and Reimann and Slaman [59]. While Reimann and Slaman gave a direct construction, the proof by Kjos-Hanssen et al. sheds light on a fascinating connection with recursion theory.

A function $g : \mathbb{N} \to \mathbb{N}$ is *diagonally non-recursive (dnr)* if for all n, $g(n) \neq \varphi_n(n)$, where $\{\varphi_n\}_{n\in\mathbb{N}}$ is some standard effective enumeration of all partial recursive functions from $\mathbb{N}$ to $\mathbb{N}$. Dnr functions play an important role in recursion theory. It is known that computing a dnr function is equivalent to computing a *fixed-point free function*, i.e. a function f such that $\varphi_{f(e)} \neq \varphi_e$ for all e [24]. The well-known *Arslanov completeness criterion* says that an r.e.

set $W \subseteq \mathbb{N}$ is Turing complete if and only if it computes a fixed-point free function.

Kjos-Hanssen et al. were able to prove the following.

THEOREM 5.9 (Kjos-Hanssen et al. [30]). *Let $x \in 2^\omega$.*

(1) *x is $\mathcal{H}^h$-random for some computable order function h if and only if it truth-table computes a dnr function.*
(2) *x is $\mathcal{H}^h$ random for some order function $h \leq_{\mathrm{T}} x$ if and only if it Turing computes a dnr function.*

Kjos-Hanssen et al. called reals which satisfy one of the equivalent conditions in (1) *complex*, those which satisfy one of the conditions in (2) *autocomplex*.

Theorem 5.9 is quite a powerful tool. For instance, together with the Arslanov completeness criterion it immediately implies that $\Omega^{(s)}$ as defined in Section 5.2, example (3), is Turing equivalent to $\emptyset'$. (Note that $\Omega^{(n)}$ is of r.e. degree.)

Furthermore, the result can be applied to prove Theorem 5.8. An intricate construction by Kumabe [34] showed the existence of a minimal degree which contains a recursively bounded dnr function g. (A function $f : \mathbb{N} \to \mathbb{N}$ is *recursively bounded* if there exists a recursive $G : \mathbb{N} \to \mathbb{N}$ such that $g(n) \leq G(n)$ for all n.) If we encode g as a real x_g (for instance, via unary representations of $g(n)$, separated by 0), the fact that g is recursively bounded implies that x_g is truth-table equivalent to g. Hence, by Theorem 5.9, x_g is complex. However, no minimal degree can contain a λ-random real, since by a theorem of Van Lambalgen [75], every recursive split of a λ-random real into two halves yields two relatively random, and hence Turing incomparable, λ-random reals.

These results are contrasted by positive results for randomness/entropy extraction if the entropy oszilllations present in a real are bounded.

Define the upper asymptotic entropy of a real x by

$$\overline{\mathrm{K}}(x) = \limsup_{n \to \infty} \frac{\mathrm{K}(x \lceil n)}{n}.$$

Note that this is a dual to the effective Hausdorff dimension of x, by Theorem 5.2. Extending earlier work by Ryabko [64, 65] and Doty [13], Bienvenu, Doty, and Stephan [4] showed the following.

THEOREM 5.10. *For all $\varepsilon > 0$ and any $x \in 2^\omega$ such that $\overline{\mathrm{K}}(x) > 0$, there exists $y \equiv_{\mathrm{wtt}} x$ such that*

$$\overline{\mathrm{K}}(y) \geq 1 - \varepsilon \quad \text{and} \quad \dim^1_{\mathrm{H}} y \geq \frac{\dim^1_{\mathrm{H}} x}{\overline{\mathrm{K}}(x)} - \varepsilon.$$

§6. **Arbitrary probability measures.** In Section 3.4 we gave a definition of randomness based on the rational representation of premeasures. While the rational representation is defined for any premeasure and hence leads to a universal notion of relative Martin-Löf-style randomness, it does not reflect the topological properties of the space of probability measures on 2^ω.

In this section we will see how, by passing to a different representation of measures, one can exploit the topological structure to prove results about randomness.

It is a classic result of measure theory (see Parthasarathy [55]) that the space of probability measures $\mathcal{P}$ on 2^ω is a compact polish space. The topology is the *weak topology*, which can be metrized by the *Prokhorov metric*, for instance. There is an *effective dense subset*, given as follows: Let Q be the set of all reals of the form $\sigma^\frown 0^\omega$. Given $\bar{q} = (q_1, \dots, q_n) \in Q^{<\omega}$ and non-negative rational numbers $\alpha_1, \dots, \alpha_n$ such that $\sum \alpha_i = 1$, let

$$\delta_{\bar{q}} = \sum_{k=1}^{n} \alpha_k \delta_{q_k},$$

where δ_x denotes the *Dirac point measure* for x. Then the set of measures of the form $\delta_{\bar{q}}$ is dense in $\mathcal{P}$.

The recursive dense subset $\{\delta_{\bar{q}}\}$ and the effectiveness of the metric d between measures of the form $\delta_{\bar{q}}$ suggests that the representation reflects the topology effectively, i.e. the set of representations should be Π^0_1. However, this is not true for the set of rational representations of probability measures. Instead, we have to resort to other representations in metric spaces, such as Cauchy sequences. Using the framework of *effective descriptive set theory*, as for example presented in Moschovakis [49], one can obtain the following.

THEOREM 6.1. *There is a recursive surjection*

$$\pi : 2^\omega \to \mathcal{P}$$

and a Π^0_1 *subset* P *of* 2^ω *such that* $\pi \lceil P$ *is one-one and* $\pi(P) = \mathcal{P}$.

The topological structure comes at price. No longer does every (pre)measure have a unique representation. In the case of Cauchy representations for instance, there are infinitely many for each measure. In particular, if x is a real and μ is a measure, we can find a Cauchy sequence representation r of μ such that x is recursive in μ. If we try to remedy this and pick out a Π^0_1 set on which the representation is one-one and onto, one could claim there is a certain arbitrariness in this.

Therefore, we either have to speak of *randomness with respect to a representation*, or try to define a notion a randomness which is *independent of the representation* of the measure.

The second path has first been followed by Levin [37, 38]. It has recently been extended by Gács [20] to a larger class of metric spaces on which random objects can be defined. It would go beyond the scope of this article to present this theory here, instead we refer to Gacs' excellent paper, which develops the theory in a mostly self-contained account. The interested reader may then pass on to Levin's much more succinct article [38].

The effective compactness of $\mathcal{P}$ has a number of remarkable properties in this theory. For instance, there exists a *neutral measure*, a measure relative to which every sequence is random.

Here we will follow the more *naive approach* and see that a result of similar nature holds. We single out a representation of $\mathcal{P}$ in the sense of Theorem 6.1. So in the following, when we speak of "measure", we will *at the same time refer to its unique representation in the* Π^0_1 *set* P given by Theorem 6.1.

6.1. Randomness of non-recursive reals. If x is an atom of some probability measure μ, it is trivially μ-random. Interestingly, only for the recursive reals this is the only way to become random.

THEOREM 6.2 (Reimann and Slaman [61]). *For any real x, the following are equivalent.*

(i) *There exists a probability measure μ such that $\mu(\{x\}) = 0$ and x is μ-random.*
(ii) *x is not recursive.*

PROOF SKETCH. A fundamental result by Kučera [33] ensures that every Turing degree above $\emptyset'$ contains a λ-random real. This result relativizes. Hence one can combine it with the *Posner-Robinson Theorem* [56], which says that for every non-recursive real x there exists a z such that $x \oplus z =_T z'$, to obtain a real R which is λ-random relative to some $z \in 2^\omega$ and which is T(z)-equivalent to x. There are Turing functionals Φ and Ψ recursive in z such that

$$\Phi(R) = x \quad \text{and} \quad \Psi(x) = R.$$

One can then use the functionals to define a Π^0_1 subset S of P, the set of representations of measures. All measures in S are consistent with the condition that it is an image measure of λ induced by Φ, and that it is non-atomic on x. In order to apply Levin's technique of conservation of randomness, one resorts to a basis result for Π^0_1 sets regarding relative randomness. ⊣

THEOREM 6.3 (Reimann and Slaman [61], Downey, et al. [15]). *Let S be $\Pi^0_1(z)$. Then, if R is λ-random relative to z, then there exists a $y \in S$ such that R is λ-random relative to $y \oplus z$.*

Theorem 6.3 is essentially a consequence of *compactness*. It seems to be quite a versatile result. For instance, it is also used in the proof of Theorem 5.5.

6.2. Randomness for continuous measures. A natural question arising regarding Theorem 6.2 is whether the measure making a real random can be ensured to have certain regularity properties; in particular, can it be chosen *continuous*? (A probability measure is *continuous* if $\mu(\{x\}) = 0$ for all $x \in 2^\omega$.)

Reimann and Slaman [61] gave an explicit construction of a non-recursive real not random with respect to any continuous measure. Call such reals 1-*ncr*. In general, let NCR_n be the set of reals which are not n-random with respect to any continuous measure.

Kjos-Hanssen and Montalban [31] observed that any member of a countable Π^0_1 class is an element of NCR_1.

PROPOSITION 6.4. *If $A \subseteq 2^\omega$ is Π^0_1 and countable, then no member of A can be in* NCR_1.

PROOF IDEA. If μ is a continuous measure, then obviously $\mu(A) = 0$. One can use a recursive tree T such that $[T] = A$ to obtain a μ-test for A. ⊣

It follows from results of Cenzer, Clote, Smith, Soare, and Wainer [9] that members of NCR_1 can be found throughout the hyperarithmetical hierarchy of Δ^1_1, whereas Kreisel [32] had shown earlier that each member of a countable Π^0_1 class is in fact hyperarithmetical.

Quite surprisingly, Δ^1_1 turned out to be the precise upper bound for NCR_1. An analysis of the proof of Theorem 6.2 shows that if x is *truth-table* equivalent to a λ-random real, then the "pull-back" procedure used to devise a measure for x yields a continuous measure. More generally, we have the following.

THEOREM 6.5 (Reimann and Slaman [60]). *Let x be a real. For any $z \in 2^\omega$, the following are equivalent.*

(i) *x is random for a continuous measure recursive in z.*

(ii) *x is random for a continuous dyadic measure recursive in z.*

(iii) *There exists a functional Φ recursive in z which is an order-preserving homeomorphism of 2^ω such that $\Phi(x)$ is λ-z-random.*

(iv) *x is truth-table equivalent to a λ-z-random real.*

Here *dyadic* measure means that the underlying premeasure is of the form $\rho(\sigma) = m/2^n$ with $m, n \in \mathbb{N}$. The theorem can be seen as an effective version of the *classical isomorphism theorem* for continuous probability measures (see for instance Kechris [29])[1].

Woodin [77], using involved concepts from set theory, was able to prove that if $x \in 2^\omega$ is not hyperarithmetic, then there is a $z \in 2^\omega$ such that $x \oplus z \equiv_{\mathrm{tt}(z)} z'$, i.e. outside Δ^1_1 the Posner-Robinson theorem holds with truth-table equivalence. Hence we have

THEOREM 6.6 (Reimann and Slaman [61]). *If a real x is not Δ^1_1, then there exists a continuous measure μ such that x is μ-random.*

It is on the other hand an open problem whether every real in NCR_1 is a member of a countable Π^0_1 class.

[1]The theorem suggests that for continuous randomness representational issues do not really arise, since there is always a measure with a computationally minimal representation.

One may ask how the complexity of NCR_n grows with n. There is some "empirical" evidence that this growth is rather fast. It it, for instance, not obvious at all whether for all n, NCR_n is countable. This, however, holds true.

THEOREM 6.7 (Reimann and Slaman [60]). *For all n, NCR_n is countable.*

PROOF IDEA. The idea is to use *Borel determinacy* to show that the complement of NCR_n contains an upper Turing cone. This follows from the fact that the complement of NCR_n contains a Turing invariant and cofinal (in the Turing degrees) Borel set. For example, we can use the set of all y that are Turing equivalent to some $z \oplus R$, where R is λ-$(n+1)$-random relative to a given z. The desired cone is given by the *Turing degree of a winning strategy* in the corresponding game (see Martin [44]).

The one can go on to show that the elements of NCR_n show up at a rather *low level of the constructible universe*. It holds that $\mathrm{NCR}_n \subseteq L_{\beta_n}$, where β_n is the least ordinal such that

$$L_{\beta_n} \models \mathsf{ZFC}^- + \text{ there exist } n \text{ many iterates of the power set of } \omega,$$

where ZFC^- is Zermelo-Fraenkel set theory without the Power Set Axiom.

To show this, given $x \notin L_{\beta_n}$, construct a set G such that $L_{\beta_n}[G]$ is a model of ZFC^-_n, and for all $y \in L_{\beta_n}[G] \cap 2^\omega$, $y \leq_{\mathrm{T}} x \oplus G$. G is constructed by *Kumabe-Slaman forcing* (see [68]). The existence of G allows to conclude: If x is not in L_{β_n}, it will belong to every cone with base in $L_{\beta_n}[G]$. In particular, it will belong to the cone given by Martin's argument (relativized to G, here one has to use absoluteness), i.e. the cone avoiding NCR_n. Hence x is random relative to G for some continuous μ, an thus in particular μ-random. ⊣

The proof of the countability of NCR_n makes essential use of Borel determinacy.

It is known from a result by Friedman [18] that the use of infinitely many iterates of the power set of ω is necessary to prove Borel determinacy. As a base for an induction on the levels of the Borel hierarchy, Friedman showed that ZFC^- does not prove the statement "All $\mathbf{\Sigma}^0_5$-games on countable trees are determined". The proof works by showing that there is a model of ZFC^- for which $\mathbf{\Sigma}^0_5$-determinacy does not hold. This model is just L_{β_0}.

Very recently, Reimann and Slaman [60] showed that for every fixed k, NCR_n is cofinal in the Turing degrees of L_{β_k}. It allowed them to infer the following result.

THEOREM 6.8 (Reimann and Slaman [60]). *For every k, the statement*

For every n, NCR_n is countable.

cannot be proven in

ZFC^- + *there exists k many iterates of the power set of ω.*

The proof uses Jensen's *master codes* [23] as witnesses for NCR_n.

This line of work indicates that questions about randomness for continuous measures formalizable in second order arithmetic (such as the one formulated in the problem above) extend far into the realm of (descriptive) set theory.

Acknowledgement. I would like to thank Barry Cooper for encouraging me to write this article, as well as for his patience as an editor throughout the process of writing it.

REFERENCES

[1] K. Ambos-Spies and A. Kučera, *Randomness in computability theory*, ***Computability Theory and its Applications (Boulder, CO, 1999)***, Contemporary Mathematics, vol. 257, AMS, Providence, RI, 2000, pp. 1–14.

[2] C. H. Bennett, *Logical depth and physical complexity*, ***The Universal Turing Machine: A Half-Century Survey***, Oxford Science Publications, Oxford University Press, New York, 1988, pp. 227–257.

[3] L. Bienvenu, *Constructive equivalence relations on computable probability measures*, ***Computer Science—Theory and Applications, First International Computer Science Symposium in Russia, CSR 2006, St. Petersburg, Russia***, Lecture Notes in Computer Science, vol. 3967, Springer, Berlin, 2006, pp. 92–103.

[4] L. Bienvenu, D. Doty, and F. Stephan, *Constructive dimension and weak truth-table degrees*, ***Computation and Logic in the Real World—Third Conference of Computability in Europe*** (S. B. Cooper, B. Löwe, and A. Sorbi, editors), Electronic Notes in Theoretical Computer Science, vol. 4497, Springer, Berlin, 2007.

[5] L. Bienvenu and W. Merkle, *Effective randomness for computable probability measures*, ***Proceedings of the Third International Conference on Computability and Complexity in Analysis (CCA 2006)***, Electronic Notes in Theoretical Computer Science, vol. 167, Elsevier, Amsterdam, 2007, pp. 117–130.

[6] P. Billingsley, ***Probability and Measure***, third ed., Wiley Series in Probability and Mathematical Statistics, John Wiley & Sons, New York, 1995.

[7] J.-Y. Cai and J. Hartmanis, *On Hausdorff and topological dimensions of the Kolmogorov complexity of the real line*, ***Journal of Computer and System Sciences***, vol. 49 (1994), no. 3, pp. 605–619.

[8] C. Calude, L. Staiger, and S. A. Terwijn, *On partial randomness*, ***Annals of Pure and Applied Logic***, vol. 138 (2006), no. 1-3, pp. 20–30.

[9] D. Cenzer, P. Clote, R. L. Smith, R. I. Soare, and S. S. Wainer, *Members of countable Π^0_1 classes*, ***Annals of Pure and Applied Logic***, vol. 31 (1986), no. 2-3, pp. 145–163, Special issue: Second Southeast Asian logic conference (Bangkok, 1984).

[10] G. J. Chaitin, *A theory of program size formally identical to information theory*, ***Journal of the Association for Computing Machinery***, vol. 22 (1975), pp. 329–340.

[11] ———, *Information-theoretic characterizations of recursive infinite strings*, ***Theoretical Computer Science***, vol. 2 (1976), no. 1, pp. 45–48.

[12] O. Demuth, *Remarks on the structure of tt-degrees based on constructive measure theory*, ***Commentationes Mathematicae Universitatis Carolinae***, vol. 29 (1988), no. 2, pp. 233–247.

[13] D. Doty, *Dimension extractors and optimal decompression*, *Theory of Computing Systems*, to appear, Special issue of invited papers from Computability in Europe 2006.

[14] R. Downey and D. R. Hirschfeldt, *Algorithmic randomness and complexity*, book, in preparation.

[15] R. Downey, D. R. Hirschfeldt, J. S. Miller, and A. Nies, *Relativizing Chaitin's halting probability*, ***J. Math. Log.***, vol. 5 (2005), no. 2, pp. 167–192.

[16] H. G. Eggleston, *The fractional dimension of a set defined by decimal properties*, ***The Quarterly Journal of Mathematics. Oxford. Second Series***, vol. 20 (1949), pp. 31–36.

[17] K. Falconer, ***Fractal Geometry: Mathematical Foundations and Applications***, John Wiley & Sons, Chichester, 1990.

[18] H. M. Friedman, *Higher set theory and mathematical practice*, ***Annals of Pure and Applied Logic***, vol. 2 (1970/1971), no. 3, pp. 325–357.

[19] P. Gács, *Every sequence is reducible to a random one*, ***Information and Control***, vol. 70 (1986), no. 2-3, pp. 186–192.

[20] ———, *Uniform test of algorithmic randomness over a general space*, ***Theoretical Computer Science***, vol. 341 (2005), no. 1-3, pp. 91–137.

[21] P. R. Halmos, ***Measure Theory***, D. Van Nostrand Company, New York, 1950.

[22] J. M. Hitchcock, J. H. Lutz, and E. Mayordomo, *Fractal geometry in complexity classes*, SIGACT News Complexity Theory Column, September 2005.

[23] R. B. Jensen, *The fine structure of the constructible hierarchy*, ***Annals of Pure and Applied Logic***, vol. 4 (1972), pp. 229–308; Erratum, ibid. vol. 4 (1972), p. 443, With a section by Jack Silver.

[24] C. G. Jockusch, Jr., M. Lerman, R. I. Soare, and R. M. Solovay, *Recursively enumerable sets modulo iterated jumps and extensions of Arslanov's completeness criterion*, ***The Journal of Symbolic Logic***, vol. 54 (1989), no. 4, pp. 1288–1323.

[25] C. G. Jockusch, Jr. and R. I. Soare, Π^0_1 *classes and degrees of theories*, ***Transactions of the American Mathematical Society***, vol. 173 (1972), pp. 33–56.

[26] D. W. Juedes, J. I. Lathrop, and J. H. Lutz, *Computational depth and reducibility*, ***Theoretical Computer Science***, vol. 132 (1994), no. 1-2, pp. 37–70.

[27] S. Kakutani, *On equivalence of infinite product measures*, ***Annals of Mathematics. Second Series***, vol. 49 (1948), pp. 214–224.

[28] S. M. Kautz, ***Degrees of Random Sequences***, Ph.D. thesis, Cornell University, Ithaca, New York, 1991.

[29] A. S. Kechris, ***Classical Descriptive Set Theory***, Graduate Texts in Mathematics, vol. 156, Springer-Verlag, New York, 1995.

[30] B. Kjos-Hanssen, W. Merkle, and F. Stephan, *Kolmogorov complexity and the recursion theorem*, ***STACS 2006***, Lecture Notes in Computer Science, vol. 3884, Springer, Berlin, 2006, pp. 149–161.

[31] B. Kjos-Hanssen and A. Montalban, *Personal communication*, March 2005.

[32] G. Kreisel, *Analysis of the Cantor-Bendixson theorem by means of the analytic hierarchy*, ***Bulletin de l'Académie Polonaise des Sciences. Série des Sciences Mathématiques, Astronomiques et Physiques***, vol. 7 (1959), pp. 621–626.

[33] A. Kučera, *Measure,* Π^0_1*-classes and complete extensions of* PA, ***Recursion Theory Week (Oberwolfach, 1984)***, Lecture Notes in Mathematics, vol. 1141, Springer, Berlin, 1985, pp. 245–259.

[34] M. Kumabe, *A fixed-point free minimal degree*, Unpublished manuscript, 51 pages, 1996.

[35] L. A. Levin, *The concept of a random sequence*, ***Doklady Akademii Nauk SSSR***, vol. 212 (1973), pp. 548–550.

[36] ———, *Laws on the conservation (zero increase) of information, and questions on the foundations of probability theory*, ***Akademiya Nauk SSSR. Institut Problem Peredachi Informatsii Akademii Nauk SSSR. Problemy Peredachi Informatsii***, vol. 10 (1974), no. 3, pp. 30–35.

[37] ———, *Uniform tests for randomness*, ***Doklady Akademii Nauk SSSR***, vol. 227 (1976), no. 1, pp. 33–35.

[38] ———, *Randomness conservation inequalities: information and independence in mathematical theories*, ***Information and Control***, vol. 61 (1984), no. 1, pp. 15–37.

[39] M. Li and P. Vitányi, ***An Introduction to Kolmogorov Complexity and Its Applications***, second ed., Graduate Texts in Computer Science, Springer-Verlag, New York, 1997.

[40] J. H. LUTZ, *Weakly hard problems*, ***SIAM Journal on Computing***, vol. 24 (1995), no. 6, pp. 1170–1189.

[41] ———, *Dimension in complexity classes*, ***Proceedings of the Fifteenth Annual IEEE Conference on Computational Complexity (Florence, 2000)***, IEEE Computer Society, Los Alamitos, CA, 2000, pp. 158–169.

[42] ———, *Gales and the constructive dimension of individual sequences*, ***Automata, Languages and Programming (Geneva, 2000)***, Lecture Notes in Computer Science, vol. 1853, Springer, Berlin, 2000, pp. 902–913.

[43] ———, *The dimensions of individual strings and sequences*, ***Information and Computation***, vol. 187 (2003), no. 1, pp. 49–79.

[44] D. A. MARTIN, *The axiom of determinateness and reduction principles in the analytical hierarchy*, ***Bulletin of the American Mathematical Society***, vol. 74 (1968), pp. 687–689.

[45] P. MARTIN-LÖF, *The definition of random sequences*, ***Information and Computation***, vol. 9 (1966), pp. 602–619.

[46] P. MATTILA, ***Geometry of Sets and Measures in Euclidean Spaces: Fractals and Rectifiability***, Cambridge Studies in Advanced Mathematics, vol. 44, Cambridge University Press, Cambridge, 1995.

[47] E. MAYORDOMO, *A Kolmogorov complexity characterization of constructive Hausdorff dimension*, ***Information Processing Letters***, vol. 84 (2002), no. 1, pp. 1–3.

[48] J. S. MILLER and A. NIES, *Randomness and computability: open questions*, ***The Bulletin of Symbolic Logic***, vol. 12 (2006), no. 3, pp. 390–410.

[49] Y. N. MOSCHOVAKIS, ***Descriptive Set Theory***, Studies in Logic and the Foundations of Mathematics, vol. 100, North-Holland, Amsterdam, 1980.

[50] A. A. MUCHNIK, A. L. SEMENOV, and V. A. USPENSKY, *Mathematical metaphysics of randomness*, ***Theoretical Computer Science***, vol. 207 (1998), no. 2, pp. 263–317.

[51] M. E. MUNROE, ***Introduction to Measure and Integration***, Addison-Wesley, Cambridge, MA, 1953.

[52] A. NIES, *Computability and randomness*, in preparation.

[53] A. NIES and J. REIMANN, *A lower cone in the wtt degrees of non-integral effective dimension*, To appear in *Proceedings of IMS workshop on Computational Prospects of Infinity*, 2006.

[54] P. G. ODIFREDDI, ***Classical Recursion Theory. Vol. II***, Studies in Logic and the Foundations of Mathematics, vol. 143, North-Holland, Amsterdam, 1999.

[55] K. R. PARTHASARATHY, ***Probability Measures on Metric Spaces***, Probability and Mathematical Statistics, No. 3, Academic Press, New York, 1967.

[56] D. B. POSNER and R. W. ROBINSON, *Degrees joining to* $\mathbf{0}'$, ***The Journal of Symbolic Logic***, vol. 46 (1981), no. 4, pp. 714–722.

[57] J. REIMANN, ***Computability and Fractal Dimension***, Doctoral dissertation, Universität Heidelberg, Heidelberg, 2004.

[58] ———, *Effectively closed sets of measures and randomness*, in preparation.

[59] J. REIMANN and T. A. SLAMAN, *Randomness, entropy, and reducibility*, in preparation.

[60] ———, *Randomness for continuous measures*, in preparation, 2007.

[61] ———, *Measures and their random reals*, To be submitted for publication, 2007.

[62] J. REIMANN and F. STEPHAN, *Hierarchies of randomness tests*, ***Mathematical Logic in Asia*** (S. S. Goncharov, R. G. Downey, and H. Ono, editors), World Science Publications, Hackensack, NJ, 2006, pp. 215–232.

[63] C. A. ROGERS, ***Hausdorff Measures***, Cambridge University Press, London, 1970.

[64] B. Y. RYABKO, *Coding of combinatorial sources and Hausdorff dimension*, ***Doklady Akademii Nauk SSSR***, vol. 277 (1984), no. 5, pp. 1066–1070.

[65] ———, *Noise-free coding of combinatorial sources, Hausdorff dimension and Kolmogorov complexity*, ***Akademiya Nauk SSSR. Institut Problem Peredachi Informatsii Akademii Nauk SSSR. Problemy Peredachi Informatsii***, vol. 22 (1986), no. 3, pp. 16–26.

[66] C.-P. SCHNORR, ***Zufälligkeit und Wahrscheinlichkeit. Eine algorithmische Begründung der Wahrscheinlichkeitstheorie***, Lecture Notes in Mathematics, vol. 218, Springer-Verlag, Berlin, 1971.

[67] A. K. SHEN, *Relationships between different algorithmic definitions of randomness*, ***Doklady Akademii Nauk SSSR***, vol. 302 (1988), no. 3, pp. 548–552.

[68] R. A. SHORE and T. A. SLAMAN, *Defining the Turing jump*, ***Mathematical Research Letters***, vol. 6 (1999), no. 5-6, pp. 711–722.

[69] R. M. SOLOVAY, *Draft of a paper on chaitin's work*, 1975, Manuscript, IBM Thomas J. Watson Research Center.

[70] L. STAIGER, *A tight upper bound on Kolmogorov complexity and uniformly optimal prediction*, ***Theory of Computing Systems***, vol. 31 (1998), no. 3, pp. 215–229.

[71] ———, *Constructive dimension equals Kolmogorov complexity*, ***Information Processing Letters***, vol. 93 (2005), no. 3, pp. 149–153.

[72] LUDWIG STAIGER, *Kolmogorov complexity and Hausdorff dimension*, ***Information and Computation***, vol. 103 (1993), no. 2, pp. 159–194.

[73] F. STEPHAN, *Hausdorff-dimension and weak truth-table reducibility*, ***Logic Colloquium 2004***, Lecture Notes in Logic, vol. 29, ASL, Chicago, IL, 2008, pp. 157–167.

[74] K. TADAKI, *A generalization of Chaitin's halting probability* Ω *and halting self-similar sets*, ***Hokkaido Mathematical Journal***, vol. 31 (2002), no. 1, pp. 219–253.

[75] M. VAN LAMBALGEN, ***Random Sequences***, Ph.D. thesis, Universiteit van Amsterdam, Amsterdam, 1987.

[76] V. G. VOVK, *On a criterion for randomness*, ***Doklady Akademii Nauk SSSR***, vol. 294 (1987), no. 6, pp. 1298–1302.

[77] W. H. WOODIN, *A tt-version of the Posner-Robinson Theorem*, Submitted for publication.

[78] A. K. ZVONKIN and L. A. LEVIN, *The complexity of finite objects and the basing of the concepts of information and randomness on the theory of algorithms*, ***Akademiya Nauk SSSR i Moskovskoe Matematicheskoe Obshchestvo. Uspekhi Matematicheskikh Nauk***, vol. 25 (1970), no. 6(156), pp. 85–127.

DEPARTMENT OF MATHEMATICS
UNIVERSITY OF CALIFORNIA
970 EVANS HALL, BERKELEY
CA 94720, USA
E-mail: reimann@math.berkeley.edu

THE DERIVED MODEL THEOREM

J. R. STEEL

§1. We shall exposit here one of the basic theorems leading from large cardinals to determinacy, a result of Woodin known as the *derived model theorem*. The theorem dates from the mid-80's and has been exposited in several sets of informally circulated lecture notes (e.g., [11, 14]), but we know of no exposition in print. We shall also include a number of subsidiary and related results due to various people.

We shall use very heavily the technique of *stationary tower forcing*. The reader should see Woodin's paper [13] or Larson's [3] for the basic facts about stationary tower forcing. The second main technical tool needed for a full proof of the derived model theorem is the theory of iteration trees. This is one of the main ingredients in the proof of Theorem 3.1 below, but since we shall simply take that theorem as a "black box" here, it is possible to read this paper without knowing what an iteration tree is.

The paper is organized as follows. In §1, we introduce homogeneity, weak homogeneity, and universal Baireness. The main result here is the Martin-Solovay theorem, according to which all weakly homogeneous sets are universally Baire. We give a reasonably complete proof of this theorem. In §2 and §3, we show that in the presence of Woodin cardinals, homogeneity, weak homogeneity, and universal Baireness are equivalent. We also give, in §3, an argument of Woodin's which shows that strong cardinals yield universally Baire representations after a collapse. In §4 we prove the *tree production lemma*, according to which sets admitting definitions with certain absoluteness properties are universally Baire. §5 contains a generic absoluteness theorem for $(\Sigma^2_1)^{\mathrm{Hom}_\infty}$ statements. In §6 we state and prove the derived model theorem. In §7 we prove that in derived models, the pointclass Σ^2_1 has the Scale Property, and in §8, we use this to produce derived models which satisfy $\mathsf{AD}_{\mathbb{R}}$.

This paper was written in Fall 2002, and circulated informally. In 2004, Larson's monograph [3] on stationary tower forcing appeared. Much of the material in §2, §3, and §4 can be found in Section 3.3 of [3]. Sections 3.2 and 3.4 of [3] use this machinery to prove important results of Woodin concerning generic absoluteness. In another direction, Neeman's recent paper [9] gives

Logic Colloquium '06
Edited by Barry Cooper, Herman Geuvers, Anand Pillay, and Jouko Väänänen
Lecture Notes in Logic, 32

a proof of $\mathsf{AD}^{L(\mathbb{R})}$ which avoids stationary tower forcing entirely, relying entirely on iteration trees instead. Finally, author has published a stationary-tower-free proof of the full derived model theorem in [10].

§2. Homogeneously Suslin and universally Baire sets. If Y is a set, then $Y^\omega = \{x \mid x\colon \omega \to Y\}$ is the set of infinite sequences of elements of Y. We often regard Y^ω as being equipped with the Baire topology, whose basic open sets are just those of the form $N_s = \{x \in Y^\omega \mid s \subseteq x\}$, for $s \in Y^{<\omega}$. We are most interested in the case $Y = \omega$. We call the elements of ω^ω reals, and write $\mathbb{R}$ for ω^ω. In this section we introduce three regularity properties a set $A \subseteq Y^\omega$ might have.

The first of these, homogeneity, derives from Martin [4], and was first explicitly isolated by Martin and Kechris. In a word, a set $A \subseteq Y^\omega$ is homogeneously Suslin just in case it is continuously reducible to wellfoundedness of towers of measures. Here is some further detail.

We shall use the terms *ultrafilter on I* and *measure on I* interchangeably; thus all our measures take values in $\{0, 1\}$.

Definition 2.1. For any Z, $\text{meas}_\kappa(Z)$ is the set of all κ-additive measures on $Z^{<\omega}$. We let $\text{meas}(Z) = \text{meas}_{\omega_1}(Z)$.

Clearly, if $\mu \in \text{meas}(Z)$, then there is exactly one $n < \omega$ such that $\mu(Z^n) = 1$. We call n the *dimension* of μ, and write $n = \dim(\mu)$.

If $\mu, \nu \in \text{meas}(Z)$, then we say that μ *projects* to ν iff for some $m \leq n < \omega$, $\dim(\mu) = n$, $\dim(\nu) = m$, and for all $A \subseteq Z^m$

$$\nu(A) = \mu(\{u \mid u \restriction m \in A\}).$$

We say μ and ν are *compatible* if one projects to the other. If μ projects to ν, then there is a natural embedding

$$\pi_{\nu,\mu}\colon \text{Ult}(V, \nu) \to \text{Ult}(V, \mu)$$

given by $\pi([f]_\nu) = [f^*]_\mu$, where $f^*(u) = f(u \restriction m)$ for all $u \in Z^n$.

A *tower of measures* on Z is a sequence $\langle \mu_n \mid n < k\rangle$, where $k \leq \omega$, such that each $\mu_n \in \text{meas}(Z)$, and whenever $m \leq n < k$, then $\dim(\mu) = n$ and μ_n projects to μ_m. If $\langle \mu_n \mid n < \omega\rangle$ is an infinite tower of measures, then

$$\text{Ult}(V, \langle \mu_n \mid n < \omega\rangle) = \text{dir lim}_{n<\omega} \text{Ult}(V, \mu_n),$$

where the direct limit is taken using the natural embeddings π_{μ_n,μ_m}, which commute with one another[1]. We say that the tower $\langle \mu_n \mid n < \omega\rangle$ is *countably complete* just in case whenever $\mu_{x\restriction n}(A_n) = 1$ for all $n < \omega$, then $\exists f\ \forall n$ $(f \restriction n \in A_n)$. It is easy to show that $\langle \mu_n \mid n < \omega\rangle$ is countably complete if and only if $\text{Ult}(V, \langle \mu_n \mid n < \omega\rangle)$ is wellfounded, and so we shall say that a tower is wellfounded just in case it is countably complete.

[1]Note that $Z^0 = \{\emptyset\}$, so in any tower, μ_0 is principal, and $\text{Ult}(V, \mu_0) = V$.

Definition 2.2. A *homogeneity system over Y with support Z* is a function

$$\bar{\mu}\colon Y^{<\omega} \to \text{meas}(Z)$$

such that, writing $\mu_s = \bar{\mu}(s)$, we have that for all $s, t \in Y^{<\omega}$,

1. $\dim(\mu_t) = \text{dom}(t)$, and
2. $s \subseteq t \Rightarrow \mu_t$ projects to μ_s.

If $\text{ran}(\bar{\mu}) \subseteq \text{meas}_\kappa(Z)$, then we say that $\bar{\mu}$ is *κ-complete*.

Definition 2.3. If $\bar{\mu}$ is a homogeneity system over Y with support Z, then for each $x \in Y^{<\omega}$, we let $\bar{\mu}_x$ be the tower of measures $\langle \mu_{x\restriction n} \mid n < \omega \rangle$, and set

$$S_{\bar{\mu}} = \{x \in Y^{<\omega} \mid \bar{\mu}_x \text{ is countably complete}\}.$$

Definition 2.4. Let $A \subseteq Y^{<\omega}$; then A is *κ-homogeneous* iff $A = S_{\bar{\mu}}$, for some κ-complete homogeneity system $\bar{\mu}$. We say A is *homogeneous* if it is κ-homogeneous for some κ.

For the collection of all κ-homogeneous sets we write

$$\text{Hom}_\kappa^Y = \{A \subseteq Y^{<\omega} \mid A \text{ is } \kappa\text{-homogeneous}\}.$$

We set

$$\text{Hom}_{<\lambda}^Y = \bigcap_{\kappa<\lambda} \text{Hom}_\kappa^Y,$$

and

$$\text{Hom}_\infty^Y = \bigcap_{\kappa\in 0R} \text{Hom}_\kappa^Y.$$

We write Hom_κ for Hom_κ^ω, etc. It is clear that Hom_κ^Y is closed downward under continuous reducibility (also called Wadge reducibility, at least if $Y = \omega$). It is not too hard to show Hom_κ^Y is closed under countable intersections. One cannot prove closure under complementation in ZFC. We don't know whether closure under union is provable in ZFC.

Homogeneity is often considered in conjunction with trees. A *tree* on a set X is a set $T \subseteq X^{<\omega}$ such that $\forall s \in T \forall k (s \restriction k \in T)$. We let $[T] = \{f \in X^\omega \mid \forall n < \omega (f \restriction n \in T)\}$ be the set of infinite branches of T, so that T is wellfounded (under reverse inclusion) iff $[T] = \emptyset$. We think of a tree on $Y \times Z$ as a set of pairs $(s, t) \in Y^{<\omega} \times Z^{<\omega}$ such that $\text{dom}(s) = \text{dom}(t)$. If T is a tree on $Y \times Z$ and $s \in Y^{<\omega}$, then

$$T_s = \{t \mid (s, t) \in T\},$$

and for $x \in Y^\omega$,

$$T_x = \bigcup_{n<\omega} T_{x\restriction n}.$$

The projection on Y^ω of $[T]$ is given by

$$\begin{aligned} x \in p[T] &\Leftrightarrow T_x \text{ is illfounded} \\ &\Leftrightarrow \exists f \in Z^\omega \forall n((x \restriction n, f \restriction n) \in T), \end{aligned}$$

for $x \in Y^\omega$. We call T a *Suslin representation* of $p[T]$, and say that $p[T]$ is Z-*Suslin* via T.

PROPOSITION 2.5 (Woodin). *Let $A = S_{\bar{\mu}}$, where $\bar{\mu}$ is a homogeneity system over Y with support Z. Suppose that $\bar{\mu}$ is $|Y|^+$-complete; then there is a tree T on $Y \times Z$ such that*

$$A = p[T],$$

and for all $s \in Y^{<\omega}$,

$$\mu_s(T_s) = 1.$$

PROOF. For each $x \in Y^\omega \setminus A$, pick a sequence of sets B_n^x witnessing that $\bar{\mu}_x$ is not countably complete, so that $\mu_{x \restriction n}(B_n^x) = 1$ for all n, but $\not\exists f \forall n$ $(f \restriction n \in B_n^x)$. Then put

$$(s, u) \in T \Leftrightarrow u \in Z^{\operatorname{dom} s} \wedge \forall x \supseteq s \left(x \notin A \Rightarrow u \in B^x_{\operatorname{dom} s}\right).$$

Since the additivity of any measure is a measurable cardinal, each μ_s is sufficiently additive that $\mu_s(T_s) = 1$. If $x \in A$, then the countable completeness of $\bar{\mu}_x$ implies there is an f such that for all n, $f \restriction n \in T_{x \restriction n}$, so that $x \in p[T]$. On the other hand, if $x \notin A$, then the choice of the B_n^x guarantees that $x \notin p[T]$. ⊣

The completeness hypothesis on $\bar{\mu}$ in Proposition 2.5 is redundant in the case Y is countable. If T is related to $\bar{\mu}$ as in Proposition 2.5, and $\bar{\mu}$ is κ-complete, then we shall say that T is a κ- *homogeneous tree*, and $\bar{\mu}$ is a homogeneity system for T. We also say that $p[T]$ is κ-*homogeneously Suslin*.

K. Windszus has proved a stronger version of Proposition 2.5, showing that if $A \subseteq Y^\omega$ is continuously reducible to wellfoundedness of direct limit systems on $|Y^\omega|$-closed models, then A is homogeneously Suslin.

REMARK 2.6. Let $Y \subseteq \operatorname{meas}_\kappa(Z)$, and let

$$A = \{\langle \mu_n \mid n < \omega \rangle \in Y^\omega \mid \langle \mu_n \mid n < \omega \rangle \text{ is wellfounded}\}.$$

It is clear that A is κ-homogeneous, the homogeneity system over Y being simply the identity function. As one might suspect, this is not a very useful fact. If in addition $|Y| < \kappa$, then by Proposition 2.5, A is κ-homogeneously *Suslin*, which is more useful. (See Remark 3.2.)

The first and most important fact about homogeneously Suslin sets is

THEOREM 2.7 (Martin [4], essentially). *If $A \subseteq Y^\omega$ is $|Y|^+$-homogeneous, then the two-person game of perfect information on Y with payoff set A is determined.*

We now consider a weakening of homogeneity, also first isolated by Kechris and Martin.

DEFINITION 2.8. A *weak homogeneity system over Y with support Z* is an injective function $\bar{\mu}\colon Y^{<\omega} \to \text{meas}(Z)$ such that for all $s \in Y^{<\omega}$

1. $\dim(\mu_s) \leq \text{dom}(s)$, and
2. if μ_s projects to ν, then $\exists i(\mu_{s\restriction i} = \nu)$.

DEFINITION 2.9. If $\bar{\mu}$ is a (κ-complete) weak homogeneity system over Y, then we set

$$W_{\bar{\mu}} = \Big\{x \in Y^\omega \mid \exists\langle i_k \mid k < \omega\rangle \in \omega^\omega(\langle \mu_{x\restriction i_k} \mid k < \omega\rangle \text{ is a wellfounded tower})\Big\},$$

and say that $W_{\bar{\mu}}$ is (κ-)*weakly homogeneous* via $\bar{\mu}$.

So a weak homogeneity system over Y associates continuously to each $x \in Y$ a countable tree of towers of measures, and x is in the set being represented iff at least one of the branches of this tree is a wellfounded tower[2]. This leads us to

PROPOSITION 2.10. *Let $A \subseteq Y^\omega$. If there is a measurable cardinal, then the following are equivalent*:

1. *A is κ-weakly homogeneous,*
2. *there is a κ-homogeneous set $B \subseteq Y^\omega \times \omega^\omega$ such that $x \in A \Leftrightarrow \exists y(x, y) \in B$, for all x.*

We leave the proof to the reader. The measurable cardinal is needed for a minor technical reason: if there are no measurables, then $Y^\omega \times Z^\omega$ is the only homogeneous $B \subseteq Y^\omega \times Z^\omega$, while the projection of any closed $B \subseteq Y^\omega \times Z^\omega$ is weakly homogeneous. This situation could be remedied by allowing *partial* $\bar{\mu}$ as homogeneity systems, but in any case, homogeneity isn't very interesting if there are no measurable cardinals.

For trees we make the following definition:

DEFINITION 2.11. A tree T on $Y \times Z$ is *κ-weakly homogeneous via $\bar{\mu}$* iff $\bar{\mu}$ is a κ-complete weak homogeneity system over Y such that

1. $p[T] = W_{\bar{\mu}}$, and
2. $\forall s \in Y^{<\omega}\exists k(\mu_s(T_{s\restriction k})) = 1$.

We say that $p[T]$ is *κ-weakly homogeneously Suslin* in this case.

---

[2]Had we taken a weak homogeneity system to be *any* function $\bar{\mu}\colon Y^{<\omega} \to \text{meas}(Z)$, we would have obtained the same class of sets $W_{\bar{\mu}}$. Our restrictions on $\bar{\mu}$ make some manipulations easier. Note that they imply $\mu_\emptyset$ is principal, and that if $\langle i_k \mid k < \omega\rangle$ is a sequence witnessing $x \in W_{\bar{\mu}}$ then $i_0 = 0$ and $k < l \Rightarrow i_k < i_l$.

Parallel to Proposition 2.5 we have:

PROPOSITION 2.12. *Let $\bar{\mu}$ be a $|Y|^+$-complete weak homogeneity system over Y with support Z; then there is a tree T on $Y \times Z$ such that which is weakly homogeneous via $\bar{\mu}$.*

So if $A \subseteq Y^\omega$ and $\kappa > |Y|$, then A is κ-homogeneous iff A is κ-homogeneously Suslin, and A is κ-weakly homogeneous iff A is κ-weakly homogeneously Suslin. In the sequel, Y will almost always be countable, so these equivalences apply. Our characterization of weak homogeneity for trees also simplifies a bit in the case Y is countable:

PROPOSITION 2.13. *A tree T on $\omega \times Z$ is κ-weakly homogeneous iff there is a countable set $\sigma \subseteq \text{meas}_\kappa(Z)$ so that $\forall\, x$*

$$x \in p[T] \Leftrightarrow \begin{array}{l}\text{There is a tower } \langle \mu_n \mid n < \omega \rangle \text{ of} \\ \text{measures from } \sigma \text{ such that} \\ \mu_n(T_{x \restriction n}) = 1 \text{ for all } n, \text{ and} \\ \langle \mu_n \mid n < \omega \rangle \text{ is countably complete.}\end{array}$$

PROOF. Given such a σ, let $\sigma = \{\nu_i \mid i < \omega\}$ be a one-one enumeration such that if ν_i projects to μ, then $\exists k \leq i(\nu_k = \mu)$. Setting $\mu_s = \nu_{\text{dom}(s)}$, it is clear that T is weakly homogeneous via $\bar{\mu}$. Conversely, if T is weakly homogeneous via $\bar{\mu}$, then take $\sigma = \text{ran}(\bar{\mu})$. Since $W_{\bar{\mu}} = p[T]$, if $x \in p[T]$, then there is a countably complete tower from σ concentrating on T_x. The converse is true in general, because of countable completeness. ⊣

A still further weakening of homogeneity is the property of being κ-*universally Baire* (see [1]).

DEFINITION 2.14. We say G is $< \kappa$-*generic* over M iff G is M-generic for some poset $\mathbb{P}$ such that $M \models |\mathbb{P}| < \kappa$.

DEFINITION 2.15. Let T on $X \times Y$ and U on $X \times Z$ be two trees; then we say T and U are κ-*absolute complements* iff whenever G is $< \kappa$-generic over V

$$V[G] \models p[T] = X^\omega \setminus p[U].$$

We say T *is* κ-*absolutely complemented* iff $\exists U$ (T and U are κ-absolute complements).

If $p[T] \cap p[U] = \emptyset$ in V, then the same is true in any generic extension of V by the absoluteness of wellfoundedness. We shall use this simple observation again and again. What absolute complementation adds is that T and U are sufficiently "fat" that in the relevant $V[G]$, we have $p[T] \cup p[U] = X^\omega$.

DEFINITION 2.16. (1) A set $A \subseteq X^\omega$ is κ-*universally Baire*, or κ-*absolutely Suslin* iff $A = p[T]$ for some κ-absolutely complemented T.

(2) $UB_\kappa = \{A \subseteq \omega^\omega \mid A \text{ is } \kappa\text{-universally Baire}\}$.

Every provably-in-ZFC Δ^1_2 set of reals is κ-universally Baire for all κ. This is the key to Solovay's proof that such sets are Lebesgue measurable and have the Baire property. Indeed, any $(2^{\aleph_0})^+$- universally Baire set of reals has these regularity properties ([1]).

It is one of the main results of Martin–Solovay [5] that every κ-weakly homogeneous Suslin set is κ-universally Baire. Here is a brief review of the Martin–Solovay construction.

For simplicity, we begin with a homogeneity systems.

DEFINITION 2.17. Let $\bar{\mu}$ be a homogeneity system over Y. For any ordinal θ, we define the Martin-Solovay tree $ms(\bar{\mu}, \theta)$ on $Y \times \mathrm{OR}$ by

$$\begin{aligned}&\big(s, \langle \alpha_n \mid n < e\rangle\big) \in ms(\bar{\mu}, \theta) \Leftrightarrow \\ &\qquad s \in Y^e \wedge \alpha_0 < \theta \wedge \forall n\big(n+1 < e \Rightarrow \pi_{\mu_{s\restriction n}, \mu_{s\restriction(n+1)}}(\alpha_n) > \alpha_{n+1}\big).\end{aligned}$$

That is, $ms(\bar{\mu}, \theta)_x$ searches for a proof that $\mathrm{Ult}(V, \bar{\mu}_x)$ is illfounded below the image of θ. (This last restriction makes it a set, rather than a proper class.) It is not hard to see that if $\bar{\mu}$ has support Z and is illfounded, then it is illfounded below the image of $|Z|^+$. It follows easily that for any $\theta \geq |Z|^+$,

$$p[ms(\bar{\mu}, \theta)] = Y^\omega \setminus S_{\bar{\mu}}.$$

Thus if $\bar{\mu}$ is a homogeneity system for T, then $ms(\bar{\mu}, \theta)$ and T complement each other in V.

Now suppose $\bar{\mu}$ is κ-complete, and G is $< \kappa$-generic over V. The measures μ_s extend to measures μ_s^* in $V[G]$, where

$$\mu_s^*(A) = 1 \Leftrightarrow \exists B \subseteq A(\mu_s(B) = 1).$$

(We shall use this *-notation in this way without much comment in the future.) Moreover, for every function $f : Z^{<\omega} \to V$ such that $f \in V[G]$ there is a function $g \in V$ such that $f(u) = g(u)$ for μ_s^*-a.e. u. Thus $\bar{\mu}^*$ is a homogeneity system in $V[G]$, whose associated embeddings, when restricted to V, are those of $\bar{\mu}$. It follows that $ms(\bar{\mu}, \theta)^V = ms(\bar{\mu}^*, \theta)^{V[G]}$, and that $p[ms(\bar{\mu}, \theta)^V] = Y^\omega \setminus S_{\bar{\mu}^*}$ in $V[G]$.

Finally, suppose in addition that $\bar{\mu}$ is a homogeneity system for T in V. In order to see that $ms(\bar{\mu}, \theta)$ is a κ-absolute complement for T, it is enough to show $p[T] = S_{\bar{\mu}^*}$ in $V[G]$[3]. Now if $x \in S_{\bar{\mu}^*}$, then the tower $\bar{\mu}_x^*$ is countably complete, and since its measures concentrate on T_x, we get that T_x is illfounded, so $x \in p[T]$. Conversely, if $x \in p[T]$, then $x \notin p[ms(\bar{\mu}, \theta)^V]$, because the projections of these two trees were disjoint in V. As we showed above, this implies $x \in S_{\bar{\mu}^*}$ in $V[G]$, as desired.

We can extend this construction to weak homogeneity systems.

[3]This also shows that $\bar{\mu}^*$ is a homogeneity system for T in $V[G]$.

DEFINITION 2.18. Let $\bar{\mu}$ be a weak homogeneity system over Y, and θ be an ordinal. We define the Martin-Solovay tree $ms(\bar{\mu}, \theta)$ on $Y \times \mathrm{OR}$ by

$$(s, \langle \alpha_e \mid e < n \rangle) \in ms(\bar{\mu}, \theta) \Leftrightarrow s \in Y^n \wedge \alpha_0 < \theta \wedge \\ \forall e < k < n\big(\mu_{s \upharpoonright k} \text{ projects to } \mu_{s \upharpoonright e} \\ \Rightarrow \pi_{\mu_{s\upharpoonright e}, \mu_{s\upharpoonright k}}(\alpha_e) > \alpha_k\big).$$

LEMMA 2.19. *Suppose that $\bar{\mu}$ is a κ-complete weak homogeneity system over Y with support Z, and $\theta \geq |Z|^+$. Let G be $< \kappa$-generic over V; then $ms(\bar{\mu}, \theta)^V = ms(\bar{\mu}^*, \theta)^{V[G]}$, moreover $p[ms(\bar{\mu}, \theta)] = Y^\omega \setminus W_{\bar{\mu}^*}$ in $V[G]$.*

PROOF. We claim first that $p[ms(\bar{\mu}, \theta)] = Y^\omega \setminus W_{\bar{\mu}}$ in V. It is clear that if $x \in p[ms(\bar{\mu}, \theta)]$ then $x \notin W_{\bar{\mu}}$, since a branch through $ms(\bar{\mu}, \theta)_x$ illfounds all the relevant towers, and in fact does so *continuously*. Conversely, suppose $x \notin W_{\bar{\mu}}$. All the relevant towers are then illfounded, but we must see this is true continuously, and below the image of θ. For that, pick, for each increasing $t \colon \omega \to \omega$ such that $\langle \mu_{x \upharpoonright t(n)} \mid n < \omega \rangle$ is a tower, a sequence A_n^t witnessing the countable incompleteness of this tower. (So $A_n^t \subseteq Z^n$ and $\mu_{x \upharpoonright t(n)}(A_n^t) = 1$.) For any k, let B_k be the intersection over all t, n such that $t(n) = k$ of the A_n^t. (n is in fact determined by k, since $\mu_{x \upharpoonright k}(Z^n) = 1$.) Then letting

$$(k, u) R (l, v) \Leftrightarrow k > l \wedge u \in B_k \wedge v \in B_l \wedge v \subseteq u,$$

we have that R is wellfounded. Set

$$f_k(u) = \text{ rank of } (k, u) \text{ in } R$$

and let

$$\alpha_k = [f_k]_{\mu_{x \upharpoonright k}}.$$

It is easy to check that $\langle \alpha_k \mid k < \omega \rangle$ is a branch through $ms(\bar{\mu}, \theta)_x$, as desired.

The remainder of the lemma is proved as it was for homogeneity systems. ⊣

THEOREM 2.20 (Martin, Solovay [5]). *Let T be κ-weakly homogeneous via $\bar{\mu}$, and $\theta > |T|^+$; then T and $ms(\bar{\mu}, \theta)$ are κ-absolute complements.*

The proof of Theorem 2.20 is just as it was for homogeneity systems, so we omit further detail. The proof also shows T remains weakly homogeneous via $\bar{\mu}^*$ in $< \kappa$-generic extensions.

COROLLARY 2.21. *If $A \subseteq \mathbb{R}$ is κ-weakly homogeneous, then A is κ-universally Baire.*

To summarize: Any κ-homogeneous set is κ-weakly homogeneous, and any κ-weakly homogeneous set is κ-universally Baire.

§3. Weak homogeneity to homogeneity. In the next two sections we show that the implications above have converses, in a certain sense. Namely, if δ is Woodin, then any δ^+-universally Baire set is $< \delta$-weakly homogeneous, and any δ^+-weakly homogeneous set is $<\delta$-homogeneous. Thus if λ is a limit of Woodins, then $\mathrm{Hom}_{<\lambda} = UB_\lambda$.

THEOREM 3.1 (Martin, Steel [6]). *Let δ be Woodin, and let $\bar{\mu}$ be a δ^+-complete weak homogeneity system over Y, where $|Y| < \delta$; then for all sufficiently large θ, $ms(\bar{\mu}, \theta)$ is κ-homogeneous for all $\kappa < \delta$.*

PROOF. Omitted. ⊣

REMARK 3.2. Our hypothesis that $|Y|$ is strictly less than the completeness of $\bar{\mu}$ implies that there is a tree T on some $Y \times Z$ which is weakly homogeneous via $\bar{\mu}$. Given T *in advance*, the construction of [6] produces a homogeneity system for some $ms(\bar{\mu}, \theta)$ in a way which is *continuous in* $\bar{\mu}$, in that finite bits $\bar{\mu} \restriction i$ are needed to determine the next measure in the homogeneity system for $ms(\bar{\mu}, \theta)$.

Woodin has observed that this has the following nice consequence. Let $Y \subseteq \mathrm{meas}_\gamma(Z)$, and $|Y| < \delta < \gamma$ for some Woodin cardinal δ. Let

$$I = \{t \in Y^\omega \mid t \text{ is an illfounded tower}\}.$$

We observed in Remark 2.6 (essentially) that $Y^\omega \setminus I$ is γ-homogeneous[4]. It then follows from the continuity implicit in the construction of [6] that I is κ-homogeneous for all $\kappa < \delta$. One should compare this with Remark 2.6.

If $A \subseteq Q \times S$, we write $\exists^Q A$ for $\{s \mid \exists q(q, s) \in A\}$, and $\forall^Q A$ for $\{s \mid \forall q \in Q(q, s) \in A\}$. If $B \subseteq Q$, then we write $\neg B$ for $Q \setminus B$, when Q is clear from context.

COROLLARY 3.3. *If $A \subseteq \mathbb{R}^2$ is δ^+-homogeneous, where δ is Woodin, then $\neg\exists^{\mathbb{R}} A$ is κ-homogeneous for all $\kappa < \delta$.*

PROOF. By Proposition 2.10, $\exists^{\mathbb{R}} A$ is δ^+-weakly homogeneous, so its complement is $<\delta$-homogeneous by Theorem 3.1. ⊣

COROLLARY 3.4. *If λ is a limit of Woodin cardinals, then* $\mathrm{Hom}_{<\lambda}$ *is closed under $\exists^{\mathbb{R}}$, negation, and continuous (i.e., "Wadge") reducibility.*

So if λ is a limit of Woodins, then projective determinacy (PD) holds, and in fact holds in all $<\lambda$-generic extensions (since λ remains a limit of Woodins in such an extension). In fact, we get projective generic absoluteness for such extensions, even with names for sets of reals in $\mathrm{Hom}_{<\lambda}$, as we now show.

The following lemma gives us our names.

LEMMA 3.5. *Let (T, U) and (R, S) be pairs of κ-absolute complements, and suppose $p[T] = p[R]$ in V. Then for any $<\kappa$-generic G, $p[T] = p[R]$ in $V[G]$.*

---

[4]One must intersect with the set of all towers in order to apply Remark 2.6, but since the set of all towers is closed in Y^ω, this is not a problem.

PROOF. Say $x \in V[G]$ and $x \in p[T]$ but $x \notin p[R]$. Since S complements R, we have $x \in p[S]$. Thus $p[T] \cap p[S] \neq \emptyset$ in $V[G]$, hence in V, a contradiction. ⊣

So we can think of a κ-absolutely complemented pair (T, U) as a name for $p[T]$, and we have that two names which agree on V also agree on any $< \kappa$-generic $V[G]$. If $A = p[T]$ for such a (T, U), then we sometimes write $A^{V[G]}$ for $p[T]^{V[G]}$ if G is $< \kappa$- generic. There is no ambiguity because $A^{V[G]}$ does not depend on which absolutely complemented name we choose.

THEOREM 3.6 (Woodin). *Let λ be a limit of Woodin cardinals, and $A \in \mathrm{Hom}_{<\lambda}$. Let G be $< \lambda$-generic over V. Then*

$$\left(HC^V, \in, A\right) \equiv \left(HC^{V[G]}, \in, A\right).$$

Notice that if x is a real, then $\{x\}$ is $\mathrm{Hom}_{<\lambda}$, and any sequence of $\mathrm{Hom}_{<\lambda}$ sets can be coded by a single one. Thus Theorem 3.6 implies a superficially stronger version of itself.

PROOF. Fix $A \in \mathrm{Hom}_{<\lambda}$. To each formula $\varphi(\vec{v}, \dot{A})$ in the language of second order arithmetic with additional predicate symbol $\dot{A}$, and each $\kappa < \lambda$, we associate a κ-homogeneous tree $T_{\varphi,\kappa}$ such that whenever G is $< \kappa$-generic

$$V[G] \models p[T_{\varphi,\kappa}] = \left\{\vec{y} \in \mathbb{R}^{<\omega} \mid \varphi\left(\vec{y}, A^{V[G]}\right)\right\}.$$

The absoluteness of wellfoundedness and the Tarski-Vaught criterion then imply that V is φ-elementary in $V[G]$, as desired.

The trees $T_{\varphi,\kappa}$ are constructed (for all κ) by induction on φ. For φ Σ^1_1, we use the given trees for the A. Now suppose $\varphi = \neg\exists\, w\psi$, and $\kappa < \lambda$. Let $\kappa < \delta < \lambda$, where δ is Woodin. Now T_{ψ,δ^+} is δ^+-homogeneous as a tree on $(\omega \times \omega) \times Z$, and hence δ^+-weakly homogeneous as a tree S on $\omega \times (\omega \times Z)$. Of course, $p[S] = \exists^{\mathbb{R}} p[T_{\varphi,\delta^+}]$ in all generic extensions. Let $\bar{\mu}$ be a δ^+-complete weak homogeneity system for S, and θ be sufficiently large, and set $T_{\varphi,\kappa} = ms(\bar{\mu}, \theta)$. This works by the Martin–Solovay and Martin–Steel theorems.

Up to logical equivalence, all Σ^1_n formulae can be built up from Σ^1_1 using $\neg\exists\, w$, so we are done. ⊣

Here is a sometimes useful observation about $\mathrm{Hom}_{<\lambda}$. It is due independently to Woodin and the author.

THEOREM 3.7. *Let λ be a limit of Woodin cardinals; then there is a $\kappa < \lambda$ such that* $\mathrm{Hom}_\kappa = \mathrm{Hom}_{<\lambda}$.

PROOF. Clearly, $\alpha < \beta \Rightarrow \mathrm{Hom}_\beta \subseteq \mathrm{Hom}_\alpha$. Each Hom_α is a boldface pointclass, that is, it is closed downward under Wadge reducibility $\leq_w$. Thus if the theorem fails, we have an infinite descending sequence $A_0 >_w A_1 >_w \ldots$ in the Wadge order. But also, we have projective-in-A_0 determinacy, and so Martin's proof that $<_w$ is wellfounded yields a contradiction. ⊣

§4. **Universally Baire to weakly homogeneous.** Our method for obtaining weak homogeneity originated in Martin's unpublished proof that $\mathsf{AD}_{\mathbb{R}}$ implies that every tree on $\omega \times \kappa$ is weakly homogeneous. Woodin extended the method to the context of large cardinals with Choice, and eventually obtained the following remarkable results.

THEOREM 4.1 (Woodin). *Let δ be Woodin, and let T and U be δ^+-absolutely complementing trees on $\omega \times Z$; then T is κ-weakly homogeneous for all $\kappa < \delta$.*

PROOF. We shall actually just use that $p[T] = \mathbb{R} \setminus p[U]$ in $V[G]$, whenever G is generic for the "countable" stationary tower $\mathbb{Q}_{<\delta}$. (Conditions are stationary $a \subseteq P_{\omega_1}(V_\alpha)$, for $\alpha < \delta$.) Names for reals modulo $\mathbb{Q}_{<\delta}$ are elements of V_δ, so if we let T^* and U^* be the subtrees of T and U consisting of all nodes definable over V_η from T, δ, U, and parameters in V_δ (where $\eta \gg \delta$ and $T, U \in V_\eta$), then T^* and U^* have size δ and are forced in $\mathbb{Q}_{<\delta}$ to project to complements. Rearranging and renaming, we may assume T and U are on $\omega \times \delta$.

Let $\kappa < \delta$ be given; we wish to show T is κ-weakly homogeneous. We may assume κ is T-reflecting in δ, since there are arbitrarily large such $\kappa < \delta$. That is, for each λ such that $\kappa < \lambda < \delta$, there is $j : V \to M$ with $\mathrm{crit}(j) = \kappa$ such that $V_\lambda \subseteq M$ and $j(T) \cap V_\lambda = T \cap V_\lambda$.

Let $\kappa < \lambda < \delta$ and j be as above. We define a continuous function $\Sigma_{\lambda,j}$ on the branches of $T \cap V_\lambda$ whose outputs are towers of measures on $T \cap V_\kappa$. Namely, for $(s, u) \in T \cap V_\lambda$, and $X \subseteq V_\kappa$, let

$$\Sigma_{\lambda,j}(s,u)(X) = 1 \text{ iff } u \in j(X).$$

Writing $\Sigma = \Sigma_{\lambda,j}$, we see

(a) $\Sigma(s, u)$ is a κ-complete measure concentrating on $T_s \cap V_\kappa$.

(b) $(s, u) \subseteq (t, v) \Rightarrow \Sigma(s, u)$ is compatible with $\Sigma(t, v)$.

(c) If $(x, f) \in [T \cap V_\lambda]$, then the tower $\Sigma(x, f) \underset{\text{df.}}{=} \langle \Sigma(x \restriction n, f \restriction n) \mid n < \omega \rangle$ is wellfounded. (Note that its ultrapower embeds into M.)

(d) In any generic extension $V[H]$ of V, if $(x, f) \in [T \cap V_\lambda]$, then $\mathrm{Ult}(V, \Sigma(x, f))$ is wellfounded. (Here the ultrapower is formed using functions in V.) This is because the tree of attempts to produce $(x, f, \vec{\alpha})$ such that $(x, f) \in [T \cap V_\lambda]$ and $\vec{\alpha}$ is a descending sequence in $\mathrm{Ult}(V, Z(x, f))$ is wellfounded in V, hence in $V[H]$.

Now let G be $\mathbb{Q}_{<\delta}$ generic over V, and

$$i : V \to N \subseteq \mathrm{Ult}(V, G)$$

be the generic embedding. It suffices to show that $i(T)$ is $i(\kappa)$-weakly homogeneous in N. ⊣

In fact, we show

Claim. $\sigma = i''\text{meas}_\kappa(\kappa^{<\omega})$ is a witness that $i(T)$ is $i(\kappa)$-weakly homogeneous in N.

Proof. Since ${}^\omega N \subseteq N$ in $V[G]$, we have $\sigma \in N$ and is countable in N. Now let $x \in p[i(T)]$ and $x \in N$. Since T and U are absolutely complementing, either $x \in p[T]$ or $x \in p[U]$. But

$$x \in p[U] \Rightarrow x \in p[i(U)] \Rightarrow p[i(U)] \cap p[i(T)] \neq \emptyset,$$

whereas $p[i(U)] \cap p[i(T)] = \emptyset$ because this is true in N, which is a wellfounded model. Thus $x \in p[T]$.

Let $\lambda < \delta$ be such that $x \in p[T \cap V_\lambda]$; note here $\delta = \omega_1^{V[G]}$, so there is such a λ. Choose $\lambda < \kappa$, and let j be such that $\Sigma = \Sigma_{\lambda,j}$ exists in V. Letting $(x, f) \in [T \cap V_\lambda]$, we have

$$N \models \text{every tower in } i(\Sigma)''\big[i(T \cap V_\lambda)\big] \text{ is wellfounded},$$

so

$$N \models \langle i(\Sigma)(x \upharpoonright n, i(f \upharpoonright n)) \mid n < \omega \rangle \text{ is wellfounded.}$$

But $\langle i(\Sigma)(x \upharpoonright n, i(f \upharpoonright n)) \mid n < \omega \rangle$ is a tower of measures from σ concentrating on $i(T_x)$. This proves the claim, and hence the theorem. ⊣

Remark 4.2. It is possible to prove the theorem using the extender algebra at δ, and iterations to make reals generic, rather than stationary tower forcing. One then uses that T and U are complementing in $V[G]$, whenever G is generic for the extender algebra at δ. Here is a sketch: let κ be T-reflecting in δ. Let $X < V_\theta$ be countable, with θ large, and $\kappa, \delta, T, U \in X$. Take $\sigma = \{\mu \in X \mid \mu$ is a κ-complete measure concentrating on some $T_s \cap V_\kappa\}$. We claim σ witnesses that T is κ-weakly homogeneous. To see this, let $\pi : N \cong X$ be the transitive collapse. Let $x \in p[T]$. We can iterate $N \stackrel{i}{\longrightarrow} M$ to make x generic over M for the extender algebra at $i(\delta)$, and we have

$$\begin{array}{ccc} N & \stackrel{\pi}{\rightarrow} & V_\theta \\ & \searrow_i & \uparrow \tau \\ & & M \end{array}$$

for some realizing map τ. Letting $\bar{T} = \pi^{-1}(T)$, etc., we can arrange $\text{crit}(i) > \bar{\kappa}$. Then $x \in p[i(\bar{T})]$, as otherwise $x \in p[i(\bar{U})]$, so $x \in p[\sigma(i(\bar{U}))]$, so $x \in p[U]$. Now then if Σ is the appropriate $\Sigma_{\lambda,j}$, and $(x, f) \in p[i(\bar{T} \cap \bar{\lambda})]$, then $(\tau \circ i)(\bar{\Sigma})(x, \tau(f))$ is a wellfounded tower from σ concentrating on T_x.

This seems to require a certain amount of iterability, but by being more careful, we can make do with a form of iterability which is provable. (Use $\delta > \kappa$ which is Woodin in $L(V_\delta, T)$, and is the least such. In the extender algebra, use only identities induced by T-strong extenders with critical point above κ.)

REMARK 4.3. Theorem 4.1 also holds for trees T and U on $Y \times Z$, where $|Y| < \delta$. Both proofs generalize easily.

The functions $\Sigma_{\lambda,j}$ of the last theorem are useful in other ways. Here is another example:

THEOREM 4.4 (Woodin). *Let δ be Woodin, and let T on $\omega \times Z$ be any tree; then there is a $\kappa < \delta$ such that*

$$V^{\mathrm{Col}(\omega,\kappa)} \models T \text{ is } \alpha\text{-weakly homogeneous, for all } \alpha < \delta.$$

PROOF. Let

$$S = \{\xi < \delta \mid \xi \text{ is } T\text{-reflecting in } \delta\},$$

and let κ be S-reflecting in δ. We shall show that in $V^{\mathrm{Col}(\omega,2^{2^\kappa})}$, T is α-weakly homogeneous for all $\alpha < \delta$. So let $\alpha < \delta$ be given. Let $\xi < \delta$ be such that $\kappa, \alpha < \xi$ and ξ is T-reflecting in δ. Let $\xi < \lambda < \delta$, where λ is large enough that

$$\overset{\mathrm{Col}(\omega,2^{2^\kappa})}{\Vdash} \quad p[T] = p[T \cap V_\lambda],$$

and pick $j : V \to M$ with critical point κ such that $j(\kappa) > \lambda$, $V_\lambda \subseteq M$, and $j(S) \cap V_\lambda = S \cap V_\lambda$. Thus

$$M \models \xi \text{ is } j(T)\text{-reflecting in } j(\delta),$$

and we can find an embedding $k : M \to N$ (with extender) in N such that

$$V^M_{j(\kappa)+2} \subseteq N \text{ and } k(j(T)) \cap V^N_{j(\kappa)} = j(T) \cap V^N_{j(\kappa)}.$$

Working still in V, we can use j and k to associate to each κ-complete measure μ on some $T_s \cap V_\kappa$ a ξ-complete measure $\nu = \nu(\mu)$ on $T_s \cap V_\xi$: we put, for $A \subseteq T_s \cap V_\xi$

$$A \in \nu \Leftrightarrow k(A) \cap V^N_{j(\kappa)} \in j(\mu).$$

Thus ν is an "average" of measures which knit together into $j(\mu)$ in the k-ultrapower. It is easy to check that ν is a ξ-complete measure on V_ξ. Note $\nu(T_s) = 1$, since

$$\begin{aligned} k(T_s \cap V_\xi) \cap V^N_{j(\kappa)} &= k(j(T_s) \cap V_\xi) \cap V^N_{j(\kappa)} \\ &= j(T_s) \cap V^N_{j(\kappa)} \in j(\mu), \end{aligned}$$

using the reflecting properties of j and k.

Now let G be V-generic over $\mathrm{Col}(\omega, 2^{2^\kappa})$, and for $\nu \in \mathrm{meas}_\kappa(V_\kappa)$, let ν^* be the canonical extension of ν to $V[G]$; note here $2^{2^\kappa} < \xi$ and ν is ξ-complete. Letting

$$\sigma = \{\nu^* \mid \nu \in \mathrm{meas}_\kappa(V_\kappa)\},$$

we have that in $V[G]$, σ is a countable family of ξ-complete measures. In order to see σ witnesses that T is ξ-weakly homogeneous in $V[G]$, fix $x \in p[T]$ in $V[G]$. By choice of λ, we have an f such that $(x, f) \in [T \cap V_\lambda]$. Now consider the tower

$$\left\langle v\left(\Sigma_{\lambda,j}(x \restriction n, f \restriction n)\right)^* \mid n \in \omega \right\rangle.$$

This is a tower concentrating on T_x, and its measures are in σ. If it is illfounded, then the tree of all attempts to produce $(y, g) \in [T \cap V_\lambda]$ together with an infinite descending sequence $\overrightarrow{\alpha}$ in $\mathrm{Ult}(V, \langle v(\Sigma_{\lambda,j}(y \restriction n, g \restriction n)) \mid n < \omega\rangle)$ has a branch $(y, g, \bar{\alpha})$ in V. (Note here it doesn't matter whether we compute the ultrapowers in V of $V[G]$, as the measures are ξ-complete.) But for such (y, g), $\Sigma_{\lambda,j}(y, g)$ is wellfounded, so $j(\Sigma_{\lambda,j}(y, g))$ is wellfounded, which easily implies $\langle v(\Sigma_{\lambda,j}(y \restriction n, g \restriction n)) \mid n < \omega\rangle$ is wellfounded, a contradiction.

Thus $\langle v(\Sigma_{\lambda,j}(x \restriction n, f \restriction n))^* \mid n < \omega\rangle$ is wellfounded in $V[G]$, and we have shown σ witnesses that T is ξ-weakly homogeneous in $V[G]$. ⊣

In fact, if κ is merely a λ-strong cardinal, as witnessed by j, then in $V^{\mathrm{Col}(\omega, 2^{2^\kappa})}$ we get an approximation to λ-weak homogeneity for $j(T)$.

Theorem 4.5 (Woodin). *Let $\lambda = |V_\lambda|$, and $j : V \to M$ witness that κ is λ-strong, with ${}^\omega M \subseteq M$, and let T be any tree on some $\omega \times Z$. Let G be V-generic over* $\mathrm{Col}(\omega, 2^{2^\kappa})$; *then*

$$V[G] \models j(T) \text{ has a } \lambda\text{-absolute complement.}$$

Proof. By a Skolem hull argument, we have a tree T^* on $\omega \times \kappa$ such that $p[T] = p[T^*]$ in any $< \kappa$-generic extension of V. Notice that $\mathrm{meas}_\kappa(\kappa^{<\omega})$ is countable in $V[G]$. Let a

$$m : \omega \stackrel{\text{onto}}{\longrightarrow} j''\mathrm{meas}_\kappa(\kappa^{<\omega})$$

be an enumeration in $V[G]$ such that each $m(e)$ concentrates on κ^n, for some $n \leq e$. Of course, the measures in $j''\mathrm{meas}_\kappa(\kappa^{<\omega})$ do not extend to $V[G]$, however, they *do* extend to $M[G]$, and in fact to $M[G][H]$ whenever H is size $< \lambda$ generic over $V[G]$, and hence over $M[G]$. This will be enough for our purpose, which is to form an analog of the Martin-Solovay tree. More precisely, we put

$$\begin{aligned}(s, \langle \alpha_0, \dots, \alpha_{n-1}\rangle) \in S \Leftrightarrow\ & s \in \omega^n \wedge \alpha_0 < j(\kappa)^+ \wedge \\ & \forall i, e\, (i < e < n \wedge m(e)(j(T^*)_s) = 1 \wedge \\ & m(e) \text{ projects to } m(i) \\ & \Rightarrow \alpha_e < \pi_{m(i),m(e)}(\alpha_i)).\end{aligned}$$

We claim S is a λ-absolute complement for $j(T)$ in $V[G]$. For let $x \in V[G][H]$ be a real, where H is size $< \lambda$ generic over $V[G]$.

For $(s, t) \in j(T^*)$, let $\Sigma(s, t)$ be the measure on T^*_s given by

$$A \in \Sigma(s, t) \Leftrightarrow t \in j(A).$$

Then if (x, f) is a branch of $j(T^*)$ in V, we have that $\Sigma(x, f)$ is a wellfounded tower of measures concentrating on T^*_x, and hence $\langle j(\Sigma(x \restriction n, f \restriction n)) \mid n < \omega \rangle$ is a wellfounded tower in M concentrating on $j(T^*)_x$. A simple absoluteness argument shows this remains true for any branch (x, f) of $j(T^*)$ in $V[G][H]$; that is, $\mathrm{Ult}(M, \langle j(\Sigma(x \restriction n, f \restriction n)) \mid n < \omega \rangle)$ is wellfounded. So if $x \in p[j(T)]^{V[G][H]} = p[j(T)]^{M[G][H]} = p[j(T^*)]^{M[G][H]}$, then one of the towers S_x is trying to prove illfounded is actually wellfounded, so that $x \notin p[S]$.

On the other hand, suppose $x \notin p[j(T)]$. Then $x \notin p[j(T^*)]$, so we have a rank function

$$f(u) = |u|_{T^*_x},$$

and $f \in M[G][H]$ because $x \in M[G][H]$. For $m(e)$ a measure in $j''\sigma$ concentrating on some $j(T^*)_{x \restriction n}$, let

$$\alpha_e = [f]_{m(e)},$$

which makes sense because f is equal modulo $m(e)$ to a function in M. Set $\alpha_e = 0$ if $m(e)$ is not such a measure. It is easy to check that $\langle \alpha_e \mid e \in \omega \rangle$ is an infinite branch of S_x, as desired. ⊣

Corollary 4.6. *Let κ be λ-strong, where $\lambda = |V_\lambda|$, and let T and U be λ-absolute complements. Let G be V-generic over* $\mathrm{Col}(\omega, 2^{2^{k}a})$; *then in $V[G]$ there are λ-absolute complements R and S such that $p[S] = \exists^{\mathbb{R}} p[T]$ in all generic extensions.*

Proof. Here T is a tree on $(\omega \times \omega) \times Z$ for some Z. Let S be T, regarded as a tree on $\omega \times (\omega \times Z)$. So $p[S] = \exists^{\mathbb{R}} p[T]$ in all generic extensions.

By the theorem, in $V[G]$ there is a λ-absolute complement R for $j(S)$, where $j : V \to M$ witnesses κ is λ-strong. It is enough then to see that $p[S] = p[j(S)]$ in $V[G][H]$, for any size $< \lambda$ generic H. For this, it is enough that $p[T] = p[j(T)]$ in $V[G][H]$. But clearly $p[T] \subseteq p[j(T)]$. If $(x, y) \notin p[T]$, then $(x, y) \in p[U]$, so $(x, y) \in p[j(U)]$, so $(x, y) \notin p[j(T)]$. Thus $p[j(T)] \subseteq p[T]$, and we are done. ⊣

Corollary 4.7 (Woodin). *If there are n strong cardinals which are $\leq \kappa$, where $1 \leq n < \omega$, then in $V^{\mathrm{Col}(\omega, 2^{2^\kappa})}$:*

(a) *For any η, there is a tree T_η such that in any $< \eta$-generic extension, $p[T_\eta]$ is the universal Σ^1_{n+3} set,*

(b) *all Σ^1_{n+2} sets are ∞-universally Baire, and*

(c) *any two set generic extensions are Σ^1_{n+3} equivalent, that is, if $x \in \mathbb{R} \cap V[G] \cap V[H]$, and φ is Σ^1_{n+3}, then $V[G] \models \varphi[x]$ iff $V[H] \models \varphi[x]$.*

The proof of this corollary is an easy induction on n, with the Martin-Solovay trees for Σ^1_3 providing the starting point in the $n = 1$ case.

Corollary 4.8. *If there are infinitely many strong cardinals below λ, then in $V^{\mathrm{Col}(\omega,\lambda)}$, projective formulae are absolute for all further set forcing.*

§5. The tree production lemma. We shall show that formulae with certain generic absoluteness properties define universally Baire sets.

Let $\varphi(v_0, v_1)$ be a Σ_n formula of the language of set theory, and let a be a parameter (not necessarily a real parameter). We are interested in the κ-universal Baireness of $\{x \in \mathbb{R} | \varphi(x, a)\}$.

Let $X \prec_{\Sigma_n} V$, with X countable and $\kappa, a \in X$. Let

$$\pi : N \cong X \prec_{\Sigma_n} V$$

be the transitive collapse, with $\pi(\bar{\kappa}) = \kappa$ and $\pi(\bar{a}) = a$. We then say that X *is (φ, a, κ)-generically correct* iff whenever g in V is N-generic over some $\mathbb{P} \in H^N_{\bar{\kappa}}$, then for all reals $x \in N[g]$,

$$N[g] \models \varphi[x, \bar{a}] \Leftrightarrow V \models \varphi[x, a].$$

Lemma 5.1. *Let $\varphi(v_0, v_1)$ be a Σ_n formula, let a be a parameter, and let M be transitive with $H_\kappa \cup \{\kappa\} \subseteq M$, and $\sigma : M \to V$ a Σ_{n+5}-elementary embedding with $a \in \mathrm{ran}(\sigma)$ and $\mathrm{crit}(\sigma) > \kappa$. The following are equivalent:*

(1) *There are club many $X \in P_{\omega_1}(M)$ such that $\sigma''X$ is (φ, a, κ)-generically correct.*

(2) *There are trees T and U such that whenever G is V-generic over some $\mathbb{P} \in H_\kappa$, then*

$$V[G] \models p[T] = \{x \mid \varphi(x, a)\} \text{ and } p[U] = \{x \mid \neg \varphi(x, a)\}.$$

Proof. Assume (1), and let $F : M^{<\omega} \to M$ be such that whenever $X \in P_{\omega_1}(M)$ and $F''X^{<\omega} \subseteq X$, then $X \prec M$ and $\sigma''X$ is (φ, a, k)-generically correct. For $x, y \in \mathbb{R}$ let

$$A(x, y) \Leftrightarrow y \text{ codes a transitive } (N, \varepsilon, \bar{\kappa}, \bar{a}) \models \mathsf{ZFC}^-, \text{ and}$$
$$\exists g \big(g \text{ is} < \bar{\kappa}\text{-generic over} N \wedge x \in N[g] \wedge N[g] \models \varphi[x, \bar{a}]\big).$$

Here $y \in {}^\omega\omega$ codes $(N, \varepsilon, \bar{\kappa}, \bar{a})$ as follows: we have $(\omega, E_y, 0, 1) \cong (N, \varepsilon, \bar{\kappa}, \bar{a})$, where $\langle n, m\rangle \in E_y$ iff $y(2^n \cdot 3^m) = 0$. By ZFC^- we mean all the Σ_5 consequences of ZFC; these are of course all true in M. The set of y which are codes is Π^1_1, so A is Σ^1_2, so there is a tree S on $\omega \times \kappa$ such that $p[S] = A$ in any size $< \kappa$ generic extension. We now define T on $\omega \times \omega \times \kappa \times M$. Let $\langle u_n \mid n < \omega\rangle$ be an enumeration of $\omega^{<\omega}$, with $\mathrm{dom}\, u_n \subset n$. Put $a^* = \sigma^{-1}(a)$, and $(u, v, r, s) \in T$ iff

(a) $(u, v, r) \in S$,

(b) $0 \in \mathrm{dom}(s) \Rightarrow s(0) = \kappa$, and $1 \in \mathrm{dom}(s) \Rightarrow s(1) = a^*$,

(c) $2k+2 \in \operatorname{dom}(s) \Rightarrow s(2k+2) = F(s \circ u_k)$, and
(d) if $2^n \cdot 3^m \in \operatorname{dom}(v)$, then $v(2^n \cdot 3^m) = 0$ iff $s(n) \in s(m)$.

Similarly, replacing φ by $\neg\varphi$, we can define a Σ^1_2 relation $B(x, y)$, and from the Shoenfield tree for B, a tree U on $\omega \times \omega \times \kappa \times M$.

Let $(x, y, f, \pi) \in [T]$. Then $(x, y, f) \in [S]$, so $A(x, y)$. By (b) and (d), π is an isomorphism between $(\omega, E_y, 0, 1)$ and $(X, \varepsilon, \kappa, a^*)$, where $X = \operatorname{ran}(\pi)$. By (c), $F``X^{<\omega} \subseteq X$. Thus $X \prec M$ and $\sigma``X$ is (φ, a, κ)-generically correct. It follows then that $\varphi(x, a)$ is true in V. Thus $p[T] \subseteq \{x \mid \varphi(x, a)\}$ in V, and similarly, $p[U] \subseteq \{x \mid \neg\varphi(x, a)\}$ in V.

Now let G be size $< \kappa$ generic over V, and suppose $V[G] \models \varphi[x, a]$. Then $M[G] \models \varphi[x, a^*]$, as σ is sufficiently elementary. We can find a countable $Z \prec M[G]$ with $G, a^*, x, \kappa \in Z$ such that setting $X = Z \cap M$, $F''X^{<\omega} \subseteq X$. Letting $X = \operatorname{ran}(\pi)$, we can find y, f such that $(x, y, f, \pi) \in [T]$. Thus $x \in p[T]^{V[G]}$. Similarly, if $V[G] \models \neg\varphi[x, a]$, then $x \in p[U]^{V[G]}$, and hence $x \notin p[T]^{V[G]}$, since $p[T]$ and $p[U]$ are disjoint in V, hence in $V[G]$. Thus $p[T] = \{x \mid \varphi(x, a)\}$ in $V[G]$, and similarly for U, as desired.

For the (2) $\Rightarrow$ (1) direction, just note that σ is sufficiently elementary that there must be trees T and U as in (2) (with a^* replacing a) such that $T, U \in M$. But then any countable $X \prec M$ such that $\kappa, a^*, T, U \in X$ is such that $\sigma``X$ is (φ, a, κ) generically correct. ⊣

Recall that $\mathbb{Q}_{<\delta}$ is Woodin's "countable" stationary tower forcing (see appendix). Conditions in $\mathbb{Q}_{<\delta}$ are stationary sets $b \subseteq P_{\omega_1}(Z)$, for some $Z \in V_\delta$.

THEOREM 5.2 (Tree production lemma, Woodin). *Let $\varphi(v_0, v_1)$ be a formula, let a be a parameter, and let δ be a Woodin cardinal. Suppose*

(1) (*Generic absoluteness*) *If G is $<\delta$-generic over V, and H is $<\delta^+$-generic over $V[G]$, then for all $x \in \mathbb{R} \cap V[G]$,*

$$V[G] \models \varphi[x, a] \text{ iff } V[G][H] \models \varphi[x, a],$$

and

(2) (*Stationary tower correctness*) *If G is $\mathbb{Q}_{<\delta}$-generic, and $j : V \to M \subseteq V[G]$ is the generic elementary embedding, then for all $x \in \mathbb{R} \cap V[G]$*

$$V[G] \models \varphi[x, a] \text{ iff } M \models \varphi[x, j(a)].$$

Then there are trees T and U such that whenever g is $<\delta$-generic over V, then

$$V[G] \models (p[T] = \{x \mid \varphi(x, a)\} \wedge p[U] = \{x \mid \neg\varphi(x, a)\}).$$

In particular, $\{x \mid \varphi(x, a)\}$ is δ-universally Baire.

PROOF. It is enough to find for each $\kappa < \delta$ trees T_κ and U_κ which work for all$<\kappa$-generic g, since then we can take $T = \bigoplus_{\kappa<\delta} T_\kappa$ and $U = \bigoplus_{\kappa<\delta} U_\kappa$. For if g is $<\kappa$-generic over V, where $\kappa < \delta$, then if $V[g] \models \varphi[x, a]$, then $x \in p[T_\kappa]$, so $x \in p[T]$. On the other hand, if $V[g] \models \neg\varphi[x, a]$, then

$x \in p[U_\kappa]$, so $x \notin p[T]$. Note here that if $p[U_\kappa] \cap p[T_\alpha] \neq \emptyset$ in some $< \delta$ generic $V[H]$, then $p[U_\kappa] \cap p[T_\alpha] \neq \emptyset$ in V, and this easily contradicts condition (1).

So fix $\kappa < \delta$. Let φ be Σ_n. Let M be transitive, $H_{\kappa^+} \subseteq M$, and $|M| < \delta$, and $\sigma : M \to V$ be Σ_{n+5} elementary, with $a = \sigma(a^*)$ and $\sigma \restriction \kappa^+ =$ identity. Let

$$b = \{X \in P_{\omega_1}(M) \mid X \prec M \text{ and } \sigma'' X \\ \text{is } (\varphi, a, k)\text{-generically correct}\}.$$

It is enough to show b contains a club in $P_{\omega_1}(M)$. If not, $P_{\omega_1}(M) \setminus b$ is a condition in $\mathbb{Q}_{<\delta}$, so we can find a $\mathbb{Q}_{<\delta}$ generic G such that $P_{\omega_1}(M) \setminus b \in G$. Let

$$j : V \to N \subseteq V[G]$$

be the generic embedding. Then $j''M \in j(P_{\omega_1}(M) \setminus b)$. Since $j''M \prec j(M)$, we have that $j(\sigma)''j''M$ is not $(\varphi, j(a), j(\kappa))$ correct in N. Since $j(\sigma)(j(z)) = j(\sigma(z))$, we see that $j(\sigma)''j''M$ collapses to M, and the image of $j(\kappa)$ under the collapse is κ, while the image of $j(a) = j(\sigma(a^*))$ is just a^*. But then for any $g \in N$ which is M-generic over poset of size $< \kappa$ in M, and any $x \in \mathbb{R} \cap M[g]$, we have

$$\begin{aligned} M[g] \models \varphi[x, a^*] &\Leftrightarrow V[g] \models \varphi[x, a] \\ &\Leftrightarrow V[G] \models \varphi[x, a] \\ &\Leftrightarrow N \models \varphi[x, j(a)]. \end{aligned}$$

The first equivalence holds because σ lifts, the second by generic absoluteness, and the third by stationary tower correctness. Thus $j(\sigma)''j''M$ is $(\varphi, j(a), j(\kappa))$-generically correct in N, a contradiction. ⊣

The Tree Production Lemma was first used by Woodin, although he did not formally state it, in the case that the parameter $a \in \mathbb{R}$, so that $j(a) = a$. The author made the trivial adaptation to the case $a \notin \mathbb{R}$ as part of the proof of the following theorem. Woodin then formally isolated the Tree Production Lemma as we have stated it.

THEOREM 5.3 (Steel). *Let λ be a limit of Woodin cardinals; then every* $\mathrm{Hom}_{<\lambda}$ *set has a* $\mathrm{Hom}_{<\lambda}$ *scale.*

For the proof, we need some elementary lemmas. The first is well-known. Let μ be a κ-complete ultrafilter on I, and g be $< \kappa$- generic over V. In $V[G]$, for $A \subseteq I$ put

$$A \in \mu^* \Leftrightarrow \exists\, B \in \mu(B \subseteq A).$$

Then μ^* is a κ-complete ultrafilter on I in $V[g]$. Moreover

PROPOSITION 5.4. *If $I \in V$, and g is $< \kappa$ generic over V, and $\nu \in V[g]$ is a κ-complete ultrafilter over I in $V[g]$, then $\nu = \mu^*$ for some $\mu \in V$.*

PROOF. Note first that if $A \in \nu$, then there is a set $B \subseteq A$ such that $B \in \nu$ and $B \in V$. (Work in $V[g]$, and let $A = \dot{A}_g$. By κ-completeness of ν, we can fix $p \in g$ so that p decides "$\check{i} \in \dot{A}$" for μ-a.e. $i \in I$. Then take $B = \{i \mid p \Vdash \check{i} \in \dot{A}\}$.)

Let $\nu = \dot{\nu}_g$. We claim there is a set $B \in \nu \cap V$ such that for all $C \subseteq B$ such that $C \in V$

$$\|\check{C} \in \dot{\nu}\| = \|\check{B} \in \dot{\nu}\| \text{ or } \|\check{C} \in \dot{\nu}\| = 0.$$

We can then define in V

$$\mu = \{C \subseteq I \mid \|\check{B} \in \dot{\nu}\| \leq \|\check{C} \in \dot{\nu}\|\},$$

and it is easy to see that $\mu = \nu \cap V$, so that $\nu = \mu^*$.

If there is no B as desired, then working in $V[g]$, we define a κ-sequence of sets $B_\alpha \in \nu \cap V$ such that

$$\alpha < \beta \Rightarrow \|\check{B}_\alpha \in \dot{\nu}\| > \|\check{B}_\beta \in \dot{\nu}\|.$$

We get $B_{\alpha+1}$ from the fact that B_α is not as desired. At limit $\lambda < \kappa$, let $A = \bigcap_{\alpha<\lambda} B_\alpha$. Since $A \in \nu$, we can find $B_\lambda \in \nu \cap V$ so that $B_\lambda \subseteq A$, and continue. But now g was generic for a poset of size $< \kappa$, so there cannot be a strictly decreasing κ-sequence of Boolean values, even in $V[g]$. ⊣

The second lemma we need is a minor variation on the well-known fact that if μ and ν are normal ultrafilters on κ and λ, with $\kappa < \lambda$, and $j : V \to M = \mathrm{Ult}(V, \mu)$ is the canonical embedding, then $j(\nu) = \nu \cap M$, and $\mathrm{Ult}(M, j(\nu))$, which is the ultrapower computed using functions in M, is the same as $\mathrm{Ult}^*(M, \nu)$, where the $*$ indicates that the ultrapower is computed using functions in V. The variation comes from letting j be a generic embedding.

LEMMA 5.5. *Let δ be Woodin, and G be $\mathbb{Q}_{<\delta}$ generic over V, with*

$$j : V \to M \subseteq \mathrm{Ult}(V, G)$$

the canonical embedding. Let μ be a δ^+-complete ultrafilter on some I, with $\mu \in V$. Then

(1) *For any $A \subseteq j(I)$ in M,*

$$A \in j(\mu) \Leftrightarrow \exists B \in \mu(j(B) \subseteq A),$$

(2) $\mathrm{Ult}(M, j(\mu)) \cong \mathrm{Ult}^*(M, \mu^*)$, *where the first ultrapower is computed using all $f : j(I) \to M$ such that $f \in M$, and the second ultrapower is computed using all $f : I \to M$ such that $f \in V[G]$, or equivalently, all $f : I \to M$ such that $f \in V$.*

(3) *Let $\nu \leq_{RK} \mu$ via $p : I \to J$, that is, let μ be the measure given by $\nu(A) = \mu(p^{-1}(A))$ for $A \subseteq J$. Let*

$$i : \mathrm{Ult}(V, \nu) \to \mathrm{Ult}(V, \mu)$$

be the canonical embedding given by $i([f]_\nu) = [f \circ p]_\mu$ *for all* $f \in V$. *Let*

$$i^* : \mathrm{Ult}(V[G], \nu^*) \to \mathrm{Ult}(V[G], \mu^*)$$

be its lift to $V[G]$, *given by* $i^*([f]_{\nu^*}) = [f \circ p]_{\mu^*}$ *for all* $f \in V[G]$. (*Thus* $i \subseteq i^*$.) *Then*

$$j(i) = i^* \restriction \mathrm{Ult}(M, j(\nu)),$$

and in particular, $j(i)$ *agrees with* i *on the ordinals.*

Proof. The $\Leftarrow$ direction of (1) is trivial. For the $\Rightarrow$ direction, let $A \in j(\mu)$, and $A = [f]_G$ where $f \in V$. We may assume $f : P_{\omega_1}(Z) \to \mu$ for some $Z \in V_\delta$. It is easy to see that

$$B = \bigcap \{f(X) \mid X \in P_{\omega_1}(Z)\}$$

is as desired.

For (2), note first

Claim. If $f : I \to M$ and $f \in V[G]$, then there is an $h \in M$ and $B \in \mu$ such that

$$f(u) = h(j(u))$$

for all $u \in I$.

Proof. Work in $V[G]$. Let $\dot{f}_G = f$. For each $u \in I$, pick $a_u \in G$ such that for some $g \in V$ with domain $P_{\omega_1}(Z_u)$

$$a_u \overset{\mathbb{Q}_{<\delta}}{\Vdash} \dot{f}(\check{u}) = [\check{g}]_{\dot{G}}.$$

We can fix $a_u = a$ and $Z_u = Z$ for μ^*-a.e. u. We then have a set $B \in \mu$ such that $a_u = a$ and $Z_u = Z$ for all $u \in B$. Going back to V, we can find for each $u \in B$ a function g_u with domain $P_{\omega_1}(Z)$ such that

$$a \overset{Q_{<\delta}}{\Vdash} \dot{f}(\check{u}) = [\check{g}_u].$$

Now, for $X \in P_{\omega_1}(Z)$ in V, set

$$h_X(u) = g_u(X)$$

for all $u \in B$. Then for all $u \in B$,

$$[\lambda X.h_X]_G(j(u)) = [g_u]_G = f(u),$$

by Los' theorem and the fact that $a \in G$. Thus $h = [\lambda X.h_X]_G$ is as desired.

Now for each $f : I \to M$ in $V[G]$, pick an $h_f \in M$ such that $h_f(j(u)) = f(u)$ for μ^*-a.e. u. Notice that if $[h_f]_{j(\mu)} \neq [h_g]_{j(\mu)}$, then by (1) we can find $B \in \mu$ s.t. $h_f(v) \neq h_g(v)$ for all $v \in j(B)$, and hence $h_f(j(u)) \neq h_g(j(u))$ for all $u \in B$, so that $f(u) \neq g(u)$ for all $u \in B$; that is, $[f]_{\mu^*} \neq [g]_{\mu^*}$. Thus the map

$$\pi([f]_{\mu^*}) = [h_f]_{j(\mu)}$$

is well-defined on equivalence classes. A similar argument shows

$$[f]_{\mu^*} \in [g]_{\mu^*} \text{ iff } [h_f]_{j(\mu)} \in [h_g]_{j(\mu)},$$

so π is an $\in$-isomorphism with its range. But if $h\colon j(I) \to M$ and $h \in M$, then letting $f(u) = h(j(u))$ for all $u \in I$, we have $f \in V[G]$ and $[h]_{j(\mu)} = [h_f]_{j(\mu)}$. Thus π is onto, and hence an isomorphism of the ultrapowers in question. ⊣

For (3), let $p : I \to J$ and $\nu(A) = 1$ iff $\mu(p^{-1}(A)) = 1$. We need to see that the diagram

$$\begin{array}{ccc} \mathrm{Ult}^*(M, \nu^*) & \xrightarrow{i^*} & \mathrm{Ult}^*(M, \mu^*) \\ \sigma \updownarrow & & \updownarrow \tau \\ \mathrm{Ult}(M, j(\nu)) & \xrightarrow{j(i)} & \mathrm{Ult}(M, j(\mu)) \end{array}$$

commutes, where σ and τ are the isomorphisms of part (2). This then means $j(i) = i^* \restriction \mathrm{Ult}^*(M, \nu^*)$ after the ultrapowers have been transitivised. So let $X = [f]_{\nu^*} \in \mathrm{Ult}^*(M, \nu^*)$. Then

$$\begin{aligned} j(i)(\sigma(x)) &= j(i)([h_f]_{j(\nu)}) \\ &= [h_f \circ j(p)]_{j(\nu)}, \end{aligned}$$

and

$$\begin{aligned} \tau(i^*(x)) &= \tau([f \circ p]_{\mu^*}) \\ &= [h_{f \circ p}]_{j(\mu)}, \end{aligned}$$

where we have adopted the notation from the proof of (2) in analyzing σ and τ. From the proof of (2), we see that it suffices to show that $h_f \circ j(p)$ and $h_{f \circ p}$ agree at all points in $j''B$, for some $B \in \mu$. But now

$$h_{f \circ p}(j(u)) = (f \circ p)(u) = f(p(u))$$

for all $u \in B_0$, where $B_0 \in \mu$. Also

$$\begin{aligned} (h_f \circ j(p))(j(u)) &= h_f(j(p)(j(u))) \\ &= h_f(j(p(u))) \\ &= f(p(u)) \end{aligned}$$

for all $u \in B_1$, where $B_1 \in \mu$. Thus $h_{f \circ p}(j(u)) = (h_f \circ j(p))(j(u))$ for all $u \in B_0 \cap B_1$, as desired. ⊣

From the last lemma we get a stationary tower correctness result for homogeneity systems.

Corollary 5.6. *Let δ be Woodin, G be V-generic over $\mathbb{Q}_{<\delta}$, and*

$$j : V \to M \subseteq \mathrm{Ult}(V, G)$$

be the generic embedding.

(1) *Let $\langle \mu_s^* \mid s \in \omega^{<\omega}\rangle$ be a homogeneity system in $V[G]$, where each μ_s is a δ^+-complete ultrafilter in V (although the* system *$\langle \mu_s \mid s \in \omega^{<\omega}\rangle$ may not be in V); then for any γ*

$$ms(\langle \mu_s^* \mid s \in \omega^{<\omega}\rangle, \gamma)^{V[G]} = ms(\langle j(\mu_s) \mid s \in \omega^{<\omega}\rangle, \gamma)^M,$$

and

$$\left(S_{\langle \mu_s^* \mid s \in \omega^{<\omega}\rangle}\right)^{V[G]} = \left(S_{\langle j(\mu_s) \mid s \in \omega^{<\omega}\rangle}\right)^M.$$

(2) *If $\lambda > \delta$ is a limit of Woodin cardinals, then $\mathrm{Hom}_{<\lambda}^{V[G]}$ is an initial segment of $j(\mathrm{Hom}_{<\lambda})$ under Wadge reducibility.*

Proof. This follows at once from the last lemma. ⊣

The corollary also holds for weak homogeneity systems, but we shall not need this fact. The corollary is due independently to Woodin and the author.

Proof of Theorem 5.3. Fix $\gamma_0 < \lambda$ such that

$$\mathrm{Hom}_{\gamma_0} = \mathrm{Hom}_{<\lambda}.$$

Let $\gamma_0 < \delta_0 < \delta_1 < \delta_2 < \lambda$, where the δ_i's are Woodin. Let $B \in \mathrm{Hom}_{<\lambda}$. It will be enough to find a scale $\{\theta_n\}$ on B such that the relation

$$S(n, x, y) \Leftrightarrow \theta_n(x) \leq \theta_n(y)$$

is δ_1^+-universally Baire. For then, S is δ_0^+-weakly homogeneous by Theorem 4.1, and hence γ_0-homogeneous by Theorem 3.1, and hence in $\mathrm{Hom}_{<\lambda}$.

Let $\langle \mu_s \mid s \in \omega^{<\omega}\rangle$ be a homogeneity system consisting of δ_2^+-complete measures such that

$$\mathbb{R} \setminus B = S_{\langle \mu_s \mid s \in \omega^{<\omega}\rangle}.$$

Let γ be a strong limit cardinal of cofinality $> \delta_2$, and let

$$T = ms(\langle \mu_s \mid s \in \omega^{<\omega}\rangle, \gamma).$$

Let $\{\theta_n\}$ be the leftmost-branch scale of T; that is, for $x \in p[T]$ define $\ell_x : \omega \to OR$ by

$$\ell_x(n) = \text{least } \alpha \text{ such that } \exists\, g(g \restriction n = \ell_x \restriction n \wedge g(n) = \alpha \wedge g \in [T_x]),$$

and put

$$\theta_n(x) \leq \theta_n(y) \text{ iff } \ell_x \restriction n \leq_{\mathrm{lex}} \ell_y \restriction n,$$

where $\leq_{\mathrm{lex}}$ is the lexicographic order. It is well-known folklore that $\{\theta_n\}$ is a scale on $p[T] = B$.

To see that the relation $S(n, x, y)$ is δ_1^+-universally Baire, we apply the tree production lemma at δ_2. Let $\varphi(v_0, v_1)$ be the natural formula defining S from the parameter T.

For generic absoluteness, let G be size $< \delta_2$ generic over V, and H size δ_2 generic over $V[G]$, and $(n, x, y) \in V[G]$. Clearly $x \in p[T]^{V[G]}$ iff $x \in p[T]^{V[G][H]}$, and similarly for y; also, $\ell_x^{V[G]} = \ell_x^{V[G][H]}$ and similarly for y. Thus

$$V[G] \models \ell_x \restriction n \leq_{\text{lex}} \ell_y \restriction n \text{ iff } V[G][H] \models \ell_x \restriction n \leq_{\text{lex}} \ell_y \restriction n,$$

as desired.

For stationary tower correctness, let $j : V \to M \subseteq V[G]$ where G is V-generic over $\mathbb{Q}_{<\delta_2}$. Clearly $j(\gamma) = \gamma$. But then

$$\begin{aligned} j(T) &= j\big(ms(\langle \mu_s \mid s \in \omega^{<\omega}\rangle, \gamma)\big) \\ &= ms\big(\langle j(\mu_s) \mid s \in \omega^{<\omega}\rangle, \gamma\big)^M \\ &= ms\big(\langle \mu_s^* \mid s \in \omega^{<\omega}\rangle, \gamma\big)^{V[G]} \\ &= T, \end{aligned}$$

using our previous lemmas. The absoluteness of wellfoundedness then tells us

$$V[G] \models \varphi[(n, x, y), T] \text{ iff } M \models \varphi[(n, x, y), j(T)],$$

as desired. ⊣

§6. $(\Sigma_1^2)^{\mathrm{Hom}_{<\lambda}}$ absoluteness. The observation that $\mathrm{Hom}_{<\lambda}^{V[G]}$ is a Wadge-initial segment of $j(\mathrm{Hom}_{<\lambda})$, recorded in Corollary 5.6, can be used to strengthen our projective generic absoluteness result. What we get is "$(\Sigma_1^2)^{\mathrm{Hom}_{<\lambda}}$ generic absoluteness". More precisely

Theorem 6.1 (Woodin). *Let $A \in \mathrm{Hom}_{<\lambda}$, where λ be a limit of Woodins, and let G be $< \lambda$-generic over V; then for any sentence φ in the language of set theory expanded by adding two new unary predicate symbols,*

$$\begin{aligned} &\exists B \in \mathrm{Hom}_{<\lambda}^V \left(HC^V, \in, A, B\right) \models \varphi \Leftrightarrow \\ &\exists B \in \mathrm{Hom}_{<\lambda}^{V[G]} \left(HC^{V[G]}, \in, A^{V[G]}, B\right) \models \varphi. \end{aligned}$$

Proof. The left-to-right direction is an immediate consequence of our projective generic absoluteness result, Theorem 3.6. For the right-to- left direction, let H be $\mathbb{Q}_{<\delta}$-generic over V for some Woodin cardinal $\delta < \lambda$, with $G \in V[H]$. (Since $\mathbb{Q}_{<\delta}$ collapses all $\eta < \delta$, general forcing theory tells us there is such an H.) By the upward absoluteness of $(\Sigma_1^2)^{\mathrm{Hom}_{<\lambda}}$ from $V[G]$ to $V[H]$, we have a $B \in \mathrm{Hom}_{<\lambda}^{V[H]}$ such that

$$\left(HC^{V[H]}, \in, A^{V[H]}, B\right) \models \varphi.$$

Letting $j: V \to M = \text{Ult}(V, H)$ be the generic embedding, we see from Corollary 5.6 that $j(A) = A^{V[H]}$ and $B \in j(\text{Hom}^V_{<\lambda})$. Of course, $HC^{V[H]} = HC^M$ as well. Thus

$$M \models \Big[\exists B \in j\Big(\text{Hom}^V_{<\lambda}\Big)(HC, \in, j(A), B) \models \varphi\Big].$$

The elementarity of j now yields the desired conclusion. ⊣

Remark 6.2. Let M_ω be the minimal iterable proper class mouse satisfying "there are infinitely many Woodin cardinals", and let λ be the supremum of the Woodin cardinals of M_ω. If $\alpha < \omega_1^{M_\omega}$, then the canonical iteration strategy for $M_\omega|\alpha$ is $\text{Hom}^{M_\omega}_{<\lambda}$, in the sense that there is a λ-absolutely complemented tree $T \in M_\omega$ which projects to this iteration strategy in all $< \lambda$-generic extensions of M_ω. It follows then that in M_ω, every real is $(\Sigma^2_1)^{\text{Hom}_{<\lambda}}$ in a countable ordinal. This statement, and the statement

$$\forall x \in \mathbb{R} \exists \Gamma \in \text{Hom}_{<\lambda} \left(\Gamma \text{ is an } \omega_1\text{-iteration strategy for the mouse } N \wedge x \in N\right)$$

have the form $\forall x \in \mathbb{R}\psi$, where ψ "is" $(\Sigma^2_1)^{\text{Hom}_{<\lambda}}$. The statements are false in $M_\omega[x]$, where x is a Cohen real over M_ω. This puts a limit on what statements about $\text{Hom}_{<\lambda}$ are provably $< \lambda$- generically absolute which lies just beyond the positive result of Theorem 6.1.

Remark 6.3. It follows from the last remark that if

$$j: M_\omega \to N = \text{Ult}(M_\omega, G)$$

the the generic embedding, where G is $\mathbb{Q}_{<\delta}$-generic over M_ω, then

$$\text{Hom}^{M_\omega[G]}_{<\lambda} \neq j\Big(\text{Hom}^{M_\omega}_{<\lambda}\Big).$$

The reason is that N satisfies both of the statements $\forall x \in \mathbb{R}\psi$ referred to in Remark 6.2, while $M_\omega[G]$ satisfies neither[5]. Since N and $M_\omega[G]$ have the same reals, it must be that $\text{Hom}^N_{<\lambda} \neq \text{Hom}^{M_\omega[G]}_{<\lambda}$.

Remark 6.4. A closely related fact is that M_ω satisfies the statement "there is a $(\Sigma^2_1)^{\text{Hom}_{<\lambda}}$ wellorder of the reals". The wellorder is not a $\text{Hom}_{<\lambda}$ set from the point of view of M_ω, since, for example, it does not have the Baire property. This shows that it is possible (consistent with the existence of infinitely many Woodins) that there are $(\Sigma^2_1)^{\text{Hom}_{<\lambda}}$ sets of reals which are not $\text{Hom}_{<\lambda}$. This in turn shows that the stationary tower correctness hypothesis of the tree production lemma is necessary, since the formula $(\Sigma^2_1)^{\text{Hom}_{<\lambda}}$ formula defining a wellorder of the reals in M_ω *is* generically absolute over M_ω.

[5]Let x be a real in $M_\omega[G] \setminus M_\omega$. If x were $(\Sigma^2_1)^{\text{Hom}_{<\lambda}}$ in $M_\omega[G]$, then by Theorem 6.1 that would remain true in $M_\omega[H]$ for some $\text{Col}(\omega, \eta)$-generic H, with $\eta < \lambda$ sufficiently large. So x would be OD in $M_\omega[H]$, and hence x would be in M_ω, a contradiction.

REMARK 6.5. The last three remarks probably generalize from M_ω to any other mouse M and ordinal λ such that $M \models \lambda$ is a limit of Woodin cardinals. This is already known for many of the more natural M. We would guess, for example that the assertion that every real is $(\Sigma_1^2)^{\mathrm{Hom}_{<\lambda}}$- definable from a countable ordinal, for all λ which are limits of Woodin cardinals, is consistent with the existence of arbitrarily large superstrong cardinals. This is because we would guess that current inner model theory, which is based on the existence of homogeneously Suslin iteration strategies, goes at least this far.

REMARK 6.6. Finally, the last remark should not be taken to mean that superstrong cardinals yield no generic absoluteness beyond that given by Theorem 6.1. Large cardinal hypotheses stronger than "there are infinitely many Woodin cardinals" will imply there are more $\mathrm{Hom}_{<\lambda}$ sets of reals, and hence that more statements can be expressed in $(\Sigma_1^2)^{\mathrm{Hom}_{<\lambda}}$ form. For example, if we add that there is a measurable cardinal above λ, then we have that $\mathbb{R}^\sharp$ is in $\mathrm{Hom}_{<\lambda}$, which implies that the first order theory of $L(\mathbb{R})$ is $< \lambda$-generically absolute. (This much generic absoluteness fails in M_ω.) One actually gets a many-one reduction of the theory of $L(\mathbb{R})$ to the set Σ of all $(\Sigma_1^2)^{\mathrm{Hom}_{<\lambda}}$ truths. Still stronger large cardinal hypotheses provide explicit provable many-one reductions to Σ of the truth sets for more powerful languages, and hence yield still stronger generic absoluteness theorems.

Woodin's Ω-conjecture implies that, granted there are arbitrarily large Woodin cardinals, all generic absoluteness theorems come through many-one reductions to $(\Sigma_1^2)^{\mathrm{Hom}_\infty}$.

§7. The derived model theorem. Let λ be a limit of Woodin cardinals. As explained in the last section, we cannot hope to show that $L(\mathbb{R}, \mathrm{Hom}_{<\lambda}) \models \mathsf{AD}$, since if V is a canonical inner model, then $L(\mathbb{R}, \mathrm{Hom}_{<\lambda})$ has a wellorder of $\mathbb{R}$ in it. (At least this is true if $V = M_\omega$, and in many other cases.) Nevertheless, one can find a model very close to $L(\mathbb{R}, \mathrm{Hom}_{<\lambda})$ which satisfies AD. This so-called *derived model* is obtained by collapsing everything below λ to be countable.

More precisely, let λ be a limit of Woodin cardinals, and let G be V-generic over $\mathrm{Col}(\omega, <\lambda)$. Let us write $G \restriction \alpha$ for $G \cap \mathrm{Col}(\omega, <\alpha)$. We set

$$\mathbb{R}^* = \mathbb{R}^*_G = \bigcup_{\alpha<\lambda} \mathbb{R} \cap V[G \restriction \alpha],$$

and

$$\mathrm{Hom}^* = \mathrm{Hom}^*_G = \{p[T] \cap \mathbb{R}^* \mid \exists \alpha < \lambda (T \in V[G \restriction \alpha) \wedge \\ V[G \restriction \alpha] \models T \text{ is } \lambda\text{-absolutely complemented})\}.$$

Put another way, for any $\alpha < \lambda$ and $A \in \mathrm{Hom}_{<\lambda}^{V[G\restriction\alpha]}$, we set

$$A^* = \bigcup_{\alpha<\beta<\lambda} A^{V[G\restriction\beta]},$$

and we have

$$\mathrm{Hom}^* = \left\{ A^* \mid \exists\alpha < \lambda \mid A \in \mathrm{Hom}_{<\lambda}^{V[G\restriction\alpha]} \right\}.$$

Then $L(\mathbb{R}^*, \mathrm{Hom}^*)$ is called a *derived model* of V at λ. Of course, it is not literally accurate to speak of *the* derived model, since the model depends not just on V and λ, but on $\mathbb{R}^*$, which can be realized in different ways with different G. However, the forcing is sufficiently homogeneous that the first order theory of $L(\mathbb{R}^*, \mathrm{Hom}^*)$ is independent of G, so there is no ambiguity if we say that "the" derived model at λ satisfies φ.

THEOREM 7.1 (Derived model theorem, Woodin). *Let λ be a limit of Woodin cardinals, and $L(\mathbb{R}^*, \mathrm{Hom}^*)$ be a derived model at λ; then*

(1) $L(\mathbb{R}^*, \mathrm{Hom}^*) \models \mathsf{AD}^+$,
(2) $\mathrm{Hom}^* = \{A \subseteq \mathbb{R}^* \mid A$ is Suslin and co-Suslin in $L(\mathbb{R}^*, \mathrm{Hom}^*)\}$.

AD^+ is the theory $\mathsf{AD} + \mathsf{DC}_{\mathbb{R}}$ + Ordinal Determinacy + "all sets of reals are ∞-Borel". These are local consequences of scales[6]. If every set in M is Suslin in some perhaps bigger model N of AD having the same reals as M, then $M \models \mathsf{AD}^+$[7]. Many of the consequences of being Suslin in a larger model of AD are theorems of AD^+. The following converse to the derived model theorem is further evidence of the significance of AD^+[8].

THEOREM 7.2 (Woodin). *Let $M \models \mathsf{AD}^+$, and let Γ be the pointclass consisting of all sets of reals which are Suslin and co-Suslin in M; then $L(\mathbb{R}^M, \Gamma)$ is a derived model of some N at some λ.*

The model N referred to in Theorem 7.2 exists in a generic extension of M. Its λ is ω_1^M, as it must be if $\mathbb{R}^M$ is to be the set of reals of a derived model at λ. We shall not prove Theorem 7.2 in these notes.

According to Theorem 7.1 and Theorem 7.2, being the pointclass of all Suslin and co-Suslin sets in a model of AD^+ is equivalent to being the pointclass of all Suslin and co-Suslin sets of a derived model (and this is equivalent to being the Hom^* of a derived model). Woodin has found a generalization of the derived model construction, and shown that the generalized derived models it produces are precisely the models of AD^+. We shall not prove this strengthening of the derived model theorem here[9].

We proceed toward the proof of the derived model theorem. The following little lemma will be useful.

---

[6] AD^+ used to be called "Within Scales".

[7] In particular, if every set of reals in M is $\mathrm{Hom}_{<\lambda}$, then $M \models \mathsf{AD}^+$.

[8] AD^+ was first isolated by Woodin. It is open whether any or all of the additional axioms of AD^+ are provable in $\mathsf{ZF} + \mathsf{AD}$.

[9] Let λ be a limit of Woodin cardinals, and $\mathbb{R}^* = \mathbb{R}^*_G$ where G is $\mathrm{Col}(\omega, <\lambda)$-generic over V. Let M be the union of all models $P \in V(\mathbb{R}^*)$ such that $P \models \mathsf{AD}^+$ (plus $V = L(P(\mathbb{R}))$?); then M is a generalized derived model of V at λ.

Lemma 7.3. *Let G be $\mathrm{Col}(\omega, <\lambda)$-generic over V, where λ is a limit of Woodin cardinals. For any $\alpha < \lambda$ and $A \in \mathrm{Hom}_{<\lambda}^{V[G\restriction\alpha]}$, $(\mathrm{HC}^{V[G\restriction\alpha]}, \in, A) \prec (\mathrm{HC}^*_G, \in, A^*)$.*

Proof. From our projective absoluteness result Theorem 3.6, we have that whenever $\alpha < \beta < \gamma < \lambda$, then $(\mathrm{HC}^{V[G\restriction\beta]}, \in, A^{V[G\restriction\beta]}) \prec (\mathrm{HC}^{V[G\restriction\gamma]}, \in, A^{V[G\restriction\gamma]})$. The lemma now follows by the Tarski-Vaught theorem on unions of elementary chains. ⊣

The heart of the matter is the following reflection result.

Lemma 7.4. *Let G be $\mathrm{Col}(\omega, <\lambda)$-generic over V, where λ is a limit of Woodin cardinals. Let $A \in \mathrm{Hom}_{<\lambda}^{V[G\restriction\alpha]}$, where $\alpha < \lambda$. Let φ be a sentence in the language of set theory with two additional unary predicate symbols, and suppose that*

$$\exists B \subseteq \mathbb{R}^*\Big(B \in L(\mathbb{R}^*, \mathrm{Hom}^*) \wedge (\mathrm{HC}^*, \in, A^*, B) \models \varphi\Big);$$

then

$$\exists B\Big(B \in \mathrm{Hom}_{<\lambda}^{V[G\restriction\alpha]} \wedge \Big(\mathrm{HC}^{V[G\restriction\alpha]}, \in, A, B\Big) \models \varphi\Big).$$

Before proving Lemma 7.4, let us use it to complete the proof of the derived model theorem. So let G be $\mathrm{Col}(\omega, <\lambda)$-generic over V, where λ is a limit of Woodins, and $\mathbb{R}^* = \mathbb{R}^*_G$ and $\mathrm{Hom}^* = \mathrm{Hom}^*_G$. We show first that $L(\mathbb{R}^*, \mathrm{Hom}^*) \models \mathsf{AD}$. For if not, there is a $B \in L(\mathbb{R}^*, \mathrm{Hom}^*)$ such that

$$(\mathrm{HC}^*, \in, B) \models \text{ the game with payoff } B \text{ is not determined.}$$

By Lemma 7.4, we can find $B \in \mathrm{Hom}^V_{<\lambda}$ such that

$$(\mathrm{HC}, \in, B) \models \text{ the game with payoff } B \text{ is not determined.}$$

This contradicts Martin's Theorem 2.7.

The remaining axioms of AD^+ are true in $L(\mathbb{R}^*, \mathrm{Hom}^*)$ for similar reasons. In each case the axiom can be expressed in the form "$\forall B \subseteq \mathbb{R}(\mathrm{HC}, \in, B) \models \varphi$", and there are no $\mathrm{Hom}_{<\lambda}$ sets B such that $(\mathrm{HC}^V, \in, B) \models \varphi$. For the axiom $\mathsf{DC}_{\mathbb{R}}$ both parts are obvious. The other two axioms have the form $\forall B \subseteq \mathbb{R}\exists C \subseteq \mathrm{OR}\ldots$, but using the Coding Lemma the quantifier on C can be reduced to a real quantifier over the field of a prewellorder which is projective in B. For Ordinal Determinacy, this is obvious, but for the assertion that B has an infinity-Borel code C, we need a preliminary argument which bounds the least size of such a code by some ordinal projective in B. This can be done[10]. Finally, the fact that there are no $\mathrm{Hom}_{<\lambda}$ counterexamples B to Ordinal Determinacy or the assertion that every set of reals is ∞-Borel

[10]The locality of ∞-Borel codes is due to Woodin.

follows from the fact that every $\text{Hom}_{<\lambda}$ set has a $\text{Hom}_{<\lambda}$ scale, together with $\text{Hom}_{<\lambda}$- determinacy[11].

To see that all Hom^* sets are Suslin in $L(\mathbb{R}^*, \text{Hom}^*)$, fix C in Hom^*. We then have $A \in \text{Hom}_{<\lambda}^{V[G\restriction\alpha]}$, for some $\alpha < \lambda$, such that $C = A^*$. By Theorem 5.3 there is $B \in \text{Hom}_{<\lambda}^{V[G\restriction\alpha]}$ which codes a scale on A. This fact can be expressed using only real quantifiers, and thus by Lemma 7.3, B^* codes a scale on A^* in $L(\mathbb{R}^*, \text{Hom}^*)$, so C is Suslin in $L(\mathbb{R}^*, \text{Hom}^*)$, as desired. Since Hom^* is closed under complement, all Hom^* sets are co-Suslin in $L(\mathbb{R}^*, \text{Hom}^*)$.

Conversely, suppose A is Suslin and co-Suslin in $L(\mathbb{R}^*, \text{Hom}^*)$, and let T and U be the trees which witness this. We can fix a set $C \in \text{Hom}^*$ such that T and U are ordinal definable over $L(\mathbb{R}^*, \text{Hom}^*)$ from C. (Every set in $L(\mathbb{R}^*, \text{Hom}^*)$ has this form.) We then have $W \in V[G \restriction \alpha]$, where $\alpha < \lambda$, such that $C = p[W] \cap \mathbb{R}^*$. It follows that T and U are definable in $V[G]$ from the parameter $\mathbb{R}^*$ and parameters in $V[G \restriction \alpha]$. But $V[G] = V[G \restriction \alpha][H]$ where H is generic for $\text{Col}(\omega, < \lambda)$, and there is a term τ such that $\tau_H = \mathbb{R}^*$ and $\text{Col}(\omega, < \lambda)$ is homogeneous with respect to τ, in that $\forall p, q \exists \pi (\pi$ is an automorphism of $\text{Col}(\omega, < \lambda)$ and $\pi(p)$ is compatible with q and $\pi\tau = \tau)$. Since T and U are subsets of $V[G \restriction \alpha]$, we have that $T, U \in V[G \restriction \alpha]$. But now T and U project to complements over $\mathbb{R}^*$, and hence in any $V[G \restriction \beta]$ for $\beta < \lambda$. Since the collapse forcing is universal, this implies that T and U are $< \lambda$-absolute complements in $V[G \restriction \alpha]$. Thus $p[T] \in \text{Hom}^*$, as desired. This completes the proof of the derived model theorem, modulo Lemma 7.4.

One key step toward the proof of Lemma 7.4 is to show that the reals of a symmetric collapse below λ can be realized as the reals of a stationary tower ultrapower. For this we use the following elementary lemma. For G $\text{Col}(\omega, < \lambda)$ generic, let HC^*_G be the collection of hereditarily countable sets having codes in $\mathbb{R}^*_G$.

Lemma 7.5. *Let N be a countable transitive model of* ZFC, *and let λ be a strong limit cardinals of N. Let X be countable. The following are then equivalent*:

(1) $X = \text{HC}^*_G$ *for some G which is* $\text{Col}(\omega, < \lambda)$*-generic over N*,

(2) $\forall y \in X(y$ is $< \lambda$ generic over N and $V_\lambda^{N[y]} \subseteq X)$, *and* $\forall y \in X \exists f \in X(f : \omega \overset{\text{onto}}{\to} y)$.

Proof. It is clear that (1) implies (2). For the converse, let $X = \{y_n \mid n < \omega\}$. We construct the desired G by defining $G \restriction \alpha_n$ by induction on n, where α_n is an increasing sequence with limit λ determined by the construction. We maintain that $G \restriction \alpha_n$ is coded by a real in X as part of the induction. Let $\langle D_n \mid n < \omega \rangle$ enumerate the dense subsets of $\text{Col}(\omega, < \lambda)$ lying in N. Given

[11] For Ordinal Determinacy, this is due independently to Moschovakis [8] and Woodin. It is folklore that all Suslin sets are ∞-Borel; see e.g. [2].

such $G \restriction \alpha_n$, we have by hypothesis that y_n is $< \lambda$-generic over N, and hence over $N[G \restriction \alpha_n]$. By general forcing theory, the complete Boolean algebra for adding y is a complete subalgebra of the collapse algebra at some $\beta < \lambda$ such that $\alpha_n < \beta$. Thus $y_n \in N[G \restriction \alpha_n][H]$ for some $\mathrm{Col}(\omega, \beta)$-generic H. We can take $H \in X$, because $V_\lambda^{N[G \restriction \alpha_n, y_n]} \subseteq X$ and every set in X has a counting in X. It is now easy to find α_{n+1} and $G \restriction \alpha_{n+1} \in X$ extending $G \restriction \alpha_n$ such that $H \in N[G \restriction \alpha_{n+1}]$ and $G \restriction \alpha_{n+1} \cap D_n \neq \emptyset$.

This completes the construction. It is clear that G is $\mathrm{Col}(\omega, < \lambda)$- generic over N, and $\mathrm{HC}^*_G = X$. ⊣

We now look at stationary tower forcing up to λ. Since λ may not itself be Woodin, $\mathbb{Q}_{<\lambda}$-generic G may be such that $\mathrm{Ult}(V, G)$ is illfounded. However, because λ is a limit of Woodins, we can find $G \subseteq \mathbb{Q}_{<\lambda}$ such that $\mathrm{Ult}(V, G)$ has wellfounded part as long as desired, and such that $\mathbb{R} \cap \mathrm{Ult}(V, G)$ is the set of reals in a symmetric collapse.

Our G will not actually be $\mathbb{Q}_{<\lambda}$-generic. However, $G \cap \mathbb{Q}_{<\delta}$ will be $\mathbb{Q}_{<\delta}$-generic for cofinally many Woodin cardinals $\delta < \lambda$. This is enough to make sense of $\mathrm{Ult}(V, G)$, since the functions used in computing this ultrapower all have domain of the form $P_{\omega_1}(V_\xi)$ for some $\xi < \lambda$ (and are in V). If $\xi < \delta$ and $G \cap \mathbb{Q}_{<\delta}$ is V-generic, then $G \cap \mathbb{Q}_{<\delta}$ measures all subsets of $P_{\omega_1}(V_\xi)$ which lie in V. Thus $\mathrm{Ult}(V, G)$ makes sense.

Let us call δ a *successor Woodin cardinal* if δ is a Woodin Cardinal which is not a limit of Woodins.

LEMMA 7.6 (Woodin). *Let λ be a limit of Woodin cardinals, let H be* $\mathrm{Col}(\omega, < \lambda)$*-generic over V, and let $\alpha \in$* OR; *then for any $b \in \mathbb{Q}_{<\lambda}$ there is a $\mathbb{Q}_{<\lambda}$-generic G over V such that $b \in G$ and*

(a) *for any successor Woodin cardinal $\delta < \lambda$ such that $\bigcup b \in V_\delta$, $G \cap \mathbb{Q}_{<\delta}$ is $\mathbb{Q}_{<\delta}$-generic over V,*
(b) *α is in the wellfounded part of* $\mathrm{Ult}(V, G)$, *and*
(c) $\mathbb{R} \cap \mathrm{Ult}(V, G) = \mathbb{R}^*_H$, *and* Hom^*_H *is a Wadge initial segment of $i_G(Hom_{<\lambda})$, where $i_G \colon V \to \mathrm{Ult}(V, G)$ is the canonical embedding.*

PROOF. There is such a G if and only if there is such a G in $V[H]^{\mathrm{Col}(\omega, \alpha)}$, because this universe is Σ^1_1-correct. Thus the existence of such a G is a first order question about $\mathbb{R}^*_H, b, V_\lambda$ and α inside $V[H]$. Because $\mathbb{R}^*_H$ is the denotation of a symmetric term, this question is decided by the empty condition in $\mathrm{Col}(\omega, < \lambda)$. So it is enough just to find *some* H and G related as in the statement of Lemma 7.6.

For this, we need the following sublemma

SUBLEMMA 7.7. There are stationarily many $X \in P_{\omega_1}(V_\lambda)$ such that $(X \cap \bigcup b) \in b$, and whenever $\delta \in X$ is a successor Woodin cardinal such that $\bigcup b \in V_\delta$, and $A \in X$ is a maximal antichain in $\mathbb{Q}_{<\delta}$, then there is an $a \in A$ such that $(X \cap \bigcup a) \in a$.

PROOF. If there is an $a \in X \cap A$ such that $(X \cap \bigcup a) \in a$, then one says that X *captures* A. In order to see that there are stationarily many X capturing all their maximal antichains at successor Woodins below λ but above $\bigcup b$, and such that $(X \cap \bigcup b) \in b$, it is enough to find one such $X \prec V_{\lambda+\omega}$ with $\lambda \in X$.

We construct X as the union of a countable elementary chain. Let X_0 be any countable elementary submodel of $V_{\lambda+\omega}$ such that $\lambda \in X_0$ and $(X_0 \cap \bigcup b) \in b$. We can find such an X_0 because b is stationary. Given $X_\alpha \prec V_{\lambda+\omega}$, let δ be the least successor Woodin in X_α not yet considered, and such that $\bigcup b \in V_\delta$. We form an elementary chain $Y_i \prec V_{\lambda+\omega}$ for $i < \omega$, setting $Y_0 = X_\alpha$ and $\gamma_0 = \bigcup\{\eta \mid \eta < \delta$ and η is Woodin $\}$. Given Y_n and γ_n, let A be the "next" maximal antichain of $\mathbb{Q}_{<\delta}$, and let $\gamma_{n+1} \in Y_n$ be such that $\gamma_n < \gamma_{n+1} < \delta$ and $A \cap \mathbb{Q}_{<\gamma_{n+1}}$ is semiproper. We can find such a γ_{n+1} since δ is Woodin. Now we get Y_{n+1} which captures A and such that $Y_n \prec Y_{n+1} \prec V_{\lambda+\omega}$ and $Y_{n+1} \cap V_{\gamma_n} = Y_n \cap V_{\gamma_n}$, as a consequence of semiproperness. The end-extension below γ_{n+1} relationship guarantees that all antichains captured at earlier stages are still captured by Y_{n+1}, and that $Y_{n+1} \cap \bigcup b = Y_n \cap \bigcup b \in b$. Let $X_{\alpha+1} = \bigcup_n Y_n$. With a little care as to the meaning of "next antichain", we shall have that $X_{\alpha+1}$ captures all maximal antichains of $\mathbb{Q}_{<\delta}$ such that $A \in X$.

At limit stages τ, set $X_\tau = \bigcup_{\alpha<\tau} X_\alpha$. It is not hard to show that there is some countable α such that X_α captures all maximal antichains at successor Woodins which it knows about, so that $X = X_\alpha$ is as desired. ⊣

We proceed to the proof of Lemma 7.6[12].

Fix an α and b; we may as well assume $\alpha > \lambda$. We claim there are G and H as desired in $V^{\mathrm{Col}(\omega,\alpha)}$.

Let $\theta = \alpha + \omega$, and let

$$\alpha, \lambda \in X \prec V_\theta,$$

where X is countable and in the stationary set given by Sublemma 7.7. Let

$$\pi\colon N \cong X$$

be the transitive collapse, and let $\pi(\langle \bar{\alpha}, \bar{\lambda}, \bar{b}\rangle) = \langle \alpha, \lambda, b\rangle$. We define $G \subseteq \mathbb{Q}^N_{<\bar{\lambda}}$ by using initial segments of X as our typical objects. More precisely, for $a \in \mathbb{Q}^N_{<\bar{\lambda}}$, let

$$a \in G \Leftrightarrow \pi``\bigcup a \in \pi(a).$$

It will be enough to show that for some H, G and H have the properties (a), (b), and (c) of Lemma 7.6 vis-a-vis N and $\bar{\alpha}, \bar{\lambda}, \bar{b}$. For then by Σ^1_1 absoluteness, $N^{\mathrm{Col}(\omega,\alpha)}$ satisfies that there are G and H with these properties (at $\bar{\alpha}$, $\bar{\lambda}$, and $\bar{b}$), and since π is elementary, we are done.

---

[12]The argument to follow is due to the author. Woodin had a somewhat different way of using the sublemma. The observation that Hom* is a Wadge initial segment of $i_G(\mathrm{Hom}_{<\lambda})$ (Lemma 7.6(c)) is due independently to the author.

It is clear that $\bar{b} \in G$. For (a), let δ be a successor Woodin cardinal of N below $\bar{\lambda}$, and let A be a maximal antichain in $\mathbb{Q}^N_{<\delta}$. Then $\pi(A)$ is a maximal antichain in $\mathbb{Q}_{<\pi(\delta)}$ and $\pi(A) \in X$, so X captures $\pi(A)$, say via $\pi(a)$. This means

$$\pi``\bigcup a = X \cap \bigcup \pi(a) \in \pi(a) \wedge \pi(a) \in \pi(A),$$

so that $a \in G \cap A$. Thus G meets all the necessary maximal antichains.

For (b), we can embed $\mathrm{Ult}(N, G)$ into V_θ as follows: let $f \in N$ be a function, and $\mathrm{dom}(f) = P_{\omega_1}(V_\gamma)^N$ where $\gamma < \bar{\lambda}$. We set

$$\sigma([f]_G) = \pi(f)\left(\pi``\bigcup \mathrm{dom}(f)\right).$$

It is easy to check that σ is well-defined and elementary (and extends π, in that $\pi = \sigma \circ \infty_G$, where i_G is the generic embedding). Thus $\mathrm{Ult}(V, G)$ is in fact fully wellfounded, and so $\bar{\alpha}$ is in its wellfounded part.

It follows immediately from (a) and Lemma 7.5 that there is a $\mathrm{Col}(\omega, <\bar{\lambda})$-generic H over N such that $\mathbb{R}^*_H = \mathbb{R} \cap \mathrm{Ult}(N, G)$. To see that $\mathrm{Hom}^*_H \subseteq i_G(\mathrm{Hom}^N_{<\bar{\lambda}})$, fix $\eta < \lambda$ and $A \in \mathrm{Hom}^{N[H\restriction\eta]}_{<\lambda}$; we must see that $A^* \in i_G(\mathrm{Hom}^N_{<\bar{\lambda}})$. Let $\gamma > \eta$ be a successor Woodin cardinal such that $H \restriction \eta \in N[G \cap \mathbb{Q}_{<\gamma}]$. Clearly, $A^* = (A^{N[G\cap\mathbb{Q}_{<\gamma}]})^*$, so to save notation, let us re-name $A = A^{N[G\cap\mathbb{Q}_{<\gamma}]}$. It follows from Corollary 5.6 that $A \in i_\gamma(\mathrm{Hom}^N_{<\lambda})$, where

$$i_\gamma \colon N \to \mathrm{Ult}\big(N, G \cap \mathbb{Q}^N_{<\gamma}\big)$$

is the canonical embedding. Let

$$\sigma \colon \mathrm{Ult}\big(N, G \cap \mathbb{Q}^N_{<\gamma}\big) \to \mathrm{Ult}(N, G)$$

be the natural embedding. (That is, $\sigma([f]_{G\cap\mathbb{Q}_{<\gamma}}) = [f]_G$.) It is enough to see that $\sigma(A) = A^*$. For that, it is enough to see that

$$i_{\gamma,\delta}(A) = A^{N[G\cap\mathbb{Q}_{<\delta}]}$$

whenever $\delta > \gamma$ is a successor Woodin cardinal and $i_{\gamma,\delta} \colon \mathrm{Ult}(N, G \cap \mathbb{Q}_{<\gamma}) \to \mathrm{Ult}(N, \mathbb{Q}_{<\delta})$ is the canonical embedding. But pick any δ^+-complete homogeneity system $\bar{\nu}$ in $N[G \cap \mathbb{Q}_{<\gamma}]$ such that

$$A = S_{\bar{\nu}}^{N[G\cap\mathbb{Q}_{<\gamma}]},$$

so that $\nu_s = \mu^*_s$ for some $\mu_s \in N$, and

$$A = S_{\langle i_\gamma(\mu_s) \mid s\in\omega^{<\omega}\rangle}$$

in $\mathrm{Ult}(N, G \cap \mathbb{Q}_{<\gamma})$. Then

$$i_{\gamma,\delta}(A) = S_{\langle i_\delta(\mu_s) \mid s\in\omega^{<\omega}\rangle}$$

in $\mathrm{Ult}(N, G \cap \mathbb{Q}_{<\delta})$, and

$$A^{N[G\cap\mathbb{Q}_{<\delta}]} = S_{\langle \mu^{**}_s \mid s\in\omega^{<\omega}\rangle}$$

in $N[G \cap \mathbb{Q}_{<\delta}]$ as a consequence of the Martin-Solovay construction. Now Corollary 5.6, applied at δ, gives the desired conclusion. ⊣

PROOF OF LEMMA 7.4. Let H be $\mathrm{Col}(\omega, <\lambda)$-generic over V, and let $A \in \mathrm{Hom}_{<\lambda}^{V[H\restriction\alpha]}$, and assume there is a $B \in L(\mathbb{R}^*, \mathrm{Hom}^*)$ such that $(\mathrm{HC}^*, \in, A^*, B) \models \varphi$. Let us call such a B a *φ-witness for A^**. What we are looking for is a φ-witness for A in $\mathrm{Hom}_{<\lambda}^{V[H\restriction\alpha]}$. By Theorem 6.1, it will suffice to find a φ-witness for $A^{V[H\restriction\beta]}$ in $\mathrm{Hom}_{<\lambda}^{V[H\restriction\beta]}$, for some $\beta < \lambda$. We consider two cases:

CASE 1. There is an $C^* \in \mathrm{Hom}^*$ such that some $B \in L(C^*, \mathbb{R}^*)$ is a φ-witness for A^*.

PROOF. By increasing α, we may as well assume that $C \in \mathrm{Hom}_{<\lambda}^{V[H\restriction\alpha]}$. We can also easily arrange that $A \leq_w C$.

Let γ_0 be least such that there is some φ-witness B for A^* with $B, A^* \in L_{\gamma_0}(C^*, \mathbb{R}^*)$[13]. Fix $x_0 \in \mathbb{R}^*$ such that some such B is ordinal definable over $L_{\gamma_0}(C^*, \mathbb{R}^*)$ from x_0 and A^*, C^*. We may as well assume $x_0 \in V[H \restriction \alpha]$. The least sequence of ordinals from which one can define a φ-witness B from $\langle x_0, A^*, C^*\rangle$ over $L_{\gamma_0}(C^*, \mathbb{R}^*)$ is definable from $\langle x_0, A^*, C^*\rangle$ over $L_{\gamma_0}(C^*, \mathbb{R}^*)$, and so we may assume that B is definable without ordinal parameters. Say

$$u \in B \Leftrightarrow L_{\gamma_0}(C^*, \mathbb{R}^*) \models \psi\big[\langle x_0, A^*, C^*\rangle, u\big].$$

Let

$$\bar{\varphi}(v_0, v_1) = \text{“}v_0\text{is a }\varphi\text{-witness for }v_1''\text{,}$$

and let $\theta(v, u)$ be the natural formula defining B from $\langle x_0, A^*, C^*\rangle$:

$$\begin{aligned}\theta(v, u) = \text{“}v \text{ is } \langle v_0, v_1, v_2\rangle &\text{ where } L(v_2, \mathbb{R}) \models \exists B \bar{\varphi}(B, v_1)\\ &\text{and if } \gamma_0 = \text{ is the least } \gamma \text{ s.t. } L_\gamma(v_2, \mathbb{R}) \models \exists B \bar{\varphi}(B, v_1),\\ &\text{then } L_{\gamma_0}(v_2, \mathbb{R}) \models \psi[v, u]''\end{aligned}$$

The key is that θ gives us an absolute definition of B. More precisely, letting $N = V[H \restriction \alpha]$ and $g \in \mathrm{HC}^*$,

CLAIM. For all $u \in \mathbb{R} \cap N[g]$

$$N[g] \models \theta[\langle x_0, A^{N[g]}, C^{N[g]}\rangle, u] \Leftrightarrow u \in B.$$

PROOF. Of course, $\mathbb{R}^*_H = \mathbb{R}^*_K$ for some $\mathrm{Col}(\omega, <\lambda)$- generic K over $N[g]$. (This follows from Lemma 7.5.) Let $G \subseteq \mathbb{Q}_{<\lambda}$ be as given by Lemma 7.6, with $N[g]$ playing the role of V, K the role of H, and γ_0 the role of α. Let

$$i\colon N[g] \to \mathrm{Ult}(N[g], G)$$

be the canonical embedding; then

$$i(A^{N[g]}) = A^* \text{ and } i(C^{N[g]}) = C^*$$

[13]We set $L_0(Z, \mathbb{R}^*) = \mathbb{R}^* \cup \{Z\}$, then iterate first order definability as usual.

by the proof of Lemma 7.6(c). Since γ_0 is in the wellfounded part of $\text{Ult}(N[g], G)$ and $\mathbb{R}^* = \mathbb{R} \cap \text{Ult}(N[g], G)$, we get

$$\text{Ult}(N[g], G) \models \big(L\big(i(C^{N[g]}), \mathbb{R}\big) \models \exists B \bar{\varphi}\big[B, i\big(A^{N[g]}\big)\big]\big)$$

and

$$\text{Ult}(N[g], G) \models \theta\big[\big\langle x_0, i(A^{N[g]}), i(C^{N[g]})\big\rangle, u\big] \Leftrightarrow u \in B.$$

Since i is elementary and the identity on reals, we have proved our claim. ⊣

Taking $g = \emptyset$ in the above, it follows from the meaning of θ that $B \cap \mathbb{R}^N$ is a φ-witness for A in the sense of N. We will be done with Case 1 if we show that $B \cap \mathbb{R}^N \in \text{Hom}^N_{<\lambda}$. This follows easily from the tree production lemma. Let (T, U) and (R, S) be $< \lambda$-absolutely complementing pairs in N such that

$$p[T] = A \text{ and } p[R] = C.$$

Let

$$\tau\big(\langle x_0, T, R\rangle, u\big) = \theta\big(\langle x_0, p[T], p[R]\rangle, u\big).$$

(The author trusts the reader will untangle the confusion of language and metalanguage here.) We apply the tree production lemma to τ, with $\langle x_0, T, R\rangle$ playing the role of the parameter a. The generic absoluteness hypothesis of the lemma is an immediate consequence of our claim and the universality of the symmetric collapse. For stationary tower correctness, let $\delta < \lambda$ be Woodin in N, and

$$i : N \to M = \text{Ult}(N, G)$$

the canonical embedding associated to a $\mathbb{Q}_{<\delta}$-generic G over N. Then

$$N[G] \models p[T] = p[i(T)] \text{ and } p[R] = p[i(R)],$$

as the reader who is still with us can easily show. Thus for $u \in \mathbb{R} \cap N[G]$,

$$\begin{aligned} M \models \tau\,\big[\langle x_0, i(T), i(R)\rangle, u\big] &\Leftrightarrow M \models \theta\,\big[\langle x_0, p[i(T)]^M, p[i(R)]^M\rangle, u\big] \\ &\Leftrightarrow N[G] \models \theta\,\big[\langle x_0, p[T]^{N[G]}, p[R]^{N[G]}\rangle, u\big] \\ &\Leftrightarrow N[G] \models \tau\,[\langle x_0, T, R\rangle, u] \end{aligned}$$

This completes the proof of Lemma 7.4 Case 1. ⊣

Case 2. Otherwise.

Proof. From Case 1 and our proof of the derived model theorem modulo Lemma 7.4, we get

$$\forall C \in \text{Hom}^*\,\big(L(C, \mathbb{R}^*) \models \mathsf{AD}^+\big).$$

For $C \in \text{Hom}^*$, let $C^\sharp$ be the type of a club class of indiscernibles for $L(C, \mathbb{R}^*)$, in the language of set theory expanded by names for each $x \in \mathbb{R}^*$. We regard $C^\sharp$, if it exists, as a subset of $\mathbb{R}^*$ under some natural coding. ⊣

CLAIM 1. $\forall C \in \mathrm{Hom}^*(C^\sharp$ exists and $C^\sharp \in \mathrm{Hom}^*)$.

PROOF. Fix C, and let $D \in \mathrm{Hom}^*$ be such that $C \notin L(D, \mathbb{R}^*)$. Now Hom^* is semi-linearly ordered by $\leq_w$ by Lemma 7.3[14], and clearly D is not Wadge-reducible to either C or $\mathbb{R}^* \setminus C$. Hence $C \leq_w D$. But now

$$\mathsf{AD}^+ \models \forall C \subseteq \mathbb{R}(\exists D \subseteq \mathbb{R}(D \notin L(C, \mathbb{R})) \Rightarrow C^\sharp \text{ exists}).$$

(See [12]; the result is due independently to the authors of that paper and to Kechris and Woodin[15].) Since $L(D, \mathbb{R}^*) \models \mathsf{AD}^+$, we get $L(D, \mathbb{R}^*) \models C^\sharp$ exists. But $L(D, \mathbb{R}^*)$ is correct about sharps because it has all the ordinals, so $C^\sharp$ exists. Finally, $C^\sharp \equiv_w \oplus_{n<\omega} B_n$, where B_n is the type of the first n indiscernibles. $B_n \in L(C, \mathbb{R}^*)$, so $B_n \leq_w D$, for all n. This implies that $\oplus_{n<\omega} B_n \leq_w D$. Since Hom^* under Wadge reducibility, $C^\sharp \in \mathrm{Hom}^*$. ⊣

CLAIM 2. For any $g \in \mathrm{HC}^*$, $\mathrm{Hom}_{<\lambda}^{V[g]}$ is closed under sharps.

PROOF. Let $C \in \mathrm{Hom}_{<\lambda}^{V[g]}$. Let $B \in \mathrm{Hom}_{<\lambda}^{V[g][h]}$, where $h \in \mathrm{HC}^*$, be such that $B^* = (C^*)^\sharp$. Such a B exists by our first claim. The relation $X = Y^\sharp$ on sets of reals is uniformly Π^1_1 in X, Y, so it follows at once from Lemma 7.3 that $B = (C^{V[g][h]})^\sharp$ holds in $V[g][h]$. But then by $(\Sigma^2_1)^{\mathrm{Hom}_{<\lambda}}$-absoluteness (cf. Theorem 6.1), $V[g]$ has a sharp for C in its $\mathrm{Hom}_{<\lambda}$. ⊣

We let $L_0(\mathbb{R}^*, \mathrm{Hom}^*) = \mathbb{R}^* \cup \mathrm{Hom}^*$, and obtain $L_\gamma(\mathbb{R}^*, \mathrm{Hom}^*)$ for $\gamma > 0$ by iterating first order definability, as usual. Let

$$\gamma_0 = \text{ least } \gamma \text{ s.t. } \exists B\Big(L_\gamma(\mathbb{R}^*, \mathrm{Hom}^*) \models \bar{\varphi}[B, A^*]\Big) \wedge \\ \forall C \in \mathrm{Hom}^* \big(|C|_w < \gamma\big).$$

Again, let $N = V[H \restriction \alpha]$.

CLAIM 3. Let $g \in \mathrm{HC}^*$, and let $G \subseteq \mathbb{Q}_{<\lambda}^{N[g]}$ be such that $G \cap \mathbb{Q}_{<\delta}^{N[g]}$ is $N[g]$-generic, for arbitrarily large Woodin cardinals $\delta < \lambda$ of $N[g]$. Suppose $\mathbb{R}^* = \mathbb{R} \cap \mathrm{Ult}(N[g], G)$ and γ_0 is in the wellfounded part of $\mathrm{Ult}(N, G)$; then letting

$$i_G \colon N \to \mathrm{Ult}(N[g], G)$$

be the canonical embedding, we have

$$i_G\Big(\mathrm{Hom}_{<\lambda}^{N[g]}\Big) = \mathrm{Hom}^*.$$

---

[14]The continuous reductions in question here are coded by reals in $\mathbb{R}^*$.

[15]Here is a short sketch. Work in $\mathsf{AD} + \mathsf{DC}_{\mathbb{R}}$. By Wadge, every set of reals in $L(C, \mathbb{R})$ is $\leq_w D$. Thus $\theta^{L(C,\mathbb{R})} < \theta$, so we can find a measurable cardinal κ such that $\theta^{L(C,\mathbb{R})} < \kappa$. Let U be a κ-complete normal ultrafilter on κ. One can use U to get the dersired indiscernibles in the usual way. The key here is that U is $\mathbb{R}$-complete over $L(C, \mathbb{R})$: is $A_x \in U$ for all $x \in \mathbb{R}$, and the function $x \mapsto A_x$ is in $L(C, \mathbb{R})$, then $\bigcap_x A_x \in U$.

PROOF. If not, let $C \in i_G(\mathrm{Hom}^{N[g]}_{<\lambda}) \setminus \mathrm{Hom}^*$ be Wadge minimal, so that

$$\mathrm{Ult}(N[g], G) \models \mathrm{Hom}^* = \{A \mid A <_w C,$$

and

$$L_{\gamma_0}(\mathbb{R}^*, \mathrm{Hom}^*) \subseteq L(C, \mathbb{R})^{\mathrm{Ult}(N[g],G)}.$$

Note $(C^\sharp)^{\mathrm{Ult}(N[g],G)}$ exists and is in $i_G(\mathrm{Hom}^{N[g]}_{<\lambda})$ by Claim 2; moreover every set Wadge reducible to it in the sense of $\mathrm{Ult}(N[g], G)$ is in $i_G(\mathrm{Hom}^{N[g]}_{<\lambda})$. It follows that

$$\mathrm{Ult}(N[g], G) \models \exists B \in i_G\Big(\mathrm{Hom}^{N[g]}_{<\lambda}\Big)\big(B \text{ is a } \varphi\text{-witness for } A^*\big).$$

Noting that $i_G(A^{N[g]}) = A^*$, we see that there is a $B \in \mathrm{Hom}^{N[g]}_{<\lambda}$ such that B is a φ-witness for $A^{N[g]}$. By Lemma 7.3, B^* is a φ-witness for A^*, and of course, $B^* \in \mathrm{Hom}^*$. This contradicts our case hypothesis. ⊣

We can now complete the proof of Lemma 7.4 just as we did in Case 1. We take a minimal-in-$L(\mathbb{R}^*, \mathrm{Hom}^*)$ φ-witness B for A^*, and show that that B has a sufficiently absolute definition that it yields a $\mathrm{Hom}^N_{<\lambda}$ witness for A. Where we used in Case 1 that $i_G(C) = C^*$ for generic embeddings induced by $\subseteq \mathbb{Q}_{<\lambda}$, we use here that $i_G(\mathrm{Hom}^{N[g]}_{<\lambda}) = \mathrm{Hom}^*$ for such embeddings. As in Case 1, this gives us the generic absoluteness of the definition of B needed in the tree production lemma. For stationary tower correctness, we use

CLAIM 4. Let $\delta < \lambda$ be a successor Woodin of N, and K be $\mathbb{Q}_{<\delta}$-generic over N; then

$$i_K\big(\mathrm{Hom}^N_{<\lambda}\big) = \mathrm{Hom}^{N[K]}_{<\lambda},$$

where i_K is the generic embedding.

PROOF. There is a φ-witness for $A^{N[K]}$ in $L(\mathbb{R}, \mathrm{Hom}_{<\lambda})^{N[K]}$ by Lemma 7.6. If Claim 4 fails, that just as in the proof of Claim 3, we get a φ-witness for A in $\mathrm{Hom}^N_{<\lambda}$, which gives us a φ-witness for A^* in Hom^*, a contradiction. ⊣

This completes the proof of Lemma 7.4. ⊣

There is a corollary worth pointing out:

COROLLARY 7.8. *Let $L(\mathbb{R}^*, \mathrm{Hom}^*)$ be a derived model, and suppose there is a φ-witness for A^* in $L(\mathbb{R}^*, \mathrm{Hom}^*)$; then there is a φ-witness for A^* in* Hom^*.

That is, in the derived model, every Σ^2_1 fact has a Suslin-co-Suslin witness. The result is an easy Corollary of 7.4 and Lemma 7.3.

§8. Scale (Σ^2_1) in derived models.

Woodin has shown

THEOREM 8.1 (Woodin). *Assume* AD^+; *then*

(1) *The pointclass Σ^2_1 has the Scale Property, and*
(2) *Every lightface Σ^2_1 collection of sets of reals has a lightface Δ^2_1 member.*

In this section we shall prove part of this theorem, namely, we shall show that (1) and (2) of Theorem 8.1 hold in any derived model. Woodin's original proof of Theorem 8.1 used this fact, together with his result that all models of AD^+ are derived models in a certain sense. (See Theorem 7.2.)[16].

In fact what we show is

LEMMA 8.2 (Woodin). *Assume* AD^+, *and suppose also that whenever* $A \subseteq \mathbb{R}$ *is Suslin and co-Suslin, and there is a* φ*-witness for* A, *then there is a* φ*-witness* B *for* A *such that* B *is Suslin and co-Suslin. Then for any Suslin-co-Suslin set* A:

(1) *The pointclass* $\Sigma_1^2(A)$ *has the Scale Property, and*
(2) *Every* $\Sigma_1^2(A)$ *collection of sets of reals has a* $\Delta_1^2(A)$ *member.*

Of course, it follows at once that if $L(\mathbb{R}^*, \mathrm{Hom}^*)$ is a derived model, and $A \in \mathrm{Hom}^*$, then in $L(\mathbb{R}^*, \mathrm{Hom}^*)$, $\Sigma_1^2(A)$ has the Scale Property, and every $\Sigma_1^2(A)$ collection of sets of reals has a $\Delta_1^2(A)$ member. These facts will be useful in the further theory of derived models which we shall develop in later sections.

The proof of Lemma 8.2 which we shall give involves techniques unlike those we have been using, and the reader without some experience with AD will no doubt find it impenetrable.

Our first lemma shows how coarse the definabilty requirement on a Σ_1^2 scale is.

LEMMA 8.3 (Woodin). *Assume* $\mathsf{AD} + \mathsf{DC}_{\mathbb{R}}$. *The following are equivalent*:

(1) Σ_1^2 *has the Scale Property,*
(2) *if* U *is a* Σ_1^2 *set of reals, then for any* $x \in U$ *there is a tree* T *on some* $\omega \times \kappa$ *such that*
    (a) $x \in p[T]$ *and* $p[T] \subseteq U$, *and*
    (b) *for some* $A \subseteq \mathbb{R}$, T *is ordinal definable in* $L(A, \mathbb{R})$[17].

PROOF. To see (1) $\Rightarrow$ (2), simply take T to be the tree of a Σ_1^2 scale on U. Clearly, T works simultaneously for all $x \in U$. Now assume (2). We define a scale $\{\psi_i\}$ on U as follows: for $x \in U$, let

$$\psi_0(x) = \langle \alpha, \beta, \gamma, \varphi \rangle,$$

where

$$\alpha = |A|_w, \text{ for } A \text{ Wadge-minimal such that}$$
$$\exists T \in \mathrm{OD}^{L(A,\mathbb{R})}(x \in p[T] \wedge p[T] \subseteq U)$$

and $\langle \beta, \gamma, \varphi \rangle$ is the lexicographically minimal tuple such that for some (equivalently all) A such that $|A|_w = \alpha$, φ defines over $L_\beta(A, \mathbb{R})$ from parameter

[16]Woodin later found a proof of Theorem 8.1 which avoids the derived model theorem.

[17]No parameters other than ordinals are allowed. In particular, then definition cannot mention A.

γ a tree T on some $\omega \times \kappa$ such that $x \in p[T] \subseteq U$. We identify the range of ψ_0 with an ordinal by ordering tuples lexicographically. It is then easy to check that ψ_0 is a Σ_1^2-norm. Let us write T^x for the tree T which arises in the definition of $\psi_0(x)$. For $x \in U$, set

$$l_x = \text{leftmost branch of } (T^x)_x,$$

and for $i > 0$ let

$$\psi_i(x) = \langle \psi_0(x), l_x \upharpoonright i \rangle.$$

Again, we use the lexicographic order to identify $\operatorname{ran}(\psi_i)$ with an ordinal. It is easy to see that the ψ_i constitute a Σ_1^2 scale on U. ⊣

REMARK 8.4. Let us write

$$P_\alpha(\mathbb{R}) = \{A \subseteq \mathbb{R} \mid |A|_w < \alpha\}.$$

Woodin has shown that AD^+ implies that for any Σ_1 formula of the language of set theory $\varphi(v_0, v_1)$ and any $A \subseteq \mathbb{R}$,

$$L(P(\mathbb{R})) \models \varphi[A, P(\mathbb{R})] \Rightarrow \exists \alpha, \beta < \Theta(L_\alpha(P_\beta(\mathbb{R})) \models \varphi[A, P_\beta(\mathbb{R})]).$$

It follows at once that for any $A \subseteq \mathbb{R}$,

$$A \in \mathrm{OD} \Leftrightarrow \exists B \subseteq \mathbb{R}(A \in \mathrm{OD}^{L(B,\mathbb{R})}).$$

We have stated Lemma 8.3 in the somewhat more complicated way we have in order to avoid using this equivalence, whose proof we do not know at the moment. In what follows, we shall often write "OD in some $L(B, \mathbb{R})$" when it might seem more natural to simply write "OD". We usually do so because the former notion is clearly Σ_1^2, and we want to avoid quoting Woodin's result that the two notions are equivalent.

We shall need to use homogeneity representations in the choiceless world of AD. The following basic theorem of Martin characterizes the sets of reals which are homogeneously Suslin via trees on some $\omega \times \kappa$. Although the proof involves some very pretty constructions of measures from games, we shall omit it, since such techniques are rather far from the other techniques we are using. See [7] for a proof.

Let Θ be the least ordinal which is not the surjective image of $\mathbb{R}$.

THEOREM 8.5 (Martin). *Assume* $\mathsf{AD} + \mathsf{DC}_\mathbb{R}$; *then for any* $A \subseteq \mathbb{R}$, *the following are equivalent*:

(1) $A = p[T]$, *for some homogeneous tree* T *on some* $\omega \times \kappa$, *where* $\kappa < \Theta$;
(2) A *is Suslin and co-Suslin.*

PROOF. Assuming (1), it is clear that A is Suslin. But the Martin-Solovay construction requires only $\mathsf{DC}_\mathbb{R}$. (One uses the Coding Lemma to code functions from κ^n to κ^+ by reals, and then $\mathsf{DC}_\mathbb{R}$ to show the appropriate ultrapowers are wellfounded.) Thus A is co-Suslin.

The author will fill in the rest later. ⊣

Part of the reason homogeneity systems yield the ordinal definable trees required in Lemma 8.3 is Kunen's theorem that all measures are ordinal definable.

THEOREM 8.6 (Kunen). *Assume* $\mathsf{AD} + \mathsf{DC}_{\mathbb{R}}$, *and let* μ *be a measure on some ordinal* $\kappa < \Theta$; *then* μ *is ordinal definable.*

PROOF. By the Coding Lemma, there is a surjective map $x \mapsto C_x$ from $\mathbb{R}$ onto $P(\kappa)$. Let D be the set of Turing degrees, and for $d \in D$, let

$$f(d) = \text{ least } \alpha \text{ such that } \alpha \in \bigcap \{C_x \mid \exists y \in d(x \leq_T y) \wedge \mu(C_x) = 1\}.$$

Since μ is countably complete, $f(d)$ exists for all d. Clearly, if $C \subseteq \kappa$, and $f(d) = g(d)$ on a Turing cone, then

$$\mu(C) = 1 \Leftrightarrow \text{ for a cone of } d, g(d) \in C.$$

This gives us a definition of μ from $[f]_\nu$, where ν is Martin's cone measure on D. But since f maps into κ, $[f]_\nu$ "is" an ordinal. ⊣

The proof of Lemma 8.2 which we shall give differs a bit from Woodin's original one. It makes use of certain observations concerning the continuous propagation of homogeneity representations which, so far as the author knows, are due to him. The first lemma in this direction elaborates on a basic construction due to Martin.

LEMMA 8.7 (Steel). *Assume* $\mathsf{AD}+\mathsf{DC}_{\mathbb{R}}$; *then for any* $\kappa < \Theta$ *there is an ordinal definable function* $F\colon \mathrm{meas}(\kappa) \to \bigcup_{\beta<\Theta} \mathrm{meas}(\beta)$ *such that*

(a) *for all* $\mu \in \mathrm{meas}(\kappa)$, $\dim(\mu) = \dim(F(\mu))$,

(b) *for all* μ, ν *in* $\mathrm{meas}(\kappa)$, μ *projects to* ν *iff* $F(\mu)$ *projects to* $F(\nu)$, *and*

(c) *for all towers of measures* $\langle \mu_n \mid n < \omega \rangle \in \mathrm{meas}(\kappa)^\omega$,

$$\langle \mu_n \mid n < \omega \rangle \text{ is wellfounded } \Leftrightarrow \langle F(\mu_n) \mid n < \omega \rangle \text{ is illfounded.}$$

PROOF. Let λ have the strong partition property $\lambda \to (\lambda)^\lambda$, and $\kappa < \lambda < \Theta$. We get the measures $F(\mu)$ we need from the strong partition property in a standard way[18]. For any $X \subseteq \lambda$ and set W equipped with a wellorder $<_W$ of order type $\leq \lambda$, let

$$[X]^W = \{f\colon W \to X \mid \forall a, b(a <_W b \Rightarrow f(a) < f(b))\}.$$

For any unbounded $X \subseteq \lambda$, let $\pi_X\colon \lambda \to X$ be the increasing enumeration of X, and set

$$X^* = \{\sup(\{\pi_X(\omega\xi + n) \mid n < \omega\}) \mid \xi < \lambda\}.$$

[18]The construction to follow is due to Martin. The new observation here is just that Martin's construction does not require a *tree* for which $\bar{\mu}$ is a homogeneity system.

The strong partition property gives us a measure σ_W on $[\lambda]^W$: for $\mathcal{A} \subseteq [\lambda]^W$, we put

$$\sigma(\mathcal{A}) = 1 \Leftrightarrow \exists C\Big(C \text{ is club in } \lambda \text{ and } [C^*]^W \subseteq \mathcal{A}\Big).$$

It is not hard to see that σ_W is a countably complete measure on $[\lambda]^W$[19].

For any $n < \omega$, let

$$W_n = \left(\bigcup_{i \leq n} \kappa^i, \leq_{\text{bk}}\right),$$

where $\leq_{\text{bk}}$ is the Brouwer-Kleene order: $s \leq_{\text{bk}} t$ iff $t \subseteq s$ or $\exists k \in \text{dom}(s) \cap \text{dom}(t)(s \restriction k = t \restriction k \wedge s(k) < t(k))$. (So W_0 consists of one point, and W_n is a suborder of W_{n+1}, for all n.) Given a measure $\mu \in \text{meas}(\kappa)$ of dimension $n > 0$, with projections μ_i to measures of dimension i for each $i \leq n$, we define a measure $F(\mu)$ on $i_\mu(\lambda)^n$ by

$$A \in F(\mu) \Leftrightarrow \text{ for } \sigma_{W_n} \text{ a.e. } f, \left\langle \left[f \restriction \kappa^1\right]_{\mu_1}, \left[f \restriction \kappa^2\right]_{\mu_2}, \ldots, \left[f \restriction \kappa^n\right]_{\mu_n}\right\rangle \in A.$$

If μ concentrates on $\kappa^0 = \{\emptyset\}$, so is principal, we let $F(\mu) = \mu$. Clearly $F(\mu) \in \text{meas}(i_\mu(\lambda))$, $i_\mu(\lambda) < \Theta$, and $\dim(\mu) = \dim(F(\mu))$.

For (b) let μ project to ν, and suppose ν has dimension i, where $i > 0$. (If $i = 0$, (b) is trivial.) Let $A \in F(\nu)$; then we can find a club C in λ such that for any $f \in [C^*]^{W_i}$, $\langle [f \restriction \kappa^0]_{\mu_0}, \ldots, [f \restriction \kappa^i]_{\mu_i}\rangle \in A$. But then for any $f \in [C^*]^{W_n}$, $f \restriction \kappa^i \in [C^*]^{W_i}$, so C witnesses that for $F(\mu)$ a.e. $\langle \alpha_1, \ldots, \alpha_n\rangle$, $\langle \alpha_1, \ldots, \alpha_i\rangle \in A$. Since A was arbitrary, we have that $F(\mu)$ projects to $F(\nu)$, as desired.

For (c), let $\langle \mu_n \mid n < \omega\rangle \in \text{meas}(\kappa)^\omega$ be a tower of measures. Notice that if $n > 0$, then $F(\mu_n)$ concentrates on tuples $\langle \alpha_1, \ldots, \alpha_n\rangle$ such that whenever $1 \leq i < n$, then $i_{\mu_i,\mu_{i+1}}(\alpha_i) > \alpha_{i+1}$. (This comes down to the fact that whenever $f \in [\lambda]^{W_{i+1}}$, then for all $s \in \kappa^{i+1}$, $f(s) < f(s \restriction i)$, because $s \leq_{\text{bk}} s \restriction i$.) Thus

$$\langle F(\mu_n) \mid n < \omega\rangle \text{ is wellfounded } \Rightarrow \langle \mu_n \mid n < \omega\rangle \text{ is illfounded},$$

since by meeting countably many measure one sets in the $F(\bar{\mu})$ tower, we produce an infinite descending chain in $\text{Ult}(V, \bar{\mu})$.

For the converse, suppose $\bar{\mu}$ is illfounded. We can then find a tree T on κ such that $\mu_n(T \cap \kappa^n) = 1$ for all n, but T is wellfounded. In order to see that

[19]For example, given $\mathcal{A} \subseteq [\lambda]^W$, partition $[\lambda]^\lambda$ by letting $F(X) = 0$ iff $g \in \mathcal{A}$ where $g \in [\lim(X)^*]^W$ is such that $\text{ran}(g)$ is an initial segment of $\lim(X)^*$. Let $H \in [\lambda]^\lambda$ be homogeneous for this partition, and $C = \lim(H)$. Then either $[C^*]^W \subseteq \mathcal{A}$ or $[C^*]^W \cap \mathcal{A} = \emptyset$. So either $\mathcal{A}$ or its complement gets measure one.

$\langle F(\mu_n) \mid n < \omega \rangle$ is countably complete, fix sets $A_n \in F(\mu_n)$, for each $n \geq 1$. We seek a "fiber" for the A_n's. Let C_n be club in λ and such that

$$f \in [C_n^*]^{W_n} \Rightarrow \left\langle \left[f \restriction \kappa^1\right]_{\mu_1}, \ldots, \left[f \restriction \kappa^n\right]_{\mu_n} \right\rangle \in A_n.$$

Let

$$C = \bigcap_{n<\omega} C_n,$$

and let $f : T \to C^*$ preserve the Brouwer-Kleene order (which is a wellorder of order type ¡ κ^+ when restricted to T), and be such that if $u, v \in T$ and $u <_{\mathrm{bk}} v$, then $C^* \cap (f(u), f(v))$ has order type at least κ^+. This spacing in C^* of the points in $\operatorname{ran}(f)$ guarantees that for any n, we can find a $g \in [C^*]^{W_n}$ such that $g \restriction (T \cap W_n) = f$. It follows that $[g \restriction \kappa^i]_{\mu_i} = [f \restriction \kappa^i]_{\mu_i}$ for $i = 1, \ldots, n$, and therefore

$$\left\langle \left[f \restriction \kappa^1\right]_{\mu_1}, \ldots, \left[f \restriction \kappa^n\right]_{\mu_n} \right\rangle \in A_n,$$

for all n. This is the desired fiber for the A's. That finishes the proof of (c). ⊣

The set of homogeneity systems over ω^k with support Z is a closed set in the topological space $\operatorname{meas}(Z)^{\omega^{k^{<\omega}}}$, where this space is given the Baire topology induced by any and all enumerations of $\omega^{k^{<\omega}}$. For any set $Y \subseteq \operatorname{meas}(Z)$, let

$$\begin{aligned}\mathcal{H}_Y^k = \{\bar{\mu} \mid \ & \bar{\mu} \text{ is a homogeneity system over } \omega^k \text{ and} \\ & \forall s \in \omega^{k^{<\omega}} (\mu_s \in Y)\end{aligned}$$

$\mathcal{H}_Y^k$ is again a closed set in the space $\operatorname{meas}(Z)^{\omega^{k^{<\omega}}}$. The topology of $\mathcal{H}_Y^k$ is generated by *finite partial homogeneity systems* (from Y, of dimension k), that is, functions $h : T \to Y$, where T is a finite tree on ω^k, such that whenever $s \in T$, then $\langle h(s \restriction i) \mid i \leq \operatorname{dom}(s) \rangle$ is a (finite) tower of measures. Given such a finite partial homogeneity system h, the set

$$N_h = \left\{\bar{\mu} \in \mathcal{H}_Y^k \mid h = \bar{\mu} \restriction T\right\}$$

is clopen in $\mathcal{H}_Y^k$, and the N_h's generate its topology. Let

$$h_Y^k = \left\{h \mid h \text{ is a finite partial homogeneity system from } Y \text{ of dimension } k\right\}.$$

For any $\pi : h_Y^k \to h_Z^n$ we let π^* be the function on $\mathcal{H}_Y^k$ given by

$$\pi^*(\bar{\mu}) = \bigcup \{\pi(h) \mid h \in h_Y^k \wedge h \subseteq \bar{\mu}\}.$$

Let us call π *good* if

$$\pi^* : \mathcal{H}_Y^k \to \mathcal{H}_Z^n$$

is a total, continuous function. (This reduces to some elementary, concrete properties of π.)

We wish to capture formulae with real quantifiers by continuous transformations π^* on homogeneity systems. More precisely, let $\mathcal{L}^*$ be the language with a unary predicate symbol $\dot{A}$, together with one k-ary relation symbol $\dot{T}$ for each k-ary recursive relation $T \subseteq (\omega^\omega)^k$. For any formula $\varphi(v_0, \ldots, v_{n-1})$ of $\mathcal{L}^*$, and any $A \subseteq \mathbb{R}$, let

$$\varphi^A = \{\bar{x} \in \mathbb{R}^n \mid (\mathbb{R}, T, A)_{T \text{ recursive}} \models \varphi[\bar{x}]\},$$

where of course A interprets $\dot{A}$ and T interprets $\dot{T}$[20]. We then have

Lemma 8.8 (Steel). *Assume* AD + DC$_\mathbb{R}$. *Let* $\varphi(v_0, \ldots, v_{n-1})$ *be a formula of* $\mathcal{L}^*$, *and let* $Y \subseteq \text{meas}(\kappa)$ *for some* $\kappa < \Theta$ *be such that* $|Y| < \Theta$, *and* $\mathcal{H}^1_Y \neq \emptyset$. *Then there is a* $\beta < \Theta$ *and* $Z \subseteq \text{meas}(\beta)$ *such that* $|Z| < \Theta$, *together with a good*

$$\pi\colon h^1_Y \to h^n_Z$$

such that

$$\forall \bar{\mu} \in \mathcal{H}^1_Y \big(\varphi^{S_{\bar{\mu}}} = S_{\pi^*(\bar{\mu})}\big).$$

Moreover, if Y *is ordinal definable, then so are* Z, π, *and* π^*.

Proof. The proof is by induction on φ.

If $\varphi = \dot{T}(v_{i_1}, \ldots, v_{i_k})$, then φ^A is recursive, hence homogeneous, and independent of A, so we let π be the appropriate constant function. If $\varphi = \dot{A}(v_i)$, then $\varphi^A = \{\bar{x} \in \mathbb{R}^n \mid x_i \in A\}$, and the desired π is a minor perturbation of π designed to accomodate the change of arity[21]. This finishes the atomic case.

If $\varphi = \neg\psi$, we can use Lemma 8.7. For let $\pi\colon h^1_Y \to h^n_Z$ witness the lemma for ψ, and let $F\colon Z \to \text{meas}(\beta)$ be an OD tower-flipping function as in Lemma 8.7. For any $h \in h^1_Y$, we let $\sigma(h)$ have the same domain as $\pi(h)$, and

$$\sigma(h)(s) = F(\pi(h)(s)).$$

It is clear that σ works for φ.

If $\varphi = \psi \wedge \rho$, and π and σ witness the lemma for φ and ρ, then it is not hard to construct a good τ with domain h^1_Y such that

$$\text{Ult}\big(V, \tau^*(\bar{\mu})\big) \cong \text{Ult}\big(\text{Ult}(V, \pi^*(\bar{\mu})), j(\sigma^*(\bar{\mu}))\big),$$

where $j\colon V \to \text{Ult}(V, \pi^*(\bar{\mu}))$ is the canonical embedding[22]. We leave the details to the reader, since the case $\varphi = \psi \wedge \rho$ can anyway be subsumed under the case $\varphi = \forall v \psi$ to follow[23].

---

[20]The notation $\varphi(v_1, \ldots, v_{n-1})$ does not presume that all v_i for $i < n$ actually occur in φ. We should therefore write $(n, \varphi)^A$, but we will allow n to be understood from context.

[21]Our induction is really on pairs n, φ such that all free variables of φ are among $v_0, \ldots, v_{n-1}$.

[22]Since we have assumed AD, j will not be elementary. However, for any $\nu \in \text{meas}(\gamma)$, $j(\nu) \in \text{meas}(j(\gamma))$, as the reader can easily check, and this is enough to make sense of the iteration.

[23]This argument in the $\wedge$ case works without AD, however, while the $\forall$ argument does not.

Finally, let $\varphi = \varphi(v_1) = \forall v_2 \psi(v_1, v_2)$, where we have taken $n = 1$ for notational simplicty. Let $\pi \colon h^1_Y \to h^2_Z$ witness the lemma for ψ. It will be enough to find a $\beta < \Theta$ and a good $\sigma \colon h^2_Z \to h^1_{\mathrm{meas}(\beta)}$ such that for all $\bar{v} \in \mathcal{H}^2_Z$ and $x \in \mathbb{R}$,

$$\sigma^*(\bar{v})_x \text{ is wellfounded } \Leftrightarrow \forall y (\bar{v})_{(x,y)} \text{ is wellfounded.}$$

For then, it is easy to find a good τ such that $\tau^* = \sigma^* \circ \pi^*$, and τ witnesses the lemma for φ.

We get σ from the standard construction, due to Martin, which obtains homogeneity from weak homogeneity, using partition cardinals. We need a little care, however, because we are not given a homogeneous *tree*.

Fix $F \colon Z \to \mathrm{meas}(\gamma)$ be a tower-flipping function as in Lemma 8.7. Given $\bar{v} \in \mathcal{H}^2_Z$, we shall define $\bar{\mu} = \sigma^*(\bar{v})$. It will be clear from the construction that $\sigma*$ is continuous. To begin with, set

$$v^*_{(s,t)} = F(v_{(s,t)})$$

for all $(s,t) \in \mathrm{dom}(\bar{v})$. Inspecting the construction of Lemma 8.7, we see that for any $(x,y) \in \mathbb{R}^2$, $\bar{v}^*_{(x,y)}$ concentrates on descending chains in $\mathrm{Ult}(V, \bar{v}_{(x,y)})$. Our tower $\bar{\mu}_x$ will concentrate on attempts to prove continuously that all $\bar{v}^*_{(x,y)}$ are illfounded. The construction generalizes that in Lemma 8.7.

Let $\langle u_i \mid i < \omega \rangle$ enumerate $\omega^{<\omega}$ so that $\forall i \forall k \exists j \leq i (u_i \restriction k = u_j)$, and let $n_i = \mathrm{dom}(u_i)$. Let

$$W_n = \left(\bigcup_{i \leq n} (\omega^i \times \gamma^i), \leq_{\mathrm{bk}} \right),$$

let λ be a strong partition cardinal such that $\gamma < \lambda < \Theta$, and for any ordered set W let σ_W be the strong partition measure on $[\lambda]^W$ defined in Lemma 8.7. For $f \in [\lambda]^{W_n}$ and $u \in \omega^k$, where $k \leq n$, let

$$f_u(t) = f(u,t)$$

for all $t \in \gamma^k$. Let

$$\beta = \sup \{ i_{v_{(s,t)}}(\lambda) \mid s,t \in \omega^{<\omega} \}.$$

For $s \in \omega^k$, where $k > 0$, we define a measure μ_s concentrating on $[\beta]^k$ by

$$\mu_s(A) = 1 \Leftrightarrow \text{ for } \sigma_{W_k}\text{-a.e. } f \ \left\langle [f_{u_1}]_{v^*_{s \restriction n_1, u_1}}, \ldots, [f_{u_n}]_{v^*_{s \restriction n_k, u_k}} \right\rangle \in A.$$

Let μ_0 be the principal measure on $\{\emptyset\}$.

$\bar{\mu}_x$ concentrates on attempts to continuously illfound all $\bar{v}^*_{(x,y)}$ below the image of λ. Thus if $\bar{\mu}_x$ is wellfounded, any fiber meeting the appropriate measure one sets witnesses that $\forall y (\bar{v}^*_{(x,y)})$ is illfounded. Conversely, let x be such that all $\bar{v}^*_{(x,y)}$ are illfounded. Let $\mu_{x \restriction n}(A_n) = 1$ for all n. Let C be club in λ, and contain all the clubs witnessing the A_n contain projections of measure one sets with respect to σ_{W_n}. Since all $\bar{\nu}_{(x,y)}$ are wellfounded, the tree T on $\omega \times \beta$ of all attempts to build a (y, g) such that g is an infinite descending chain in $\mathrm{Ult}(V, \bar{\nu}_{(x,y)})$ below the image of λ is wellfounded. Let

$$f : T \to C^*$$

preserve $\leq_{\mathrm{bk}}$ on T. Then f determines a fiber for the A_n's, as in the proof of Lemma 8.7, and we are done.

It is easy to check that the continuous homogeneity transformations we have *defined* are ordinal definable. ⊣

Proof of Lemma 8.2. We shall prove the result for Σ_1^2, and leave it to the reader to provide the easy generalization to $\Sigma_1^2(A)$, where A is Suslin and co-Suslin.

We begin by showing Σ_1^2 has the Scale Property. For this, we shall show that (2) of Lemma 8.3 holds. So let U be Σ_1^2; say

$$x \in U \Leftrightarrow \exists B \subseteq \mathbb{R} x \in \neg\varphi^B,$$

where $\varphi = \varphi(v_1)$ is an $\mathcal{L}^*$ formula. Fix $x \in U$. By hypothesis, there is a Suslin, co-Suslin B such that $x \in \neg\varphi^B$. By the theorems of Martin and Kunen, we can fix $A \subseteq \mathbb{R}$ such that for some $\kappa < \Theta^{L(A,\mathbb{R})}$, there is a homogeneity system $\bar{\mu}$ over ω, with support κ, such that $\bar{\mu} \in L(A, \mathbb{R})$ and

$$x \in (\neg\varphi)^{S_{\bar{\mu}}},$$

and

$$\forall s \in \omega^{<\omega} \big(\mu_s \in \mathrm{OD}^{L(A,\mathbb{R})}\big).$$

We now work in $L(A, \mathbb{R})$. Fix $\alpha_0 < \Theta$ such that

$$\forall s \big(\mu_s \in L_{\alpha_0}(A, \mathbb{R})\big),$$

and let

$$Y = \mathrm{meas}(\kappa) \cap L_{\alpha_0}(A, \mathbb{R}).$$

Y and each of its elements are OD. Let

$$\pi : h_Y^1 \to h_Z^1$$

be the OD good function for $\neg\varphi$ given by Lemma 8.8, where $Z \subseteq \mathrm{meas}(\lambda)$. Thus

$$\exists \bar{\mu} \in \mathcal{H}_Y^1 \big(\pi^*(\bar{\mu})_x \text{ is illfounded}\big),$$

and for any $z \in \mathbb{R}$ and $\bar{\mu} \in \mathcal{H}_Y^1$,

$$\pi^*(\bar{\mu})_z \text{ is illfounded} \Rightarrow z \in U.$$

Let $\beta = \sup\{i_\nu(\lambda) \mid \nu \in Z\}$, and let T be the tree on $\omega \times (Y \times \beta)$ which attempts to build a triple $(z, \bar{\mu}, g)$ such that $\bar{\mu} \in \mathcal{H}_Y^1$ and g is an infinite descending chain in $\mathrm{Ult}(V, \pi^*(\bar{\mu}))$ below the image of λ. It follows from the statements just displayed that $x \in p[T]$, and $p[T] \subseteq U$. Since Y has an OD wellorder of length $< \Theta$, we may regard T as a tree $\omega \times \gamma$ for some $\gamma < \Theta$. Thus T witnesses (2) of Lemma 8.3, as desired.

We now show that Δ^2_1 is a basis for Σ^2_1. It will be enough to show every non-empty projective collection of sets of reals has a Δ^2_1 member[24]. Fix then $\varphi(v_1)$ an $\mathcal{L}^*$ formula, and put

$$\mathcal{S}(B) \Leftrightarrow 0 \in \varphi^B.$$

Then $\mathcal{S}$ is a typical projective collection of sets of reals. Assume $\mathcal{S} \neq \emptyset$. It will be enough to show

$$\exists A, B \subseteq \mathbb{R}\big(\mathcal{S}(B) \wedge B \in \mathrm{OD}^{L(A,R)}\big).$$

For the we can let A be Wadge-minimal as above, and let B be the first $\mathrm{OD}^{L(A,\mathbb{R})}$ set in $\mathcal{S}$ in some natural wellorder of $\mathrm{OD}^{L(A,\mathbb{R})}$, and it is easy to see that B is Δ^2_1.

But now let's look at the proof of Scale(Σ^2_1), in the case our real $x = 0$ and our formula is $\neg\varphi$. Let T be the tree on $\omega \times (Y \times \beta)$ produced there, and $A \subseteq \mathbb{R}$ such that T is $\mathrm{OD}^{L(A,\mathbb{R})}$. Then $0 \in p[T]$, so we can set

$$(\bar{\mu}, g) = \text{ leftmost branch of } T_0,$$

using the $\mathrm{OD}^{L(A,\mathbb{R})}$ wellorder of Y to help make sense of "leftmost". Then $\bar{\mu} \in \mathcal{H}^1_Y$, moreover $\bar{\mu}$, and hence

$$B = S_{\bar{\mu}},$$

are $\mathrm{OD}^{L(A,\mathbb{R})}$. But g witnesses that $\pi^*(\bar{\mu})_0$ is illfounded, which in turn means that $0 \notin (\neg\varphi)^{S_{\bar{\mu}}}$, that is, $0 \in \varphi^B$, as desired. ⊣

§9. **Derived models of** $\mathsf{AD}_{\mathbb{R}}$. It is not too hard to see that if V is the minimal fully iterable canonical inner model with ω Woodin cardinals (i.e., $V = M_\omega$), then the derived model at the unique limit of Woodin cardinals has the form $L(\mathbb{R}^*)$[25]. In this case, $\mathrm{Hom}^* = (\Delta^2_1)^{L(\mathbb{R}^*)}$. That is, if we start with the weakest ground model which has a derived model, we get the weakest model of AD. It is natural to ask whether stronger large cardinal properties in V yield stronger forms of determinacy in its derived models. In fact, there seems to be a systematic, detailed correspondence, much of which has yet to be mapped out. In this section, we consider one very natural strengthening of AD, namely $\mathsf{AD}_{\mathbb{R}}$. We shall show that if λ is a limit of Woodin cardinals and of cardinals which are $< \lambda$-strong, then the derived model at λ satisfies $\mathsf{AD}_{\mathbb{R}}$. This result is due to Hugh Woodin. As a corollary, one has that the consistency of ZFC together with the existence of such a cardinal λ implies the consistency of

---

[24]If $\mathcal{S}(B) \Leftrightarrow \exists A \mathcal{R}(A, B)$, where $\mathcal{R}$ is projective, then a Δ^2_1 set $A \oplus B$ such that $\mathcal{R}(A, B)$ yields a Δ^2_1 set B such that $\mathcal{S}(B)$.

[25]This observation is probably due to Woodin and the author.

$\mathsf{ZF} + \mathsf{AD}_{\mathbb{R}}$. The author has recently proven the converse relative consistency theorem, and thus the existence of such a λ is in fact equiconsistent with $\mathsf{AD}_{\mathbb{R}}$.

$\mathsf{AD}_{\mathbb{R}}$ is a bit of a red herring here, as explained by the following unpublished results (from the early 80's?).

THEOREM 9.1 (Martin, Woodin). *Assume* AD. *If every set of reals is Suslin, then* $\mathsf{AD}_{\mathbb{R}}$ *holds.*

THEOREM 9.2 (Woodin). *If* $\mathsf{AD}_{\mathbb{R}}$ *holds, then all sets of reals are Suslin.*

So in the presence of AD, $\mathsf{AD}_{\mathbb{R}}$ is equivalent to the assertion that every set of reals is Suslin. A derived model $L(\mathbb{R}^*, \mathrm{Hom}^*)$ will therefore satisfy $\mathsf{AD}_{\mathbb{R}}$ if and only if $P(\mathbb{R}^*) \cap L(\mathbb{R}^*, \mathrm{Hom}^*) \subseteq \mathrm{Hom}^*$. Our main goal in this section is to prove

THEOREM 9.3 (Woodin). *Let λ be a limit of Woodin cardinals and of cardinals which are $< \lambda$-strong, and let $L(\mathbb{R}^*, \mathrm{Hom}^*)$ be a derived model at λ; then*

- $P(\mathbb{R}^*) \cap L(\mathbb{R}^*, \mathrm{Hom}^*) = \mathrm{Hom}^*$*,so*
- $L(\mathbb{R}^*, \mathrm{Hom}^*) \models \mathsf{AD}_{\mathbb{R}}$

First, an well-known basic lemma:

LEMMA 9.4 (Kechris, Solovay). *Assume* $\mathsf{AD} + \mathsf{DC}_{\mathbb{R}}$, *and let* $A \subseteq \mathbb{R}$. *For* $x, y \in \mathbb{R}$, *put*

$$R(x, y) \Leftrightarrow \forall B \subseteq \mathbb{R}\big(y \notin \mathrm{OD}^{L(B,\mathbb{R})}(A, x)\big);$$

then

(a) *R is a $\Pi^2_1(A)$ relation,*
(b) $\forall x \exists y R(x, y)$,
(c) $\neg(\exists f : \mathbb{R} \to \mathbb{R} \exists B \subseteq \mathbb{R} \exists x_0 \in \mathbb{R}(f \in \mathrm{OD}^{L(B,\mathbb{R})}(A, x_0) \wedge \forall x \in \mathbb{R}\, R(x, f(x))))$.

PROOF. (a) is obvious. (b) holds because $\{y \mid \neg R(x, y)\}$ is wellordered (all its members being $\mathrm{OD}(A, x)$), hence countable. For (c), suppose f, B, x_0 were a counterexample. Then $f(x_0)$ is $\mathrm{OD}^{L(B,\mathbb{R})}(A, x_0)$, so $\neg R(x_0, f(x_0))$, a contradiction. ⊣

So $\mathsf{AD} + \mathsf{DC}_{\mathbb{R}}$ implies that for any $A \subseteq \mathbb{R}$, there is a $\Pi^2_1(A)$ relation with no uniformization, and hence no scale, which is ordinal definable from A and a real in some $L(B, \mathbb{R})$. This leads at once to the following corollary:

COROLLARY 9.5. *Let $L(\mathbb{R}^*, \mathrm{Hom}^*)$ be a derived model; then the following are equivalent:*

(1) $P(\mathbb{R}^*) \cap L(\mathbb{R}^*, \mathrm{Hom}^*) = \mathrm{Hom}^*$,
(2) $\forall A \in \mathrm{Hom}^*$, *every* $\Sigma^2_1(A)^{L(\mathbb{R}^*,\mathrm{Hom}^*)}$ *set of reals is in* Hom^*.

PROOF. (1) $\Rightarrow$ (2) is trivial. Now suppose toward a contradiction that (2) holds and (1) fails, and let $B \subseteq \mathbb{R}^*$ be in $L(\mathbb{R}^*, \mathrm{Hom}^*) \setminus \mathrm{Hom}^*$. Note that every set in Hom^* is Wadge reducible (in the sense of $L(\mathbb{R}^*, \mathrm{Hom}^*)$) to B, so that $L(\mathbb{R}^*, \mathrm{Hom}^*) = L(B, \mathbb{R}^*)$. Let us work now in this universe. Since

$B \in L(\mathbb{R}^*, \mathrm{Hom}^*)$ we can fix $A \in \mathrm{Hom}^*$ such that $B \in \mathrm{OD}(A, \mathrm{Hom}^*)$. But $\mathrm{Hom}^* = P_\alpha(\mathbb{R}^*)$ for some α, so $\mathrm{Hom}^* \in \mathrm{OD}$, and $B \in \mathrm{OD}(A)$. Letting R be the $\Pi^2_1(A)$ relation of Lemma 9.4, we have by (2) that $R \in \mathrm{Hom}^*$, so that R has a scale in Hom^*, and hence a uniformizing function $f \in \mathrm{Hom}^*$. But the $f \in L(B, \mathbb{R}^*)$, so f is ordinal definable in $L(B, \mathbb{R}^*)$ from B and some $x \in \mathbb{R}^*$, so f is ordinal definable in $L(B, \mathbb{R}^*)$ from A and some $x \in \mathbb{R}^*$, which contradicts property (c) of Lemma 9.4. ⊣

Proof of Theorem 9.3. Let λ be a limit of Woodins, and of cardinals which are $< \lambda$-strong, and let $L(\mathbb{R}^*, \mathrm{Hom}^*)$ be a derived model at λ. By Corollary 9.5 and Theorem 9.1, it will suffice to show that whenever $A \in \mathrm{Hom}^*$, then every $\Sigma^2_1(A)^{L(\mathbb{R}^*, \mathrm{Hom}^*)}$ set of reals is in Hom^*. We shall give the proof for $A = \emptyset$; the proof in general is a simple relativization of the one we give. (One must replace V by some intermediate extension having a λ-absolutely complemented tree projecting to A.)

So let $U \subseteq \mathbb{R}^*$ be Σ^2_1 in $L(\mathbb{R}^*, \mathrm{Hom}^*)$, say

$$U(x) \Leftrightarrow \exists B \subseteq \mathbb{R}^* \big(B \in L(\mathbb{R}^*, \mathrm{Hom}^*) \wedge (\mathrm{HC}^*, \in, B) \models \varphi[x]\big).$$

By Lemma 8.2, U has a Σ^2_1 scale in $L(\mathbb{R}^*, \mathrm{Hom}^*)$. Let T be the tree of such a scale. Since T is OD in $L(\mathbb{R}^*, \mathrm{Hom}^*)$, $T \in V$. We must find a λ-absolute complement for T in some $V[g]$, for $g \in \mathrm{HC}^*$. But let $\kappa < \lambda$ be η-strong for all $\eta < \lambda$, and let $g \in \mathrm{HC}^*$ be $\mathrm{Col}(\omega, |\mathrm{meas}(\kappa)|)$-generic over V. We shall show that T has such an absolute complement in $V[g]$. The key to this is Theorem 4.5.

It will be enough to find, for each $\eta < \lambda$, an η-absolute complement S_η for T in $V[g]$, for then a simple amalgamation $\oplus_\eta S_\eta$ is a λ-absolute complement for T. So fix $\eta < \lambda$. We may as well assume $\eta = V_\eta$, and $\kappa < \eta$. Let γ be such that $\eta < \gamma < \lambda$, and

$$V[g] \models \overset{\mathrm{Col}(\omega,\eta)}{\Vdash} \mathrm{Hom}_\gamma = \mathrm{Hom}_{<\lambda}.$$

Let δ be the 5th Woodin cardinal above γ. Now we apply Theorem 4.5: going back to V, let

$$j\colon V \to M \wedge \mathrm{crit}(j) = \kappa \wedge V_\delta \subseteq M$$

be such that in $V[g]$ we have a tree S such that

$$V[g] \models S \text{ is a } \delta\text{-absolute complement for } j(T).$$

We claim that S is the desired η-absolute complement for T in $V[g]$.

Since $p[T] \subseteq p[j(T)]$, it is clear that $p[S] \cap p[T] = \emptyset$. Now let $x \in \mathbb{R}^*$ be $< \eta$-generic over $V[g]$. We can find a $\mathrm{Col}(\omega, \eta)$-generic h over $V[g]$ such that $x \in V[g][h]$. We must see $x \in p[S] \cup p[T]$. Suppose $x \notin p[S]$. Since x is $< \delta$-generic over $V[g]$, we have $x \in p[j(T)]$. Now let's look at what this means.

Note $x \in M[g][h]$, where $j(T)$ has its meaning. Moreover, $M[g][h]$ satisfies the statement that

$$p[j(T)] \cap \mathbb{R}^* = \{x \mid \exists B \in L(\mathbb{R}^*, \mathrm{Hom}^*)(\mathrm{HC}^*, \in, B) \models \varphi[x]\},$$

where this statement is phrased as a statement about the collapse up to λ over $M[g][h]$. Thus for our particular x,

$$M[g][h] \models \exists B \in L(\mathbb{R}^*, \mathrm{Hom}^*)(\mathrm{HC}^*, \in, B) \models \varphi[x],$$

where again this is a statement about the collapse up to λ. But now, applying Lemma 7.4 inside $M[g][h]$, we get a B such that

$$M[g][h] \models B \in \mathrm{Hom}_{<\lambda} \wedge (\mathrm{HC}, \in, B) \models \varphi[x].$$

Moving back to $V[g][h]$, which agrees up to δ with $M[g][h]$, we see B is δ-absolutely complemented in $V[g][h]$. But there are enough Woodin cardinals between γ and δ that this implies B is Hom_γ in $V[g][h]$, and by our choice of γ, that B is $\mathrm{Hom}_{<\lambda}$ in $V[g][h]$. But then, using Lemma 7.3, we can push up the Σ_1^2 fact that B is witnessing in $V[g][h]$ to $L(\mathbb{R}^*, \mathrm{Hom}^*)$, and we conclude that $U(x)$. That is, $x \in p[T]$, as desired. $\dashv$

REFERENCES

[1] Q. Feng, M. Magidor, and W. H. Woodin, *Universally Baire sets of reals*, ***Set Theory of the Continuum (Berkeley, CA, 1989)*** (H. Judah, W. Just, and W. H. Woodin, editors), Publications of the Research Institute for Mathematical Sciences, vol. 26, Springer, New York, 1992, pp. 203–242.

[2] A. S. Kechris and Y. N. Moschovakis, *Notes on the theory of scales*, ***Cabal Seminar 76–77 (Proc. Caltech-UCLA Logic Seminar, 1976–77)*** (A. S. Kechris and Y. N. Moschovakis, editors), Lecture Notes in Mathematics, vol. 689, Springer, Berlin, 1978, pp. 1–53.

[3] P. Larson, ***The Stationary Tower***, University Lecture Series, vol. 32, AMS, Providence, RI, 2004.

[4] D. A. Martin, *Measurable cardinals and analytic games*, ***Polska Akademia Nauk. Fundamenta Mathematicae***, vol. 66 (1969/1970), pp. 287–291.

[5] D. A. Martin and R. M. Solovay, *A basis theorem for $\sum_3^1$ sets of reals*, ***Annals of Mathematics. Second Series***, vol. 89 (1969), pp. 138–159.

[6] D. A. Martin and J. R. Steel, *A proof of projective determinacy*, ***Journal of the American Mathematical Society***, vol. 2 (1989), no. 1, pp. 71–125.

[7] ———, *The tree of a Moschovakis scale is homogeneous*, ***Games, Scales, and Suslin Cardinals, The Cabal Seminar, Vol. I*** (B. Loewe, A. Kechris, and J. R. Steel, editors), Cambridge University Press, Cambridge, 2008, pp. 401–421.

[8] Y. N. Moschovakis, *Ordinal games and playful models*, ***Cabal Seminar 77–79 (Proc. Caltech-UCLA Logic Seminar, 1977–79)*** (A. S. Kechris, D. A. Martin, and Y. N. Moschovakis, editors), Lecture Notes in Mathematics, vol. 839, Springer, Berlin, 1981, pp. 169–201.

[9] I. Neeman, *Determinacy in $L(\mathbb{R})$*, to appear in the Handbook of Set Theory, M. Foreman and A. Kanamori (Editors).

[10] J. R. Steel, *A stationary-tower-free proof of the derived model theorem*, ***Advances in Logic: The North Texas Logic Conference*** (S. Gao, S. Jackson, and Y. Zhang, editors), Contemporary Mathematics, vol. 425, AMS, Providence, RI, 2007, pp. 1–8.

[11] ———, *unpublished lecture notes*, UCLA 1994.

[12] J. R. Steel and R. Van Wesep, *Two consequences of determinacy consistent with choice*, ***Transactions of the American Mathematical Society***, vol. 272 (1982), no. 1, pp. 67–85.

[13] W. H. Woodin, *Supercompact cardinals, sets of reals, and weakly homogeneous trees*, ***Proceedings of the National Academy of Sciences of the United States of America***, vol. 85 (1988), no. 18, pp. 6587–6591.

[14] ———, *unpublished lecture notes*, Berkeley 1993-94.

DEPARTMENT OF MATHEMATICS
UNIVERSITY OF CALIFORNIA
BERKELEY, CA 94720, USA
E-mail: steel@math.berkeley.edu

FORCING AXIOMS AND CARDINAL ARITHMETIC

BOBAN VELIČKOVIĆ

Abstract. We survey some recent results on the impact of strong forcing axioms such as the Proper Forcing Axiom PFA and Martin's Maximum MM on cardinal arithmetic. We concentrate on three combinatorial principles which follow from strong forcing axioms: stationary set reflection, Moore's Mapping Reflection Principle MRP and the P-ideal dichotomy introduced by Abraham and Todorčević which play the key role in these results. We also discuss the structure of inner models of PFA and MM and present some open problems.

§1. **Introduction.** Cardinal arithmetic has been one of the main fields of research in set theory since the foundational works by Cantor in the last quarter of the 19-th century [10]. The central question is the celebrated continuum problem which asks for the specific value of $2^{\aleph_0}$. A more general version of this problem is to determine all the rules which govern the exponential function $\kappa \mapsto 2^\kappa$ on infinite cardinals. Since the seminal work of Gödel [19] and Cohen [11] it has been known that the usual axioms ZFC of set theory do not decide the value of the continuum. Moreover, soon after Cohen introduced the method of forcing Easton [14] generalized Cohen's result and showed that the exponential function $\kappa \mapsto 2^\kappa$ on regular cardinals can be arbitrary modulo some mild restrictions. The situation for singular cardinals is much more subtle. Silver [33] showed that the Generalized Continuum Hypothesis (GCH) cannot first fail at a singular cardinal of uncountable cofinality. Subsequently, Shelah developed a rich and powerful theory of possible cofinalities (PCF theory) with important applications in cardinal arithmetic. One of the most striking of Shelah's results is that $\aleph_\omega^{\aleph_0} < \max\{\aleph_{\omega_4}, (2^{\aleph_0})^+\}$ holds in ZFC (see [31] or [2]).

In 1947 Gödel ([20] and [21]) speculated on the ontological status of set theory and conjectured correctly that ZFC would be too weak to settle the continuum problem. He concluded that it was necessary to seek new natural axioms that could give a satisfactory solution to the continuum problem as

2000 *Mathematics Subject Classification.* Primary 03Exx.

Key words and phrases. cardinal arithmetic, continuum problem, singular cardinal hypothesis, PFA, MM, stationary set reflection, MRP, P-ideal dichotomy.

Logic Colloquium '06
Edited by Barry Cooper, Herman Geuvers, Anand Pillay, and Jouko Väänänen
Lecture Notes in Logic, 32

well as other natural problems arising in the field. Gödel stressed that these axioms should satisfy some form of maximality property which he opposed to the minimality condition satisfied by the constructible universe L (see [21], Section 4, p. 478, note 19). This became later known as Gödel's program. Over the years two types of new axioms satisfying Gödel's condition emerged: *large cardinal axioms* and *forcing axioms*. While large cardinals give rise to a very rich theory and decide many natural questions about the reals they have very little to say about the continuum problem. Forcing axioms on the other hand have strong influence on cardinal exponentiation and decide many natural questions about uncountable cardinals left open by ZFC. It should be pointed out that these two types of axioms are very closely intertwined. Typically one needs large cardinals to prove the consistency of strong forcing axioms.

One way to motivate forcing axioms is as generalizations of Baire's category theorem. Suppose $\mathcal{K}$ is a class of partial orderings and κ an infinite cardinal. Then $\mathrm{FA}(\mathcal{K}, \kappa)$ is the following statement.

> *Suppose $\mathcal{P}$ is in $\mathcal{K}$ and $\mathcal{D}$ is a family of κ dense subsets of $\mathcal{P}$. Then there is a filter G in $\mathcal{P}$ such that $G \cap D \neq \emptyset$, for all $D \in \mathcal{D}$.*

The study of these axioms was started by Martin and Solovay [26] who introduced Martin's axiom (MA) as an abstraction of Solovay and Tennenbaum's approach to solving Suslin's problem [34], a question about uncountable trees. MA is the statement that $\mathrm{FA}(\mathrm{ccc}, \kappa)$ holds for all $\kappa < \mathfrak{c}$, where ccc denotes the class of forcing notions having the countable chain condition. It was soon realized that MA together with the negation of the Continuum Hypothesis (CH) provides a rich structure theory for the reals. As the method of forcing was further developed generalizations of $\mathrm{MA} + \neg\mathrm{CH}$ to larger classes of forcing notions were considered as well. One of the most successful of these axioms is the Proper Forcing Axiom (PFA) introduced by Baumgartner and Shelah in the early 1980s (see, for example, Baumgartner's survey paper [7]). PFA says that $\mathrm{FA}(\mathrm{Proper}, \aleph_1)$ holds, where Proper is the class of proper forcing notions. In the mid 1980s Foreman, Magidor and Shelah [16] formulated Martin's Maximum (MM) the provably strongest forcing axiom and showed that it is relatively consistent with ZFC. MM is the statement that $\mathrm{FA}(\mathcal{K}, \aleph_1)$ holds, where $\mathcal{K}$ is the class of forcing notions preserving stationary subsets of ω_1. It should be pointed out that while the consistency of $\mathrm{MA} + \neg\mathrm{CH}$ does not require any large cardinals, the proofs of the consistency of PFA and MM use a supercompact cardinal.

In recent years, bounded versions of traditional forcing axioms have received a considerable amount of attention as they have many of the same consequences, yet require much smaller large cardinal assumptions. These statements were first considered by Goldstern and Shelah in [22] who showed that the Bounded Proper Forcing Axiom (BPFA) is equiconsistent with a relatively modest large cardinal axiom, the existence of a Σ_1-reflecting cardinal.

An appealing formulation of bounded forcing axioms as principles of generic absoluteness was provided by Bagaria [5]. Namely, suppose $\mathcal{K}$ is a class of forcing notions. The bounded forcing axiom BFA($\mathcal{K}$) is the statement asserting that for every $\mathcal{P} \in \mathcal{K}$,

$$(H_{\aleph_2}, \in) \prec_{\Sigma_1} (V^{\mathcal{P}}, \in).$$

Here $H_{\aleph_2}$ denotes the collection of all sets whose transitive closure has size at most $\aleph_1$. Thus, BFA($\mathcal{K}$) states that for every Σ_0 formula $\psi(x, a)$ with parameter $a \in H_{\aleph_2}$, if some forcing notion from $\mathcal{K}$ introduces a witness x for $\psi(x, a)$, then such an x already exists. For example, $\text{MA}_{\aleph_1}$ is BFA(ccc), BPFA is BFA(Proper) and BMM is BFA($\mathcal{K}$) where $\mathcal{K}$ is the class of forcing notions that preserve stationary subsets of ω_1.

Martin's Axiom does not decide the value of the continuum, but PFA and MM have strong influence on cardinal arithmetic. Thus, Foreman, Magidor and Shelah [16] showed that MM implies that $\mathfrak{c} = \aleph_2$. In fact, they showed that it implies that $\kappa^{\aleph_1} = \kappa$, for all regular $\kappa \geq \aleph_2$. As a consequence of this and Silver's theorem [33] it follows that MM implies the Singular Cardinal Hypothesis (SCH). By a more involved argument Todorčević and the author [38] showed that the weaker Proper Forcing Axiom (PFA) also implies that $\mathfrak{c} = 2^{\aleph_1} = \aleph_2$. In [42] Woodin identified a statement ψ_{AC} which follows from both Woodin's $\mathbb{P}_{\max}$-axiom $(*)$ and from MM, and implies that $\mathfrak{c} = 2^{\aleph_1} = \aleph_2$ and that there is a well-ordering of the reals definable with parameters in $(H_{\aleph_2}, \in)$. Moreover, Woodin showed that BMM together with the existence of a measurable cardinal implies that the continuum is $\aleph_2$. The assumption of the existence of a measurable cardinal was later eliminated by Todorčević [36] who deduced these consequences of ψ_{AC} from a statement he called θ_{AC} that he showed follows from BMM. Recently, Moore [27] introduced the Mapping Reflection Principle (MRP) and deduced it from PFA. Although MRP does not follow from BPFA, Moore [27] used a bounded version of MRP to show that BPFA implies a certain statement υ_{AC} which in turn implies that there is a well ordering of the reals, and in fact of $\mathcal{P}(\omega_1)$, of order type ω_2 which is Δ_2-definable in the structure $(H_{\aleph_2}, \in)$ with parameter a subset of ω_1. This result was later improved by Caicedo and the author [9] who showed that BPFA implies the existence of a Δ_1-definable well ordering of the reals. Finally, in 2005 Viale [40] resolved a long standing problem by showing that PFA implies SCH. In fact, he produced two proofs of this result. In one he deduced SCH from MRP and in the other he obtained the same conclusion from the P-ideal dichotomy PID introduced by Abraham and Todorčević [3]. It was known previously that PID follows from PFA.

In this paper we survey the recent results concerning the impact of forcing axioms on cardinal arithmetic. Rather than working with forcing axioms directly we will mostly concentrate on three combinatorial principles: stationary set reflection, the Mapping Reflection Principle and the P-ideal dichotomy.

They are all consequences of MM and the last two follow from PFA as well. These principles express a certain form of reflection and can be used to get bounds on the value of the continuum and more generally $\kappa^{\aleph_1}$, for $\kappa > \aleph_1$. Therefore, most of the arguments we present will be purely combinatorial and the use of forcing axioms is just to prove the relevant reflection principle.

The paper is organized as follows. In §2 we discuss various versions of stationary set reflection and present a simplified proof of a result of Shelah [32] saying that stationary set reflection implies the Singular Cardinal Hypothesis. In §3 we turn our attention to Moore's Mapping Reflection Principle and its consequences. In §4 we discuss the P-ideal dichotomy PID introduced by Abraham and Todorčević [3] and present a result of Viale [40] saying that PID implies the Singular Cardinal Hypothesis SCH. §5 is devoted to presenting the main ideas of [9] where it is shown that the Bounded Proper Forcing Axiom implies that there is a well ordering of the reals Δ_1 definable with parameter a subset of ω_1. In §6 we consider inner models of strong forcing axioms and present some rigidity results due to Viale. Finally, in §7 we present some open questions and directions for further research. We point out that §2, §3 and §4 are independent of each other, §5 requires §3, while §6 requires all the previous sections.

Our notation is mostly standard. If κ is an infinite cardinal, then H_κ is the family of sets whose transitive closure has cardinality smaller than κ, i.e., $H_\kappa = \{x \colon |\mathrm{tc}(x)| < \kappa\}$, where $\mathrm{tc}(\cdot)$ denotes transitive closure. We use $V^{\mathcal{P}}$ to denote the Boolean-valued extension of V by the forcing notion $\mathcal{P}$, equivalently, and abusing language, $V^{\mathcal{P}}$ denotes any extension $V[G]$ where G is $\mathcal{P}$-generic over V. For notation and concepts from PCF theory we refer the reader to [2], for all other notation not explicitly defined in this paper as well as an introduction to set theory and forcing we refer the reader to [23].

§2. Stationary set reflection. The purpose of this section is to present some basic results on reflection of stationary sets and present a proof of a recent result of Shelah [32] saying that stationary set reflection implies the Singular Cardinal Hypothesis. We start by recalling some definitions. Recall that an *algebra* on an infinite set I is just a function $F : I^{<\omega} \to I$. For a subset X of I the closure $cl_F(X)$ of X under F is the smallest subset Y of I containing X and such that $F[Y^{<\omega}] \subseteq Y$. A subset C of $\mathcal{P}(I)$ is *club* if there is an algebra F such that $C = \{X \subseteq I : \mathrm{cl}_F(X) = X\}$. A subset S of $\mathcal{P}(I)$ is *stationary* if $S \cap C \neq \emptyset$, for all club $C \subseteq \mathcal{P}(I)$. These notions generalize the well known notions of club and stationary subsets of a regular cardinal κ. In particular, we have a version of Fodor's theorem for stationary subsets of $\mathcal{P}(I)$. For some basic information concerning club and stationary subsets of $\mathcal{P}(I)$ see [23, chapter 8 and 38].

EXAMPLE 2.1. Given an infinite cardinal $\kappa \leq |I|$ the following set is stationary

$$[I]^\kappa = \{X \subseteq I : |X| = \kappa\}.$$

In what follows we shall restrict our attention to the space $[I]^{\omega}$ and the notions of club and stationary will be interpreted as relativized to this space. In many applications I will be H_{θ}, for some fixed regular cardinal θ. In this case stationary sets are those $S \subseteq [H_{\theta}]^{\omega}$ such that for any model $(H_{\theta}, \in, \dots)$ there is $M \in S$ with $M \prec (H_{\theta}, \in, \dots)$. We shall be mostly concerned with the following reflection principle.

Definition 2.2 (Reflection Principle (RP)). For every regular $\lambda \geq \aleph_2$ the following principle RP(λ) holds:

> If S is a stationary subset of $[\lambda]^{\omega}$ then there is $X \subseteq \lambda$ of cardinality $\aleph_1$ such that $S \cap [X]^{\omega}$ is stationary in $[X]^{\omega}$.

RP follows from Martin's Maximum and has a number of interesting consequences. In particular, it implies that every forcing notion preserving stationary subsets of ω_1 is semiproper and this in turn implies that the nonstationary ideal NS_{ω_1} is precipitous, see [16]. Therefore RP has large cardinal strength. However, many applications of MM require an even stronger reflection principle for stationary sets. In order to state this principle we will need the following definition.

Definition 2.3. A set $S \subseteq [\lambda]^{\omega}$ is *projective stationary* if for every stationary set $T \subseteq \omega_1$, the set the set $\{X \in S : X \cap \omega_1 \in T\}$ is stationary.

Definition 2.4 (Strong Reflection Principle (SRP)). If $\lambda \geq \aleph_2$ is regular then the following principle SRP(λ) holds:

> If S is projective stationary in $[H_{\lambda}]^{\omega}$ then there is continuous increasing $\in$-chain $\langle M_{\alpha} : \alpha < \omega_1 \rangle$ of countable elementary submodels of H_{λ} such that $M_{\alpha} \in S$, for all α.

The following result is implicitly proved in [30].

Theorem 2.5. MM *implies* SRP. ⊣

SRP is a very strong combinatorial principle and it implies many of the key consequences of MM, such as Strong Chang's Conjecture, the saturation of the nonstationary ideal NS_{ω_1} on ω_1, etc. (see [23], Chapter 37).

Proposition 2.6. SRP(λ) *implies* RP(λ), *for every regular* $\lambda \geq \omega_2$.

Proof. We show that SRP(λ) implies the following stronger version RP*(λ) of stationary set reflection:

> *If S is a stationary subset of $[H_{\lambda}]^{\omega}$ then there is a continuous increasing $\in$-chain $\langle M_{\alpha} : \alpha < \omega_1 \rangle$ of countable elementary submodels of H_{λ} such that $\{\alpha < \omega_1 : M_{\alpha} \in S\}$ is stationary.*

Let $S \subseteq [H_{\lambda}]^{\omega}$ be stationary. Since NS_{ω_1} is saturated by SRP($\aleph_2$) there is a stationary $A \subseteq \omega_1$ such that for every stationary $B \subseteq A$ the set $\{M \in S : M \cap \omega_1 \in B\}$ is stationary. Therefore, $T = \{M : M \in S \text{ or } M \cap \omega_1 \notin A\}$ is projective stationary. By SRP(λ), T contains a continuous increasing chain

$\langle M_\alpha : \alpha < \omega_1 \rangle$ of countable elementary submodels of H_λ. It follows that $M_\alpha \in S$, for all $\alpha \in A$. ⊣

Concerning the impact of stationary reflection principles on cardinal arithmetic, Foreman, Magidor and Shelah's proof from [16] that MM implies the Singular Cardinal Hypothesis, SCH, can be factored through SRP. Todorčević (unpublished) showed that RP($\aleph_2$) implies $2^{\aleph_0} \leq \aleph_2$ and the author showed in [38] that RP* implies SCH. The same conclusion was later obtained in [17] form simultaneous reflection of 3 stationary sets in $[\lambda]^\omega$. In the remainder of this section we present a simplified proof of the main result of Shelah [32] saying that RP implies that SCH. We begin by recalling the following.

DEFINITION 2.7. The Singular Cardinal Hypothesis (SCH) states that, for any singular cardinal κ, if $2^{\mathrm{cof}(\kappa)} < \kappa$ then $\kappa^{\mathrm{cof}(\kappa)} = \kappa^+$.

In order to prove that RP implies SCH we shall need Lemma 2.9 in order to set off the induction in the main proof (Lemma 2.8 is used to prove Lemma 2.9). Then we state and prove the main result in Theorem 2.10. The fact that stationary set reflection implies SCH comes in Corollary 2.17 as a consequence of the main theorem. Finally, in Corollary 2.18, we show that the previous results still hold without any constraint on the size of the reflecting sets.

LEMMA 2.8. RP($\aleph_2$) *implies that for every stationary subset* S *of* $[\omega_2]^\omega$ *there is* $\alpha < \omega_2$ *such that* $S \cap [\alpha]^\omega$ *is stationary in* $[\alpha]^\omega$.

PROOF. Let S be a stationary subset of $[\omega_2]^\omega$, and suppose that S does not reflect in any $\alpha \in \omega_2$. Then for each $\alpha \in \omega_2$ there is a club set $C(f_\alpha)$ in $[\alpha]^\omega$ containing the closure points in $[\alpha]^\omega$ of some function $f_\alpha : [\alpha]^{<\omega} \to \alpha$, and such that $S \cap [\alpha]^\omega \cap C(f_\alpha) = \emptyset$. Letting $f_\alpha(e) = \min(e)$ for $e \in [\omega_2]^{<\omega} \setminus [\alpha]^{<\omega}$, we can extend f_α to a function from $[\omega_2]^{<\omega}$ into ω_2. We denote this new function also by f_α.

By making simple definitions by cases, we can build two functions f and g from $[\omega_2]^{<\omega}$ into ω_2 such that for all $X \subseteq \omega_2$:

(1) if X is closed by f, then for all $\alpha \in X$, X is closed by f_α;
(2) if X is closed by g and $Card(X) = \aleph_1$, then either $X \in \omega_2$ or $\mathrm{ot}(X) = \omega_1$.

The definition of g, for instance, goes as follows.

(1) For $e \in [\omega_2]^n$ with $n > 2$, $g(e) = n - 3$. Thus, if $X \subseteq \omega_2$ is closed by g, then $\omega \subseteq X$.
(2) For $n \in \omega$ and $\xi \in \omega_1 - \omega$, $g(\{n, \xi\}) = h_\xi(n)$, where $h_\xi : \omega \to \xi$ is a fixed bijection. Thus, if $X \subseteq \omega_2$ is closed by g, then $X \cap \omega_1 \in \omega_1 + 1$.
(3) For $\xi \in \omega_1$ and $\alpha \in \omega_2 - \omega_1$, $g(\{\xi, \alpha\}) = i_\alpha(\xi)$, where $i_\alpha : \omega_1 \to \alpha$ is a fixed bijection. Thus, if $X \in [\omega_2]^{\aleph_1}$ is closed by g and $\omega_1 \subseteq X$, then $X \in \omega_2$.

(4) For $\alpha < \beta$ in $\omega_2 - \omega_1$, $g(\{\alpha, \beta\}) = i_\beta^{-1}(\alpha)$. Thus, if $X \in [\omega_2]^{\aleph_1}$ is closed by g and $\mathrm{ot}(X) > \omega_1$, then $X \cap \omega_1$ is unbounded, hence $\omega_1 \subseteq X$ by point (2), hence $X \in \omega_2$ by point (3).

(5) In all other cases $g(e)$ equals 0.

Let $C(f)$ and $C(g)$ be the respective club sets of closure points of f and g in $[\omega_2]^\omega$.

Finally, let $A \in [\omega_2]^{\aleph_1}$ such that $S \cap C(f) \cap C(g)$ reflects in A. A is closed by g, but $A \notin \omega_2$ by hypothesis, so $\mathrm{ot}(A) = \omega_1$. Then the set of proper initial segments of A is a club in $[A]^{\aleph_0}$, so there exists $\alpha \in A$ such that $A \cap \alpha \in S$. Since A is closed under f and $\alpha \in A$, by the choice of f we know that A is closed by f_α, hence, due to the definition of f_α, so is $A \cap \alpha$. On the other hand, since $S \cap [\alpha]^\omega \cap C(f_\alpha) = \emptyset$ and $A \cap \alpha \in S \cap [\alpha]^\omega$, $A \cap \alpha$ cannot be closed under f_α, a contradiction. ⊣

LEMMA 2.9. $\mathrm{RP}(\aleph_2)$ *implies* $\aleph_2^{\aleph_0} = \aleph_2$.

PROOF. For each $\alpha \in \omega_2$, let us pick $\langle X_\xi^\alpha : \xi < \omega_1 \rangle$ a continuous increasing sequence of countable subsets of α such that $\bigcup\{X_\xi^\alpha : \xi < \omega_1\} = \alpha$. Let $C = \{X_\xi^\alpha : \xi < \omega_1, \alpha < \omega_2\}$. Notice that $[\omega_2]^\omega \setminus C$ cannot reflect in any $\alpha \in \omega_2$ by the choice of C, so by Lemma 2.8 it does not reflect at all, hence it is not stationary, hence C contains a club set. Since club sets in $[\omega_2]^\omega$ are of size $\aleph_2^{\aleph_0}$ (see [8], Theorem 3.2) and $Card(C) = \aleph_2$, we conclude that $\aleph_2^{\aleph_0} = \aleph_2$. ⊣

The following theorem is due to Shelah [32].

THEOREM 2.10. RP *implies that* $\lambda^{\aleph_0} = \lambda$, *for any regular cardinal* $\lambda \geq \aleph_2$.

PROOF. Assume that the theorem does not hold, and let λ be the least counterexample. Basic cardinal arithmetic (along with Lemma 2.9) shows that λ is the successor of some κ of cofinality $\aleph_0$, and $\kappa^{\aleph_0} > \lambda$. Furthermore, Lemma 2.9 implies that $2^{\aleph_0} < \kappa$. Our goal is to show that $\mathrm{RP}(\lambda)$ does not hold. To that end we will need the following notion from PCF theory [31]. We borrow the terminology from [12]. ⊣

DEFINITION 2.11. Given a sequence $\langle \mu_\alpha : \alpha < \beta \rangle$ of regular ordinals and an ideal I on β, an I-*scale* on $\langle \mu_\alpha : \alpha < \beta \rangle$ is a $<_I$-strictly increasing and cofinal sequence $\langle f_\xi : \xi < \gamma \rangle$ in $\prod_{\alpha<\beta} \mu_\alpha$. The scale $\langle f_\xi : \xi < \gamma \rangle$ is said to be *better* if and only if, for every ordinal $\alpha < \gamma$ with $\mathrm{cof}(\alpha) > \beta$, there exists a club set $C \subseteq \alpha$, and $Z_i \in I$, for each $i \in C$, such that if $i, j \in C$ and $i < j$ then $f_i \restriction (\beta \setminus (Z_i \cup Z_j)) < f_j \restriction (\beta \setminus (Z_i \cup Z_j))$.

PCF theory shows that we can choose an increasing sequence $\langle \kappa_n : n < \omega \rangle$ of regular cardinals in κ with limit κ so as to have a *better scale* $\langle f_\xi : \xi < \lambda \rangle$ on $\langle \kappa_n : n < \omega \rangle$, with respect to the ideal *FIN* of finite subsets of ω (see [31, II, Claim 1.5A]). In fact, for the purpose of this proof, we shall only need the *better* scale property to hold for α of cofinality $\aleph_1$.

For $X \subseteq \mathrm{ORD}$, let $\delta(X) = \sup(X \cap \lambda)$, and $\chi(X)(n) = \sup(X \cap \kappa_n)$. Most of the proof will hinge on the comparison, for $X \in [\lambda]^\omega$, between $\chi(X)$ and $f_{\delta(X)}$. Let us then define:

$$E_X = \{n < \omega : \chi(X)(n) \leq f_{\delta(X)}(n)\}.$$

Let $\{A_\xi : \xi < \omega_1\}$ be a family of pairwise almost-disjoint infinite subsets of ω, that is, for all $\xi \neq \zeta$ in ω_1, $A_\xi \cap A_\zeta$ is finite; and let $\phi : \mathcal{P}(\omega) \to \omega_1$ be a partial function defined by

$$\phi(E) = \min\{\xi < \omega_1 : Card(A_\xi \cap E) = \aleph_0\},$$

if such an ordinal ξ exists. Finally, let us consider the set:

$$\mathcal{S} = \{X \in [\lambda]^\omega : \phi(E_X) \text{ is defined and } \phi(E_X) \geq \mathrm{ot}(X), \; X \text{ is closed by } x \mapsto f_x(n) \text{ for all } n\}.$$

We are going to show that $\mathcal{S}$ is stationary, yet does not reflect in any $A \in [\lambda]^{\aleph_1}$.

Claim 2.12. $\mathcal{S}$ does not reflect in any $A \in [\lambda]^{\aleph_1}$.

Proof. Let us assume to the contrary that $\mathcal{S}$ reflects in some $A \in [\lambda]^{\aleph_1}$. Let $\langle X_i : i < \omega_1\rangle$ be a continuous cofinal increasing sequence of countable subsets of A, and let $R = \{i < \omega_1 : X_i \in \mathcal{S}\}$. Since $\{X_i : i < \omega_1\}$ is club in $[A]^\omega$, saying that $\mathcal{S}$ reflects in A is the same as saying that $\{X_i : i \in R\}$ is stationary in $[A]^\omega$, or that R is stationary in ω_1. First, we show that $\mathrm{cof}(\sup(A)) = \aleph_1$.

Let us assume towards contradiction that $\mathrm{cof}(\sup(A)) < \aleph_1$. Then there exists $\gamma \in \omega_1$ such that $\sup(X_\gamma) = \sup(A)$. Let $\delta = \sup(A)$. Now for any $n \in \omega$, let $\alpha(n) = \min\{\alpha \in R : \chi(X_\alpha)(n) > f_\delta(n)\}$, if such α exists, and $\alpha(n) = \min(R)$ otherwise. Letting $\alpha = \max\{\gamma, \sup_{n\in\omega} \alpha(n)\}$, for all $\beta \geq \alpha$, $E_{X_\alpha} = E_{X_\beta}$, and in particular $\phi(E_{X_\beta}) = \phi(E_{X_\alpha})$. However, since $X_\beta \in \mathcal{S}$, we also have $\phi(E_{X_\beta}) \geq \mathrm{ot}(X_\beta)$, and $\mathrm{ot}(X_\beta)$ converges to ω_1, so there is a contradiction.

Since $\mathrm{cof}(\sup(A)) = \aleph_1$, we may assume that $\delta(X_i) = \sup(X_i)$ is strictly increasing, trimming $\langle X_i : i < \omega_1\rangle$ if necessary. Let $\delta_i = \delta(X_i)$, and let $\beta_i = \min(A \setminus \delta_i)$. Trimming $\langle X_i : i < \omega_1\rangle$ two more times, we can ensure that:

$$\forall i < j \in R, (\beta_i < \delta_j) \wedge (\beta_i \in X_j). \tag{2.1}$$

Now, by applying the better scale property of $\langle f_\xi : \xi < \gamma\rangle$ to $\delta(A)$, there exists a club set $C \subseteq \omega_1$, along with a sequence $\langle n_i : i \in C\rangle$ of elements of ω such that for $i < j \in R \cap C$ and $n \geq n_i, n_j$, we have $f_{\delta_i}(n) < f_{\delta_j}(n)$. As $i \mapsto n_i$ yields a partition of $C \cap R$ into $\aleph_0$ subsets, one of them is stationary, let us rename it R. Thus, we may assume that there exists $k \in \omega$ such that for all $i < j$ in R, $f_{\delta_i} \restriction [k, \omega) < f_{\delta_j} \restriction [k, \omega)$.

Because of (2.1), we know that for i in R if $j = \min(R \setminus (i+1))$ then $f_{\delta_i} \leq_{FIN} f_{\beta_i} <_{FIN} f_{\delta_j}$, so there exists $m_i \in \omega$ such that for all $n \geq m_i$, $f_{\delta_i}(n) \leq f_{\beta_i}(n) < f_{\delta_j}(n)$. By the same reasoning as before, we can thin R so as to have $m_i = m$ a constant, and increase k so that $k \geq m$. As a result:

$$(2.2) \qquad \forall i < j \in R,\ f_{\delta_i} \restriction [k,\omega) \leq f_{\beta_i} \restriction [k,\omega) < f_{\delta_j} \restriction [k,\omega).$$

Now let $f \in \prod_{n<\omega} \kappa_n$ be defined by $f(n) = \sup_{i \in R} f_{\beta_i}(n)$ if $n \geq k$ and 0 otherwise. Because of (2.1) and the closure properties of $\mathcal{S}$, for $i < j \in R$ and $n \in \omega$ we have $f_{\beta_i}(n) \in X_j$, so $f(n) \leq \sup(\bigcup_{i\in R}(X_i \cap \kappa_n)) = \chi_A(n)$. Let $B = \{n \in [k,\omega) : f(n) = \chi_A(n)\}$ and $\overline{B} = \{n \in [k,\omega) : f(n) < \chi_A(n)\} = [k,\omega) \setminus B$. We are going to prove that, for all i in some stationary subset of R, $f_{\delta_i} \restriction B \geq \chi(X_i) \restriction B$ and $f_{\delta_i} \restriction \overline{B} < \chi(X_i) \restriction \overline{B}$.

Let $n \in B$. Since $f(n) = \chi(A)(n)$ and $\mathrm{cof}(f(n)) = \aleph_1$, we can define a club set C_n in $\chi(A)(n)$ such that for all $i < j$ in C_n, $\chi(X_i)(n) < f_{\delta_j}(n)$, and also $f_{\delta_i}(n) \in X_j$. As a result, for l a limit point of C_n, we get $f_{\delta_l}(n) \geq \bigcup_{i\in C_n \cap l} f_{\delta_i}(n) = \bigcup_{i \in C_n \cap l} \chi(X_i)(n) = \chi(X_l)(n)$. Let D_n be the club set of limit points of C_n, as $R \cap \bigcap\{D_n : n < \omega\}$ is stationary, we can rename it R again, and so we may assume that for all $i \in R$ we have $f_{\delta_i} \restriction B \geq \chi(X_i) \restriction B$.

Let $n \in \overline{B}$. Since $f(n) < \chi(A)(n)$, there exists $i(n) \in \omega_1$ such that $f(n) < \chi(X_{i(n)}(n))$. As $\sup_{n\in B}(i(n)) < \omega_1$, $R \setminus \sup_{n \in B}(i(n))$ is stationary and we can rename it again R. Thus, for all $i \in R$ we have $f_{\delta_i} \restriction \overline{B} < \chi(X_i) \restriction \overline{B}$.

We have shown that for $i \in R$, $\{n \in [k,\omega) : \chi(X_i)(n) \leq f_{\delta_i}(n)\} = B$, so $E_{X_i} =_{FIN} B$. In particular, $\phi(E_{X_i})$ remains constant on R. That is a contradiction, since $\phi(E_{X_i}) \geq \mathrm{ot}(X_i)$ and $\mathrm{ot}(X_i)$ converges to ω_1. ⊣

Claim 2.13. $\mathcal{S}$ is stationary.

Proof. Let C be a club in $[\lambda]^\omega$ and fix a function $f_C : [\lambda]^{<\omega} \to \lambda$ such that $C = \{X \in [\lambda]^\omega : cl_{f_C}(X) = X\}$. We need to find $X \in \mathcal{S}$ such that $f_C[X^{<\omega}] \subseteq X$. In order to build such a set $X \in \mathcal{S}$, the main issue is to control both E_X and $\mathrm{ot}(X)$, so that $\phi(E_X) \geq \mathrm{ot}(X)$. For that purpose, we consider a closed two-player game G_ε, for each choice of $\varepsilon \in \omega_1$. Player 1 sets up constraints that will later on allow us to control E_X and ensure that $\phi(E_X) \geq \varepsilon$; meanwhile, player 2 tries to meet these constraints, build the set X, bound $\chi(X)$, as well as prove that $\mathrm{ot}(X) \leq \varepsilon$. ⊣

In the first part of the proof, we show that player 2 has a winning strategy for some $\varepsilon \in \omega_1$. In the second part, we show how player 1 should play against that strategy in order to obtain X as required. We begin by describing G_ε.

Let θ be a sufficiently large regular cardinal, say $\theta = (2^\lambda)^+$, and let us fix a well-ordering $\lhd$ of H_θ. For $X \subseteq \mathrm{ORD}$, let $sk(X)$ be the Skolem hull of X in $(H_\theta, \in, \lhd)$, $sk_\lambda(X) = sk(X) \cap \lambda$, and $cl(X) = sk_\lambda(X \cup \{\langle f_\xi : \xi < \lambda\rangle, f_C\})$. Then we can find functions $t_n : \lambda^n \to \lambda$ such that if $X = \{\alpha_n : n < \omega\}$ is a subset of λ then $cl(X) = \{t_n(\alpha_0, \dots, \alpha_n) : n < \omega\}$.

At stage n the game G_ε proceeds as follows.

(1) (a) If n is of the form $2l$: player 1 picks an ordinal $\xi_{2l} \in \kappa_l$. Player 2 then picks α_{2l} and γ_{2l} in κ_l such that $\xi_{2l} \leq \alpha_{2l} \leq \gamma_{2l}$.
 (b) If n is of the form $2l+1$: player 1 picks an ordinal $\xi_{2l+1} \in \lambda$. Player 2 then picks α_{2l+1} in λ such that $\xi_{2l+1} \leq \alpha_{2l+1}$.

(2) Player 2 chooses an ordinal ζ_n in ε.

Once the game is over, let $X = cl(\{\alpha_n : n \in \omega\})$ and $\tau_n = t_n(\alpha_0, \dots, \alpha_n)$, for all n. Then $\{\tau_n : n < \omega\}$ constitutes an enumeration of X.

We say that player 2 wins the game iff:

(1) For all $n \in \omega$, $X \cap \kappa_n \subseteq \gamma_{2n}$.
(2) The mapping $g : X \to \varepsilon$ given by $\tau_n \overset{g}{\mapsto} \zeta_n$, for all n, is well-defined and order preserving. As such g witnesses $\mathrm{ot}(X) \leq \varepsilon$.

FACT 2.14. There exists $\varepsilon \in \omega_1$ such that player 2 has a winning strategy for G_ε.

PROOF. Let $\varepsilon \in \omega_1$. The first point is that the game G_ε is closed, because if player 2 loses, that loss is apparent after a finite number of moves. Indeed, if at the end of the game, for some $n \in \omega$, $X \cap \kappa_n \not\subseteq \gamma_n$, then some element τ_n of X witnesses it and the value τ_n can be computed at stage n of the game. The same goes for the second winning condition.

Since G_ε is closed, the Gale-Stewart theorem [18] guarantees that one of the two players has a winning strategy. Let us assume towards contradiction that player 1 has a winning strategy σ_ε, for all $\varepsilon \in \omega_1$. The crux of the matter here is that player 1's best interest is always to play ξ_n as high possible. In particular, if we modify σ_ε to increase player 1's answer ξ_n to some sequence of moves by player 2, we still get a winning strategy.

Assuming that player 1 follows the strategy σ_ε, for any given sequence s of moves by player 2 up to step n of the game, let $\sigma_\varepsilon(s)$ be the answer ξ_n dictated to player 1 by his strategy σ_ε (letting $\sigma_\varepsilon(s) = 0$ if s is not a possible sequence of moves for player 2 when player 1 applies σ_ε). We can define a new strategy σ for player 1 by $\sigma(s) = \sup\{\sigma_\varepsilon(s) : \varepsilon \in \omega_1\}$. Due to the remark above, σ is a winning strategy for player 1 for all the games G_ε.

Let us assume for a while that player 2 always plays $\alpha_n = \xi_n$, i.e. as the first part of his move he copies the move of player 1 and then chooses some γ_n (if n is even), and some $\zeta_n < \varepsilon$. We let player 1 follow his winning strategy σ. Thus, up to step $2n$, this subgame is determined by player 2's choices of the $\gamma_{2i} < \kappa_i$, for $i \leq n$, and the $\zeta_n < \varepsilon < \omega_1$. As a result, there are only κ_n possible sequences of moves up to step $2n+1$ (we may assume $\omega_1 < \kappa_0$), so the set of all possible plays ξ_{2n+2} by player 1 is bounded in κ_{n+1}. Thus, improving σ if necessary, we can assume that player 1's moves are independent of all the previous moves. Let $\langle \xi_n : n < \omega \rangle$ be the sequence of player 1's moves.

Let us now turn back to the regular games G_ε, and play as player 2 against strategy σ using the following strategy of our own. We are going to play $\alpha_n = \xi_n$ at every turn, so we know from the start the set $X = cl(\{\alpha_n : n < \omega\}) = cl(\{\xi_n : n < \omega\})$. Let $\varepsilon = \mathrm{ot}(X)$. We play in the game G_ε.

Since we know X we can fix an order preserving mapping $g : X \to \varepsilon$ in advance. At each turn, we play $\alpha_n = \xi_n$ and $\gamma_n = \sup(X \cap \kappa_n)$ in case n is even. We can also compute $\tau_n = t_n(\alpha_0, \dots, \alpha_n)$ and we play $\zeta_n = g(\tau_n)$. By playing in this way player 2 clearly wins the game, so σ is not a winning strategy, which is a contradiction. ⊣

Let then $\varepsilon \in \omega_1$ be such that player 2 has a winning strategy τ for G_ε. If we play against τ, we know that we will get a set X such that $\mathrm{ot}(X) \leq \varepsilon$, $X \in C$, and X is closed by all the relevant functions, thus the last remaining point is to ensure $\phi(E_X) \geq \varepsilon$. Letting $A = A_\varepsilon$, we are going to achieve this by having $E_X =_{FIN} A$.

First, we need a countable $M \prec H_\theta$ such that $\chi(M) \leq_{FIN} f_{\delta(M)}$, and M contains all the relevant objects: $\langle f_\xi : \xi < \lambda\rangle$, $\langle \kappa_i : i < \omega\rangle$, τ, f_C. To obtain such an M we build a continuous $\in$-chain $\langle M_\zeta : \zeta < \omega_1\rangle$ of countable elementary submodels of H_θ such that the aforementioned objects belong to M_0. Let $\delta_\zeta = \sup(M_\zeta \cap \lambda)$, for all ζ, and let $\delta = \sup\{\delta_\zeta : \zeta < \omega_1\}$. Then $\chi(M_\zeta) <_{FIN} f_{\delta_{\zeta+1}} < \chi(M_{\zeta+2})$, for all ζ. Since $\{\delta_\zeta : \zeta < \omega_1\}$ is a club in δ we can use the better scale property of $\langle f_\xi : \xi < \lambda\rangle$ at δ to find a club D in ω_1 and l_ζ, for each $\zeta \in D$, such that if $\zeta, \eta \in D$ and $n \geq l_\zeta, l_\eta$ then $f_{\delta_\zeta}(n) < f_{\delta_\eta}(n)$. Moreover, by increasing the l_ζ if necessary, we may assume that if $\zeta \in D$ and we let $\zeta^+ = \min(D \setminus (\zeta + 1))$ then $\chi(M_\zeta)(n) < f_{\delta_{\zeta^+}}(n)$, for all $n \geq l_\zeta$. Let $k_\zeta = \max(l_\zeta, l_{\zeta^+})$, for each $\zeta \in D$. The map $\zeta \mapsto k_\zeta$ yields a partition of D into countably many pieces. Therefore, there is k such that the set $E = \{\zeta \in D : k_\zeta = k\}$ is stationary in ω_1. Fix a limit point η of E, an increasing sequence $\langle \eta(i) : i < \omega\rangle$ of elements of E converging to η and let $M = M_\eta$. We now have that for every every i and every $n \geq k$,

$$f_{\delta_{\eta(i)}}(n) < \chi(M_{\eta(i+1)})(n) < f_{\delta_{\eta(i+2)}}(n) < f_{\delta_\eta}(n).$$

Since $\chi(M) = \sup\{\chi(M_{\eta(i)}) : i < \omega\}$ and $\delta(M) = \delta_\eta$ it follows that $\chi(M) \restriction [k, \omega) \leq f_{\delta(M)} \restriction [k, \omega)$, as required.

Now, for simplicity of notation let us assume that $\chi(M)(n) \leq f_{\delta(M)}(n)$, for all n, i.e. that $k = 0$.

FACT 2.15. $\delta(sk_\lambda(M \cup \kappa)) = \delta(M)$. Similarly, $\chi(sk_\lambda(M \cup \kappa_n))(n+1) = \chi(M)(n+1)$, for all $n < \omega$.

PROOF. Let $\alpha \in sk_\lambda(M \cup \kappa)$. Then here exists a Skolem term z_1 with parameters $x_1, \dots, x_m$ in M and $y_1, \dots, y_n$ in κ such that

$$\alpha = z_1(\beta_1, \dots, \beta_m, \gamma_1, \dots, \gamma_n).$$

Now if we consider the Skolem function $z_2(\beta_1, \dots, \beta_m)$ representing

$$\sup\{z_1(\beta_1, \dots, \beta_m, x_1, \dots, x_n) : x_1, \dots, x_n \in \kappa\}$$

it is clear that $z_2(\beta_1, \dots, \beta_m) \geq \alpha$. Then $z_2(\beta_1, \dots, \beta_m) \in M$, since all the β_i are in M. Necessarily $z_2(\beta_1, \dots, \beta_m) < \lambda$, for cofinality reasons. Since $\alpha \in sk_\lambda(M \cup \kappa)$ was arbitrary, it follows that $\delta(sk_\lambda(M \cup \kappa)) \leq \delta(M)$. The converse also holds since $M \cap \lambda \subset sk_\lambda(M \cup \kappa)$, so we have an equality. We can apply the same reasoning to obtain the second equality. ⊣

Let $\langle m_i : i < \omega \rangle$ be an enumeration of $M \cap \lambda$. Now we play the game G_ε as player 1 against player 2's winning strategy τ as follows.

(1) At step $2l$:
 (a) if $l \in A$, we play $\xi_{2l} = 0$;
 (b) if $l \notin A$, we play $\xi_{2l} = f_{\delta_M}(l) + 1$.
(2) At step $2l + 1$, we play $\xi_{2l+1} = m_l$.

At stage $2l$, player 2 played $\alpha_{2l}, \gamma_{2l}, \zeta_{2l}$ such that $\xi_{2l} \leq \alpha_{2l} \leq \gamma_{2l} < \kappa_l$, and at stage $2l + 1$ player 2 played $\alpha_{2l+1}, \zeta_{2l+1}$ such that $\xi_{2l+1} \leq \alpha_{2l+1} < \lambda$. All the ζ_n are less than ε. As before let $\tau_n = t_n(\alpha_0, \dots, \alpha_n)$ and let $X = cl(\{\alpha_n : n < \omega\}) = \{\tau_n : n < \omega\}$. Since player 2 played following his winning strategy we know that the map $\tau_n \mapsto \zeta_n$ witnesses that X is of order type at most ε and that $\chi(X)(l) \leq \gamma_{2l}$, for all l.

Fact 2.16. $E_X = A$.

Proof. Since player 1's move at stage $2l + 1$ is $\xi_{2l+1} = m_l$ and $\alpha_{2l+1} \geq \xi_{2l+1}$ it follows that $\delta(M) \leq \delta(X)$. Conversely, $X \subset sk_\lambda(M \cup \kappa)$, so Fact 2.15 yields $\delta_M = \delta_X$.

Let us look at step $2l$ of the game we have described. First, if $l \in A$, player 1 played $\xi_{2l} = 0$. Since all the moves of the game up to that point as well as the winning strategy τ belong to $M \cup \kappa_{l-1}$ the move played by player 2 at this stage belongs to $sk(M \cup \kappa_{l-1})$, as well. Therefore

$$\chi(X)(l) \leq \gamma_{2l} \leq \chi(sk_\lambda(M \cup \kappa_{l-1}))(l) = \chi(M)(l) \leq f_{\delta(M)}(l) = f_{\delta(X)}(l).$$

Therefore $l \in E_X$.

Second, if $l \notin A$, player 1 played $\xi_{2l} = f_{\delta(M)}(l) + 1$. Since $\xi_{2l} \leq \alpha_{2l} < \kappa_l$ and $\delta(X) = \delta(M)$ we have that $\chi(X)(l) > f_{\delta(X)}(l)$, i.e. $l \notin E_X$. ⊣

Corollary 2.17. RP *implies* SCH.

Proof. Assume to the contrary that SCH does not hold and let κ be the first cardinal for which SCH fails. Silver's theorem [33] implies that $\mathrm{cof}(\kappa) = \aleph_0$, so we have $\kappa^{\aleph_0} > \kappa^+$. Theorem 2.10, on the other hand, states that $(\kappa^+)^{\aleph_0} = \kappa^+$, so there is a contradiction. ⊣

Corollary 2.18. *Assume for all regular $\lambda \geq \aleph_2$ every stationary subset of $[\lambda]^\omega$ reflects in some subset of λ of cardinality $< \lambda$. Then $\lambda^{\aleph_0} = \lambda$, for all regular $\lambda \geq \aleph_2$. In particular,* SCH *holds.*

PROOF. That is, Theorem 2.10 still holds if we allow reflection in any $A \in [\lambda]^{<\lambda}$. This stems from the fact that, in the proof of Theorem 2.10, the set S does not actually reflect in any $A \in [\lambda]^{<\lambda}$.

Assume towards contradiction that S reflects in some $A \subseteq \lambda$ of cardinality $< \lambda$. First, suppose that $Card(A) < \kappa$. Then we can collapse $Card(A)$ to ω_1 by the usual forcing with countable conditions $Coll(\omega_1, \kappa)$. Since this forcing does not add new ω-sequences and preserves stationary sets, using the same definition in the generic extension $V[G]$ we get the same set S. Moreover, S still reflects in A, while $Card(A)$ becomes ω_1. However, this contradicts Theorem 2.10 applied in $V[G]$. As a side effect, however, since ordinals in λ that were formerly of cofinality greater than $\aleph_1$ may end up after the forcing with cofinality $\aleph_1$ and we still need to apply the *better* scale property to them, we now have to use the better scale property on all ordinals of cofinality greater than or equal to $\aleph_1$ and not just on those of cofinality $\aleph_1$, as was the case in Theorem 2.10.

Now suppose that $Card(A) = \kappa$. Let $\delta = \sup(A)$ and $\gamma = \mathrm{cof}(\delta)$. Since κ is singular, we have $\gamma < \kappa$. Recall that in Claim 2.12, we have shown that $\mathrm{cof}(\sup(A)) > \aleph_0$. The reasoning we used does not really depend on $Card(A)$ and still holds. Thus we can apply the better scale property of $\langle f_\xi : \xi < \lambda \rangle$ to δ in order to conclude that the set $\{n \in \omega : \mathrm{cof}(f_\delta(n)) = \gamma\}$ is cofinite. Let then $m \in \omega$ be such that $n \geq m$ implies $\mathrm{cof}(f_\delta(n)) = \gamma$.

Let $U_0 = \{n \geq m : \chi(A)(n) = f_\delta(n)\}$. For each $n \in U_0$ choose $B_n \subseteq A \cap [\kappa_{n-1}, \kappa_n)$ cofinal in $\chi(A)(n)$ and of cardinality γ. Let $U_1 = \{n \geq m : \chi(A)(n) > f_\delta(n)\}$. For each $n \in U_1$ let $\alpha_n = \min(A \setminus (f_\delta(n) + 1))$ and let $R = \{\alpha_n : n \in U_1\}$. Finally, let $U_3 = (\omega \setminus m) \setminus (U_0 \cup U_1)$. Since $\mathrm{cof}(\delta) > \aleph_0$, using the better scale property at δ we can find $\alpha < \delta$ such that $f_\alpha > \chi(A)(n)$, for all but finitely many elements of U_3. Let $B_\lambda \subseteq A \cap [\kappa, \delta)$ be cofinal in δ and of cardinality γ. Finally, let

$$B = \bigcup \{B_n : n \in U_0\} \cup B_\lambda \cup R \cup \{\alpha\}.$$

Increasing B if necessary, we may assume that B is closed by $x \mapsto f_x(n)$, for all n (because A satisfies this condition).

The construction of B ensures that the set

$$D = \{X \in [A]^\omega : R \cup \{\alpha\} \subseteq X, \delta(X) = \delta(X \cap B) \\ \wedge n \in U_0 \Longrightarrow \chi(X)(n) = \chi(X \cap B)(n)\}$$

is club in $[A]^\omega$. The construction of D, in turn, ensures that for each $X \in S \cap D$, $E_X =_{FIN} E_{X \cap B}$ (recall that $E_X = \{n < \omega : \chi(X)(n) \leq f_{\delta(X)}(n)\}$), therefore $\phi(X \cap B) = \phi(X)$. On the other hand, obviously $\mathrm{ot}(X \cap B) \leq \mathrm{ot}(X)$. By the definition of S this means that if $X \in S \cap D$ then $X \cap B \in S$. As a result S reflects in B. Indeed, let C be a club set in $[B]^\omega$. Then $C_A = \{X \in [A]^\omega : X \cap B \in C\}$ is also a club. Let $X \in S \cap D \cap C_A$. It follows that $X \cap B \in \mathcal{S} \cap C$. Therefore, $\mathcal{S}$ reflects in B, as well. However, B is of cardinality $\gamma < \kappa$ and by the previous case this give a contradiction. ⊣

§3. Mapping Reflection Principle. The Mapping Reflection Principle was introduced by Moore in [27] and used by him and others to obtain a number of interesting consequences. It can be thought of as a localized version of stationary set reflection, but is in fact independent of the statements RP and SRP discussed in the previous section. While PFA does not imply RP it does imply Mapping Reflection Principle and, indeed, proofs of some important consequences of PFA can be factored through MRP. In particular, MRP plays a key role in Moore's proof [28] that PFA implies the existence of a five element basis for the class of uncountable linear orderings. In this section we present this principle and discuss some of its consequences. We start by recalling the relevant definitions from [27].

DEFINITION 3.1. Let θ be a regular cardinal, let X be uncountable, and let M be a countable elementary submodel of H_θ such that $[X]^\omega \in M$. A subset Σ of $[X]^\omega$ is *M-stationary* iff for all $E \in M$ such that $E \subseteq [X]^\omega$ is club, $\Sigma \cap E \cap M \neq \emptyset$.

Recall that the Ellentuck topology on $[X]^\omega$ is obtained by declaring a set open iff it is the union of sets of the form

$$[x, N] = \{Y \in [X]^\omega : x \subset Y \subseteq N\},$$

where $N \in [X]^\omega$ and $x \subset N$ is finite.

DEFINITION 3.2. Σ is an *open stationary set mapping* if and only if there is an uncountable set $X = X_\Sigma$ and a regular cardinal $\theta = \theta_\Sigma$ such that $[X]^\omega \in H_\theta$, $\mathrm{dom}(\Sigma)$ is a club in $[H_\theta]^\omega$ consisting of elementary submodels and for all $M \in \mathrm{dom}(\Sigma)$, $X \in M$ and $\Sigma(M) \subseteq [X]^\omega$ is open in the Ellentuck topology on $[X]^\omega$ and M-stationary.

The Mapping Reflection Principle MRP is the following statement.

If Σ is an open stationary set mapping, there is a continuous $\in$-chain $\vec{N} = \langle N_\xi : \xi < \omega_1 \rangle$ of elements in the domain of Σ such that for all limit ordinals $\xi < \omega_1$ there is $\nu < \xi$ such that $N_\eta \cap X \in \Sigma(N_\xi)$ for all η such that $\nu < \eta < \xi$.

If $\langle N_\xi : \xi < \omega_1 \rangle$ satisfies the conclusion of MRP for Σ then it is said to be a *reflecting sequence* for Σ. The following was proved in [27].

THEOREM 3.3. PFA *implies* MRP. ⊣

Moore also showed in [27] that MRP implies the failure of the principle $\square(\kappa)$, for all regular $\kappa > \aleph_1$. Thus, MRP has considerable large cardinal strength. Moreover, Sharon [29] showed that MRP implies the failure of $\square_{\kappa,\omega}$, for all regular $\kappa \geq \aleph_1$, and that MRP together with $\mathrm{MA}_{\aleph_1}$ implies

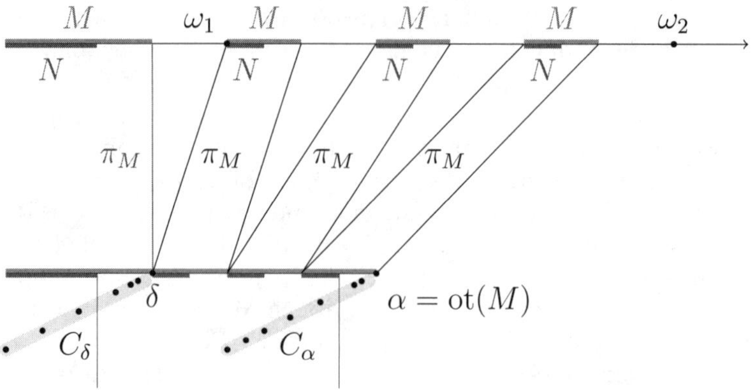

FIGURE 1. $w(N \cap \omega_1, M \cap \omega_1) = 3$, $w(N, M) = 5$.

the failure of $\square_{\kappa,\omega_1}$, for all regular $\kappa \geq \aleph_1$. On the other hand, MRP is *not* equivalent to PFA, and in fact it does not even imply MA. This is a consequence of the fact that the posets used to force the individual instances of MRP are proper and do not add reals and therefore they are ω^ω-bounding and the fact that ω^ω-bounding is preserved by countable support iteration of proper forcing notions, see [1].

Concerning the impact of MRP on cardinal arithmetic, Moore also showed in [27] that MRP implies $2^{\aleph_0} = \aleph_2$. In fact, for this result one only needs a bounded version of MRP and thus obtains that even BPFA implies $2^{\aleph_0} = \aleph_2$. Somewhat later, Viale [40] showed that MRP and therefore also PFA implies SCH, and thus resolved a long standing open problem. We now sketch Moore's proof that MRP implies $2^{\aleph_0} = \aleph_2$. We begin with the following definition.

DEFINITION 3.4. A sequence $\vec{C} = \langle C_\xi : \xi < \omega_1 \text{ and } \lim(\xi)\rangle$ is called a *C-sequence* if C_ξ is an unbounded subset of ξ of order type ω, for all limit $\xi < \omega_1$.

Fix a C-sequence $\vec{C}$. For $N \subseteq M$ countable sets of ordinals such that $\mathrm{ot}(M)$ is a limit ordinal and $\sup(N) < \sup(M)$, set

$$w(N, M) = \left|\sup(N) \cap \pi^{-1}[C_\alpha]\right|,$$

where $\alpha = \mathrm{ot}(M)$ and $\pi : M \to \alpha$ is the transitive collapse of M. See Figure 1

DEFINITION 3.5. Let $A \subseteq \omega_1$. Then $\upsilon_{\mathrm{AC}}(A)$ holds iff there is an uncountable $\delta < \omega_2$ and an increasing continuous sequence $\langle N_\xi : \xi < \omega_1\rangle$ of countable subset of δ whose union is δ such that for all limit ordinals $\nu < \omega_1$ there is a $\nu_0 < \nu$ such that for all $\xi \in (\nu_0, \nu)$,

$$N_\nu \cap \omega_1 \in A \text{ iff } w(N_\xi \cap \omega_1, N_\nu \cap \omega_1) < w(N_\xi, N_\nu).$$

The principle $\upsilon_{\rm AC}$ states that $\upsilon_{\rm AC}(A)$ holds for all $A \subseteq \omega_1$.

THEOREM 3.6. MRP *implies that* $\upsilon_{\rm AC}$ *holds. This also follows from* BPFA.

PROOF. It suffices to prove the following.

LEMMA 3.7. *Let* M *be a countable elementary submodel* $H_{(2^{\aleph_1})^+}$. *Then* $\Sigma_<(M)$ *and* $\Sigma_\geq(M)$ *are open in the Ellentuck topology on* $[\omega_2]^\omega$ *and* M-*stationary, where*

$$\Sigma_<(M) = \{\, N \in [M \cap \omega_2]^\omega : w(N \cap \omega_1, M \cap \omega_1) < w(N, M)\,\}$$

and

$$\Sigma_\geq(M) = \{\, N \in [M \cap \omega_2]^\omega : w(N \cap \omega_1, M \cap \omega_1) \geq w(N, M)\,\}.$$

To see that the theorem follows from the lemma, let $A \subseteq \omega_1$ and notice that Lemma 3.7 implies immediately that Σ_A is open stationary where, for a countable elementary submodel M of $H_{(2^{\aleph_1})^+}$ we let

$$\Sigma_A(M) = \begin{cases} \Sigma_<(M) & \text{if } M \cap \omega_1 \in A \\ \Sigma_\geq(M) & \text{if } M \cap \omega_1 \notin A. \end{cases}$$

Let $\langle N_\xi : \xi < \omega_1 \rangle$ be a reflecting sequence for Σ_A and let $\delta = \omega_2 \cap \bigcup\{N_\xi : \xi < \omega_1\}$. Then δ and $\langle N_\xi \cap \omega_2 : \xi < \omega_1\rangle$ witness $\upsilon_{\rm AC}(A)$.

To see that BPFA suffices, notice that for each $A \subseteq \omega_1$, $\upsilon_{\rm AC}(A)$ is a Σ_1 statement in the parameters A and $\vec{C}$. By the proof of Theorem 3.3 there is a proper poset $\mathcal{P}$ that forces $\upsilon_{\rm AC}(A)$. It follows that $\upsilon_{\rm AC}(A)$ holds under BPFA.

PROOF OF LEMMA 3.7. To see that $\Sigma_<(M)$ is open in the Ellentuck topology let $N \in \Sigma_<(M)$. Let $\beta \in N$ be such that

$$\sup(N) \cap \pi^{-1}[C_{ot(M\cap\omega_2)}] \subset \beta,$$

and let $\gamma \in N \cap \omega_1$ be such that $\sup(N \cap \omega_1) \cap \pi^{-1}[C_{ot(M\cap\omega_1)}] \subset \gamma$, where $\pi : M \cap \omega_2 \to ot(M \cap \omega_2)$ is the transitive collapse of $M \cap \omega_2$. β and γ exist because $N \in M$ is countable so $\sup(N) < \sup(M \cap \omega_2)$ and $\sup(N \cap \omega_1) < \sup(M \cap \omega_1)$, and each C_α has order type ω. Then $[\{\beta, \gamma\}, N] \subseteq \Sigma_<(M)$. Exactly the same argument works for $\Sigma_\geq(M)$.

To see that $\Sigma_<(M)$ is M-stationary, let $E \in M$ be club in $[\omega_2]^\omega$. By the pigeonhole principle there is $\gamma < \omega_1$ such that

$$\{\sup(N) : N \in E \text{ and } N \cap \omega_1 \subseteq \gamma\}$$

is unbounded in ω_2. By elementarity of M, there is $\gamma < M \cap \omega_1$ such that $\{\sup(N) : N \in E \cap M \text{ and } N \cap \omega_1 \subseteq \gamma\}$ is unbounded in $M \cap \omega_2$, so we can find $N \in E \cap M$ such that $N \cap \omega_1 \subseteq \gamma$ and

$$w(N, M \cap \omega_2) = \left|\sup(N) \cap \pi^{-1}[C_{ot(M\cap\omega_2)}]\right| > |C_{M\cap\omega_1} \cap \gamma|.$$

As before, N is countable and belongs to M, so $N \subset M$ and $\sup(N) < \sup(M)$. It follows that $N \in E \cap \Sigma_<(M) \cap M$.

Finally, we show that $\Sigma_{\geq}(M)$ is M-stationary. Let $E \in M$ be club in $[\omega_2]^\omega$. Let $\delta \in M$ be such that $\delta < \omega_2$, δ is uncountable, and $E \cap [\delta]^\omega$ is club in $[\delta]^\omega$. Now let $N \in M$ be such that $N \in E \cap [\delta]^\omega$,

$$|\sup(N \cap \omega_1) \cap C_{M \cap \omega_1}| \geq \left|\delta \cap \pi^{-1}[C_{ot(M \cap \omega_2)}]\right|$$

and $\delta \cap \pi^{-1}[C_{ot(M\cap\omega_2)}] \subseteq N$. Then $N \in E \cap \Sigma_{\geq}(M) \cap M$. This completes the proof of Lemma 3.7 and Theorem 3.6. ⊣

The following theorem was also proved in [27].

THEOREM 3.8. υ_{AC} *implies that* $L(\mathcal{P}(\omega_1))$ *satisfies the Axiom of Choice and that* $2^{\aleph_0} = 2^{\aleph_1} = \aleph_2$.

PROOF. For $A \subseteq \omega_1$ let $[A]$ denote the equivalence class of A in the quotient $\mathcal{P}(\omega_1)/\mathrm{NS}_{\omega_1}$ and let δ_A denote the smallest uncountable ordinal such that some sequence $\vec{N}_A = \langle N^A_\xi : \xi < \omega_1\rangle$ and δ_A witness $\upsilon_{AC}(A)$. Notice that δ_A depends only the equivalence class $[A]$ of A. The assignment $[A] \mapsto \delta_A$ is in fact an injection of $\mathcal{P}(\omega_1)/\mathrm{NS}_{\omega_1}$ into ω_2, since $\delta_A = \delta_B$ implies that for any witnessing sequences $\vec{N}_A$ and $\vec{N}_B$ as above, the set of $\xi < \omega_1$ such that $N^A_\xi = N^B_\xi$ is club in ω_1 and therefore for any ξ limit point of this club, $\xi \in A$ iff $\xi \in B$.

It follows that $\mathcal{P}(\omega_1)/\mathrm{NS}_{\omega_1}$ is well-orderable in $L(\mathcal{P}(\omega_1))$ in order type ω_2. But this implies that $\mathcal{P}(\omega_1)$ itself is well-orderable, and therefore $L(\mathcal{P}(\omega_1)) \models$ AC. In effect, fix a sequence $\langle S_\alpha : \alpha < \omega_1\rangle$ of disjoint stationary subsets of ω_1 and notice that the map $A \mapsto [\bigcup_{\alpha\in A} S_\alpha]$ is an injection of $\mathcal{P}(\omega_1)$ into $\mathcal{P}(\omega_1)/\mathrm{NS}_{\omega_1}$. In particular, $2^{\aleph_1} = \aleph_2$.

To show that $2^{\aleph_0} = 2^{\aleph_1}$, we refute the weak diamond principle of Devlin and Shelah, see [13]. Suppose then that υ_{AC} holds and let $F : 2^{<\omega_1} \to 2$ be given by $F(f) = 0$ iff f codes in some reasonable fashion an increasing continuous $\langle N_\alpha : \alpha < \beta\rangle$ with β limit such that if $N = \bigcup_{\alpha<\beta} N_\alpha$, then there is a $\nu < \beta$ such that for all $\alpha \in (\nu, \beta)$, $w(N_\alpha, N) < w(N_\alpha \cap \omega_1, N \cap \omega_1)$. Let $g \in 2^{\omega_1}$ witness weak diamond for F, let $S = g^{-1}(1)$ and let $f \in 2^{\omega_1}$ code $\langle N_\alpha : \alpha < \omega_1\rangle$ witnessing $\upsilon_{AC}(S)$. There is a club $C \subseteq \omega_1$ such that $f \restriction \nu$ codes $\langle N_\alpha : \alpha < \nu\rangle$, for all $\nu \in C$. Let $A = \{\nu : F(f \restriction \nu) = g(\nu)\}$, so A is stationary. If $A \cap S$ is stationary, let $\nu \in A \cap S \cap C$. Then $F(f \restriction \nu) = 0$, contradicting that $g(\nu) = 1$. It follows that $A \setminus S$ is stationary. Let $\nu \in C \cap (A \setminus S)$. Then $F(f \restriction \nu) = 1$, contradicting that $g(\nu) = 0$. This contradiction implies that weak diamond fails and therefore $2^{\aleph_0} = 2^{\aleph_1}$. ⊣

REMARK. This well-ordering is $\Delta_2(\vec{C})$, since being stationary is Π_1 and we need this to state that two sets are assigned different ordinals under this well-ordering. In §5 we show that this can be improved and that MRP and BPFA imply that there is a $\Delta_1(\vec{C})$ well-ordering of $\mathcal{P}(\omega_1)$, where $\vec{C}$ is a given C-sequence.

Our next goal is to present a result of Viale [40] saying that MRP implies SCH.

THEOREM 3.9. MRP *implies that* $\kappa^{\aleph_0} = \kappa$, *for all regular cardinals* $\kappa > \aleph_1$.

We will need the following definition.

DEFINITION 3.10. Let κ be an infinite cardinal. A *covering matrix* for κ^+ is a sequence $\mathcal{C} = \{K(n, \beta) : n < \omega, \beta < \kappa^+\}$ such that

(1) $|K(n, \beta)| < \kappa$, for all n and β,
(2) $K(n, \beta)$ is a closed subset of β in the order topology, for all n,
(3) $K(n, \beta) \subseteq K(m, \beta)$, for all β and $n < m$,
(4) $\beta = \bigcup\{K(n, \beta) : n < \omega\}$, for all β,
(5) for all $\alpha < \beta$ and n there is m such that $K(n, \alpha) \subseteq K(m, \beta)$.

LEMMA 3.11. *Assume* κ *is a singular cardinal of cofinality* ω. *Then there exists a covering matrix for* κ^+.

PROOF. Fix a sequence $\langle \kappa_n : n < \omega \rangle$ of regular cardinals converging to κ and for each $\xi < \kappa^+$ a surjection $\varphi_\xi : \kappa \to \xi$. We define a sequence $\{K(n, \beta) : n < \omega\}$ by induction on $\beta < \kappa^+$. In order to guarantee condition (1) we also arrange that $|K(n, \beta)| \leq \kappa_n$, for all β and n. Assume we have defined $K(n, \xi)$, for all n and $\xi < \beta$. We let

$$K(n, \beta) = \overline{\varphi_\beta[\kappa_n] \cup \bigcup\{K(n, \xi) : \xi \in \varphi_\beta[\kappa_n]\}} \cap \beta,$$

where the closure is taken in the order topology. It is easy to see that this sequences satisfies the required conditions. ⊣

PROOF OF THEOREM 3.9. We prove the theorem by induction on κ. The only problem occurs at successors of singular cardinals of cofinality ω. So, assume $\mathrm{cof}(\kappa) = \omega$ and fix a covering matrix for κ^+, say $\{K(n, \beta) : n < \omega, \beta < \kappa^+\}$. By the inductive hypothesis we have that $|[K(n, \beta)]^\omega| < \kappa$, for all n and β. We also fix a C-sequence $\langle C_\xi : \xi < \omega_1 \rangle$. For a countable subset X of κ^+ let

$$\alpha_X = \sup(X \cap \omega_1) \text{ and } \delta_X = \sup(X).$$

If $\alpha < \gamma < \omega_1$ let $\mathrm{ht}_\gamma(\alpha) = |C_\gamma \cap \alpha|$. If $\xi < \beta < \kappa^+$ let

$$w(\xi, \beta) = \min\{n : \xi \in K(n, \beta)\}.$$

Now, fix a sufficiently large regular cardinal θ and let M be a countable elementary submodel of H_θ containing all the relevant information. Fix an ordinal $\beta_M \geq \delta_M$ such that for all $\gamma < \kappa^+$ and all $n < \omega$ there is m such that $K(n, \gamma) \cap M \subseteq K(m, \beta_M)$. To see that such an ordinal β_M exists note that $\{K(n, \gamma) \cap M : n < \omega, \gamma < \kappa^+\}$ is of cardinality at most $2^{\aleph_0}$ and $\mathrm{cof}(\kappa^+) > 2^{\aleph_0}$, and use property (5) of the covering matrix. Let $\Sigma(M)$ be the set of all $X \subseteq M \cap \kappa^+$ such that

$$\alpha_X < \alpha_M, \delta_X < \delta_M \text{ and } \mathrm{ht}_{\alpha_M}(\alpha_X) < w(\delta_X, \beta_M).$$

CLAIM 3.12. $\Sigma(M)$ is open in the Ellentuck topology.

PROOF. If $X \in \Sigma(M)$ we have to find a finite $x \subseteq X$ such that $[x, X] \subseteq \Sigma(M)$. First note that if $\delta_X \in X$ then $[\{\delta_X\}, X] \subseteq \Sigma(M)$. So, assume now that X does not have a maximal element and let $n = w(\delta_X, \beta_M)$. By definition of w and the fact that the $K(n, \beta_M)$ are closed in the order topology it follows that $K(n-1, \beta_M) \cap \delta_X$ is bounded in δ_X. So, we can find $\gamma < \delta_X$ such that $[\gamma, \delta_X) \cap K(n-1, \beta_M) = \emptyset$. It follows now that $[\{\gamma\}, X] \subseteq \Sigma(M)$, as desired. ⊣

In general we cannot show that $\Sigma(M)$ is M-stationary, but we do have the following.

CLAIM 3.13. Assume $\kappa^{\aleph_0} > \kappa^+$. Then $\Sigma(M)$ is M-stationary.

PROOF. Suppose $f : [\kappa^+]^{<\omega} \to \kappa^+$ belongs to M. We need to find $Y \in M \cap \Sigma(M)$ which is closed under f. First, find a countable elementary submodel N of $H_{\kappa^{++}}$ containing f and all other relevant objects and such that $N \in M$. Let $m = \mathrm{ht}_{\alpha_M}(\alpha_N)$. Let E be the set of $\gamma < \kappa^+$ of cofinality ω which are closed under f. Then E is of cardinality κ^+ and so, by our assumption $|[E]^{\aleph_0}| > \kappa^+$. By the inductive hypothesis, $\mu^{\aleph_0} < \kappa$, for all $\mu < \kappa$. Therefore $[E]^{\aleph_0}$ is not a subset of

$$\bigcup\{[K(n, \beta)]^{\aleph_0} : n < \omega, \beta < \kappa^+\}.$$

By elementarity of N there exists $X \in [E]^{\aleph_0} \cap N$ such that $X \not\subseteq K(n, \beta)$, for all n and β. In particular, $X \not\subseteq K(m, \beta_M)$. Since $X \in N$ and X is countable it follows that $X \subseteq N$. We conclude that there exists $\gamma \in E \cap N$ such that $w(\gamma, \beta_M) > m$. Since γ is of cofinality ω and closed under f, we can pick a countable subset Y of γ which is cofinal in γ, closed under f and belongs to N. Then we have that $\delta_Y = \gamma$. Since $Y \subseteq N$ we have that $\alpha_Y \leq \alpha_N$. Therefore

$$\mathrm{ht}_{\alpha_M}(\alpha_Y) \leq \mathrm{ht}_{\alpha_M}(\alpha_N) = m < w(\delta_Y, \beta_M).$$

It follows that $Y \in M \cap \Sigma(M)$, as desired. ⊣

Now apply MRP to Σ and let $\langle N_\xi : \xi < \omega_1 \rangle$ be a reflecting sequence. Let $\delta_\xi = \sup(N_\xi \cap \kappa^+)$, for all $\xi < \omega_1$. Then $C = \{\delta_\xi : \xi < \omega_1\}$ is a club in $\delta = \sup C$. Since δ is of cofinality ω_1 there is n such that $K(n, \delta)$ is unbounded in δ. Since $K(n, \delta)$ is also closed it follows that there exists a club D in ω_1 such that $\{\delta_\xi : \xi \in D\} \subseteq K(n, \delta)$. Let ν be any limit point of D and let $M = N_\nu$. By the definition of β_M and properties (3) and (5) of the covering matrix there exists m such that $K(n, \delta) \cap M \subseteq K(m, \beta_M)$. It follows that $w(\delta_\xi, \beta_M) \leq m$, for all $\xi \in D \cap \nu$. On the other hand, $\mathrm{ht}_{\alpha_M}(\alpha_{N_\xi})$ converges to ω, as ξ converges to ν. It follows that $\mathrm{ht}_{\alpha_M}(\alpha_{N_\xi}) > w(\delta_\xi, \beta_M)$, for eventually all $\xi \in D \cap \nu$, which contradicts the fact that $\langle N_\xi : \xi < \omega_1 \rangle$ is a reflecting sequence for Σ. This completes the proof of Theorem 3.9. ⊣

Using the same argument as in Corollary 2.17 we have the following.

COROLLARY 3.14. MRP *implies* SCH. ⊣

§4. P-ideal dichotomy. In this section we introduce a combinatorial principle which captures some of the essential features of PFA yet is compatible with CH. A restricted version of this principle was introduced and studied by Abraham and Todorčević in [3] and the principle was later extended and generalized by Todorčević in [35].

We start with some basic definitions concerning P-ideals of countable sets. We fix an infinite set X. For $A, B \subseteq X$ let us write $A \subseteq_* B$ if $A \setminus B$ is finite. We say that A and B are *orthogonal* and write $A \perp B$ if and only if $A \cap B$ is finite.

DEFINITION 4.1. Let X be a set. Suppose $\mathcal{I} \subseteq [X]^{\leq\aleph_0}$ is an ideal containing all finite sets. We say that $\mathcal{I}$ is a P-*ideal of countable sets* if for every sequence $\{X_n : n < \omega\} \subseteq \mathcal{I}$ there is $X \in \mathcal{I}$ such that $X_n \subseteq_* X$, for all n.

DEFINITION 4.2. The *P-ideal dichotomy* PID is the statement that for every P-ideal $\mathcal{I}$ of countable sets on an uncountable set X, either

(1) there is an uncountable $Y \subseteq X$ such that $[Y]^{\leq\aleph_0} \subseteq \mathcal{I}$, or
(2) we can write $X = \bigcup\{X_n : n < \omega\}$, where $X_n \perp \mathcal{I}$, for all n.

The following was shown for P-ideals of countable sets on ω_1 in [3] and for general P-ideals of countable sets in [35].

THEOREM 4.3. PFA *implies* PID. ⊣

The following was also shown in [35].

THEOREM 4.4. *Assume there is a supercompact cardinal* κ. *Then there is generic extension satisfying* GCH *in which* PID *holds.* ⊣

It was shown in [3] that PID refutes some standard consequences of $\diamondsuit$ and therefore these statements do not follow from CH. For instance, PID implies that there are no ω_1-Souslin trees and that all (ω_1, ω_1^*)-gaps are Hausdorff. In [35] it was shown, among other things, that PID implies that $\square(\kappa)$ fails, for all regular $\kappa > \aleph_1$. Thus, PID has considerable large cardinal strength. For many more applications of PID see [6], [39] and [15].

We will primarily be interested in the consequences of PID to cardinal arithmetic. Before we start we make some definitions which will be useful in the future. Given a subset $\mathcal{A}$ of $\mathcal{P}(X)$ let

$$\mathcal{A}^{\perp} = \{B \subseteq X : B \perp A, \text{ for all } A \in \mathcal{A}\}.$$

Notice that $\mathcal{A}^{\perp}$ is always an ideal, but not necessarily a P-ideal.

DEFINITION 4.5. An ideal $\mathcal{J}$ on a set X is *locally countably generated* if $\mathcal{J} \restriction Y$ is countably generated, for every countable $Y \subseteq X$.

LEMMA 4.6. *Assume an ideal $\mathcal{J}$ on a set X is locally countably generated, then $\mathcal{J}^\perp \cap [X]^{\leq\aleph_0}$ is a* P*-ideal.*

PROOF. Let $\mathcal{I} = \mathcal{J}^\perp \cap [X]^{\aleph_0}$. Assume $Y_n \in \mathcal{I}$, for all n. Let $Y^* = \bigcup_n Y_n$. Fix a generating family $\{Z_n : n < \omega\}$ for $\mathcal{J} \restriction Y^*$. We may assume $Z_n \subseteq Z_{n+1}$, for all n. Let $Y = \bigcup\{Y_n \setminus Z_n : n < \omega\}$. It follows that $Y \in \mathcal{I}$ and $Y_n \subseteq_* Y$, for all n. Therefore $\mathcal{I}$ is a P-ideal. ⊣

LEMMA 4.7. *Assume $\mathcal{J}$ is a locally countably generated ideal on a set X. Let $\mathcal{I} = \mathcal{J}^\perp \cap [X]^{\leq\aleph_0}$. Then $\mathcal{I}^\perp \cap [X]^{\aleph_0} \subseteq \mathcal{J}$.*

PROOF. Let $Y \in \mathcal{I}^\perp$ be countable. Let $\{Z_n : n < \omega\}$ be a generating family for $\mathcal{J} \restriction Y$. We may assume $Z_n \subseteq Z_{n+1}$, for all n. We need to show that $Y \subseteq Z_n$, for some n. Assume otherwise and pick distinct points $x_n \in Y \setminus Z_n$, for $n < \omega$. Then $U = \{x_n : n < \omega\}$ is an infinite subset of Y and is orthogonal to all the Z_n, so $U \in \mathcal{I}$. But this contradicts the fact that $Y \in \mathcal{I}^\perp$. ⊣

We now turn to consequences of PID to cardinal arithmetic. As a warm up let us first prove the following result from [3].

THEOREM 4.8. PID *implies there are no ω_1-Suslin tree.*

PROOF. Assume $(T, \leq)$ is a Souslin tree. For each $t \in T$ let $\hat{t} = \{s \in T : s <_T t\}$. Let $\mathcal{J}$ be the ideal generated by $\{\hat{t} : t \in T\}$. Then it is easy to see that $\mathcal{J}$ is locally countably generated and so $\mathcal{I} = \mathcal{J}^\perp \cap [T]^{\leq\aleph_0}$ is a P-ideal. Consider the two alternatives of PID.

Suppose first that we can write $T = \bigcup\{X_n : n < \omega\}$, where $X_n \in \mathcal{I}^\perp$, for all n. Fix any n such that X_n is uncountable. By Lemma 4.7 we have that $[X_n]^{\leq\aleph_0} \subseteq \mathcal{J}$. Therefore X_n has no infinite antichains, and so by a standard argument has an uncountable chain, a contradiction.

Suppose now that there is an uncountable $X \subseteq T$ such that $[X]^{\leq\aleph_0} \subseteq \mathcal{I}$. Then $\hat{t} \cap X$ is finite, for every $t \in T$. It follows that X, considered as a tree with the inherited ordering has at most $\leq \omega$ levels. Therefore X contains an uncountable antichain, again a contradiction. ⊣

It is an open question if PID bounds the value of the continuum, in particular, if it implies that $2^{\aleph_0} \leq \aleph_2$. However it follows from some results of Todorčević [35] that PID implies that the bounding number $\mathfrak{b}$ is at most $\aleph_2$. In order to discuss this result let us recall the following.

DEFINITION 4.9. Let S be a set. A *pregap* on S is a pair $(\mathcal{A}, \mathcal{B})$ of two orthogonal families of countable subsets of S.

A pregap $(\mathcal{A}, \mathcal{B})$ is a *gap* iff there does not exist $x \subseteq S$ such that $a \subseteq_* x$, for all $a \in \mathcal{A}$, and $b \perp x$, for all $b \in \mathcal{B}$.

Furthermore, two families $\{a_\xi : \xi < \omega_1\}$ and $\{b_\xi : \xi < \omega_1\}$ form a *Hausdorff gap* if the sets $\{\xi < \alpha : a_\alpha \cap b_\xi \subseteq a_\alpha[n]\}$ are finite for all $\alpha < \omega_1$ and $n < \omega$, where $a_\alpha[n]$ denotes the set of the first n elements of a_α.

Notice that a Hausdorff gap is a gap. Indeed, it is easily seen that for every $x \subseteq \bigcup_{\xi<\omega_1}(a_\xi \cup b_\xi)$, either $\{\xi < \omega_1 : a_\xi \perp x\}$ or $\{\xi < \omega_1 : b_\xi \subseteq_* x\}$ is countable.

For a pregap $(\mathcal{A}, \mathcal{B})$ we consider the ideal $\mathcal{I}_{\mathcal{A},\mathcal{B}}$ of countable subsets B of $\mathcal{B}$ for which there is $a \in \mathcal{A}$ such that the set $B(a, n) = \{b \in B : a \cap b \subseteq a[n]\}$ is finite, for all n. The purpose of this condition is to capture the Hausdorff property, in the following sense.

Lemma 4.10. *If there is an uncountable $X \subseteq \mathcal{B}$ such that $[X]^\omega \subseteq \mathcal{I}_{\mathcal{A},\mathcal{B}}$ then $(\mathcal{A}, \mathcal{B})$ contains a Hausdorff subgap.*

Proof. Simply pick a sequence $\{b_\xi : \xi < \omega_1\}$ of distinct elements of X and $\{a_\xi : \xi < \omega_1\}$ such that a_α witnesses $\{b_\xi : \xi < \alpha\} \in \mathcal{I}_{\mathcal{A},\mathcal{B}}$, for all α. ⊣

Lemma 4.11. *If $\mathcal{A}$ is σ-directed, then $\mathcal{I}_{\mathcal{A},\mathcal{B}}$ is a P-ideal.*

Proof. Let $\{B_i : i < \omega\}$ be a family of elements of $\mathcal{I}_{\mathcal{A},\mathcal{B}}$, and let $a_i \in \mathcal{A}$ witness that $B_i \in \mathcal{I}_{\mathcal{A},\mathcal{B}}$, for all i. Since $\mathcal{A}$ is σ-directed, we can choose $a \in \mathcal{A}$ such that $a_i \subseteq_* a$, for all i. Notice that $B_i(a, n)$ is finite, for all i and n, since as soon as $m \geq n$ is large enough that $a_i \setminus a$ and $a[n] \cap a_i$ are included in $a_i[m]$, we get $B_i(a, n) \subseteq B_i(a_i, m)$. Finally, letting $B = \bigcup\{B_n \setminus B_n(a, n) : n < \omega\}$, it is clear that $B_n \subseteq_* B$, for all $n < \omega$, and a witnesses $B \in \mathcal{I}_{\mathcal{A},\mathcal{B}}$. ⊣

Thus, as long as $\mathcal{A}$ is σ-directed we can apply the P-ideal dichotomy to $\mathcal{I}_{\mathcal{A},\mathcal{B}}$. We have seen in Lemma 4.10 how the first alternative of the dichotomy translates in terms of gaps and we are now left to examine the second alternative.

Lemma 4.12. *Let $X \subseteq \mathcal{B}$ be orthogonal to $\mathcal{I}_{\mathcal{A},\mathcal{B}}$. Then $\bigcup X$ is orthogonal to $\mathcal{A}$.*

Proof. Assume to the contrary that there exists $a \in A$ such that $a \cap (\bigcup X)$ is infinite. Then for each $n < \omega$ we can choose $b_n \in X$ such that $b_n \cap a \not\subseteq a[n]$. The family of all such b_n's is an infinite subset of X belonging to $\mathcal{I}_{\mathcal{A},\mathcal{B}}$, which is a contradiction. ⊣

This concludes the proof of the following theorem from [35].

Theorem 4.13. *Assume* PID *and let $(\mathcal{A}, \mathcal{B})$ be a pregap on some set S such that $\mathcal{A}$ is σ-directed under $\subseteq_*$. Then one of the following holds.*

(1) *$(\mathcal{A}, \mathcal{B})$ contains a Hausdorff subgap.*
(2) *There exists a countable family $\{S_n : n < \omega\}$ of subsets of S such that S_n is orthogonal to $\mathcal{A}$, for all n, and every element of $\mathcal{B}$ is included in S_n, for some n.* ⊣

Given two regular cardinals κ and λ recall that a (κ, λ)-*gap* is a gap $(\mathcal{A}, \mathcal{B})$ such that $\mathcal{A}$ is well ordered by $\subseteq_*$ in order type κ and $\mathcal{B}$ is well ordered by $\subseteq_*$ in order type κ. We now have the following.

Corollary 4.14. *Assume* PID *and let κ and λ be two uncountable regular cardinals such that there exists a (κ, λ)-gap. Then $\kappa = \lambda = \aleph_1$.*

PROOF. Assume otherwise and let $(\mathcal{A}, \mathcal{B})$ be a (κ, λ)-gap on a set S. Since $\mathcal{A}$ is well ordered by $\subseteq_*$ in order type κ and κ is regular and uncountable, it follows that $\mathcal{A}$ is σ-directed by $\subseteq_*$. Notice that the second alternative of Theorem 4.13 cannot hold. Namely, suppose $\{S_n : n < \omega\}$ is a family of subsets of S such that S_n is orthogonal to $\mathcal{A}$, for all n, and for every $b \in \mathcal{B}$ there is n such that $b \subseteq_* S_n$. Since λ is also regular and uncountable there is n such that $\{b \in \mathcal{B} : b \subseteq_* S_n\}$ is cofinal in $\mathcal{B}$ under $\subseteq_*$. It follows that $S_n \perp a$, for all $a \in \mathcal{A}$, and $b \subseteq_* S_n$, for all $b \in \mathcal{B}$, a contradiction. Therefore, $(\mathcal{A}, \mathcal{B})$ has a Hausdorff subgap $(\mathcal{A}^*, \mathcal{B}^*)$. But then $\mathcal{A}^*$ is a cofinal subset of $\mathcal{A}$ and $\mathcal{B}^*$ is a cofinal subset of $\mathcal{B}^*$. Since both $\mathcal{A}^*$ and $\mathcal{B}^*$ have cardinality $\aleph_1$ it follows that $\kappa = \lambda = \aleph_1$. ⊣

Recall that the *bounding number* $\mathfrak{b}$ is the least cardinal κ such that there is a subset $\mathcal{F}$ of ω^ω of cardinality κ which is unbounded under eventual dominance $\leq_*$. We now have the following.

COROLLARY 4.15. PID *implies that* $\mathfrak{b} \leq \aleph_2$.

PROOF. Considering the structure $(\omega^\omega, \leq_*)$ one can define a similar notion of a (κ, λ^*)-gap and show, assuming PID, that if κ and λ are uncountable and regular and there is a (κ, λ^*)-gap in $(\omega^\omega, \leq_*)$ then $\kappa = \lambda = \aleph_1$. Now, assuming $\mathfrak{b} > \aleph_2$ one can build an (ω_2, λ^*)-gap in $(\omega^\omega, \leq_*)$ for some regular $\lambda \geq \aleph_1$, see for instance [23, Theorem 29.8, pp.559]. This is a contradiction. ⊣

We now turn to the following result of Viale [41].

THEOREM 4.16. *Assume* PID. *Then* $\kappa^{\aleph_0} = \kappa$, *for all regular* $\kappa \geq 2^{\aleph_0}$.

PROOF. We prove this by induction on κ. The only problem occurs at successors of singular cardinals of countable cofinality. Thus, assume κ is singular of countable cofinality and the theorem holds for all regular $\mu < \kappa$. By Lemma 3.11 we can fix a covering matrix for κ^+, say $\mathcal{C} = \{K(n, \beta) : n < \omega, \beta < \kappa^+\}$. Let $\mathcal{J}$ be the ideal generated by $\mathcal{C}$ and let $\mathcal{I} = \mathcal{J}^\perp \cap [\kappa^+]^{\leq\aleph_0}$.

CLAIM 4.17. $\mathcal{J}$ is locally countably generated and thus $\mathcal{I}$ is a P-ideal.

PROOF. Let Y be a countable subset of κ^+. For each $\alpha < \kappa^+$ let $\mathcal{J}_\alpha$ be the ideal generated by $\{K(n, \alpha) : n < \omega\}$. By property 4. of the covering matrix $\mathcal{J}_\alpha \subseteq \mathcal{J}_\beta$, for $\alpha < \beta$. Since $|\mathcal{P}(Y)| = 2^{\aleph_0} < \kappa^+$ there is $\beta < \kappa^+$ such that

$$\mathcal{J}_\gamma \restriction Y = \mathcal{J}_\beta \restriction Y$$

for all $\gamma \geq \beta$. So, $\mathcal{J} \restriction Y$ is generated by $\{K(n, \beta) : n < \omega\}$. ⊣

CLAIM 4.18. There is no uncountable set $X \subseteq \kappa^+$ such that $[X]^{\leq\aleph_0} \subseteq \mathcal{I}$.

PROOF. Suppose X is of size $\aleph_1$ and let $\alpha = \sup(X)$. Then for some n $K(n, \alpha) \cap X$ is uncountable. Let Z be an infinite subset of $K(n, \alpha)$. Then $Z \notin \mathcal{I}$. ⊣

By the P-ideal dichotomy one gets a decomposition $\kappa^+ = \bigcup_n X_n$ such that $X_n \perp \mathcal{I}$, for all n. By Lemma 4.7 we have the following.

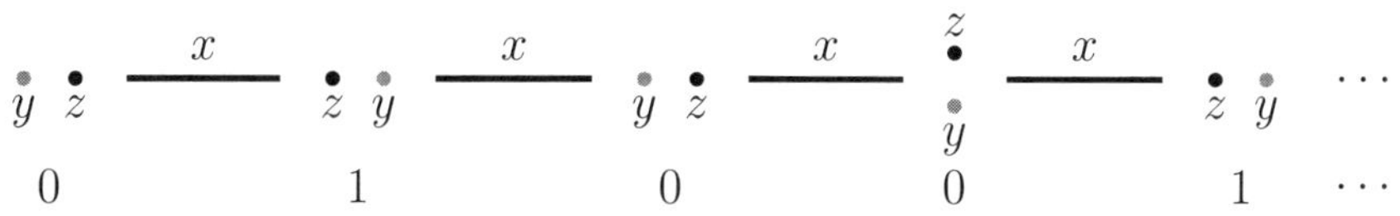

FIGURE 2. $o(x, y, z) = 01001\ldots.$

CLAIM 4.19. $[X_n]^{\aleph_0} \subseteq \bigcup\{[K(m, \beta)]^{\aleph_0} : m < \omega, \beta < \kappa^+\}$. ⊣

Since there is n such that $|X_n| = \kappa^+$ and the $K(m, \beta)$ are all of size $< \kappa$ it follows that

$$(\kappa^+)^{\aleph_0} = \kappa^+ \cdot \sup_{\lambda<\kappa} \lambda^{\aleph_0} = \kappa^+.$$

This completes the proof of Theorem 4.16. ⊣

As in Corollary 2.17 we have the following.

COROLLARY 4.20. PID *implies* SCH. ⊣

§5. Definable well-orderings of the reals. In this section we show that BPFA implies the existence of a definable well ordering of the reals of optimal complexity. This result comes from the paper of Caicedo and the author [9] and refines Theorem 3.8 due to Moore [27]. The interest of this result is that it suggests that there is a structure theory for models of strong forcing axioms. We will say more about this in §6.

The idea is to design a robust coding of reals by triples of ordinals smaller than ω_2 which guarantees that if M is an inner model, BPFA holds in both M and V, and $\aleph_2^M = \aleph_2$, then $\mathcal{P}(\omega_1) \subset M$. This allows us to show that BPFA implies the existence of a well-ordering of the reals which is Δ_1 with parameter a subset of ω_1. Obviously, under $\neg$CH there cannot be a Δ_0-definable well ordering of the reals with parameter a subset of ω_1. Also, a well ordering of the reals cannot be Δ_1 with real parameter r, since then, by a result of Mansfield (see [23, Theorem 25.39]), $\mathbb{R} \subseteq L[r]$, which satisfies CH.

We start by describing the coding machinery which is reminiscent of Moore's coding from §3. We fix a C-sequence $\vec{C} = \langle C_\xi : \xi < \omega_1 \text{ and } \lim(\xi)\rangle$. Given x, y, z sets of natural numbers, define an equivalence relation $\sim_x$ on $\omega \setminus x$ by setting $n \sim_x m$ (for $n \leq m$) iff $[n, m] \cap x = \emptyset$. Thus the equivalence classes of $\sim_x$ are simply the intervals between the consecutive members of x. Let $(I_k)_{k\leq t}$ $(t \leq \omega)$ be the natural enumeration of those equivalence classes which intersect both y and z. Let the *oscillation* of x, y, z be the function $o(x, y, z) : t \to 2$ defined by letting for all $k < t$,

$$o(x, y, z)(k) = 0 \text{ iff } \min(I_k \cap y) \leq \min(I_k \cap z).$$

In Figure 2 we show an example of sets x, y and z such that $o(x, y, z) = 01001\ldots.$

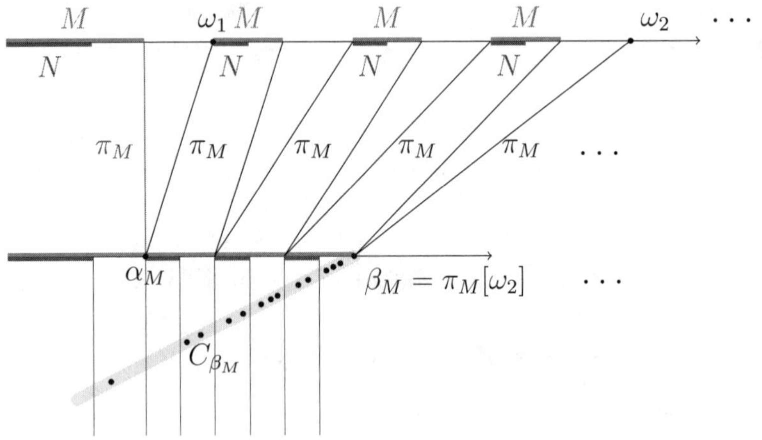

FIGURE 3. $x = \{0, 1, 3, 4, 5, 8, 9, 10\}$.

Let $\omega_1 < \beta < \gamma < \delta$ be fixed limit ordinals and suppose $N \subseteq M \subseteq \delta$ are countable sets of ordinals. Assume that $\{\omega_1, \beta, \gamma\} \subset N$, that $\sup(\xi \cap N) < \sup(\xi \cap M)$ and $\sup(\xi \cap M)$ is a limit ordinal, for every $\xi \in \{\omega_1, \beta, \gamma, \delta\}$. Then the pair (N, M) codes a finite binary sequence as follows. Take the transitive collapse $\bar{M}$ of M and let π be the collapsing map. Let $\alpha_M = \pi(\omega_1), \beta_M = \pi(\beta), \gamma_M = \pi(\gamma), \delta_M = \bar{M}$, each of these is a countable limit ordinal. Let the *height* of $\alpha_N = \sup(\pi[\omega_1 \cap N])$ in α_M be the integer

$$n = n(N, M) = |\alpha_N \cap C_{\alpha_M}|.$$

Define three sets x, y and z of integers by

$$x = \{\, |\pi(\xi) \cap C_{\beta_M}| \colon \xi \in \beta \cap N \,\}$$

and similarly for y and z with γ and δ, respectively, instead of β. See Figure 3 for an example.

Notice that x, y, z are finite by our assumption on N and M. Now, we look at the oscillation of $x \setminus n$, $y \setminus n$ and $z \setminus n$, which is a binary sequence, and if its length is at least n then we let

$$s_{\beta\gamma\delta}(N, M) = o(x \setminus n, y \setminus n, z \setminus n) \restriction n.$$

Otherwise we let $s_{\beta\gamma\delta}(N, M) = *$. We similarly write

$$s_{\beta\gamma\delta}(N, M) \restriction l = *$$

if $l > n(N, M)$.

REMARK. Notice that there is a finite $T \subset N$ such that for any $S \subset N$, if $T \subset S$ then $s_{\beta\gamma\delta}(S, M) = s_{\beta\gamma\delta}(N, M)$. In effect, it suffices that T contains $\{\omega_1, \beta, \gamma\}, \pi^{-1}[\alpha_N \cap C_{\alpha_M}]$, one point of N for each interval in $\beta \cap M$ determined by $\pi^{-1}[C_{\beta_M}]$ that N meets, and similarly for γ and δ.

Finally, we say that the triple (β, γ, δ) *codes* a real r iff there is a continuous increasing sequence $\langle N_\xi : \xi < \omega_1 \rangle$ of countable sets whose union is δ such that for every countable limit ordinal ξ there is $\nu < \xi$ such that

$$r = \bigcup_{\nu < \eta < \xi} s_{\beta\gamma\delta}(N_\eta, N_\xi).$$

It was shown in [9] that BPFA implies the following two facts:

(1) Given ordinals $\omega_1 < \beta < \gamma < \delta < \omega_2$ of cofinality ω_1 there is an increasing continuous sequence $\langle N_\xi : \xi < \omega_1 \rangle$ of countable sets whose union is δ such that for every limit ordinal $\xi < \omega_1$ and every integer n there is $\nu < \xi$ and $s_\xi^n \in \{0, 1\}^n \cup \{*\}$ such that $s_{\beta\gamma\delta}(N_\eta, N_\xi) \restriction n = s_\xi^n$ for every η such that $\nu < \eta < \xi$.

(2) For each real r there are ordinals $\omega_1 < \beta < \gamma < \delta < \omega_2$ of cofinality ω_1 such that the triple (β, γ, δ) codes r.

We will not prove these two facts here and refer the interested to [9] for details.

Remark. Notice that in (1) we are not claiming that for a fixed n the values of s_ξ^n cohere as ξ varies. In fact, this is not necessarily true. However, by the way we have defined our coding it follows that for a fixed limit ordinal ξ the values of s_ξ^n (other than $*$) cohere as n varies. Also, if $s_\xi^n = *$ then $s_\xi^m = *$ for all $m \geq n$.

We now show how facts (1) and (2) are used to obtain the desired consequences.

Theorem 5.1. *Assume M is an inner model,* BPFA *holds in both M and V, and $\omega_2^M = \omega_2$. Then $\mathcal{P}(\omega_1) \subset M$.*

Proof. Assume that M is an inner model of V, that BPFA holds in both M and V and $\omega_1^M = \omega_1$. Fix a C-sequence in M to carry out the codings just described. Suppose $\omega_1 < \beta < \gamma < \delta < \omega_2^M$ are ordinals of cofinality ω_1 and the triple (β, γ, δ) codes in V a real r. Let $\langle N_\xi : \xi < \omega_1 \rangle$ be a continuous increasing sequence of countable sets with union δ witnessing this. In M there is a sequence $\langle P_\xi : \xi < \omega_1 \rangle$ witnessing (1) for (β, γ, δ). There is a club $C \subseteq \omega_1$ in V such that $N_\xi = P_\xi$, for every $\xi \in C$. Then it follows that for any ξ which is a limit point of C, $r = \bigcup_n s_\xi^n$, as computed in M relative to the sequence $\langle P_\xi : \xi < \omega_1 \rangle$. It follows that $r \in M$. If, moreover, $\omega_2 = \omega_2^M$, then any real is coded in V by some triple of ordinals less than ω_2^M and thus all reals are in M. But then $\mathcal{P}(\omega_1) \subset M$, since, given any ω_1-sequence in M of almost disjoint reals, BPFA, and in fact $\mathrm{MA}_{\aleph_1}$, allows us to code any subset of ω_1 via this sequence and a real. ⊣

Since BPFA implies $\mathrm{MA}_{\aleph_1}$ this also implies that BPFA implies that $2^{\aleph_0} = 2^{\aleph_1} = \aleph_2$.

THEOREM 5.2. BPFA *implies that there is a* Δ_1 *well-ordering of* $\mathcal{P}(\omega_1)$ *with parameter a subset of* ω_1. *The length of the well-ordering is* ω_2.

PROOF. Fix as parameter a C-sequence $\vec{C} = \langle C_\xi : \xi < \omega_1 \text{ and } \lim(\xi)\rangle$. Let $\mathrm{T} = \mathrm{T}_{\vec{C}}$ be the theory

$$\text{“ZFC}-\text{Power set}+\mathrm{MA}_{\aleph_1}+\text{(1) and (2) hold with respect to } \vec{C} \\ +\forall x\,(|x| \leq \aleph_1)\text{”}.$$

Notice that any transitive model M of T which contains $\vec{C}$ is uniquely determined by $\mathrm{ORD} \cap M$. In effect, notice that since $\vec{C} \in M$, M computes ω_1 correctly. Suppose a real r is coded by some triple (β, γ, δ) of ordinals in M. Then, arguing as in the proof of Theorem 5.1, we see that $r \in M$. Notice that we are not claiming that M knows that (β, γ, δ) codes r, just that $r \in M$. Since M also satisfies (2) it follows that the reals in M are precisely the reals coded by some triple of ordinals which belong to M. Since $\mathrm{MA}_{\aleph_1}$ holds in M it follows that $\mathcal{P}(\omega_1)^M$ is completely determined as well. Namely, from $\vec{C}$ we can define a canonical ω_1-sequence $\vec{r}$ of almost disjoint reals, and we can use the standard almost disjoint coding to code a subset of ω_1 by the sequence $\vec{r}$ and a real. Now, for an ordinal $\theta < \omega_2$, let M_θ be the unique transitive model M of T containing $\vec{C}$ such that $\mathrm{ORD} \cap M = \theta$, if it exists; otherwise let $M_\theta = \emptyset$. Notice that the function $\theta \mapsto M_\theta$ is Δ_1 in the parameter $\vec{C}$: $M = M_\theta$ if and only if for every transitive model N of enough set theory that contains $\vec{C}$, θ and M, $N \models M = M_\theta$, if and only if there is one such model N.

Let $<_*$ be the antilexicographic ordering on the class $[\mathrm{ORD}]^3$ of increasing triples of ordinals. For a real r let θ_r be the least θ such that $r \in M_\theta$ and let $(\beta_r, \gamma_r, \delta_r)$ be the $<_*$-least triple of ordinals smaller than θ_r such that $M_{\theta_r} \models (\beta_r, \gamma_r, \delta_r)$ codes r. Finally, let $r \lhd s$ iff either $\theta_r < \theta_s$ or $\theta_r = \theta_s$ and $(\beta_r, \gamma_r, \delta_r) <_* (\beta_s, \gamma_s, \delta_s)$.

We can now define a well-ordering $\prec$ of $\mathcal{P}(\omega_1)$ as follows. For a and b subsets of ω_1, we define $a \prec b$ iff the $\lhd$-least real coding a (from the sequence $\vec{r}$ of almost disjoint reals defined from $\vec{C}$) is $\lhd$-smaller than the $\lhd$-least real coding b. By an argument similar to the above, $\prec$ is also Δ_1 in the parameter $\vec{C}$. ⊣

REMARK. It is open whether even MM implies that there is a definable well ordering of the reals without parameters. On the other hand, Asperó [4] showed that it is consistent with PFA that there is a well-ordering of the reals definable over $H_{\aleph_2}$ without parameters and Larson [24] showed the same result with MM in place of PFA.

§6. Inner models of forcing axioms. The results of the previous section suggest that there should be a structure theory for models of strong forcing axioms such as PFA or MM. In order to make this question precise the following conjecture was formulated by Caicedo and the author.

CONJECTURE 6.1. Assume $V \subseteq W$ are two models of PFA or MM which have the same cardinals. Then $\mathrm{ORD}^{\omega_1} \cap V = \mathrm{ORD}^{\omega_1} \cap W$.

We fix the cardinals to avoid situations like:

- If V satisfies MM and there is a supercompact cardinal then we can collapse $\aleph_2$ and force MM again. If W is the resulting generic extension then V and W have little in common.
- If V has a measurable cardinal κ and W is the result of iterating ω-many times a normal measure on κ then $V \subset W$ are elementarily equivalent, but V is not closed under ω-sequences.
- If V has a proper class of completely Jónsson cardinals, $\mathcal{P}_\infty$ is the class stationary tower forcing (see [25]), G is $\mathcal{P}_\infty$-generic over V, and $W = V[G]$, then there is an elementary embedding $j : V \to W$ and we can arrange that $\mathrm{cp}(j)$ is arbitrarily high and $\mathrm{cof}^W(\mathrm{cp}(j)) = \omega$.

While Conjecture 6.1 is still open, in this section we present some partial results. We start with the following observation. The proof is essentially the same as the proof that MM implies $\kappa^{\aleph_1} = \kappa$, for all regular $\kappa \geq \aleph_2$, see [16].

PROPOSITION 6.2. *Assume W is a set generic extension of V and W satisfies* SRP. *Assume moreover that V and W have the same ordinals of cofinality ω and ω_1 and $\mathcal{P}(\omega_1)^V = \mathcal{P}(\omega_1)^W$. Then* $\mathrm{ORD}^{\omega_1} \cap V = \mathrm{ORD}^{\omega_1} \cap W$.

PROOF. Let $\mathcal{P} \in V$ be a forcing notion such that $W = V[G]$, for some V-generic filter over $\mathcal{P}$. Let $\kappa = |\mathcal{P}|^+$. Then obviously $\mathcal{P}$ satisfies the κ-chain condition. Let $S = \{\alpha < \kappa : \mathrm{cof}(\alpha) = \omega\}$ and fix in V a partition $S = \bigcup\{S_\xi : \xi < \kappa\}$ into disjoint stationary sets. By the κ-cc the S_ξ are still stationary in W. Fix, also in V, a partition $\omega_1 = \bigcup\{E_\xi : \xi < \omega_1\}$ into disjoint stationary sets. Then, by our assumption, the E_ξ are also stationary in W. Now, work in W and suppose $A \in W$ is a subset of κ of cardinality ω_1. We want to show that $A \in V$. To this end let $f : \omega_1 \to \kappa$ be a $1-1$ enumeration of A and consider the following set

$$T = \{X \in [\kappa]^{\aleph_0} : \sup(X \cap \omega_1) \in E_\xi \text{ iff } \sup(X) \in S_{f(\xi)}, \text{ for all } \xi\}.$$

Let θ be a sufficiently large regular cardinal and let T^* be the set of all countable elementary submodels M of H_θ such that $M \cap \kappa \in T$.

CLAIM 6.3. T^* is projective stationary.

PROOF. Let E be a stationary subset of ω_1. We need to show that $T^*_E = \{M \in T^* : \sup(M \cap \omega_1) \in E\}$ is stationary in $[H_\theta]^\omega$. Let $F : [H_\theta]^{<\omega} \to H_\theta$ be an algebra on H_θ. We need to find $M \in T^*$ which is closed under F. First we find $\xi < \omega_1$ such that $E \cap E_\xi$ is stationary. Since $S_{f(\xi)}$ is a stationary subset of κ we can find an elementary submodel N of $(H_\theta, \in, F)$ such that $\delta = N \cap \kappa \in S_{f(\xi)}$. Build a continuous increasing chain $\langle M_\eta : \eta < \omega_1 \rangle$ of countable elementary submodels of N such that $M_0 \cap \delta$ is cofinal in δ (we can do this since $\mathrm{cof}(\delta) = \omega$) and such that if we let $\alpha_\eta = \sup(M_\eta \cap \omega_1)$ the sequence $\langle \alpha_\eta : \eta < \omega_1 \rangle$ is strictly increasing. Then $D = \{\alpha_\eta : \eta < \omega_1\}$ is a club in ω_1 and since $E \cap E_\xi$ is stationary there is η such that $\alpha_\eta \in D \cap E \cap E_\xi$. Then $M_\eta \in T^*_E$, as desired. ⊣

Now, by SRP we can find a continuous $\in$-chain $\langle M_\eta : \eta < \omega_1 \rangle$ of countable elementary submodels of H_θ such that $M_\eta \in T^*$, for all η. Let $\alpha_\eta = \sup(M_\eta \cap \omega_1)$ and $\delta_\eta = \sup(M_\eta \cap \kappa)$. Then $\{\alpha_\eta : \eta < \omega_1\}$ is a club in ω_1 and $\{\delta_\eta : \eta < \omega_1\}$ is a club in some $\delta < \kappa$ with $\mathrm{cof}(\delta) = \omega_1$. Since $M_\eta \in T^*$ we have that $\alpha_\eta \in E_\xi$ iff $\delta_\eta \in S_{f(\xi)}$, for all ξ. By our assumption $\mathrm{cof}^V(\delta) = \omega_1$ and the notion of stationary subset of δ is the same in V and W. Since A is equal to the set of $\zeta < \kappa$ such that S_ζ is stationary in δ it follows that $A \in V$. ⊣

In [41] Viale defined a family of covering principles which he used to address problems related to Conjecture 6.1. We will need a generalization of the notion of a covering matrix introduced in §3.

Definition 6.4. Let θ and κ be two regular cardinals with $\theta < \kappa$. an infinite cardinal. A *θ-covering matrix* for κ is a sequence $\mathcal{K} = \{K(\xi, \beta) : \xi < \theta, \beta < \kappa\}$ such that

(1) $K(\xi, \beta)$ is a closed subset of β in the order topology, for all ξ and β,
(2) $K(\xi, \beta) \subseteq K(\eta, \beta)$, for all $\xi < \eta$ and all β,
(3) $\beta = \bigcup\{K(\xi, \beta) : \xi < \theta\}$, for all β,
(4) for all $\alpha < \beta$ and ξ there is η such that $K(\xi, \alpha) \subseteq K(\eta, \beta)$.

$\tau_{\mathcal{K}}$ is the least ordinal τ such that $\mathrm{ot}(K(\xi, \beta)) < \tau$, for all ξ and β. $\mathcal{K}$ is *trivial* if $\tau_{\mathcal{K}} = \kappa$.

Thus, what we called a covering matrix for κ^+ in Definition 3.10 is an ω-covering matrix $\mathcal{K}$ for κ^+ such that $\tau_{\mathcal{K}} = \kappa$.

We say that a θ-covering matrix for κ $\mathcal{K}$ is *downward coherent* if in addition it satisfies the following coherence property.

(5) *For every $\alpha < \beta < \kappa$ and $\xi < \theta$ there is η such that $K(\xi, \beta) \cap \alpha \subseteq K(\eta, \alpha)$.*

Given a family $\mathcal{A}$ of subsets of κ we say that $\mathcal{A}$ is *covered* by $\mathcal{K}$ if for every $X \in \mathcal{A}$ there is $K \in \mathcal{K}$ such that $X \subseteq K$.

Definition 6.5 (Covering Property (CP)). Given two regular cardinal $\theta < \kappa$, the *covering property* $\mathrm{CP}(\theta, \kappa)$ is the following statement.

For every θ-covering matrix $\mathcal{K} = \{K(\xi, \beta) : \xi < \theta, \beta < \kappa\}$ for κ there is an unbounded subset A of κ such that $[A]^\theta$ is covered by $\mathcal{K}$.

$\mathrm{CP}(\kappa)$ abbreviates $\mathrm{CP}(\omega, \kappa)$ and CP states that $\mathrm{CP}(\kappa)$ holds, for all regular $\kappa > \mathfrak{c}$.

It should be pointed out that the proofs of both Theorem 3.9 and Theorem 4.16 can be factored through CP. Namely, in [41] Viale proved the following.

THEOREM 6.6. (1) *Both* MRP *and* PID *imply* CP.

(2) CP *implies* $\kappa^{\aleph_0} = \kappa$, *for all regular* $\kappa \geq \mathfrak{c}$. *In particular,* CP *implies* SCH. ⊣

PROPOSITION 6.7. *Let* κ *be a regular cardinal. Then there is a downward coherent* κ*-covering matrix* $\mathcal{K}$ *for* κ^+ *with* $\tau_{\mathcal{K}} = \kappa$.

PROOF. For each limit ordinal $\alpha < \kappa^+$ fix a club C_α in α of minimal order type. If $\alpha = \beta + 1$ is a successor ordinal let $C_\alpha = \{\beta\}$. Recall the definition of Todorčević's ρ function, see for instance [37, pp. 271]. The function $\rho : [\kappa^+]^2 \to \kappa$ is defined recursively as follows:

$$\rho(\alpha, \beta) = \sup\{\mathrm{ot}(C_\beta \cap \alpha), \rho(\alpha, \min(C_\beta \setminus \alpha)), \rho(\xi, \alpha) : \xi \in C_\beta \cap \alpha\},$$

where we let $\rho(\alpha, \alpha) = 0$, by convention. The function ρ has the following properties, see [37, Lemmas 9.1.1 and 9.1.2].

(1) For every $\nu < \kappa$ and $\alpha < \kappa^+$ the set $P_\nu(\alpha) = \{\xi < \alpha : \rho(\xi, \alpha) < \nu\}$ has cardinality at most $|\nu| + \aleph_0$.

(2) For every $\alpha < \beta < \gamma < \kappa^+$,
   (a) $\rho(\alpha, \gamma) \leq \max\{\rho(\alpha, \beta), \rho(\beta, \gamma)\}$,
   (b) $\rho(\alpha, \beta) \leq \max\{\rho(\alpha, \gamma), \rho(\alpha, \gamma)\}$.

Let now $K(\nu, \alpha) = \overline{P_\nu(\alpha)}$, for all $\nu < \kappa$ and $\alpha < \kappa^+$. Then the sequence $\mathcal{K} = \{K(\nu, \alpha) : \nu < \kappa, \alpha < \kappa^+\}$ is a downward coherent κ-covering matrix for κ^+ and $\tau_{\mathcal{K}} = \kappa$. ⊣

In [41] Viale also showed the following.

THEOREM 6.8. *Let* $V \subseteq W$ *be two models of set theory with the same reals and cardinals. Assume* W *satisfies* CP. *Then* V *and* W *have the same ordinals of cofinality* ω.

PROOF. Assume otherwise and let κ be the least uncountable cardinal which is regular in V and has cofinality ω in W. Then by our assumption $\kappa > \mathfrak{c}$. In V fix a downward coherent κ-covering matrix $\mathcal{K} = \{K(\xi, \alpha) : \xi < \kappa, \alpha < \kappa^+\}$ for κ^+ such that $\tau_{\mathcal{K}} = \kappa$. Now, since $\mathrm{cof}^W(\kappa) = \omega$ we can fix an increasing ω-sequence $\{\xi_n : n < \omega\}$ of ordinals converging to κ. Let $L(n, \alpha) = K(\xi_n, \alpha)$, for all n and α. Since $(\kappa^+)^V = (\kappa^+)^W$ it follows that $\mathcal{L} = \{L(n, \alpha) : n < \omega, \alpha < \kappa^+\}$ is a downward coherent ω-covering matrix for κ^+ in W. Now, in W, apply $\mathrm{CP}(\omega, \kappa^+)$ to $\mathcal{L}$ to get an unbounded subset A of κ^+ such that $[A]^{\aleph_0}$ is covered by $\mathcal{L}$. Let α be such that $\mathrm{ot}(A \cap \alpha) = \kappa$. Since $\mathrm{ot}(L(n, \alpha)) < \kappa$ we can pick $\xi_n \in A \setminus L(n, \alpha)$, for all n. Let $X = \{\xi_n : n < \omega\}$. By our construction X is not covered by $L(n, \alpha)$, for any n. Since $\mathcal{L}$ is a downward coherent ω-covering matrix it follows that X is not covered by any member of $\mathcal{L}$, a contradiction. ⊣

Concerning agreement on ordinals of cofinality ω_1 Viale [41] also proved the following.

THEOREM 6.9. *Let $V \subseteq W$ be two models of set theory with the same cardinals. Assume W satisfies* MM *and that all limit cardinals are strong limits. Then V and W have the same ordinals of cofinality ω_1.* ⊣

§7. Open problems. In §4 we showed that PID implies that the bounding number $\mathfrak{b}$ is at most $\aleph_2$. Assuming the existence of a supercompact cardinal one can produce models of PID in which the continuum is either $\aleph_1$ or $\aleph_2$. However, the following is still open.

QUESTION 1. Does PID imply that $\mathfrak{c} \leq \aleph_2$?

In §5 we showed that BPFA implies the existence of a well ordering of the reals which is Δ_1-definable in parameter a subset of ω_1. Concerning definable (without parameters) well orderings of the reals, as we already mentioned, Asperó [4] and Larson [24] showed that the existence of such well orderings is compatible with PFA and MM. However, it is possible that the existence of such well orderings as an outright consequence of MM.

QUESTION 2. Does MM imply the existence of a well ordering of the reals which is definable without parameters?

Finally, we restate Conjecture 6.1 we discussed in §6.

QUESTION 3. Assume $V \subseteq W$ are two models of set theory with the same cardinals and W satisfies MM. Is $\mathrm{ORD}^{\omega_1} \cap V = \mathrm{ORD}^{\omega_1} \cap W$?

Acknowledgment. The author would like to thank Andres Caicedo, Brice Minaud, and Matteo Viale for their help in the preparation of this paper.

REFERENCES

[1] U. ABRAHAM, *Proper forcing*, ***Handbook of Set Theory*** (M. Foreman and A. Kanamori, editors), Springer-Verlag, Berlin, New York, 2008.

[2] U. ABRAHAM and M. MAGIDOR, *Cardinal arithmetic*, ***Handbook of Set Theory*** (M. Foreman and A. Kanamori, editors), Springer-Verlag, Berlin, New York, 2008.

[3] U. ABRAHAM and S. TODORČEVIĆ, *Partition properties of ω_1 compatible with CH*, ***Fundamenta Mathematicae***, vol. 152 (1997), pp. 165–180.

[4] D. ASPERÓ, *Guessing and non-guessing of canonical functions*, ***Annals of Pure and Applied Logic***, vol. 146 (2007), no. 2-3, pp. 150–179.

[5] J. BAGARIA, *Bounded forcing axioms as principles of generic absoluteness*, ***Archive for Mathematical Logic***, vol. 39(6) (2000), pp. 393–401.

[6] B. BALCAR, T. JECH, and T. PAZÁK, *Complete ccc boolean algebras, the order sequential topology, and a problem of von neumann*, ***Bulletin of the London Mathematical Society***, vol. 37 (2005), pp. 885–898.

[7] J. E. BAUMGARTNER, *Applications of the proper forcing axiom*, ***Handbook of Set-Theoretic Topology***, North-Holland, Amsterdam, 1984, pp. 913–959.

[8] J. E. BAUMGARTNER and A. D. TAYLOR, *Saturation properties of ideals in generic extensions. I*, ***Transactions of the American Mathematical Society***, vol. 270 (1982), no. 2, pp. 557–574.

[9] A. Caicedo and B. Veličković, *Bounded proper forcing axiom and well orderings of the reals*, ***Mathematical Research Letters***, vol. 13 (2006), no. 2-3, pp. 393–408.

[10] G. Cantor, *Ein Beitrag zur Mannifaltigkeitslehre*, ***Journal für die reine und angewandte Mathematik***, vol. 84 (1878), pp. 242–258.

[11] J. P. Cohen, *The Independence of the Continuum Hypothesis*, ***Proceedings of the National Academy of Sciences of the United States of America***, vol. 50, 1963, pp. 1143–1148.

[12] J. Cummings, *Notes on singular cardinal combinatorics*, ***Notre Dame Journal of Formal Logic***, vol. 46 (2005), no. 3, pp. 251–282.

[13] K. Devlin and S. Shelah, *A weak form of diamond which follows from* $2^{\aleph_0} < 2^{\aleph_1}$, ***Israel Journal of Mathematics***, vol. 29 (1978), pp. 239–247.

[14] W. B. Easton, *Powers of regular cardinals*, ***Annals of Mathematical Logic***, vol. 1 (1970), pp. 139–178.

[15] T. Eisworth and P. Nyikos, *Antidiamond principles and topological applications*, ***Transactions of the American Mathematical Society***, to appear.

[16] M. Foreman, M. Magidor, and S. Shelah, *Martin's Maximum, saturated ideals and nonregular ultrafilters*, ***Annals of Mathematics. Second Series.***, vol. 127(1) (1988), no. 1, pp. 1–47.

[17] M. Foreman and S. Todorčević, *A new löwenheim-skolem theorem*, ***Transactions of the American Mathematical Society***, vol. 357 (2005), no. 5, pp. 1693–1715.

[18] D. Gale and F. M. Stewart, *Infinite games with perfect information*, ***Annals of Mathematics Studies***, vol. 28 (1953), pp. 245–266.

[19] K. Gödel, ***The Consistency of the Axiom of Choice and of the Generalized Continuum Hypothesis with the Axioms of Set Theory***, Princeton University Press, 1940.

[20] ———, *What is Cantor's continuum problem?*, ***American Mathematical Monthly***, vol. 54 (1947), pp. 515–525.

[21] ———, *What is Cantor's continuum problem?*, ***Philosophy of Mathematics Selected Readings*** (P. Benacerraf and H. Putnam, editors), Cambridge University press, 1983, pp. 470–485.

[22] M. Goldstern and S. Shelah, *The bounded proper forcing axiom*, ***The Journal of Symbolic Logic***, vol. 60 (1995), no. 1, pp. 58–73.

[23] T. Jech, ***Set Theory, the Third Millennium Edition, Revised and Expanded***, Springer, New York, Berlin, 2002.

[24] P. B. Larson, *Martin's maximum and definability in* $H(\aleph_2)$., ***Annals of Pure and Applied Logic***, to appear.

[25] ———, ***The Stationary Tower: Notes on a Course by W. Hugh Woodin***, AMS, 2004.

[26] D. A. Martin and R.M. Solovay, *Internal Cohen extensions*, ***Annals of Mathematical Logic***, vol. 2 (1970), pp. 143–178.

[27] J. T. Moore, *Set mapping reflection*, ***Journal of Mathematical Logic***, vol. 5 (2005), no. 1, pp. 87–97.

[28] J. T. Moore, *A five element basis for the uncountable linear orders*, ***Annals of Mathematics***, vol. 163 (2006), no. 2, pp. 669–688.

[29] A. Sharon, MRP *and square principles*, unpublished, 23 pages, 2006.

[30] S. Shelah, *Semiproper forcing axiom implies Martin's maximum but not* PFA$^+$, ***The Journal of Symbolic Logic***, vol. 52 (1987), no. 2, pp. 360–367.

[31] ———, ***Cardinal Arithmetic***, Oxford University Press, 1994.

[32] ———, *Reflection implies* SCH, ***Fundamenta Mathematicae***, vol. 198 (2008), pp. 95–111.

[33] J. H. Silver, *On the singular cardinals problem*, ***Proceedings of the International Congress of Mathematicians (Vancouver, B. C., 1974), Vol. 1***, Canad. Math. Congress, 1975, pp. 265–268.

[34] R. M. Solovay and S. Tennenbaum, *Iterated Cohen extensions and Souslin's problem*, ***Annals of Mathematics. Second Series***, vol. 94 (1971), pp. 201–245.

[35] S. Todorčević, *A dichotomy for P-ideals of countable sets*, ***Fundamenta Mathematicae***, vol. 166(3) (2000), pp. 251–267.

[36] ———, *Generic absoluteness and the continuum*, ***Mathematical Research Letters***, vol. 9 (2002), no. 4, pp. 465–471.

[37] S. Todorčević, ***Walks on Ordinals and Their Characteristics***, Progress in Mathematics, vol. 263, Birkhäauser Verlag AG, 2007.

[38] B. Veličković, *Forcing axioms and stationary sets*, ***Advances in Mathematics***, vol. 94(2) (1992), no. 2, pp. 256–284.

[39] ———, *CCC forcing and splitting reals*, ***Israel Journal of Mathematics***, vol. 147 (2005), pp. 209–220.

[40] M. Viale, *The proper forcing axiom and the singular cardinal hypothesis*, ***The Journal of Symbolic Logic***, vol. 71 (2006), no. 2, pp. 473–479.

[41] ———, *A family of covering properties*, ***Mathematical Research Letters***, vol. 15 (2008), no. 2, pp. 221–238.

[42] W. H. Woodin, ***The Axiom of Determinacy, Forcing Axioms, and the Nonstationary Ideal***, Walter de Gruyter and Co., 1999.

ÉQUIPE DE LOGIQUE MATHÉMATIQUE
UNIVERSITÉ PARIS DIDEROT PARIS 7
UFR DE MATHÉMATIQUES CASE 7012, SITE CHEVALERET
75205 PARIS CEDEX 13, FRANCE
E-mail: boban@logique.jussieu.fr

HRUSHOVSKI'S AMALGAMATION CONSTRUCTION

FRANK O. WAGNER

Abstract. An overview is given of the various structures obtained by means of Hrushovski's amalgamation method and variants thereof.

§1. Introduction. In 1986, Ehud Hrushovski modified Fraïssé's construction of a universal homogeneous countable relational structure from the class of its finite substructures, in order to obtain *stable* structures with particular properties. In particular, he constructed

(i) an $\aleph_0$-categorical stable pseudoplane (4.1.1),
(ii) a strongly minimal set with an exotic geometry which is not disintegrated, but does not interpret any group (4.2.3),
(iii) the fusion of two strongly minimal sets in disjoint languages in a third one,

obtaining counter-examples to conjectures by Lachlan and Zilber. In particular, a stable $\aleph_0$-categorical structure need not be ω-stable (i), and the trichotomy suggested by Zilber for strongly minimal sets (disintegrated/vector-space-like/field-like) does not hold in general (ii); moreover, every strongly minimal set (with an additionnal condition, the definable multiplicity property DMP) has an essential strongly minimal extension, so there is no maximal one (iii). In particular, an algebraically closed field has proper strongly minimal extensions, even interpreting another algebraically closed field of possibly different characteristic, answering a question of Zilber.

His method was taken up by Baldwin, Baudisch, Evans, Hasson, Hils, Holland, Martín Pizarro, Poizat, Ziegler, Zilber and others, who constructed a variety of strange objects, most notably

(iv) a disintegrated stable structure with a reduct which is not one-based (3.1.1),
(v) a new $\aleph_1$-categorical nilpotent group of class 2 and exponent p,
(vi) a field of Morley rank 2 with an additive subgroup of Morley rank 1 in characteristic $p > 0$ (6.2),

---

Membre junior de l'Institut Universitaire de France.

Logic Colloquium '06
Edited by Barry Cooper, Herman Geuvers, Anand Pillay, and Jouko Väänänen
Lecture Notes in Logic, 32

(vii) a field of Morley rank 2 with a multiplicative subgroup of Morley rank 1 in characteristic 0 (7.2),

(viii) the fusion of two strongly minimal sets over a common $\aleph_0$-categorical reduct.

(iv) answers negatively the question of Poizat whether a reduct of a disintegrated structure must be disintegrated, or a reduct of a one-based structure must be one-based. (v) yields another counterexample to the trichotomy conjecture, which is not only combinatorial, but carries algebraic structure. (vi) and (vii) answer questions by Zilber and Poizat about the existence of such structures. The problem arose in the context of the analysis of simple groups of finite Morley rank (which were conjectured by Zilber and Cherlin to be algebraic); a soluble non-nilpotent subgroup can be reduced to the form $K^+ \rtimes T$ for some non-trivial $T \leq K^\times$, and in order to prove the existence of elements of finite order, one would have liked to show $T = K^\times$ (but the current approach to the conjecture manages to circumvent this difficulty). Finally, Hrushovski remarks in [11] that it should be possible to do the fusion of (iii) not in disjoint languages, but over a common vector space over a finite field; this is the essential case of (viii).

We shall sketch how these structures are obtained by variations of the original construction.

§2. The original construction of Fraïssé. Let $\mathcal{C}$ be a class of finite structures in a finite relational language, closed under substructures, and with the *amalgamation property* (where we allow $A = \emptyset$)

AP: For all injective $\sigma_i : A \to B_i$ in $\mathcal{C}$ for $i = 1, 2$ there are injective $\rho_i : B_i \to D \in \mathcal{C}$ with $\rho_1\sigma_1 = \rho_2\sigma_2$.

Then there is a unique countable structure $\mathfrak{M}$ satisfying

Richness: For all finite $A \subset \mathfrak{M}$ and $A \subset B \in \mathcal{C}$ there is an embedding $B \to \mathfrak{M}$ which is the identity on A.

The proof is by successive amalgamation over all possible situations, using AP.

We call $\mathfrak{M}$ the *generic* model; it is ultrahomogeneous, and hence $\aleph_0$-categorical (since the language is finite). It is axiomatized by the universal axioms for $\mathcal{C}$ (which proscribe the sets not in $\mathcal{C}$) and the inductive axioms for richness.

§3. Modification I. Rather than considering all inclusions $A \subset B \in \mathcal{C}$, we only consider *certain* inclusions $A \leq B$, which we call *closed*. We require $\leq$ to be transitive and preserved under intersections: If $A \leq B$ and $C \subseteq B$, then $A \cap C \leq C$. We only demand AP for closed inclusions, and obtain a generic structure $\mathfrak{M}$ satisfying

Closed Richness: For any finite $A \leq \mathfrak{M}$ and $A \leq B \in \mathcal{C}$ there is a *closed* embedding $B \hookrightarrow \mathfrak{M}$ which is the identity on A.

Moreover, $\mathfrak{M}$ is ultrahomogeneous for finite *closed* subsets.

Let $\bar{\mathcal{C}}$ be the class of structures whose finite substructures are in $\mathcal{C}$. For infinite embeddings $A \subseteq B \in \bar{\mathcal{C}}$ we put $A \leq B$ if $A \cap C \leq C$ for all finite $C \subseteq B$. For $A \subset \mathfrak{M} \in \bar{\mathcal{C}}$ we define the *closure* $\mathrm{cl}_{\mathfrak{M}}(A)$ to be the smallest $B \leq \mathfrak{M}$ containing A. It is easily seen to be unique, but it can be infinite (for instance $\mathfrak{M}$ itself).

In order to axiomatize, we need to express $A \leq \mathfrak{M}$. If this can be done by a first-order formula, and if there is a finite bound on the size of the closure of a finite set in terms of size of the set itself (and hence a finite bound on the number of n-types for all $n < \omega$, since the language is finite), the generic model is $\aleph_0$-categorical.

However, even in a finite relational language, closedness need not be a definable property. If $A \leq \mathfrak{M}$ is only *type*-definable, or if closures can be infinite, we need definability of

Approximate Richness: If $A \leq B \in \mathcal{C}$ and A is sufficiently closed in $\mathfrak{M}$, then there is an embedding of B into $\mathfrak{M}$ over A whose image has a pre-described leved of closedness.

This yields homogeneity for closures of finite subsets in an $\aleph_0$-saturated model, or for countable closed subsets in an $\aleph_1$-saturated model.

3.1. Example. [7] Consider the class $\mathcal{C}$ of finite directed graphs without directed cycles and out-valency 2. Define $A \leq B$ if all descendants of A in B are already in A. Note that the closure is *disintegrated*:

$$\mathrm{cl}_{\mathfrak{M}}(A) = \bigcup_{a \in A} \mathrm{cl}_{\mathfrak{M}}(a).$$

This class has AP (namely the free amalgam, i.e. the disjoint union over the intersection, with no edges added). Clearly, richness (for finite sets) is definable; let T be the theory of (finitely) rich directed graphs without directed cycles and out-valency 2. By compactness, an $\aleph_0$-saturated model of T is rich even for closures of finite sets, so T is complete, and isomorphic closed sets have the same type.

3.1.1. *A non-disintegrated reduct.* Now consider the undirected reduct of models of T. By induction, one sees that a finite undirected graph has an orientation without directed cycles and of out-valency 2 iff every subgraph has a vertex of valency at most 2. For these graphs we define $A \leq B$ if B has such an orientation in which A is closed in the directed sense. This class is closed under free amalgamation; its generic model is the reduct of a generic model of the directed class.

However, the closure of A in $\mathfrak{M}$ is the intersection of all directed closures of A in $\mathfrak{M}$, for all possible orientations. This closure is no longer disintegrated; since no group is definable, it is not even one-based. Counting types, one can show that both theories are stable.

§4. **Modification II.** For every relation $R \in \mathcal{L}$ we choose a weight $\alpha_R > 0$ and define a *predimension* on finite $\mathcal{L}$-structures:

$$\delta(A) = |A| - \sum_{R \in \mathcal{L}} \alpha_R |R(A)|, \qquad \text{as well as}$$

$$\delta(A/B) = \delta(AB) - \delta(B) = |A \setminus B| - \text{weights of the new relations;}$$

the latter makes sense even if B is infinite. Define

$$A \leq B \quad \Leftrightarrow \quad \delta(B'/A) \geq 0 \text{ for all } B' \subseteq B.$$

Let $\mathcal{C}$ be the universal class of all finite $\mathcal{L}$-structures whose substructures have non-negative predimension (*ab initio*).

It is closed under free amalgamation, and thus has AP. Closedness is type-definable (this uses finiteness of the language), richness is approximately definable, and $\aleph_1$-saturated models are rich. The generic model is ω-stable if all the α_R are rational, and stable otherwise.

Since we have free amalgamation, cl is equal to algebraic closure (in the model-theoretic sense).

If A is a finite subset of the generic model, then $\text{cl}_{\mathfrak{M}}(A)$ is the unique smallest superset of A with minimal predimension (for irrational α_R this may be infinite, and we interpret its predimension as a suitable limit). Define the *dimension* of A in $\mathfrak{M}$ to be $d_{\mathfrak{M}}(A) = \delta(\text{cl}_{\mathfrak{M}}(A))$.

Then two closed sets are independent in the forking sense iff they are freely amalgamated over their intersection, and the amalgam is closed in $\mathfrak{M}$. It is now easy to see that the generic model has weak elimination of imaginaries.

Finally, for rational α_R Evans [8] has defined a notion of *directed hypergraph* such that $A \leq B$ in the predimension sense iff there is an orientation of B in which A is closed in the directed sense. So 3.1.1 above generalizes, and all ω-stable ab initio constructions arise as reducts of disintegrated geometries.

4.1. $\aleph_0$-categoricity. If we want the resulting structure to be $\aleph_0$-categorical, there must be a bound on the size of the closure of a finite set (as the closure is always contained in the algebraic closure).

Note that the free amalgam of A and B over $A \cap B$ forms the fourth point of a parallelogram:

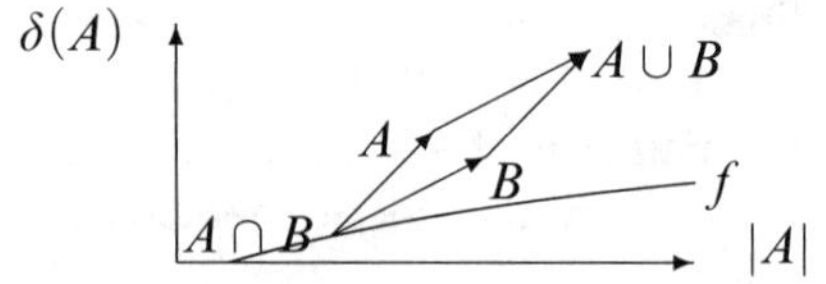

We choose an unbounded increasing convex cut-off function f and consider the subclass $\mathcal{C}_f = \{A \in \mathcal{C} : \delta(A) \geq f(|A|)\}$.

If the slope of f at $|B|$ is at most the minimal slope of an arrow at B, then $\mathcal{C}_f$ has free amalgamation. This requires in particular $\delta(B/A)$ to be strictly positive, and the α_R to be linearly independent over $\mathbb{Q}$. Axiomatizability and $\aleph_0$-categoricity follow.

4.1.1. *Pseudoplanes.* An incidence system $I \subset P \times L$ is a *pseudoplane* if

- every point $p \in P$ lies on infinitely many lines $\ell \in L$,
- every line contains infinitely many points,
- every two points are incident with finitely many lines,
- every two lines intersect in finitely many points.

The pseudoplane is *complete* if I is a complete type.

Hrushovki has shown that for a single binary relation the set of (irrational) α which allow an unbounded increasing convex cut-off function is co-meagre. Putting $P = L = \mathfrak{M}$ and $I = R$, this yields an $\aleph_0$-categorical stable complete pseudoplane (answering a question of Lachlan).

It is easy to see that a disintegrated stable theory cannot type-define a complete pseudoplane. So if an ω-stable complete pseudoplane has a disintegrated expansion à la Evans, the pseudoplane obviously still exists in the expansion, but is no longer complete.

4.2. Strong minimality.

4.2.1. *Geometry.* Suppose all the α_R are integers. For a single point a and a set B (which we may assume to be closed) of parameters in a generic structure $\mathfrak{M}$, there are two possibilities:

- $d_{\mathfrak{M}}(a/B) = \delta(a/B) = 1$. Then $aB \leq \mathfrak{M}$, so this determines a unique type, the *generic* type.
- $d_{\mathfrak{M}}(a/B) = 0$. So a is in the *geometric* closure $\mathrm{gcl}(B)$.

Clearly gcl is increasing and idempotent, hence a closure operator, which in addition satisfies the exchange rule:

If $a \in \mathrm{gcl}(Bc) \setminus \mathrm{gcl}(B)$, then $c \in \mathrm{gcl}(Ba)$.

We should like to restrict the class $\mathcal{C}$ so that gcl becomes algebraic closure, thus yielding a *strongly minimal* set (every definable subset is uniformly finite or cofinite). For that, we have to bound the number of possible realisations of any $a \in \mathrm{gcl}(B)$ uniformly and definably.

4.2.2. *Minimal and pre-algebraic extensions, codes.* A proper closed extension $A \leq B \in \mathcal{C}$ is *minimal* if $A \leq A' \leq B$ implies $A' = A$ or $A' = B$. Equivalently, $\delta(B/A') < 0$ for all $A \subset A' \subset B$.

It is *pre-algebraic* if $\delta(B/A) = 0$. In this case, minimality is equivalent to $\delta(B'/A) > 0$ for all $A \subset B' \subset B$.

For a minimal pre-algebraic extension $A \leq B$ let $A_0 \leq A$ be the closure of the points in A related to some points in $B \setminus A$. This is the unique minimal closed subset of A over which $B \setminus A$ is pre-algebraic (and in fact minimal).

We call $A_0 \leq B$ *bi-minimal pre-algebraic*; its *code* $\varphi(\bar{x}, \bar{y})$ is the quantifier-free diagram of $(B \setminus A, A_0)$.

Clearly it is sufficient to bound the number of realisations for each bi-minimal pre-algebraic code.

4.2.3. *New strongly minimal sets.* [12] Let μ be a function from the set of codes to the integers, satisfying $\mu(\varphi(\bar{x},\bar{y})) \geq \delta(\bar{y})$. Let $\mathcal{C}^{\mu}$ be the class of $A \in \mathcal{C}$ such that for any code φ and any A_0 there are at most $\mu(\varphi)$ disjoint realizations of $\varphi(\bar{x}, A_0)$. Again, this is a universal class.

Hrushovski has shown that $\mathcal{C}^{\mu}$ has *thrifty amalgamation*:

> If $A \leq B \in \mathcal{C}^{\mu}$ is minimal and $A \leq M \in \mathcal{C}^{\mu}$, then either the free amalgam of B and M over A is still in $\mathcal{C}^{\mu}$, or B embeds closedly into M over A.

So a generic model exists; since approximate richness is still definable, its theory is strongly minimal.

§5. Extension to general languages. If the language is infinite or non-relational (but still countable), one considers a class $\mathcal{C}$ of finitely generated structures with AP, and a predimension δ on $\mathcal{C}$ satisfying

(1) $\delta(\emptyset) = 0$, and
(2) *Submodularity:* $\delta(A) + \delta(B) \geq \delta(A \cup B) + \delta(A \cap B)$ for closed A, B.

Two main difficulties arise:

- Axiomatization of the class $\mathcal{C}$ (the excluded structures need not be finite).
- Axiomatization of the richness condition, or even of approximate richness.

In practice, they are usually overcome by requiring definability of rank (and possibly also degree), and *modularity* of the negative part of the predimension (i.e. the inequality above is strengthened to equality).

5.1. Poizat's black field of rank $\omega \cdot 2$. [13] A *black field* is an ω-stable algebraically closed field K with a predicate N for a subset of comparable rank. According to Berline and Lascar [6] the Lascar rank of an infinite superstable field must be of the form $\omega^{\alpha} \cdot n$ for some ordinal α and some positive integer n; they conjectured that $n = 1$. Since $U(K) > U(N)$, black fields yield a counter-example to this conjecture.

Let $\mathcal{C}$ be the class of finitely generated fields k of fixed characteristic with a unary predicate N, such that for all finitely generated subfields k'

$$\delta(k') = 2\,\mathrm{tr.deg}(k') - \left|N(k')\right| \geq 0,$$

and for $k' \in \bar{\mathcal{C}}$ finitely generated over $k \in \bar{\mathcal{C}}$ put

$$\delta(k'/k) = 2\,\mathrm{tr.deg}(k'/k) - \left|N(k') \setminus N(k)\right|,$$

where again $\bar{\mathcal{C}}$ is the class of structures whose finitely generated substructures are in $\mathcal{C}$.

Since C has free amalgamation, a generic model $\mathfrak{M}$ exists. The class C is universally axiomatized by formulas of the form $\forall \bar{x}\, \neg[\varphi(\bar{x}) \wedge \nu(\bar{x})]$, where φ is a quantifier-free formula in the field language explicitly making strictly more than half of $\bar{x}$ algebraic over the rest, and $\nu(\bar{x})$ states that all elements of $\bar{x}$ are distinct and black. In a similar way, modulo the axioms for algebraically closed fields, richness is approximately definable by inductive formulas

$$\forall \bar{y}\left[\vartheta(\bar{y}) \rightarrow \left\{\exists \bar{x}\left[\varphi(\bar{x}, \bar{y}) \wedge \nu(\bar{x})\right]\right\}\right],$$

where φ and ϑ are quantifier-free formulas in the pure language of fields, $\varphi(\bar{x}, \bar{y})$ implies that $\mathrm{tr.deg}(\bar{x}'/\bar{x}''\bar{y}) < 2|\bar{x}'|$ for every non-trivial partition $\bar{x} = \bar{x}'\bar{x}''$, and $\vartheta(\bar{y})$ states that $\varphi(\cdot, \bar{y})$ has Morley rank $|\bar{x}|/2$ and Morley degree one (which are definable properties in pure algebraically closed fields). So $\aleph_0$-saturated models of $\mathrm{Th}(\mathfrak{M})$ are rich.

For a point a and a set B in $\mathfrak{M}$ there are three possibilities:

(1) $d_{\mathfrak{M}}(a/B) = 2$. Then a is not black, and aB is closed (*white generic*).
(2) $d_{\mathfrak{M}}(a/B) = 1$. There is a black point a' interalgebraic with a over B, and $a'B$ is closed (*black generic*).
(3) $d_{\mathfrak{M}}(a/B) = 0$. Then either a is algebraic over B, or pre-algebraic.

One can show that minimal pre-algebraic extensions have rank 1, that the black generic is the limit of pre-algebraic types of arbitrarily big finite rank (and hence of rank ω), and that the white generic has rank $\omega \cdot 2$ (all other types have rank strictly less than $\omega \cdot 2$).

5.2. The collaps. [13, 1, 2] If $\varphi(\bar{x}, \bar{y})$ and $\vartheta(\bar{y})$ are as in one of the inductive axioms, a *code* is given by a formula $\varphi \wedge \vartheta$ (where we assume in addition $\bar{y}$ to be the canonical parameter). For a code φ we fix a quantifier-free definable function $f_\varphi(\bar{x}_1, \dots, \bar{x}_{n_\varphi}, \bar{y})$ which explicitly defines $\bar{y}$ over n_φ (independent) generic realizations $\bar{x}_i$ of $\varphi(\cdot, \bar{y})$. Let μ be a function from the set of codes to ω with $\mu(\varphi) \geq (2n+2)n_\varphi$, and C^μ the class of structures in C which for any code φ contain at most $\mu(\varphi)$ disjoint black tuples such that any n_φ of them have the same image $\bar{y}$ under f, and they all satisfy $\varphi(\cdot, \bar{y})$. Then C^μ has thrifty amalgamation; if μ is finite-to-one, a generic model is ω-saturated, and hence of Morley rank 2. A black generic has Morley rank 1; a white (field) generic has Morley rank 2, and is the sum of two independent generic black points.

Since the axiomatization is inductive and any complete theory of fields of finite Morley rank is $\aleph_1$-categorical, Lindström's theorem implies model-completenes of the theory of the generic model.

§6. Fields with an additive subgroup.

6.1. Poizat's red field of rank $\omega \cdot 2$. [14] A *red field* is an ω-stable algebraically closed field K with a predicate R for a connected additive subgroup of comparable rank. Note that in characteristic 0 this gives rise to an infinite definable subfield $\{a \in K : aR \leq R\}$, so the structure has rank at least ω.

Let $\mathcal{C}$ be the class of finitely generated fields k of characteristic $p > 0$ with a predicate R for an additive subgroup, such that for all finitely generated subfields k'

$$\delta(k') = 2\,\text{tr.deg}(k') - \text{lin.dim}_{\mathbb{F}_p}(R(k')) \geq 0.$$

This condition is universal, since we have to say that $2n$ linearly independent red points do not lie in any variety of dimension $< n$.

For $k \subseteq k' \in \bar{\mathcal{C}}$ with k' finitely generated over k put

$$\delta(k'/k) = 2\,\text{tr.deg}(k'/k) - \text{lin.dim}_{\mathbb{F}_p}(R(k')/R(k)).$$

Since $\mathcal{C}$ has free amalgamation, a generic model $\mathfrak{M}$ exists; as $\mathcal{C}$ is (universally) axiomatizable and richness is approximately definable, $\aleph_0$-saturated models of $\text{Th}(\mathfrak{M})$ are rich.

For a point a and a set B in $\mathfrak{M}$ there are three possibilities:

(1) $d_{\mathfrak{M}}(a/B) = 2$. Then a is not red, and aB is closed. $RM(a/B) = \omega \cdot 2$.
(2) $d_{\mathfrak{M}}(a/B) = 1$. There is a red point a' interalgebraic with a over B, and $a'B$ is closed. $\omega \cdot 2 > RM(a/B) \geq RM(a'/B) = \omega$.
(3) $d_{\mathfrak{M}}(a/B) = 0$. Then a is algebraic or pre-algebraic over B, of rank $< \omega$.

6.2. The collaps. [5] We have to restrict the number of bi-minimal pre-algebraic extensions. A *code* is a formula $\varphi(\bar{x}, \bar{y})$ in the field language with $n = |\bar{x}|$ such that

(1) For all $\bar{b}$ either $\varphi(\bar{x}, \bar{b})$ is empty, or has Morley degree 1.
(2) $\text{tr.deg}(\bar{a}/\bar{b}) = n/2$ and $\text{lin.dim}_{\mathbb{F}_p}(\bar{a}/\bar{b}) = n$ for generic $\bar{a} \models \varphi(\bar{x}, \bar{b})$, and $2\,\text{tr.deg}(\bar{a}/U\bar{b}) < n - \text{lin.dim}_{\mathbb{F}_p}(U)$ for all non-trivial subspaces U of $\langle \bar{a} \rangle$.
(3) If $\text{tr.deg}(\varphi(\bar{x}, \bar{b}) \cap \varphi(\bar{x}, \bar{b}')) = n/2$, then $b = b'$.
(4) If $\varphi(\bar{x}, \bar{b})$ is disintegrated for some $\bar{b}$, it is disintegrated (or empty) for all $\bar{b}$.
(5) For any $H \in \text{GL}_n(\mathbb{F}_p)$, $\bar{m}$ and $\bar{b}$ there is $\bar{b}'$ with $\varphi(H\bar{x} + \bar{m}, \bar{b}) \equiv \varphi(\bar{x}, \bar{b}')$.

Remark. (1) says that $\varphi(\bar{x}, \bar{b})$ determines a unique generic type $p_{\varphi(\bar{x},\bar{b})}$ (or is empty).

(2) says that $\bar{b} \leq \bar{a}\bar{b}$ with $\bar{a}$ red is minimally pre-algebraic. Moreover, $\delta(\bar{a}'/B) < 0$ for any $B \ni \bar{b}$ and non-generic red $\bar{a}' \notin \text{acl}(B)$ realizing $\varphi(\bar{x}, \bar{b})$.
(3) says that $\bar{b}$ is the canonical base for $p_{\varphi(\bar{x},\bar{b})}$, so the extension $\bar{b} \leq \bar{a}\bar{b}$ is bi-minimal.

(4) says that φ fixes the type of the extension: disintegrated, or generic in a group coset (minimal pre-algebraic types are locally modular).
(5) says that affine transfomations preserve the code.

Inductively one constructs a set $\mathcal{S}$ of codes such that every minimal pre-algebraic extension is coded by a unique $\varphi \in \mathcal{S}$.

6.2.1. *Difference sequences.* For a code φ and some $\bar{b}$ consider a Morley sequence $(\bar{a}_0, \bar{a}_1, \ldots, \bar{a}_k, f)$ for $p_{\varphi(\bar{x},\bar{b})}$, and put $\bar{e}_i = \bar{a}_i - \bar{f}$.

We can then find a formula $\psi_\varphi \in \text{tp}(\bar{e}_0, \ldots, \bar{e}_k)$ in the field language such that

- Any realization $(\bar{e}'_0, \ldots, \bar{e}'_k)$ of ψ_φ is $\mathbb{F}_p$-linearly independent, and there is a unique $\bar{b}'$ definable over sufficiently large finite subsets of the $\bar{e}'_i$ such that $\models \varphi(\bar{e}'_i, \bar{b}')$.
- ψ_φ is invariant under the finite group of *derivations* generated by

$$\partial_i : \bar{x}_j \mapsto \begin{cases} \bar{x}_j - \bar{x}_i & \text{if } j \neq i, \\ -\bar{x}_i & \text{if } j = i \end{cases}$$

for $0 \leq i \leq k$.
- Some condition ensuring dependence of affine combinations, and invariance under the stabiliser of the group for coset codes.

A *difference sequence* for a code φ is any realization of ψ_φ.

6.2.2. *A counting lemma.* Given a code φ and natural numbers m, n, there is some λ such that for every $M \leq N \in \mathcal{C}$ and red difference sequence $(\bar{e}_0, \ldots, \bar{e}_\lambda)$ in N with canonical parameter $\bar{b}$, either

- the canonical parameter for some derived sequence lies in M, or
- for every $A \subset N$ of size m the sequence $(\bar{e}_0, \ldots, \bar{e}_\lambda)$ contains a Morley subsequence in $p_{\varphi(\bar{x},\bar{b})}$ over MA of length n.

Let μ be a sufficiently fast-growing finite-to-one function from $\mathcal{S}$ to ω, and $\mathcal{C}^\mu$ the class of $A \in \mathcal{C}$ which do not contain a red difference sequence for φ of length $\mu(\varphi)$ for any $\varphi \in \mathcal{S}$.

The above lemma allows us to characterize when a minimal pre-algebraic extension of some $M \in \mathcal{C}^\mu$ is no longer in $\mathcal{C}^\mu$, and to prove thrifty amalgamation for $\mathcal{C}^\mu$.

6.2.3. *Axiomatization.* It follows that there is a generic model $\mathfrak{M}$; since approximate richness remains definable, $\aleph_0$-saturated models of $\text{Th}(\mathfrak{M})$ are rich, $\mathfrak{M}$ has Morley rank 2, and $R^{\mathfrak{M}}$ has Morley rank 1.

Alternatively, the field can be inductively axiomatized as follows:

- Finitely generated subfields are in $\mathcal{C}^\mu$.
- ACF_p.
- The extension of the model generated by a red generic realization of some code instance $\varphi(\bar{x}, \bar{b})$ is not in $\mathcal{C}^\mu$.

So Lindström's theorem again implies model-completeness of $\text{Th}(\mathfrak{M})$.

§7. Fields with a multiplicative subgroup.

7.1. Poizat's green field of rank $\omega \cdot 2$. [14] A *green field* is an ω-stable algebraically closed field K with a predicate $\ddot{U}$ for a connected multiplicative subgroup of comparable rank. Note that in characteristic $p > 0$ a green field of finite rank implies that there are only finitely many *p-Mersenne primes* $\frac{p^n-1}{p-1}$, and $\tilde{\mathbb{F}}_p$ is a prime model.

Let $\mathcal{C}$ be the class of finitely generated fields k of characteristic 0 with a predicate $\ddot{U}$ for a torsion-free multiplicative subgroup, such that for all finitely generated subfields k'

$$\delta(k') = 2\,\mathrm{tr.deg}(k') - \mathrm{lin.dim}_{\mathbb{Q}}(\ddot{U}(k')) \geq 0,$$

where the linear dimension is taken multiplicatively. For $k' \in \bar{\mathcal{C}}$ finitely generated over $k \subseteq k'$ put

$$\delta(k'/k) = 2\,\mathrm{tr.deg}(k'/k) - \mathrm{lin.dim}_{\mathbb{Q}}(\ddot{U}(k')/\ddot{U}(k)).$$

7.1.1. *The weak CIT.* While linear dimension over a finite field is definable, this is no longer true for dimension over $\mathbb{Q}$, as there are infinitely many scalars (exponents). Poizat used Zilber's weak CIT, a consequence of Ax' differential Schanuel conjecture:

For any uniform family $V_{\bar{z}}$ of varieties there is a finite set $T_0, \dots, T_r$ of tori, such that for any torus T, any $V_{\bar{b}}$ and any irreducible component $W \ni \bar{a}$ of $V_{\bar{b}} \cap \bar{a} \cdot T$, for some $i \in [0, r]$ $W \subseteq \bar{a} \cdot T_i$ and $\dim(T_i) - \dim(V \cap \bar{a} \cdot T_i) = \dim T - \dim W$, or $\dim T - \dim W \geq n - \dim V$, where $V \subseteq (K^{\times})^n$.

This specifies finitely many possibilities for $\mathbb{Q}$-linear relations on a family of varieties which could render δ negative. Hence $\mathcal{C}$ is again universal, approximate richness axiomatizable, the generic model exists, and $\aleph_0$-saturated models of its theory are rich. It has Morley rank $\omega \cdot 2$, and a generic green point has rank ω.

7.2. The collaps. [3] A *code* is a formula $\varphi(\bar{x}, \bar{y})$ in the field language with $n = |\bar{x}|$ such that

1. For all $\bar{b}$ either $\varphi(\bar{x}, \bar{b})$ is empty, or has Morley degree 1.
2. $\mathrm{tr.deg}'(\bar{a}/\bar{b}) = n/2$ and $\mathrm{lin.dim}_{\mathbb{Q}}(\bar{a}/\bar{b}) = n$ for generic $\bar{a} \models \varphi(\bar{x}, \bar{b})$, and for $i = 2, \dots, r$ and any W irreducible component of $V \cap \bar{a}T_i$ of maximal dimension, $\dim(T_i) > 2 \cdot \dim(W)$ if $V \cap \bar{a}T_i$ is infinite.
3. If $\mathrm{tr.deg}(\varphi(\bar{x}, \bar{b}) \cap \varphi(\bar{x}, \bar{b}')) = n/2$, then $b = b'$.
4. For any multiplicatively invertible $\bar{m}$ and $\bar{b}$ there is $\bar{b}'$ with $\varphi(\bar{x} \cdot \bar{m}, \bar{b}) \equiv \varphi(\bar{x}, \bar{b}')$.

7.2.1. *Toric correspondences.* This time $\mathrm{GL}_n(\mathbb{Q})$ acts on the codes, which is infinite. Hence we cannot put invariance under $\mathrm{GL}_n(\mathbb{Q})$ into the axioms, but have to deal with it outside the codes. Using weak CIT we obtain:

There exists a collection $\mathcal{S}$ of codes such that for every minimal prealgebraic definable set X there is a unique code $\varphi \in \mathcal{S}$ and finitely many tori T such that $T \cap (X \times \varphi(\bar{x}, \bar{b}))$ projects generically onto X and $\varphi(\bar{x}, \bar{b})$ for some $\bar{b}$. We call such a T a *toric correspondence*. In particular, for any code φ only finitely many tori can induce a toric correspondence between instances of φ.

7.2.2. *Difference sequences.* For every code φ there is some formula

$$\psi(\bar{x}_0, \dots, \bar{x}_k) \in \operatorname{tp}(\bar{e}_0 \cdot \bar{f}^{-1}, \dots, \bar{e}_k \cdot \bar{f}^{-1})$$

for some Morley sequence $(\bar{e}_0, \dots, \bar{e}_k, \bar{f})$ in $p_{\varphi(\bar{x}, \bar{b})}$ such that:

(1) Any realization $(\bar{e}'_0, \dots, \bar{e}'_k)$ of ψ is disjoint, and $\models \varphi(\bar{e}'_i, \bar{b}')$ for some unique $\bar{b}'$ definable over sufficiently large finite subsets of the $\bar{e}'_i$.

(2) If $\models \psi(\bar{e}_0, \dots, \bar{e}_k)$, then $\models \psi(\bar{e}_0, \dots, \bar{e}_{k'})$ for each $k' \leq k$, and ψ is invariant under derivations.

(3) Let $i \neq j$ and $(\bar{e}_0, \dots, \bar{e}_k)$ realize ψ with canonical parameter $\bar{b}$. If there is some toric correspondence T on φ and $\bar{e}'_j$ with $(\bar{e}_j, \bar{e}'_j) \in T$, then $\bar{e}_i \not\mathrel{\unicode{x2AEB}}_{\bar{b}} \bar{e}'_j \cdot \bar{e}_i^{-1}$ in case $\bar{e}_i$ is a generic realization of $\varphi(\bar{x}, \bar{b})$.

7.2.3. *Counting.* Miraculously, this is enough to obtain the same counting lemma as before.

Since $\mathbb{Q}$-linear dependence need not be uniform, we have to use the weak CIT in order to uniformize dependencies in a non-generic difference sequence, and then the finite Ramsey theorem to obtain a long derived sequence inside the original model.

Again, we choose a fast-growing finite-to-one function μ from $\mathcal{S}$ to ω, and define $\mathcal{C}^\mu$ to be the class of all $A \in \mathcal{C}$ who do not have a difference sequence for φ of length $\mu(\varphi)$. We obtain a characterisation when a structure in $\mathcal{C}^\mu$ has a minimal pre-algebraic extension not in $\mathcal{C}^\mu$ similar to the red case, and can prove thrifty amalgamation. Hence there is a generic model $\mathfrak{M}$.

7.2.4. *Axiomatization.* When we want to axiomatize approximate richness, we have to say that for all $\bar{a} \in A \leq \mathfrak{M}$, a code instance $\varphi(\bar{x}, \bar{a})$ has an A-generic realization in $\mathfrak{M}$, *unless* for a generic realization B we would have $AB \notin \mathcal{C}^\mu$. The weak CIT allows us to limit the possible $\mathbb{Q}$-linear dependencies we have to consider, with an extra twist: We may first have to extend by finitely many green generic points.

It follows that $\aleph_0$-saturated models of $\operatorname{Th}(\mathfrak{M})$ are rich, $\mathfrak{M}$ has Morley rank 2, and $\ddot{U}^{\mathfrak{M}}$ has Morley rank 1.

Moreover, there also is an alternative inductive axiomatization analogous to the red case, which yields model-completeness by Lindström's theorem.

§8. Fusion. Using similar techniques as for the expansions of fields, one also obtains:

(1) Two strongly minimal sets with definable multiplicity property (DMP) in disjoint languages can be amalgamated freely with predimension

$\delta(A/B) = RM_1(A/B) + RM_2(A/B) - |A \setminus B|$, and collapsed to a strongly minimal set [11].

(2) Two strongly minimal sets with DMP with a common $\aleph_0$-categorical reduct, one expansion preserving multiplicities, can be amalgamated freely with predimension

$$\delta(A/B) = RM_1(A/B) + RM_2(A/B) - RM_0(A/B),$$

and collapsed to a strongly minimal set. See [9] for the one-based case, [4] for the fusion over vector spaces, which implies the full result according to [10].

(3) Two theories of finite and definable Morley rank with DMP and of the same Morley degree can be amalgamated freely with predimension

$$\delta(A/B) = n_1 \cdot RM_1(A/B) + n_2 \cdot RM_2(A/B) - n \cdot |A \setminus B|,$$

where $n_1 \cdot RM(T_1) = n_2 \cdot RM(T_2) = n$, and collapsed to a structure of Morley rank n [15]. (It may be necessary to extend the language.)

In all cases except for the one-based collapse in (2), the language has to be countable in order to collapse.

REFERENCES

[1] J. Baldwin and K. Holland, *Constructing ω-stable structures: Rank 2 fields*, ***The Journal of Symbolic Logic***, vol. 65 (2000), no. 1, pp. 371–391.

[2] J.. Baldwin and K. Holland, *Constructing ω-stable structures: Computing rank*, ***Fundamenta Mathematicae***, vol. 170 (2001), no. 1-2, pp. 1–20.

[3] A. Baudisch, M. Hils, A. Martín Pizarro, and F. O. Wagner, *Die Böse Farbe*, ***Journal of the Institut de Mathématiques de Jussieu***, to appear.

[4] A. Baudisch, A. Martin-Pizarro, and M. Ziegler, *Fusion over a vector space*, ***Journal of Mathematical Logic***, vol. 6 (2006), no. 2, pp. 141–162.

[5] ———, *Red fields*, ***The Journal of Symbolic Logic***, vol. 72 (2007), no. 1, pp. 207–225.

[6] C. Berline and D. Lascar, *Superstable groups*, ***Annals of Pure and Applied Logic***, vol. 30 (1986), no. 1, pp. 1–43.

[7] D. Evans, *Ample dividing*, ***The Journal of Symbolic Logic***, vol. 68 (2003), no. 4, pp. 1385–1402.

[8] ———, *Trivial stable structures with non-trivial reducts*, ***Journal of the London Mathematical Society. Second Series***, vol. 72 (2005), no. 2, pp. 351–363.

[9] A. Hasson and M. Hils, *Fusion over sublanguages*, ***The Journal of Symbolic Logic***, vol. 71 (2006), no. 2, pp. 361–398.

[10] M. Hils, ***Fusion libre et autres constructions génériques***, Ph.D. thesis, Université Paris-7, Paris, 2006.

[11] E. Hrushovski, *Strongly minimal expansions of algebraically closed fields*, ***Israel Journal of Mathematics***, vol. 79 (1992), no. 2-3, pp. 129–151.

[12] ———, *A new strongly minimal set*, ***Annals of Pure and Applied Logic***, vol. 62 (1993), no. 2, pp. 147–166.

[13] B. Poizat, *Le carré de l'égalité*, ***The Journal of Symbolic Logic***, vol. 64 (1999), no. 3, pp. 1339–1355.

[14] ———, *L'égalité au cube*, ***The Journal of Symbolic Logic***, vol. 66 (2001), no. 4, pp. 1647–1676.

[15] M. Ziegler, *Fusion of structures of finite Morley rank*, preprint.

UNIVERSITÉ DE LYON; UNIVERSITÉ LYON 1; CNRS
INSTITUT CAMILLE JORDAN UMR 5208
VILLEURBANNE F-69622
FRANCE
E-mail: wagner@math.univ-lyon1.fr